Annika Grafschafter, Susanne Lux, Matthäus Siebenhofer

Off-Gas Purification

Also of Interest

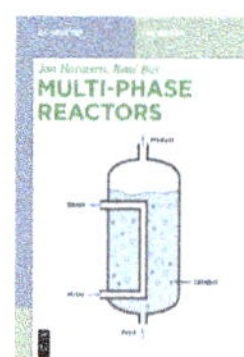

Multiphase Reactors.
Reaction Engineering Concepts, Selection, and Industrial Applications
Harmsen, Bos, 2023
ISBN 978-3-11-071376-3, e-ISBN (PDF) 978-3-11-071377-0,
e-ISBN (EPUB) 978-3-11-071384-8

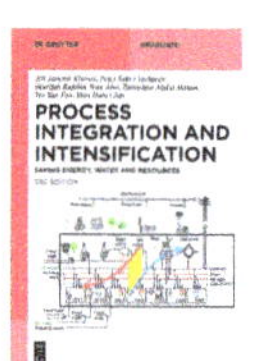

Sustainable Process Integration and Intensification.
Saving Energy, Water and Resources
3rd Edition
Klemes, Varbanov, Wan Alwi, Manan, Fan, Chin (Eds.), 2023
ISBN 978-3-11-078283-7, e-ISBN (PDF) 978-3-11-078298-1,
e-ISBN (EPUB) 978-3-11-078300-1

Energy, Environment and New Materials Vol. 1-3
van de Voorde (Ed.), 2021
[Set ISBN 978-3-11-075497-1]
Vol. 1 Hydrogen Production and Energy Transition
ISBN 978-3-11-059622-9, e-ISBN (PDF) 978-3-11-059625-0,
e-ISBN (EPUB) 978-3-11-059405-8
Vol. 2 Hydrogen Storage for Sustainability
ISBN 978-3-11-059623-6, e-ISBN (PDF) 978-3-11-059628-1,
e-ISBN (EPUB) 978-3-11-059431-7
Vol. 3 Utilization of Hydrogen for Sustainable Energy and Fuels
ISBN 978-3-11-059624-3, e-ISBN (PDF) 978-3-11-059627-4,
e-ISBN (EPUB) 978-3-11-059410-2

Green Chemistry and Technologies
Long, Changsheng, Dai (Eds.), 2018
ISBN 978-3-11-047861-7, e-ISBN (PDF) 978-3-11-047931-7,
e-ISBN (EPUB) 978-3-11-047978-2

Annika Grafschafter, Susanne Lux,
Matthäus Siebenhofer

Off-Gas Purification

Basics, Exercises and Solver Strategies

DE GRUYTER

Authors
Dipl.-Ing. Dr. techn. Annika Grafschafter
Hammerweg 7
9562 Himmelberg
Austria
a.grafschafter@tugraz.at

Assoc. Prof. Dr. Dipl.-Ing. Susanne Lux
Institute of Chemical Engineering and Environmental Technology
TU Graz
Inffeldgasse 25/C/II
8010 Graz
Austria
susanne.lux@tugraz.at

Prof. Dr. Dipl.-Ing. Matthäus Siebenhofer
Institute of Chemical Engineering and Environmental Technology
TU Graz
Inffeldgasse 25/C/II
8010 Graz
Austria
m.siebenhofer@tugraz.at

ISBN 978-3-11-076390-4
e-ISBN (PDF) 978-3-11-076392-8
e-ISBN (EPUB) 978-3-11-076400-0

Library of Congress Control Number: 2023931745

Bibliographic information published by the Deutsche Nationalbibliothek
The Deutsche Nationalbibliothek lists this publication in the Deutsche Nationalbibliografie; detailed bibliographic data are available on the Internet at http://dnb.dnb.de.

Cover image: Annika Grafschafter – Droplets at a glance
Typesetting: Integra Software Services Pvt. Ltd.
Printing and binding: CPI books GmbH, Leck

www.degruyter.com

Preface

This off-gas purification book is intended to offer graduate students and practicians basic tools for process and equipment design in off-gas purification. It has been adjusted to the needs of engineers in the industry. The content is based on practical experience in industrial off-gas purification for more than two decades.

We will start the discussion with off-gas specification and develop the thermodynamic basics for discussion of off-gas specification. Then we have to identify the need of measures before we develop a strategy for appropriate purification processes. This introductory topic is followed by the detailed presentation of off-gas purification technologies, covering dedusting, thermal and catalytic technologies as well as absorption processes and adsorption.

Precipitation of particles is still the main challenge in off-gas purification. Several boundaries, whether it is the technological progress in general, the necessity of increased attention of safety and health affairs or the demand for a balanced environment, all need the highest possible standards with regard to prevention of particle emission and emission control. Therefore, particle precipitation is a high priority. Absorptive precipitation techniques, catalytic processes and adsorptive measures will be discussed in further topics.

The main objective of this book is to discuss calculation and design strategies and to identify the limits of application. The reader will just need an appropriate hand calculator and scale paper. Whenever possible, we have tried to explain the basics with an example. The strategy is simple: Learning by doing.

The book has been written and compiled to the best of the authors' knowledge. Nevertheless, freedom from any errors and mistypings cannot be guaranteed. No liability can be taken for errors in equipment design based on the material presented.

https://doi.org/10.1515/9783110763928-202

Contents

Part II: **Technologies**

Symbols and abbreviations

Abbreviations

ACF	Activated carbon fiber
AF	Auxiliary fuel
ECA	Excess combustion air
EROM	European reference odor mass
ESP	Electrostatic precipitator
ESU	Electrostatic unit
CA	Combustion air
CE	Combustion efficiency
CFD	Computational fluid dynamics
CG	Combustion gas
CMS	Carbon molecular sieving
CRE	Chemical reaction engineering
DA	Daily average
DALR	Dry adiabatic lapse rate
FIC	Flow indication control
FGD	Flue gas gypsum
FR	Flow recorder
GAC	Granular activated carbon
HM	Hourly mean
HHM	Half-hourly mean
IUPAC	International Union of Pure and Applied Chemistry
MHHM	Maximum half-hourly mean
NG	Natural gas
ODT	Odor detection threshold
OG	Off-gas
PAC	Powdered activated carbon
PFR	Plug flow reactor
PIR	Pressure indication control
PM	Particulate matter
PO	Pollutant
ppm	Part per million
PSD	Particle size distribution
SCR	Selective Catalytic Reduction
TI	Temperature indication
TOF	Turnover frequency
TON	Turnover number

Symbols

A	Cross-sectional area (see also CSA) (m^2)
A_{free}	Free cross-sectional area of trays (absorption) (m^2)

https://doi.org/10.1515/9783110763928-204

A	Frequency factor (also pre-exponential Arrhenius factor) (same unit as the reaction rate constant, depending on the reaction kinetics)
AA	Annual average
AV	Area velocity (s^{-1})
a	Actual (prefix, specifying the gaseous state)
a	Activity
a	Specific mass transfer area (m^2 m^{-3})
a	Maximum immission concentration
B	Dust load (kg kg^{-1})
BP	Barometric pressure (hPa)
b	Width, distance (m)
b_e	Inlet width (m)
b_{VdW}	Van der Waals constant
$C(x,y,z)$	Mass concentration of the pollutant at the Cartesian level (x,y,z) (mg m^{-3})
C, c	Concentration Mass concentration (g m^{-3}) Molar concentration (mol L^{-1}, mol m^{-3})
CE	Combustion efficiency (%)
CSA	Cross-sectional area (m^2)
Cp	Specific heat (kJ kg^{-1} K^{-1})
C_w	Drag coefficient
D, d	Diameter (m)
D	Diffusion coefficient (m^2 s^{-1})
D	Rate of transmission
D	Column diameter (m)
De	Dean number
d_{dr}	Droplet diameter (m)
d_G	Bubble cap diameter (mm)
$d_{1,2}$	Sauter mean diameter (m)
E	Separation efficiency
E_A	Activation energy (kJ mol^{-1})
E_0	Corona onset field intensity (kV m^{-1} or $kg^{0.5}$ $m^{-0.5}$ s^{-1})
E_p	Precipitation field strength ($kg^{0.5}$ $m^{-0.5}$ s^{-1})
%EA	Percent excess air
e	Diameter of the interception area (m)
F	Capacity factor (absorption)
F_A	Molar flow rate of reactant A (mol $time^{-1}$)
F_e	Cross-sectional area of the inlet tube (cyclone) (m^2)
F_i	Cross-sectional area of the vortex finder (cyclone) (m^2)
F_M, F_n	Molar flow rate (mol $time^{-1}$) (also F_A for reactant A in chemical reaction engineering)
F_m	Mass flow rate (g $time^{-1}$)
F_V	Volumetric flow rate (m^3 $time^{-1}$)
$F_{V,g}$	Gas flow rate (m^3 $time^{-1}$)
$F_{V,l}$	Liquid (absorbent) flow rate (m^3 $time^{-1}$)
g	Acceleration due to gravity (9.81 m s^{-2})
G	Gas flow rate (in absorption) Mass gas flow rate (kg h^{-1}) Molar gas flow rate (mol h^{-1}) Volumetric gas flow rate (m^3 h^{-1})

ΔG	Gibbs free enthalpy
$\Delta_f G^0$	Gibbs free standard enthalpy of formation (kJ mol^{-1})
$\Delta_R G^0$	Gibbs free standard enthalpy of reaction (kJ mol^{-1})
Gz	Graetz number
H	Henry constant (MPa, bar, hPa)
H^*	Henry constant
H, h	Height (m)
HCV	Higher colorific value (kJ mol^{-1})
$\Delta_f H^0$	Standard enthalpy of formation (kJ mol^{-1})
$\Delta_R H^0$	Standard enthalpy of reaction (kJ mol^{-1})
h	Effective height of emission (m)
h	Specific enthalpy (kJ kg^{-1})
h	Adhesivity
h_{1+x}	Specific enthalpy of humid gas (kJ kg^{-1} dry gas)
h_{St}	Height of the stack (m)
h_e	Inlet height (cyclone) (m)
h_{ex}	Excess height due to plume rise (m)
h_i	Level of the vortex finder (m)
h_t	Height of the vortex finder (cyclone) (m)
h_Z	Height of the cylindrical part (cyclone) (m)
I	specific corona current ($kg^{0.5}$ $m^{0.5}$ s^{-2})
K	Ion mobility ($m^{1.5}$ $kg^{-0.5}$)
K	Adsorption constant
K	(Thermodynamic) equilibrium constant
K	Dissociation constant
K_c	Concentration-based equilibrium constant
K_p	Partial pressure-based equilibrium constant
k	Reaction rate constant (also k', k'', k''', k'''', k_s)
k_B	Boltzmann constant ($1.380649 \cdot 10^{-23}$ J K^{-1})
k_s	Reaction rate constant of the surface reaction
k_{total}	Overall reaction rate constant
k_V	Load factor (absorption)
k'	Equilibrium constant (pressure $concentration^{-1}$)
L	Absorbent flow rate (in absorption)
	Absorbent mass flow rate (kg h^{-1})
	Absorbent molar flow rate (mol h^{-1})
	Absorbent volumetric flow rate (m^3 h^{-1})
LCV	Lower calorific value (kJ kg^{-1})
lg, $\log_{10}$	Logarithm to the base of ten
M, m	Mass (kg)
MM	Molar mass (g mol^{-1})
MV	Molar volume (dm^3 $mol^{-1)}$
MOV	Measurement oxygen value (vol%)
m	H/P
m	Specific purification volume
N	Number (of particles, separation stages, etc.)
N, n	Number (of moles)
OU	Odor unit (n m^{-3})
OUE	European odor unit (n m^{-3})

P	(Total) pressure (Pa)
P	Penetration factor (–)
P	Total pressure, system pressure (Pa, hPa, bar)
P_{theor}	Theoretically required power (W)
P_v	Velocity pressure (Pa)
p	Partial pressure (Pa)
P_s	Static pressure (Pa)
p_0	Vapor pressure, saturation pressure (Pa)
Δp	Pressure drop (Pa)
Q	Energy (kJ)
$\dot{Q}$	Heat energy flux of the off-gas (kJ s^{-1})
$\dot{Q}$	Energy input (kJ h^{-1})
Q_{diff}	Differential heat of adsorption (J mol^{-1})
Q_{st}	Isosteric heat of adsorption (J mol^{-1})
Q	Cumulative count (number, volume, mass)
$Q\Delta$	Distributive count
q	Probability, frequency
q	Equilibrium concentration in the solid (kg kg^{-1}, kmol $kmol^{-1}$)
q_m	Mass load of adsorbate to the mass of adsorbent (kg kg^{-1})
q_0	Adsorption capacity (kg kg^{-1}, kmol $kmol^{-1}$)
$q_{0,m}$	Mass load capacity of the adsorbent (kg kg^{-1})
R	Rate of retention
R	Universal gas constant (8.314 J mol^{-1} K^{-1})
Re	Reynolds number (–)
RV	Reference value
ROV	Reference oxygen value (vol%)
r	Equivalent rate of reaction (mol s^{-1} $basis^{-1}$)
r_A	Rate of reaction regarding reactant A; reference basis: reaction volume (mol s^{-1} m^{-3})
r_A'	Rate of reaction regarding reactant A; reference basis: mass of catalyst (mol s^{-1} kg^{-3})
r_A''	Rate of reaction regarding reactant A; reference basis: surface of catalyst (mol s^{-1} m^{-2})
r_A'''	Rate of reaction regarding reactant A; reference basis: volume of catalyst (mol s^{-1} m^{-3})
r_A''''	Rate of reaction regarding reactant A; reference basis: reactor volume (mol s^{-1} m^{-3})
r_a	Radius of the cyclone (m)
r_e	Radius of the inlet opening (cyclone) (m)
r_i	Radius of the vortex finder (cyclone) (m)
S	Surface (m^2)
ΔS	Change of entropy (J K^{-1} mol^{-1})
S	Selectivity (%)
Sc	Schmidt number (–)
Sh	Sherwood number (–)
SV	Space velocity (s^{-1})
s	Stability parameter
T	Temperature (K)
TC	Total carbon (mg m^{-3})
t	Time (s, h)
$T(x)$	Grade separation efficiency
U	Voltage (V)
U_0	Corona onset voltage (V or $kg^{0.5}$ $m^{0.5}$ s^{-1})
U_b	Breakdown voltage (V)

u_h	Wind speed (m s^{-1})
u_r	Mean wind speed, determined at height z_a (m s^{-1})
V	Volume (m^3)
v, w	Velocity (m s^{-1})
v_g, v_{gas}	Gas velocity (m s^{-1})
v_F	Rate of filtration (m s^{-1})
$v_{\phi i}$	Circumferential velocity (m s^{-1})
v_t	Terminal velocity of droplets (absorption) (m s^{-1})
W	Resistance force
W	Areal dust density (kg m^{-2})
W_{MP}	Volume of micropores filled with adsorbate (m^3 kg^{-1})
$w(x)$	Rate of particle migration (m s^{-1})
x,y,z	Cartesian coordinates of the investigated location (stack coordinates: 0, 0, 0)
x	Molar fraction of a substance in liquid (–)
x	Particle diameter (µm, m)
x_T	Cut-off particle size (diameter) (µm, m)
x'	Statistical mean particle diameter (µm, m)
X	Load Molar load (mol mol^{-1}, mol kg^{-1}) Mass load (kg kg^{-1})
X	Specific humidity, specific mass load (kg kg^{-1})
X_A	Relative conversion of reactant A (%)
y	Molar fraction of gaseous substances (–)
Y	Load Molar load (mol mol^{-1}, mol kg^{-1}) Mass load (kg kg^{-1})
Y	Yield (%)
z	Height above the ground level (m)
z	Filter layer thickness (µm)
z_G	Height of a bubble cap (mm)
z_a	Height of speed determination (m)

Greek letters

α	Inlet obstruction factor (–)
α	Bunsen absorption coefficient (m^3 solute m^{-3} absorbent)
β_g	Mass transfer coefficient in the gas phase (m^{-1})
γ_+, γ_-	Ion-specific activity coefficient
$\gamma^{\pm}$	Mean activity coefficient (–)
ε	Cone angle (°)
ε	Void fraction (–)
ε_A	Volumetric factor (with respect to reactant A) (–)
δ	Relative gas density (–)
δ	Thickness of the phase boundary layer (µm)
Δ	Delta (differential or fractional amount) (–)
θ	Temperature (°C)

θ	Fractional coverage (–)
ν	Stoichiometric coefficient (–)
ν	Kinematic viscosity ($m^2\ s^{-1}$)
ϕ	Relative humidity (%)
ϕ	Separation efficiency of single droplet/fiber (–)
λ	Air number (–)
μ	Chemical potential (–)
μ	Dynamic viscosity (Pa s, cP)
ρ	Density ($kg\ m^{-3}$)
τ	Residence time (s)
σ_A	Change in total number of moles in relation to the number of moles of key component A (–)
σ_y, σ_z	Horizontal and vertical dispersion parameter (–)
Ω	Angular velocity ($m\ s^{-1}$)
λ	Wall friction factor (–)
ξ	Pressure drop coefficient (–)
η	Dynamic viscosity (Pa s)
η	Degree of impaction/probability of interception (–)
η	Stokes number (–)

Subscripts

A	Any substance, reactant
Ad	Adsorbent
ads	Adsorbate
acl	Adiabatic cooling limit
B	Bottom (of the column)
cat	Catalyst
cl	Cooling limit
comb	Combustion
D	Dust
d (dry)	Dry
des	Desorption
diff	Differential
dp	Dew point
dr	Droplet
eff	Effective
F	Fiber
f	Formation
G, g, gas	Gas, gaseous
H_2O	Aqueous, water
Hg	Mercury
h	Horizontal
humid	Humid
I	Inerts, inert substance
I	Constituent
L	Length
m	Mass

m	Maximum (absorption)
m, *n*	Stoichiometric coefficients of elements in a molecule (also *a*, *b*, *c*, *d*)
max	Maximum
min	Minimum
mon	Monomolecular
OG	Off-gas
Off-gas	Off-gas
S	Surface
op	Operation
PG	Purge gas
p	Particle
prec	Precipitation/precipitated
R	Reactor, reaction
STP	Standard temperature and pressure
s	Static
s	Settling
s	Surface
sat	Saturated
sp	Specific
st	Isosteric
T	Temperature
T	Top (of the column)
tot	Total
ul	Upper limit
v	Vaporous, volumetric
vap	Vaporization
W	Water
0	Starting, initial
1,2	State

Superscripts

n	Distribution (smoothness) parameter
n	Reaction order
m	Power exponent
0	Standard

Part I: **Basics**

1 Do we need off-gas purification?

1.1 Pollutants

Human-made sources of air pollutants cover a wide variety of chemicals and physical activities. They are the major contributors to urban air pollution. The main classes of pollutants are particulates, carbon monoxide, hydrocarbons, nitrogen oxides, sulfur oxides and hydrogen halides. Classification may also consider the origin (primary or secondary).

Specification of gaseous substances has to distinguish between inorganic and organic pollutants.

The group of inorganic pollutants mainly covers:
- Sulfur oxides
- Hydrogen sulfide
- Nitrogen oxides
- Halogenes
- Halides
- Ozone and oxidants
- Cyanides
- Ammonium compounds

The group of organic pollutants mainly considers
- Hydrocarbons (paraffins, olefins, aromatics)
- Oxygenated aliphatic compounds (alcohols, aldehydes, ketones, acids, peroxides)
- Organic halides
- Organic sulfides and amines

Particulates include solid and liquid matter whose diameter is larger than a molecule but smaller than about 100 µm. Particulates dispersed in gaseous carrier are termed an "aerosol."

Aerosols are difficult to classify on a scientific basis in terms of their fundamental properties such as settling rate, optical activity, reaction activity or physiological impact.

1.2 Effects of air pollutants

Air pollution can have a significant effect on human health, animals, vegetation and materials [1]. Every endeavor has been made in the last decades to raise ambient air quality. However, reports of environmental organizations still figure out the necessity of improvement, exemplarily underlined by the 2019 Report on Air Quality in Europe of the European Environment Agency [2].

https://doi.org/10.1515/9783110763928-001

Although acute problems may underline the significance of pollution control they are actually the lesser of the health problems. People exposed to polluted atmosphere over extended periods of time suffer from several ailments and a reduction in life span. Air pollution contributes to increased morbidity and accelerated onset of chronic respiratory diseases. The EEA report 15/2021 [3] on air quality in Europe attributes 307,000 premature deaths to chronic exposure to fine particulate matter (PM), 40,400 premature deaths to chronic nitrogen dioxide exposure and 16,800 premature deaths to acute ozone exposure in 2019 [4], although death toll was cut by 30% since 2005 due to huge efforts made in pollution control.

To encourage the member states to intensify activities in air quality improvement the European Parliament and the Council have promulgated Directive 2016/2284 [5], with the objective, as specified in Article 1:

> In order to move towards achieving levels of air quality that do not give rise to significant negative impacts on and risks to human health and the environment, this Directive establishes the emission reduction commitments for the Member States' anthropogenic atmospheric emissions of sulfur dioxide (SO_2), nitrogen oxides (NO_x), non-methane volatile organic compounds (NMVOC), ammonia (NH_3) and fine particulate matter (PM2.5) and requires that national air pollution control programs be drawn up, adopted and implemented and that emissions of those pollutants and the other pollutants referred to in Annex I, as well as their impacts, be monitored and reported.

Annex I lists detailed monitoring and reporting requirements and a list of national emission reduction commitments of the member states in Annex II, as well as guidelines for national air pollution programs in Annex III and national emission inventories in Annex IV.

EEA report 15/2021 [3] mentions a decline of 29% of PM 2.5, 36% of NO_x and 76% of SO_2 related to 2005. In 2019 the manufacturing and extractive industry still contributed 9% to black carbon emissions, 21% to CO emissions, 47% to non-methane volatile organic carbon emissions (NMVOCs), 14% of NO_x emissions, 22% of PM10 emissions, 17% of PM2.5 emissions, 35% of SO_2 emissions, 13% of CH_4 emissions, 63% of Pb emissions, 55% of Cd emissions, 44% of Hg emissions and 36% of As emissions, leaving a huge potential for future efforts in off-gas purification. (It is probably worth mentioning the 94% share of agriculture in NH_3 emissions according to EEA report 15/2021 [3].)

The acute toxicological effects are reasonably well understood. But the effects of exposure to heterogeneous mixtures of gases and particulates at low concentrations are only beginning to be comprehended and they need extended epidemiological and laboratory research. The mechanism by which an animal can become poisoned is completely different from that by which humans are affected. Inhalation is an important route of entry in acute air pollution exposures. Most common exposure for herbivorous animals is ingestion of feed contaminated with air pollutants. Vegetation is more sensitive to many air contaminants than humans and animals. The effects of air pollution on vegetation can appear as death, stunted growth, reduced crop yield and degradation of color.

The damage that air pollution can do to materials is well known. Among the most important effects are discoloration, corrosion, soiling and impairment of atmospheric visibility.

1.3 How can we get it done?

Let us start the discussion of the basic aspects with an off-gas specification, shown in Tab. 1.1. We may for example find such specifications in off-gas measurement reports for off-gas from combustion of a residual solvent dispersion. We will quickly find out that seemingly a combustion process is the origin of this off-gas when comparing the concentration values of nitrogen, oxygen and carbon dioxide. Our job will be to design a complete off-gas treatment line, including the technologies, the equipment design and the expected performance, to comply with the legislative needs. But how? Do we actually need any off-gas purification equipment, and what kind of equipment? What is the expected/needed performance? What do the readings of Tab. 1.1 tell us?

Tab. 1.1: Our off-gas specification.

Specification	Value	Unit	Dust, PSD	Q_3
P_{stat}	980	hPa		
P_{dyn}	981.095	hPa	x_{ul}	
θ	200	°C	0.2	0.004
Stack diameter	1	m	0.4	0.018
N_2	78	%	0.6	0.039
O_2	12	%	0.8	0.069
CO_2	10	%	1.0	0.105
X_{H_2O}, volume-based	100	$g\ m^{-3}{}_{STP,dry}$	2.0	0.359
ROV	3	%	4.0	0.831
NO_x	300	ppm	6.0	0.982
SO_x	300	ppm	8.0	0.999
HCl	30	ppm	10.0	1.000
Dust	3.0	$g\ m^{-3}{}_{STP,dry}$		

Before we jump into this activity, let us explain the last question with an example, listed in Tab. 1.2. Table 1.2 shows the clean gas specification of waste incineration experiments in a fluidized bed incinerator. The off-gas purification line was equipped with a Venturi scrubber and three spray columns in series operated in co-current mode. Except PM the concentration values are referred to the reference oxygen value (ROV) of 11%.

Some concentration values, shown in bold letters, failed the emission limit, offering important background information about the fuel quality. Seemingly sewage sludge produces a huge amount of fines when fed to a fluidized bed incinerator. Wooden railroad ties had a lead pin seal, and garage waste provides all ingredients for PCDD (polychlorinated dibenzodioxine) synthesis. These comments just intend to

Tab. 1.2: Measurement report of incineration experiments in a fluidized bed incinerator.

Parameter	Waste wood	Sewage sludge	Railroad ties	Garage waste	Emission limit
ROV (vol%)	11	11	11	11	11
Particulate (mg $m^{-3}{}_{STP}$)	1.5	**28**	6	1	10
CO (mg $m^{-3}{}_{STP}$)	6	10	12	8	50
NO_2 (mg $m^{-3}{}_{STP}$)	188	277	213	121	300
SO_2 (mg $m^{-3}{}_{STP}$)	1	57	6	5	100
VOC (mg $m^{-3}{}_{STP}$)	0.3	11	16	1	20
Cl^- (mg $m^{-3}{}_{STP}$)	6	0.9	0.9	1	15
F^- (mg $m^{-3}{}_{STP}$)	0.05	0.1	0.1	0.01	0.7
Cd (mg $m^{-3}{}_{STP}$)	0.001	0.03	0.03	0.01	0.05
Hg (mg $m^{-3}{}_{STP}$)	0.01	0.001	0.004	0.03	0.05
As, Co, Ni (mg $m^{-3}{}_{STP}$)	0.004	0.07	0.02	0.03	0.7
Pb, Zn, Cr (mg $m^{-3}{}_{STP}$)	0.01	0.8	**3.7**	0.7	3
PCDD-TE (mg $m^{-3}{}_{STP}$)	0.02	0.03	0.007	**0.98**	0.1

encourage the reader to spend sufficient time with analyzing and understanding a measurement report, and to try to collect as much background information about a specific emission source as possible.

1.4 Our job

We start with discussing and interpreting the specific off-gas measurement report of Tab. 1.1. This measurement report will be our guide for developing the basics and the technologies. Before discussing technologies we will need the complete off-gas specification and the link to the appropriate legislative framework, which will help us in specifying the need of action and the expected outcome. Then we discuss technologies, the arrangement of equipment and the design of equipment. For interpreting the report, we make use of the physical–chemical basics.

Before discussing any technical measures, we have to complete the specification. We have to develop solver strategies for the following specification details:

First	Mean molar mass (MM_{mean}) of the carrier gas
Second	Gas density ($\rho_{g,STP,dry}$)
	Gas density ($\rho_{g,STP,humid}$)
	Gas density ($\rho_{g,actual}$)
Third	Actual gas flow rate ($am^3\ h^{-1}$)
	Gas flow rate of the humid gas ($m^3{}_{STP,humid}\ h^{-1}$)
	Flow rate of the carrier gas ($m^3{}_{STP,dry}\ h^{-1}$)
Fourth	Partial pressure of vaporous water (p_{H_2O})

Fifth	Water dew point temperature (θ_{dp})
Sixth	Volumetric water content (vol%)
Seventh	Sulfuric acid dew point temperature (θ_{dp, H_2SO_4})
Eighth	Cooling limit temperature (θ_{cl}) and adiabatic cooling limit temperature (θ_{acl})

If you are familiar with these calculations, you may directly move to the off-gas purification technologies. If you are not familiar with these calculations, we will discuss these questions step by step. Off-gas purification technologies need the complete specification of the actual and the standard state of gases to be able to work out mass balances and hydraulic design data. Specification of dry gas, adiabatic change of state, humid gas and gas in actual state needs to be performed correctly. We of course have to care for PM too. We will discuss the basics we need for PM control separately. Dust collection is concerned with the removal or collection of (mainly solid) dispersoids from gaseous carrier for purposes of

- air pollution control (fly ash),
- reduction of equipment-maintenance (intake-air of engines),
- elimination of safety- or health hazards (siliceous dust or flour dust),
- improvement of product quality (pharmaceutical industry),
- recovery of valuable products (smelters and dryers), and
- collection of powdered products (pneumatic conveying and spray drying).

2 The basics (for successful design work)

The physical–chemical home base of any discussion of the gaseous state is the perfect gas law:

$$P \cdot V = n \cdot R \cdot T \tag{2.1}$$

It is based on the observation of a constant ratio of any change of volume V, pressure P or temperature T in K:

$$\frac{P_1 \cdot V_1}{T_1} = \frac{P_2 \cdot V_2}{T_2} = \text{constant} \tag{2.2}$$

Any change of the gaseous state may be quantified by this noble correlation.

Definition of the standard state:

Pressure P: 1,013 hPa (= 1,013 mbar)

Temperature T: 273.15 K

According to eq. (2.1) the molar volume (MV) of a perfect gas (= the same number of molecules needs the same amount of volume, independent of the molar mass of the substance) in the standard state is MV = 22.418 m^3_{STP} $kmol^{-1}$. (Remember: The quantity of 1 mol is based on the equivalent mass of a defined number of molecules in comparison with the equivalent number of molecules of the isotope C-12.)

2.1 Volume

In off-gas purification we have to distinguish between the following properties.

2.1.1 Standard volume, dry (in $m^3_{STP,dry}$)

The standard dry volume of a gas mixture defines the amount of the carrier gas at standard conditions (STP) but without water. The carrier gas consists of the constituents N_2, O_2 and CO_2. We assume that the amount of these constituents does not change during off-gas purification, although we meanwhile have to consider carbon dioxide capture technologies.

2.1.2 Standard volume, humid (in $m^3_{STP,humid}$)

The standard humid volume of a gas mixture defines the amount of carrier gas including vaporous (gaseous) water at standard conditions.

https://doi.org/10.1515/9783110763928-002

2.1.3 Actual volume (in am^3)

Any change in temperature and/or pressure of an ideal gas will cause a change in the volume, as explained in the perfect gas law in eq. (2.2).

The specification of the actual state of our off-gas is a necessity for correct hydraulic design work, while dry standard conditions form the base for any mass balance. Let us do some calculations to become used to these specifications.

Exercise 2.1: Mean molar mass, gas density and gas flow rate
Air with a flow rate of $F_V = 1{,}000\ \text{m}^3{}_{\text{STP,dry}}\ \text{kmol}^{-1}$ has a water load of $X = 10\ \text{g kg}^{-1}$:
- Determine the flow rate of the humid gas at STP.
- Determine the actual gas flow rate at $\theta = 60\ °\text{C}$ and a pressure of 900 hPa.

Solver:
In the first step, the mass flow rate of vaporous water has to be specified. For that purpose, we need the density of dry air at STP and the mass flow rate of air.

The molar mass of air is
$\text{MM}_{\text{air}} = 0.79 \cdot 28 + 0.21 \cdot 32 = 28.82\ \text{kg kmol}^{-1}$ (also: g mol^{-1})
The MV of air (an ideal gas) at standard temperature and pressure is

$$\text{MV} = \frac{n \cdot R \cdot T}{P} = \frac{1{,}000 \cdot 8.314 \cdot 273.15}{101{,}300} = 22.418\ \text{m}^3{}_{\text{STP}}\ \text{kmol}^{-1} \left(\text{also L mol}^{-1}{}_{\text{STP}} \text{ or m}^3{}_{\text{STP}} \text{ per } 1{,}000 \text{ mol}\right)$$

And the density of air at standard temperature and pressure ρ_{STP} is

$$\rho_{\text{STP}} = \frac{\text{MM}_{\text{air}}}{\text{MV}} = \frac{28.82}{22.418} = 1.286\ \text{kg m}^{-3}{}_{\text{STP}}$$

The mass flow rate F_m of dry air is then

$$F_m = F_V \cdot \rho = 1{,}000 \cdot 1.286 = 1{,}286\ \text{kg h}^{-1}$$

The mass flow rate of water vapor F_{m,H_2O} is derived from the mass flow rate of dry air $F_{m,\text{air}}$ multiplied with the water load X:

$$F_{m,H_2O} = F_{m,\text{air}} \cdot X = F_V \cdot \rho \cdot X = 1{,}000 \cdot 1.286 \cdot 10 = 1{,}286 \cdot 10\ \text{g h}^{-1} = 12.86\ \text{kg h}^{-1}$$

Now the volumetric flow rate of vaporous water can be determined.

The molar mass of water is MM = 18 kg kmol^{-1}. The MV of vaporous water at STP is MV = 22.418 m^3 kmol^{-1}.
The mass flow rate of water F_{m,H_2O} = 12.86 kg h^{-1} corresponds with a water vapor flow rate of

$$F_{V,H_2O} = \frac{F_{m,H_2O} \cdot \text{MV}_{H_2O}}{\text{MM}_{H_2O}} = \frac{12.86 \cdot 22.418}{18} = 16.016\ \text{m}^3{}_{\text{STP}}\ \text{h}^{-1}$$

The total volumetric flow rate of humid air $F_{V,\text{STP,humid}}$ (including the air and vaporous water) is

$$F_{V,\text{air, STP, humid}} = F_{V,\text{air, STP, dry}} + F_{V,\text{water vapor, STP}} = 1{,}000 + 16.016 = 1{,}016.016\ \text{m}^3{}_{\text{STP,humid}}\ \text{kmol}^{-1}$$

At STP we have specified the total gas flow rate $F_{V,1}$ for state (1) with
$P_1 = 1{,}013$ hPa
$T_1 = 273.15$ K
$F_{V,1} = 1{,}016.016\ \text{m}^3{}_{\text{STP,humid}}\ \text{kmol}^{-1}$

According to eq. (2.2), the perfect gas law, we get for the actual state (state 2) with $P_2 = 900$ hPa and

$T_2 = (273.15 + 60) = 333.16$ K:

$$\frac{P_1 \cdot F_{V,1}}{T_1} = \frac{P_2 \cdot F_{V,2}}{T_2} = \frac{1{,}013 \cdot 1{,}016.016}{273.15} = \frac{900 \cdot F_{V,2}}{(273.15 + 60)} \text{ and } F_{V,2} = 1{,}394.82 \text{ am}^3 \text{ h}^{-1}$$

2.2 Concentration units

Concentration of gaseous substances is specified differently. The choice of units mainly depends on the amount of the individual substance in a gas mixture.

2.2.1 Volume fraction

The volume fraction is mainly used for highly concentrated gaseous substances (the carrier gas), e.g., the volume fraction of oxygen in air is 0.21 and the volume fraction of nitrogen is 0.79.

2.2.2 ppm

The amount of dilute gaseous constituents is preferably specified by the volumetric concentration ppm in the off-gas.

Exercise 2.2: Definition of ppm

Off-gas is contaminated with 1 cm³ NO in 1 m³ of gas at STP. The volumetric concentration of NO is then:

$$1\left[\frac{\text{cm}^3\text{ NO}}{\text{m}^3}\right] = \frac{1[\text{cm}^3\text{ NO}]}{1,000,000[\text{cm}^3\text{gas}]} = 1 \text{ part of NO in } 1 \text{ million parts} = 1 \text{ ppm}$$

2.3 Conversion from ppm into mg m^{-3}_{STP}

By multiplying the volumetric concentration ppm with the molar density of the pollutant $\rho_{Pollutant}$ we obtain the mass concentration. Several constituents of an off-gas may occur in different form (e.g., SO_2 and SO_3 have to be specified by the equivalent SO_2 content, or NO, NO_2, N_2O_3, N_2O_5 have to be expressed by the equivalent NO_2 content).

Exercise 2.3: Conversion from ppm into mg $m^{-3}{}_{STP}$

We start with a volumetric NO concentration of 1,000 ppm or 1,000 cm^3 NO in 1 m^3 at STP. The molar mass of NO is 30 g mol^{-1} or 30,000 mg mol^{-1}. The molar mass of NO_2 is 46 g mol^{-1} or 46,000 mg mol^{-1}, and the MV of either constituents is MV = 22,418 $cm^3{}_{STP}$ mol^{-1}.

Solver:

We calculate the molar concentration of NO by dividing the volume of 1,000 ppm (equivalent to 1,000 cm^3 m^{-3}) with the MV,

$$\text{MV} = 22{,}418\ \text{cm}^3\ \text{mol}^{-1}$$

and the mass concentration by multiplying the molar concentration with the molar mass:

$$1{,}000\ \text{ppm NO} = 1{,}000\left[\text{cm}^3\ \text{NO m}^{-3}\right]\cdot\frac{1}{22{,}418}\left[\text{mol cm}^{-3}\right]\cdot 30{,}000\left[\text{mg NO mol}^{-1}\right] \equiv 1{,}338\ \text{mg NO m}^{-3}$$

If we have to specify the equivalent NO_2 mass concentration, we just have to exchange molar masses

$$1{,}000\ \text{ppm NO}_2 = 1{,}000\left[\text{cm}^3\ \text{NO}_2\ \text{m}^{-3}\right]\cdot\frac{1}{22{,}418}\left[\text{mol cm}^{-3}\right]\cdot 46{,}000\left[\text{mg NO}_2\ \text{mol}^{-1}\right]$$
$$\equiv 2{,}052\ \text{mg NO}_2\ \text{m}^{-3}$$

2.4 Reference oxygen value (ROV)

Concentration of oxygen in off-gas from combustion processes can vary widely. For comparability of specifications we have to define reference oxygen concentration values (ROVs), a necessity for the comparability of emission data from incineration processes, to avoid falsification of concentration results by dilution. With regard to emission data from incineration processes, the ROV is specified by legislation. Table 2.1 shows the selected ROVs for different fuels.

Tab. 2.1: Selected reference oxygen values for different fuels (and incineration facilities with <50 MW fuel heat duty).

Fuel type	ROV (%)
Solid fuels	6
Fuel oils, vegetable oils, gaseous fuels	3
Wood	11

Annex V of EU Directive 2010/75 [6] has specified emission limit values for combustion plants. Referring to Article 30(2),

> all permits for installations containing combustion plants which have been granted a permit before 7 January 2013, or the operators of which have submitted a complete application for a permit before that date, provided that such plants are put into operation no later than 7 January 2014, shall include conditions ensuring that emissions into air from these plants do not exceed the emission limit values set out in Part 1 of Annex V.

In particular Annex V precises emission limits:

> All emission limit values shall be calculated at a temperature of 273.15 K, a pressure of 101.3 kPa and after correction for the water vapor content of the waste gases and at a standardized O_2 content (= ROV) of 6% for solid fuels, 3% for combustion plants, other than gas turbines and gas engines using liquid and gaseous fuels and 15% for gas turbines and gas engines.

To briefly explain the background of ROVs we compare off-gas data from magnesite sintering in a rotary kiln in Fig. 2.1. During burner optimization the kiln was fueled with natural gas. The air/fuel ratio was varied to find a correlation between the residual oxygen content of the off-gas, NO_x load and CO load.

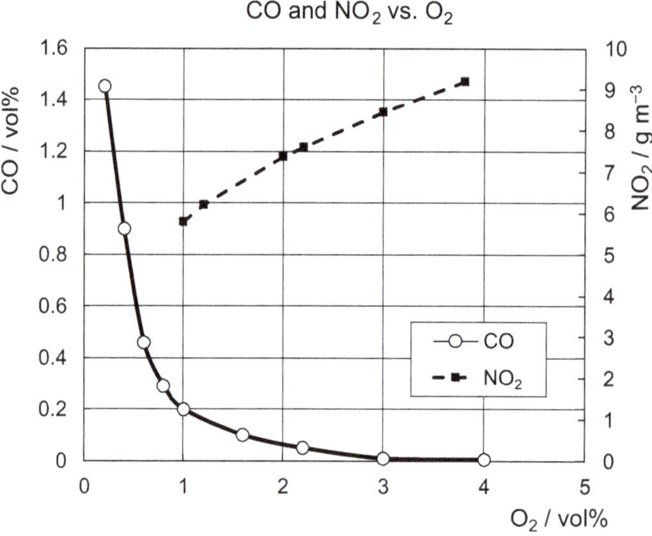

Fig. 2.1: Oxygen concentration, CO concentration and NO_x concentration of off-gas from magnesite sintering in a rotary kiln, fueled with natural gas.

Figure 2.1 clearly indicates the onset of increasing CO concentration in the off-gas at 3% residual oxygen content. This example may explain the ROV value of 3% oxygen for fueling combustion facilities with natural gas.

Exercise 2.4: Conversion of concentration data and the background of the ROV

Spent lubrication oil is incinerated. The off-gas record shows an oxygen content of 14 vol% and a NO_x concentration of 1,000 ppm. According to legislative needs the emission data for NO_x has to be recorded in the concentration unit mg NO_2 $m^{-3}{}_{STP}$ based on a reference oxygen content of 3 vol%; therefore, mg NO_2 $m^{-3}{}_{STP,O_2 = 3\%}$.

Solver:
In the first step, we transfer from ppm NO_x to the mass concentration mg NO_2 m^{-3} at STP (have in mind that different ideal gases do have the same MV but different molar mass):

$$1{,}000 \text{ ppm } NO_x \equiv 1{,}000 \cdot \frac{46}{22.418} = 2{,}052 \text{ mg } NO_2 \text{ m}^{-3}$$

When relating the calculated emission value of a pollutant at given oxygen content MOV (measurement oxygen value) of 14 vol% in the carrier gas V_1 to a reference oxygen content ROV of 3 vol%, we must by means remove a portion of air ($V_1 - V_2$) with an oxygen content of 21 vol% from the off-gas. The pollutant is left in the residual off-gas V_2 with an oxygen content of 3 vol%. We set up the oxygen balance

$$V_1 \cdot \text{MOV} - (V_1 - V_2) \cdot 21 = V_2 \cdot \text{ROV}$$

and insert numbers for $V_1 = 1 \text{ m}^3$

$$1 \cdot 14\% - (1 - V_2) \cdot 21\% = V_2 \cdot 3\%$$

MOV Measurement oxygen value (= 14 vol% O_2)
ROV Reference oxygen value (= 3 vol% O_2)

resulting in $V_1 - V_2 = 0.611 \text{ m}^3$ of air to be removed per m^3 of dry off-gas, and to leave a residual off-gas volume of $V_2 = 1 - 0.611 = 0.389 \text{ m}^3$ with all the NO_x of 2,052 mg NO_2 m^{-3} in the original off-gas being kept in the residual volume V_2.

Now we perform the NO_x balance.

Since all the pollutant is left in the residual gas V_2 the concentration must rise to suffice the mass balance:

$$V_1 \cdot c_1 = V_2 \cdot c_2 \text{ and } c_2 = \frac{V_1 \cdot c_1}{V_2} = \frac{2{,}052 \cdot 1}{0.389} = 5{,}275 \text{ mg } NO_2 \text{ m}^{-3}{}_{\text{ROV}=3\%}$$

With this procedure in mind we may simplify:

First step: Oxygen balance

$$V_1 \cdot \text{MOV} - (V_1 - V_2) \cdot 21 = V_2 \cdot \text{ROV and } \frac{V_1}{V_2} = \frac{21 - \text{ROV}}{21 - \text{MOV}} \tag{2.3}$$

$$V_1 \cdot c_1 = V_2 \cdot c_2 \quad \text{and} \quad c_2 = \frac{V_1}{V_2} \cdot c_1 \tag{2.4}$$

Second step: Pollutant balance

$$c_{NO_2, 3\%\, O_2} = 2{,}052 \cdot \frac{21 - 3}{21 - 14} = 5{,}275 \text{ mg } NO_2 \text{ m}^{-3}{}_{\text{ROV}=3\%}$$

To avoid confusion: Whenever you are faced with the specification mg m^{-3} without any additional information you may assume that a mass concentration at STP and dry gas and actual oxygen content is specified.

2.5 Gas velocity

The velocity of off-gas in a ducted system can be determined by anemometric measurement, by hot wire detection or by determination of the velocity pressure P_V. Latter is performed by carrying out traverses with a Prandtl probe at a plane in a section of the duct and at right angles to the walls of the duct. The section chosen should be at least six duct diameters or widths downstream of any bend or obstruction. We will discuss velocity determination via P_V monitoring in detail.

As shown in Fig. 2.2, the measurement equipment for P_V monitoring consists of the Prandtl probe tube and a manometer. The Prandtl probe tube consists of the probe head, the stem and the tail. The construction of the head includes the dip for the detection of the total pressure P_1 and holes on the outer tube, which are needed for detecting the static pressure P_2. Both concentrically arranged tubes are led to pipe couplings at the tail end. The manometer is connected with the pipe plugs with flexible hoses.

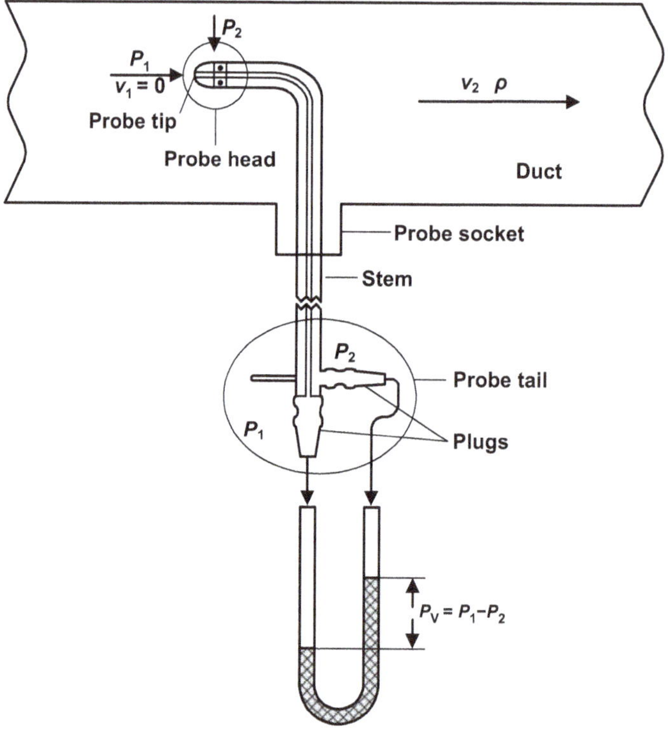

Fig. 2.2: Measurement setup for determining the gas velocity with a Prandtl probe tube in a duct.

During measurement, the static (duct) pressure P_2 is determined by connecting the static pressure plug with the manometer. The velocity pressure P_V is measured by connecting both plugs with the manometer.

The Prandtl static tube is inserted into the duct in two rectangular traverses via holes at appropriate positions. It is moved to fixed positions. The distance between different measurement positions is indicated with sliding clip markers fitted to the stem of the Prandtl probe tube. We have to monitor the gas velocity over the whole cross section to obtain a reliable result. Carefully read the guidelines for correct monitoring of the gas velocity (e.g., DIN EN ISO16911-1 (2013–06): Stationary source emissions – Manual and automatic determination of velocity and volume flow rate in ducts – Part 1: Manual reference method) before you perform these measurements.

It is very important to position the head of the Prandtl probe tube parallel to the wall of the duct. The direction of the head is indicated by a pointer fixed to the tail of the Prandtl probe tube.

The bulk of the gas flow shall be parallel to the walls. Eddies and cross-flow have to be considered by finding the angle at which maximum velocity pressure reading is obtained. For that purpose, the head of the Prandtl probe tube has to be turned in both pitch and yaw. The angle should be less than 15° from the parallel position to the wall. The gas velocity is then derived from Bernoulli's law (at constant height) via $P_V = (P_1 - P_2)$:

$$P_1 + \frac{\rho}{2} \cdot v_1^2 = P_2 + \frac{\rho}{2} \cdot v_2^2$$

With $v_1 = 0$ we get for the velocity v_2 of the gas:

$$v_2 = \sqrt{\frac{2 \cdot (P_1 - P_2)}{\rho}} \tag{2.5}$$

The average velocity in a duct is calculated from averaging the square root of several velocity measurement readings.

Exercise 2.5: Gas velocity, gas flow rate and state

A storehouse with a volume of 2,000 m^3 is aerated with fresh air. The air is released to the environment via a stack installed on the roof. We have to check by airflow measurement with a Prandtl probe tube whether a recommended air exchange rate of 2 (twice the air volume of the storehouse per hour) is obtained. The reading says:

Stack diameter D = 400 mm

Temperature θ = 28 °C

Static pressure P_2 = 1,002 hPa (= 100,200 Pa)

Total pressure P_1 = 1,002.6 hPa (= 100,260 Pa)

Solver:

$$\mathrm{MM_{air}} = (0.79 \cdot 28 + 0.21 \cdot 32) = 28.84 \text{ g mol}^{-1}$$

$$\mathrm{MV}_{\mathrm{air},T,P_2} = \frac{22.418 \cdot 1{,}013 \cdot (273.15 + 28)}{1{,}002 \cdot 273.15} = 25.00 \text{ m}^3\text{ kmol}^{-1} = 25.00 \text{ L mol}^{-1}$$

$$\rho_{T,P_2} = \frac{\mathrm{MM_{air}}}{\mathrm{MV}_{\mathrm{air},T,P_2}} = \frac{28.84}{25.00} = 1.154 \text{ kg m}^{-3}$$

$$v_2 = \sqrt{\frac{2 \cdot (P_1 - P_2)}{\rho}} = \sqrt{\frac{2 \cdot (100{,}260 - 100{,}200)}{1.154}} = 10.2 \text{ m s}^{-1}$$

With the stack cross-sectional area A

$$A = \frac{D^2 \cdot \pi}{4} = 0.126 \text{ m}^2$$

we can calculate the actual flow rate of air $F_{V,\mathrm{air}}$ and with eq. (2.2), we obtain the flow rate at STP

$$F_{V,\,\text{actual}} = 10.2 \cdot 0.126 \cdot 3{,}600 = 4{,}611.74\ \text{am}^3\ \text{h}^{-1}$$

$$F_{V,\,\text{STP}} = \frac{4{,}611.74 \cdot 273.15 \cdot 1{,}002}{(273.15 + 28) \cdot 1{,}013} = 4{,}137.5\ \text{m}^3{}_{\text{STP}}\ \text{h}^{-1}$$

!

Discussion: The air exchange rate is above the set point (even for STP; the specification for the gas exchange rate is actually not correct, because it misses the indication of the state of the gas flow rate). We therefore offer both values.

2.6 Dew point, cooling limit, adiabatic cooling limit and sulfuric acid dew point

If we want to make use of off-gas energy for heating purposes, we have to transfer this portion of energy from the off-gas to the energy user via heat exchanger. The maximum possible heat we may withdraw from the heat source is limited by the water dew point and by the sulfuric acid dew point of our off-gas. We therefore have to determine the corresponding dew points. When we have to quench off-gas by evaporating water, we have to determine how much water we need to either achieve a specific temperature or to achieve complete water saturation. In latter case we also have to determine the corresponding off-gas temperature. For doing the balances we briefly address the definition of the thermodynamic properties.

2.6.1 Definition

$h_{g,\,273.15} = 0\ \text{kJ kg}^{-1}$	Enthalpy of the gas at 273.15 K
$h_{H_2O,\,273.15} = 0\ \text{kJ kg}^{-1}$	Enthalpy of water added at 273.15 K
$h_{vap,\,H_2O,\,273.15} = 2{,}501\ \text{kJ kg}^{-1}$	Enthalpy of evaporation of water at 273.15 K
$h_T\ [\text{kJ kg}^{-1}] = Cp_g \cdot (T - 273.15)$	Specific enthalpy of the gas at temperature T
$Q_T\ [\text{kJ}] = m_g \cdot Cp_g \cdot (T - 273.15)$	Thermal energy of gases at temperature T

$$h_{1+X}\left[\text{kJ kg}^{-1}\,\text{dry air}\right] = \frac{\left(m_g \cdot Cp_g + X \cdot Cp_{H_2O,\,vap}\right) \cdot (T - 273.15) + m_g \cdot X \cdot h_{vap,\,H_2O,\,273.15}}{m_g}$$

$$h_{1+X}\left[\text{kJ kg}^{-1}\,\text{dry air}\right] \frac{\left(m_g \cdot Cp_g + X \cdot Cp_{H_2O,\,vap}\right) \cdot \theta + m_g \cdot X \cdot h_{vap,\,H_2O,\,273.15}}{m_g} \tag{2.6}$$

where h_{1+X} is the specific enthalpy of the humid gas at temperature θ (= $T - 273.15$) (kJ kg^{-1} dry gas), θ is the temperature (°C), m_g is the mass of dry gas (carrier gas) (kg), X is the specific humidity (kg water vapor kg^{-1} dry gas), Cp_g is the specific heat of the gas at constant pressure and $Cp_{H_2O,\,v}$ is the specific heat of vaporous water at constant pressure.

We may collect the thermodynamic data from any data source (e.g., VDI Wärmeatlas). Tables 2.2 and 2.3 show selected data for some constituents.

We will not fail much, when we simplify by assuming $Cp_{N_2, O_2, CO_2, \text{mean}} = 1\ \text{kJ kg}^{-1}\ \text{K}^{-1}$ and $Cp_{H_2O, \text{vap}, \text{mean}} = 1.91\ \text{kJ kg}^{-1}\ \text{K}^{-1}$.

Tab. 2.2: Specific heat of selected off-gas constituents at constant pressure [7].

Constituent	$Cp_{g, 273\ K}$ (kJ kg^{-1} K^{-1})	$Cp_{g, 373\ K}$ (kJ kg^{-1} K^{-1})	$Cp_{g, 473\ K}$ (kJ kg^{-1} K^{-1})	$Cp_{g, \text{mean}}$ (kJ kg^{-1} K^{-1})
N_2	1.036	1.044	1.056	1.05
O_2	0.923	0.943	0.966	0.94
CO_2	0.849	0.920	0.985	0.92
H_2O_{vap}	1.871	1.906	1.95	1.91

We will not fail much, when we simplify by assuming $Cp_{H_2O, l, \text{mean}} = 4.19\ \text{kJ kg}^{-1}\ \text{K}^{-1}$.

Tab. 2.3: Selected thermodynamic data of water in the liquid state [7].

	$T = 273.15$ K	$T = 323.15$ K	$T = 373.15$ K
Cp_l (kJ kg^{-1} K^{-1})	4.219	4.18	4.216
Δh_{vap} (kJ kg^{-1})	2,500.9	2,382.3	2,256.5

2.6.2 Vapor pressure of water (Antoine equation)

$$\log p_{H_2O, \theta, 0} = A - \frac{B}{C + \theta} \tag{2.7}$$

$p_{H_2O, \theta, 0}$ is the vapor pressure of water at temperature θ, and θ is the temperature (°C).

After rearranging we get the temperature θ for given vapor pressure $p_{H_2O, \theta, 0}$:

$$\theta = \frac{B}{A - \log(p_{H_2O, \theta, 0})} - C \tag{2.8}$$

For pressure unit hPa we may find the Antoine parameters A, B and C for water in several data series:

$A = 8.19625$

$B = 1730.462$

$C = 233.426$

Comment: Be careful with the Antoine parameters. They may be listed for different pressure units and different temperature units. You may simply check that, when inserting the boiling point temperature of the constituent in eq. (2.7) (e.g., water: $P = 1{,}013$ hPa, $bp = 99.96$ °C).

Boundary condition for water saturation of the gas phase: The partial pressure $p_{H_2O,\theta}$ and the vapor pressure $p_{H_2O,\theta,0}$ are same:

$$p_{H_2O,\theta} = p_{H_2O,\theta,0}$$

Relative humidity:

$$\varphi = \frac{p_{H_2O,\theta}}{p_{H_2O,\theta,0}} \tag{2.9}$$

Dalton's law:

$$p = y \cdot P_{tot} \tag{2.10}$$

2.6.3 The specific water load and the partial pressure of water vapor

Based on the perfect gas law we may derive the specific water load X in kg H_2O $kg^{-1}_{STP,dry}$ of a binary gas mixture (water vapor plus gas) for given

- System pressure: P_{tot}
- Partial pressure of water vapor: p_{H_2O}
- Partial pressure of the gas: $p_{gas} = P_{tot} - p_{H_2O}$

$$p_{H_2O} \cdot V = n_{H_2O} \cdot R \cdot T = \frac{m_{H_2O}}{MM_{H_2O}} \cdot R \cdot T \text{ and } p_{H_2O} = \frac{m_{H_2O}}{MM_{H_2O}} \cdot \frac{R \cdot T}{V}$$

After rearranging we get

$$m_{H_2O} = p_{H_2O} \cdot MM_{H_2O} \cdot \frac{V}{R \cdot T} \quad \text{and}$$

$$m_{gas} = p_{gas} \cdot MM_{gas} \cdot \frac{V}{R \cdot T} = \left(P_{tot} - p_{H_2O}\right) \cdot MM_{gas} \cdot \frac{V}{R \cdot T} \quad \text{and}$$

$$\frac{m_{H_2O}}{m_{gas}} = X = \frac{p_{H_2O}}{\left(P_{tot} - p_{H_2O}\right)} \cdot \frac{MM_{H_2O}}{MM_{gas}} \cdot \frac{V}{R \cdot T} \cdot \frac{R \cdot T}{V}$$

with

$$X = \frac{p_{H_2O}}{\left(P_{tot} - p_{H_2O}\right)} \cdot \frac{MM_{H_2O}}{MM_{gas}} \tag{2.11}$$

and after rearranging for p_{H_2O}:

$$p_{H_2O} = \frac{X \cdot P_{tot}}{\frac{MM_{H_2O}}{MM_{gas}} + X} \tag{2.12}$$

The dimension of the partial pressure of water p_{H_2O} just depends on the dimension of the system pressure P_{tot}.

After determining the partial pressure of water vapor in the off-gas, we can finally calculate the temperature θ at which the partial pressure equals the vapor pressure according to the condition for water saturation:

$$p_{H_2O,\theta} = p_{H_2O,\theta,0}$$

To avoid condensation, we must keep the gas temperature in dry off-gas purification equipment, ducting and heat transfer equipment well above the water dew point temperature.

Exercise 2.6: Water load, partial pressure, dew point temperature and specific enthalpy

The temperature of air in a ducted system is $\theta = 200$ °C, the specific water load is $X = 0.12$ kg kg^{-1} and the system pressure is $P_{tot} = 1{,}025$ hPa. We have to determine

- the partial pressure of water vapor p_{H_2O} in hPa,
- the dew point temperature θ_{dp} and
- the specific enthalpy h_{1+X} in kJ kg^{-1},

and we have to determine how much heat energy we may withdraw from this portion of air without slipping into a water vapor condensation disaster.

Solver:

The molar mass of air is $MM_{air} = (0.79 \cdot 28 + 0.21 \cdot 32) = 28.84$ g mol^{-1}

The molar mass of water is $MM_{H_2O} = 18$ g mol^{-1}.

The partial pressure of the water vapor p_{H_2O} is

$$p_{H_2O} = \frac{X \cdot P_{tot}}{\frac{MM_{H_2O}}{MM_{gas}} + X} = \frac{0.12 \cdot 1{,}025}{\frac{18}{28.84} + 0.12} = 165.29\,\text{hPa}$$

When we withdraw energy from our gas via heat exchanger, the temperature will drop and finally we will end up at the dew point temperature when the partial pressure of the humidity becomes equivalent the vapor pressure at corresponding (dew point) temperature $p_{H_2O,\theta} = p_{H_2O,\theta,0}$, and

$$\theta_{dp,H_2O} = \frac{B}{A - \log(p_{H_2O})} - C = \frac{1{,}730.462}{8.196 - \log(165.29)} - 233.426 = 56.05\,°\text{C}$$

Discussion: If we make use of some specific heat energy Δh of the air as specified via heat exchanger, we must not go below a temperature of 56.05 °C to avoid condensation of water vapor.

The specific enthalpy of our gas is made up of the specific enthalpy of the dry gas

$$h_{dry\,gas} = m_{gas} \cdot Cp_g \cdot \theta$$

the energy for water evaporation at $T = 273.15$ K

$$Q_{H_2O,\,evap,\,273.15} = m_{gas} \cdot X \cdot h_{vap,\,H_2O,\,273.15}$$

and the energy of vaporous water

$$Q_{vapor} = m_{gas} \cdot X \cdot Cp_{H_2O,\,vap} \cdot \theta$$

$$h_{1+X}\left[\text{kJ kg}^{-1}\text{ dry gas}\right] = \frac{\left(m_{gas} \cdot Cp_g + m_{gas} \cdot X \cdot Cp_{H_2O,\,vap}\right) \cdot \theta + m_{gas} \cdot X \cdot h_{vap,\,H_2O,\,273.15}}{m_{gas}} \quad (2.13)$$

$$h_{1+X,\ 200\,°C} = \frac{(1 \cdot 1 + 1 \cdot 0.12 \cdot 1.91) \cdot 200 + 1 \cdot 0.12 \cdot 2,501}{1} = 545.96\ \text{kJ kg}^{-1}\text{ dry gas}$$

$$h_{1+X,\ 56.05\,°C} = \frac{(1 \cdot 1 + 1 \cdot 0.12 \cdot 1.91) \cdot 56.05 + 1 \cdot 0.12 \cdot 2,501}{1} = 369.01\ \text{kJ kg}^{-1}\text{ dry gas}$$

We may therefore make use of as much as $\Delta Q = (545.96 - 369.01) = 176.95$ kJ kg^{-1}dry air.

! Discussion: In dry off-gas purification it is recommended to keep the temperature well above any dew point temperature.

2.6.4 The adiabatic cooling limit

Off-gas purification by absorption is carried out by cooling the gaseous phase with the solvent (water), followed by the absorption process. Hydraulic design and the evaluation of the mass balance need the quantitative determination of the evaporative transfer of the solvent (water) from the liquid state to the gas phase to achieve complete (water) solvent saturation. The determination of the adiabatic cooling limit is performed by solving the energy balance and the physical–chemical equilibrium iteratively.

Adding a certain portion of water $F_{m,W}$, $\theta_W = 0$ °C and $h = 0$ kJ kg^{-1}, to a given quantity of gas with the temperature $\theta_{g,in}$ will cause evaporative cooling of the gaseous phase, according to the heat needed for evaporation of the added amount of water $F_{m,W} \cdot h_{vap}$, and the temperature $\theta_{g,out}$ will drop according to eq. (2.14). The water load in the outlet X_{out} will increase according to eq. (2.15), and the partial pressure of water $p_{H_2O,\,out}$ will increase according to eq. (2.12). Then the vapor pressure of water $p_{H_2O,\,\theta,\,0}$ is calculated for the outlet temperature $\theta_{g,out}$ via eq. (2.7). Water saturation is achieved, when the partial pressure $p_{H_2O,\,\theta}$, derived from the mass balance, equals the vapor pressure $p_{H_2O,\,\theta,\,0}$, derived from the Antoine equation. Figure 2.3 shows the mass balance. We have to add just as much water $F_{m,W}$, as we need for complete water saturation of the gas.

If you miss the energy of evaporation for the vaporous water fed to the adiabatic black box with the entering gas $F_{m,g} \cdot X_{in} \cdot h_{vap}$, do not worry. The same amount of energy also exits the adiabatic black box. As a consequence, this amount of energy makes itself vanish. If you complain the missing energy of the water fed to the gas cooling black box $Q_{W,\,in} = F_{m,W} \cdot Cp_{H_2O,l} \cdot \theta_{H_2O,\,in}$, you are right, but kindly consider the rules of the game for the adiabatic cooling limit ($h_{H_2O,\,273.15}$ = 0 kJ kg^{-1}). However, do not worry, we will discuss it. Have in mind that you add just us much water to the gas as needed for obtaining the adiabatic cooling limit temperature (but not any further droplet).

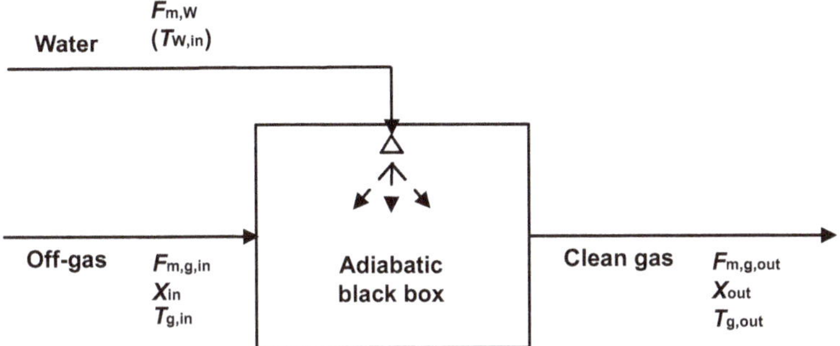

Fig. 2.3: Schematic of the setup for adiabatic gas cooling.

Solver:
First step: Energy balance for an assumed amount of water $F_{m,W}$ added, and calculation of θ_{out} and X_{out}:

$$\left(F_{m,g} \cdot Cp_g + F_{m,g} \cdot X_{in} \cdot Cp_{H_2O,vap}\right) \cdot (T_{in} - 273.15) - F_{m,W} \cdot h_{vap}$$
$$= \left(F_{m,g} \cdot Cp_g + F_{m,g} \cdot X_{in} \cdot Cp_{H_2O,vap} + F_{m,W} \cdot Cp_{H_2O,vap}\right) \cdot (T_{out} - 273.15)$$

$$\text{or } \left(F_{m,g} \cdot Cp_g + F_{m,g} \cdot X_{in} \cdot Cp_{H_2O,vap}\right) \cdot \theta_{in} - F_{m,W} \cdot h_{vap}$$
$$= \left(F_{m,g} \cdot Cp_g + F_{m,g} \cdot X_{in} \cdot Cp_{H_2O,vap} + F_{m,W} \cdot Cp_{H_2O,vap}\right) \cdot \theta_{out}$$

$$\theta_{g,out} = \frac{\left(F_{m,g} \cdot Cp_g + F_{m,g} \cdot X_{in} \cdot Cp_{H_2O,vap}\right) \cdot \theta_{in} - F_{m,W} \cdot h_v}{\left(F_{m,g} \cdot Cp_g + F_{m,g} \cdot X_{in} \cdot Cp_{H_2O,vap} + F_{m,W} \cdot Cp_{H_2O,vap}\right)} \tag{2.14}$$

and

$$X_{out} = \frac{F_{m,g} \cdot X_{in} + F_{m,W}}{F_{m,g}} \tag{2.15}$$

Second step: Calculation of the partial pressure of water $p_{H_2O,out}$ with X_{out}

$$p_{H_2O,out} = \frac{X_{out} \cdot P_{tot}}{\frac{MM_{H_2O}}{MM_{gas}} + X_{out}}$$

Third step: Comparison of $p_{H_2O,\,out}$ (partial pressure, as derived from the inflow water load plus the water fed to the system) and $p_{H_2O,\,\theta,0}$ (vapor pressure, calculated with temperature $\theta_{g,\,out}$)

Comparison:

$p_{H_2O,\,out} < p_{H_2O,\,\theta,0}$ the adiabatic cooling limit (acl) has not been obtained; additional water must be added ($\theta_{g,out} = (T_2 - 273.15) > T_{acl}$)

$p_{H_2O,\,out} = p_{H_2O,\,\theta,0}$ the adiabatic cooling limit (acl) has been obtained; the gas is water saturated ($\theta_{g,out} = (T_2 - 273.15) = T_{acl}$)

$p_{H_2O,\,out} > p_{H_2O,\,\theta,0}$ too much water has been added; the amount of water must be decreased ($\theta_{g,out} = (T_2 - 273.15) < T_{acl}$)

Alternatively, we may also compare temperatures (probably the better route):
First step: We determine the exit temperature $\theta_{g,out}$ from the energy balance.
Second step: We determine X_{out} (from the water load of 1 kg gas and the water $F_{m,W}$ fed to the system.
Third step: We determine $p_{H_2O,\,out}$ via X_{out}.
Fourth step: We determine

$$\theta_{dp} = \frac{B}{A - \log\left(p_{H_2O,\,out}\right)} - C$$

Fifth step: We compare $\theta_{g,out}$ from the energy balance and $\theta_{g,dp}$

$\theta_{g,out} > \theta_{dp}$ water saturation is not obtained

$\theta_{g,out} = \theta_{dp}$ water saturation is obtained

$\theta_{g,out} < \theta_{dp}$ too much water was added

Exercise 2.7: The adiabatic cooling limit

The temperature of air in a ducted system is $\theta = 200$ °C, humidity is $X = 0.12$ kg kg^{-1} and the system pressure is $P_{tot} = 1{,}025$ hPa. We have to find out how much water $F_{m,W}$ we have to add to the gas $F_{m,g}$ to obtain complete water saturation under adiabatic conditions ($h_{H_2O,\,273.15} = 0$ kJ kg^{-1}), and of course we would like to get a result for the gas temperature $\theta_{g,out}$ in the outflow. We compare different solver approaches.

Solver 2.7.1: The "incidence" approach
For convenience we decide for 1 kg h^{-1} of dry gas, $\theta = 200$ °C, specific water load $X = 0.12$ kg kg^{-1} and system pressure $P_{tot} = 1{,}025$ hPa. By guessing we assume $F_{m,W} = 0.064$ kg h^{-1} (comment: This is a prepared "lucky punch"; by guessing you will need several trials. You may rather make use of a solver, if your hand calculator offers this service.)
Then we do calculations step by step.

First step: With the assumed (guessed) amount of water $F_{m,W} = 0.064$ kg h^{-1} evaporated with the heat energy of our gas we can calculate the temperature $\theta_{g,out}$ of the humidified gas via eq. (2.14):

$$\theta_{g,\,out} = \frac{\left(F_{m,g} \cdot Cp_g + F_{m,g} \cdot X_{in} \cdot Cp_{H_2O,\,vap}\right) \cdot \theta_{in} - F_{m,W} \cdot h_{vap}}{\left(F_{m,g} \cdot Cp_g + F_{m,g} \cdot X_{in} \cdot Cp_{H_2O,\,vap} + F_{m,W} \cdot Cp_{H_2O,\,vap}\right)}$$

$$= \frac{(1 \cdot 1 + 1 \cdot 0.12 \cdot 1.91) \cdot 200 - 0.064 \cdot 2{,}501}{(1 \cdot 1 + 1 \cdot 0.12 \cdot 1.91 + 0.064 \cdot 1.91)} = 63.49\ °C$$

Second step: We determine the outflow water load X_{out} via eq. (2.15):

$$X_{out} = \frac{F_{m,g} \cdot X_{in} + F_{m,W}}{F_{m,g}} = \frac{1 \cdot 0.12 + 0.064}{1} = 0.184 \text{ kg } H_2O \text{ kg}^{-1} \text{ air}$$

Third step: We determine the partial pressure of water p_{H_2O} via eq. (2.12):

$$p_{H_2O,out} = \frac{X_{out} \cdot P_{tot}}{\frac{MM_{H_2O}}{MM_g} + X_{out}} = \frac{0.184 \cdot 1{,}025}{\frac{18}{28.84} + 0.184} = 233.37 \text{ hPa}$$

Fourth step: We determine the dew point temperature θ_{dp} and compare it with the temperature θ_{out} we have obtained from solving the energy balance via eq. (2.8):

$$\theta_{dp} = \frac{B}{A - \log(p_{H_2O,out})} - C = \frac{1{,}730.462}{8.19625 - \log(233.37)} - 233.426 = 63.49\ °C$$

Discussion of results: Seemingly we scored with the guess (by incidence?). You will for sure get the same result in a single step, when working with a solver.

Solver 2.7.2: The “lazy student” approach
When we do not feed any water to the off-gas ($F_{m,g} = 1$ kg h^{-1}, $\theta = 200$ °C, $X = 0.12$ kg kg^{-1} and $P_{tot} = 1{,}025$ hPa), we get the dew point temperature for $X = 0.12$ kg kg^{-1} and $p_{H_2O} = 165.29$ hPa

$$\theta_{dp} = \frac{B}{A - \log(p_{H_2O})} - C = \frac{1{,}730.462}{8.196 - \log(165.29)} - 233.426 = 56.05\ °C$$

Now we let the “lazy student” evaporate an (assumed) amount of $F_{m,w} = 0.1$ kg h^{-1} water in our gas. The off-gas temperature will drop to $\theta = -3$ °C (the energy balance will permit it, although it is complete nonsense from a thermodynamics point of view!).

The corresponding dew point temperature for $X = (0.12 + 0.1) = 0.22$ kg kg^{-1} is $\theta_{dp} = 66.506$ °C.

The “serious” engineer will immediately stop and revise the amount of water to be fed because of the cooling temperature miracle, while the “lazy student” wants to get a quick solution, but how?

By constructing two straight lines with the data,

Line 1: $y_1 = k_1 \cdot x + d$, with $d = 200$ and $k_1 = (200 - (-3))/0.1$
Line 2: $y_2 = k_2 \cdot x + d$, with $d = 56.05$ and $k_2 = (66.506 - (-56.05))/0.1$

and intersect them at $y_1 = y_2$ to get: $200 - 2{,}030 \cdot x = 56.05 + 104.6 \cdot x$, with $x = 0.067$, resulting in $X_{out} = 0.187$ kg kg^{-1}, $p_{H_2O,out} = 236.37$ hPa and $\theta_{g,out} = 63.8$ °C.

To avoid confusion, we let the “lazy student” show his solver in a sketch in Fig. 2.4.

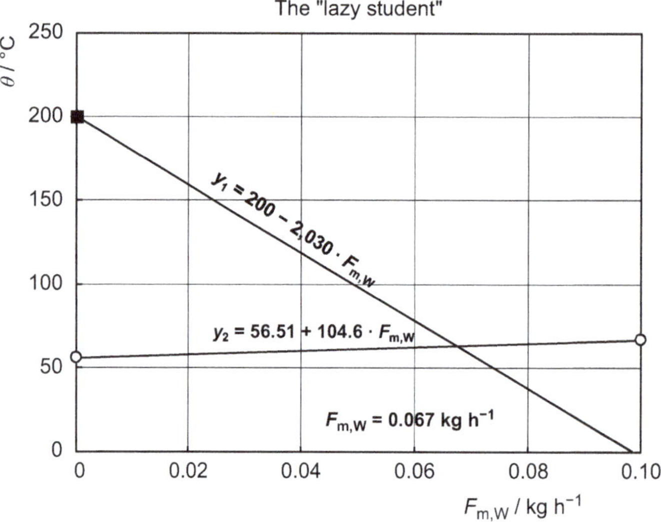

Fig. 2.4: The "lazy student" solver.

Discussion of results: The "lazy student" (we should rather address her/him the "efficient student") comes very close to the correct adiabatic cooling limit result (θ_{dp}= 63.8 °C). However, the difference in water consumption is significant compared to the "incidence approach" (just have in mind the difference of 343.9 kg or 5.4% in water consumption, e.g., 100,000 kg of air).

2.6.5 The cooling limit (or "the correct engineer" approach)

Again we start with 1 kg h^{-1} of air (θ = 200 °C, X = 0.12 kg kg^{-1} and P_{tot} = 1,025 hPa), but consider that the water is fed to the system with temperature $\theta_{W,in}$. Per definition for adiabatic cooling we assumed that the water $F_{m,W}$ fed to our cooling box in the sketch in Fig. 2.3 does not contribute to the energy balance, when actually the energy content $Q_{W,in}$ of $F_{m,W}$ is $Q_{W,in} = F_{m,W} \cdot Cp_{H_2O,l} \cdot \theta_{H_2O,in}$.

If we consider the amount of energy $Q_{W,in} = F_{m,W} \cdot Cp_{H_2O,l} \cdot \theta_{H_2O,in}$ in an extended version of eq. (2.14), we have to adjust our energy balance according to the following equation:

$$\theta_{out} = \frac{\left(F_{m,gas} \cdot Cp_g + F_{m,gas} \cdot X_{in} \cdot Cp_{H_2O,vap}\right) \cdot \theta_{in} - F_{m,W} \cdot h_{vap} + F_{m,W} \cdot Cp_{H_2O,l} \cdot \theta_{W,in}}{\left(F_{m,gas} \cdot Cp_g + F_{m,gas} \cdot X_{in} \cdot Cp_{H_2O,vap} + F_{m,W} \cdot Cp_{H_2O,vap}\right)} \tag{2.16}$$

Exercise 2.8: The cooling limit

We repeat the "incidence approach," but we assume that $\theta_{W,in} = 25$ °C, resulting in

$\theta_{g,out} = 63.72$ °C for $F_{m,W} = 0.067$ kg H_2O kg^{-1} air, and $X_{out} = 0.187$ kg H_2O kg^{-1} air, $p_{H_2O} = 235.85$ hPa, $\theta_{dp} = 63.72$ °C.

Discussion of results: The cooling limit temperature and water consumption will (must) rise, and also h_{1+X} must rise (because of the additional energy we feed to the system with the water).

Check and compare:

h_{1+X} of 1 kg h^{-1} of air before adding water for $\theta_{air} = 200$ °C and $X = 0.12$ kg kg^{-1} is

$$h_{1+X,200\,°C} = \frac{(1\cdot 1 + 0.12\cdot 1.91)\cdot 200 + 1\cdot 0.12\cdot 2{,}501}{1} = 546 \text{ kJ kg}^{-1} \text{ dry air}$$

h_{1+X} of 1 kg h^{-1} of air after adiabatic cooling for $\theta_{dp} = 63.49$ °C with $X_{out} = 0.184$ kg kg^{-1} is

$$h_{1+X,dp} = \frac{(1\cdot 1 + 0.184\cdot 1.91)\cdot 63.49 + 1\cdot 0.184\cdot 2{,}501}{1} = 546 \text{ kJ kg}^{-1} \text{ dry air}$$

and for the cooling limit:

h_{1+X} of 1 kg h^{-1} of air after cooling for $\theta_{W,in} = 25$ °C, $\theta_{dp} = 63.72$ °C with $X_{out} = 0.187$ kg kg^{-1} is

$$h_{1+X,dp} = \frac{(1\cdot 1 + 0.187\cdot 1.91)\cdot 63.72 + 1\cdot 0.187\cdot 2{,}501}{1} = 553 \text{ kJ kg}^{-1} \text{ dry air}$$

Discussion: As expected, the energy fed to our cooling box with the water (for $\theta_{W,in} = 25$ °C) will increase the water consumption, the cooling limit temperature and the specific enthalpy.

Exercise 2.9: The "application" case (in absorption)

We rarely just want to saturate the off-gas with water for cooling purposes. In off-gas purification we may for example make use of water quenching, to rapidly drop the gas temperature. In these applications we do not target complete water saturation but a specific temperature above the cooling limit temperature. In absorption, gas cooling by water saturation is just part of the technology (the technology of absorption will be discussed in detail separately). For our purposes we need to adapt the sketch of Fig. 2.3, because we add fresh water for gas cooling as well as the absorption process, as shown in Fig. 2.5. For latter application we adjust a continuous flow of water, fixed by the so-called liquid to gas ratio ($L/G = F_{m,W,abs,in}/F_{m,g,in}$). When performing the energy balance, we have to consider the energy inflow (gas plus water) and the energy outflow (again gas plus water). The cooling limit temperature of the clean gas $T_{g,out}$ ($\theta_{g,out}$) and the temperature of the wastewater $T_{W,out}$ ($\theta_{W,out}$) have the same value. The temperatures of the gas inflow and the water inflow are different.

Solver:

First step: We start with the energy balance (we distinguish between the amount of water $F_{m,W,cool,in}$ needed for evaporative cooling and the amount of water $F_{m,W,abs,in}$ needed for absorption).

$$\left(F_{m,g}\cdot Cp_g + m_g\cdot X_{in}\cdot Cp_{H_2O,vap}\right)\cdot\theta_{in} - F_{m,W,cool,in}\cdot h_{vap} + \left(F_{m,W,abs,in} + F_{m,W,cool,in}\right)\cdot Cp_{H_2O,l}\cdot\theta_{W,in}$$
$$= \left(F_{m,g}\cdot Cp_g + F_{m,g}\cdot X_{in}\cdot Cp_{H_2O,vap} + F_{m,W,cool,in}\cdot Cp_{H_2O,vap} + F_{m,W,abs,out}\cdot Cp_{H_2O,l}\right)\cdot\theta_{out}$$

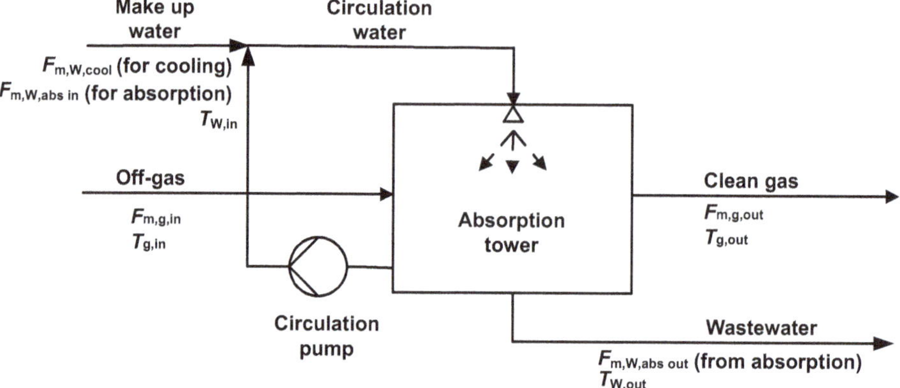

Fig. 2.5: The "application case" (absorption setup).

and rearrange for

$$\theta_{\text{out}} = \frac{\left(F_{\text{m,g}} \cdot Cp_{\text{g}} + m_{\text{g}} \cdot X_{\text{in}} \cdot Cp_{\text{H}_2\text{O,vap}}\right) \cdot \theta_{\text{in}} - F_{\text{m,W,cool,in}} \cdot h_{\text{vap}} + \left(F_{\text{m,W,abs,in}} + F_{\text{m,W,cool,in}}\right) \cdot Cp_{\text{H}_2\text{O,l}} \cdot \theta_{\text{W,in}}}{\left(F_{\text{m,g}} \cdot Cp_{\text{g}} + F_{\text{m,g}} \cdot X_{\text{in}} \cdot Cp_{\text{H}_2\text{O,vap}} + F_{\text{m,W,cool,in}} \cdot Cp_{\text{H}_2\text{O,vap}} + F_{\text{m,W,abs,out}} \cdot Cp_{\text{H}_2\text{O,l}}\right)} \tag{2.14}$$

Second step: For $F_{m,W,cool,in} = 0.014$ kg kg^{-1} and $\theta_{W,in} = 25$ °C at $L/G = 1$ ($F_{m,g} = 1$ kg and $F_{m,W,abs,in} = 1$ kg) we will get $\theta_{out} = 58$ °C for the clean gas as well as the wastewater.

! Discussion of results: Compared to the "cooling limit case" and the adiabatic cooling limit case we remove a lot of energy from the feed gas with the wastewater $F_{m,W,abs,out}$ in the (absorption) application case. Much less water will therefore be needed for cooling the feed gas. The reader is encouraged to find out how much water we need for evaporative gas cooling for $L/G = 0.5$ and $L/G = 2$. (Consider in the latter case that excess water may even cause condensation of water vapor from the gas phase.)

2.6.6 Estimation of the sulfuric acid dew point of SO_2-laden off-gas

We have already discussed the problems humid off-gas may cause in dry off-gas purification when temperature falls below the water vapor dew point. However, damage to equipment may become a serious issue, when humid off-gas also does contain sulfur dioxide. Emission of sulfur dioxide from incineration processes is mainly originating in fuel composition. Sulfur dioxide is partially oxidized to sulfur trioxide to form liquid sulfuric acid aerosols of characteristic bluish plume with vaporous water. Oxidation of sulfur dioxide with residual oxygen is an exothermic reversible process. Figure 2.6 shows the temperature dependency of SO_2 conversion for exemplarily 3,000 mg SO_2 m^{-3} and 11 vol% of oxygen [8].

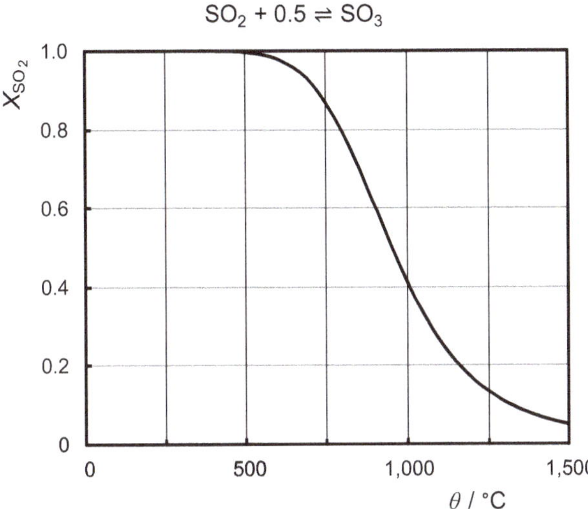

Fig. 2.6: Conversion $_x$ of SO_2 (3000 mg SO_2 m^{-3}) with oxygen (11 vol%) at different temperatures.

Figure 2.6 suggests complete conversion of SO_2 below temperatures of $\theta = 500$ °C. SO_2 is also converted to SO_3 with NO_x. Figure 2.7 shows the standard free enthalpy $\Delta_R G^0$ for SO_2 conversion with different oxidizers [8].

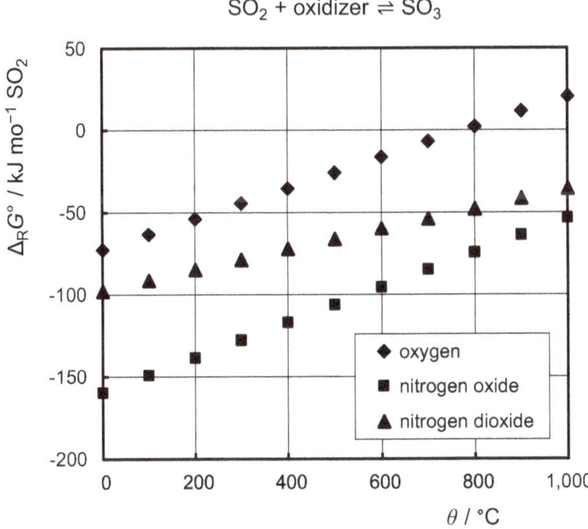

Fig. 2.7: Temperature dependency of SO_2 conversion with different oxidizers.

Gladly kinetics of SO_2 conversion does not follow the recommendations of thermodynamics. Catalysts such as V_2O_5 or Fe_2O_3 speed up conversion, but the technology is very sophisticated [9]. We may therefore expect less conversion of SO_2 in off-gas purification, but the question is, how much conversion we have to expect, because SO_3 immediately reacts with vaporous water to form liquid sulfuric acid aerosols. This property becomes very important in off-gas purification, because the boiling point of concentrated sulfuric acid (98.3 wt%) is as high as $\theta = 339$ °C, and condensation of sulfuric acid in technical equipment is a serious issue. Figure 2.8 shows condensation isotherms for corresponding sulfuric acid/vaporous water load in air [9].

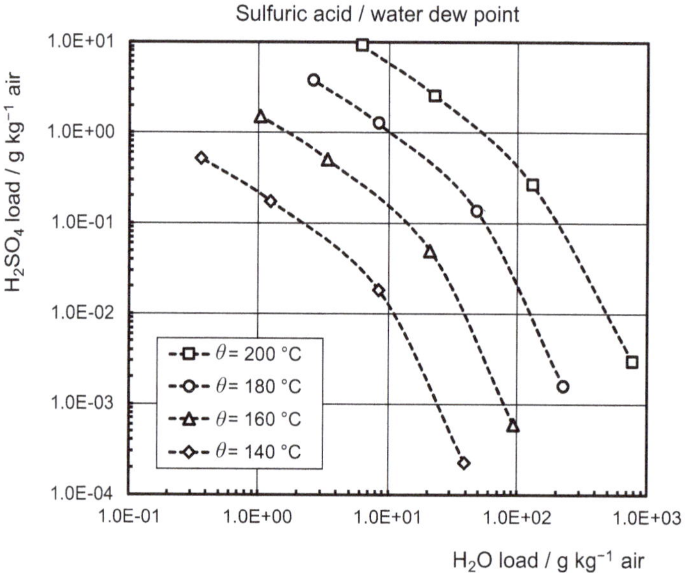

Fig. 2.8: Dew point isotherms for different sulfuric acid/water load of air.

The reading of Fig. 2.8 tells us that for given temperature and given sulfuric acid load of air you must not go beyond the corresponding maximum water load to avoid condensation of sulfuric acid (e.g., at θ = 160 °C and H_2SO_4 load = 0.05 g kg^{-1} H_2O load must not exceed 20.8 g kg^{-1}).

We have discussed some problems SO_2 may cause in dry off-gas purification, but without much help in estimating the hazard potential of sulfuric acid condensation. Several proposals (about 17) for sulfuric acid dew point estimation in off-gas were developed in the past. If interested, you may check the proposals of [10, 11]. Based on empirical experience, Pierce [12] has developed a correlation for the estimation of the sulfuric acid dew point. We will discuss this two-parameter approach, because you can easily apply it with your hand calculator:

$$\theta_{dp,\,H_2SO_4} = \frac{1{,}000}{1.7842 + 0.0269 \cdot B - 0.1029 \cdot A + 0.0329 \cdot A \cdot B} - 273 \tag{2.17}$$

$\theta_{dp,\,H_2SO_4}$ Sulfuric acid dew point temperature

$$A = \log\left(\frac{SO_2 \cdot BP}{10^9}\right) \tag{2.18}$$

where SO_2 is sulfur dioxide (ppm) and BP is the barometric pressure (hPa)

$$B = \log\left(\frac{H_2O \cdot BP}{10^5}\right) \tag{2.19}$$

where H_2O is the water content (vol%).

Exercise 2.10: Application of the basics

Table 1.1 shows our basic off-gas measurement report with several specification questions. With our toolbox we are now ready to answer these questions.

First is the mean molar mass MM_{mean} of the carrier gas:

The carrier gas is made up of the "inerts" oxygen (MM = 32 kg kmol^{-1}), nitrogen (MM = 28 kg kmol^{-1}) and carbon dioxide (MM = 44 kg kmol^{-1}), and the mean molar mass MM_{mean} is

$$MM_{mean} = \frac{(78 \cdot 28 + 12 \cdot 32 + 10 \cdot 44)}{100} = 30.08\ \text{kg kmol}^{-1}$$

Second is the gas density of the dry gas $\left(\rho_{STP,\,dry}\right)$

$$\rho_{STP,\,dry} = \frac{30.08}{22.418} = 1.34\ \text{kg m}^{-3}{}_{STP,\,dry}$$

Gas density of the humid gas $\rho_{g,\ STP,humid}$

Now we have to consider the amount of vaporous water in the inflow gas, specified as volume based load of 100 g m$^{-3}{}_{STP,\,dry}$. We start with 1 m^3 of dry gas and determine the mass of gas, and we repeat the procedure with 0.1 kg m$^{-3}{}_{STP,\,dry}$ vaporous water.

$$\rho_{STP,\,humid} = \frac{m_{dry\ gas} + m_{H_2O}}{V_{dry\ gas} + V_{water\ vapor}} = \frac{1.34 + 0.1}{1 + \frac{0.1}{18} \cdot 22.418} = 1.281\ \text{kg m}^{-3}{}_{STP,\,humid}$$

Gas density in actual state $\left(\rho_{gas,\,actual}\right)$

We make use of eq. (2.2):

$$\rho_{actual} = \rho_{STP,\,humid} \cdot \frac{T_{standard\ state}}{T_{actual\ state}} \cdot \frac{P_{actual\ state}}{P_{standard\ state}} = 1.28 \cdot \frac{273.15}{(273.15 + 200)} \cdot \frac{980}{1{,}013} = 0.715\ \text{kg am}^{-3}$$

Third is the gas flow rates.

We have to determine the correct gas flow rate for hydraulic design purposes (actual gas flow rate) and for balance purposes (dry gas flow rate at STP).

- Actual gas flow rate in am^3 h^{-1}:

with eq. (2.5) we obtain

$$v = \sqrt{\frac{2 \cdot (98{,}109.5 - 98{,}000)}{0.715}} = 17.5\ \text{m s}^{-1}$$

and with the cross-sectional area A

$$A = \frac{D^2 \cdot \pi}{4} = 0.785\ \text{m}^2$$

we get

$$F_{V,\text{actual}} = 17.5 \cdot 0.785 \cdot 3{,}600 = 49{,}453.4\ \text{am}^3\ \text{h}^{-1}$$

– Humid gas flow rate $\left(F_{V,\text{STP, humid}}\ \text{in}\ \text{m}^3_{\text{STP, humid}}\ \text{h}^{-1}\right)$:
with eq. (2.2) we get

$$F_{V,\text{STP, humid}} = 49{,}453.4 \cdot \frac{980}{1{,}013} \cdot \frac{273.15}{(273.15 + 200)} = 27{,}619.4\ \text{m}^3_{\text{STP, humid}}\ \text{h}^{-1}$$

– Carrier gas flow rate in $\text{m}^3_{\text{STP,dry}}\ \text{h}^{-1}$
We have to convert the volume-based water load from Tab. 1.1 into mass-based water load X:

$$X_{H_2O} = \frac{X_{H_2O,\ \text{volume-based}}}{\rho_{\text{dry gas}}} = \frac{100}{1{,}000 \cdot 1.34} = 0.075\ \text{kg H}_2\text{O kg}^{-1}\ \text{dry gas}$$

Then we perform a volume balance

$$F_{V,\text{STP, dry}} + F_{V,\text{STP, dry}} \cdot \rho_{\text{dry}} \cdot X_{H_2O} \cdot \frac{22.418}{18} = F_{V,\text{STP, humid}}$$

and rearrange for $F_{V,\text{STP, dry}}$

$$F_{V,\text{STP, dry}} = \frac{27{,}619.4}{1 + 0.075 \cdot 1.34 \cdot \frac{22.418}{18}} = 24{,}560.6\ \text{m}^3\ \text{h}^{-1}{}_{\text{STP, dry}}$$

Fourth is the partial pressure of water p_{H_2O}:

With eq. (2.9) we determine the partial pressure of water p_{H_2O}

$$p_{H_2O} = \frac{0.075 \cdot 980}{\frac{18}{30.08} + 0.075} = 108.5\ \text{hPa}$$

Fifth is the water dew point temperature $(\theta_{\text{dp},\,H_2O})$

With eq. (2.7) we determine the dew point temperature (corresponding to $p_{H_2O} = p_{H_2O,\theta,0}$)

$$\theta_{\text{dp},\,H_2O} = \frac{1{,}730.462}{8.19625 - \log(108.5)} - 233.426 = 47.5\ °\text{C}$$

Sixth is the volumetric water content in vol%.

The measurement reading of Tab. 1.2 says: $X_{H_2O,\ \text{volume-based}} = 100\ \text{g m}^{-3}{}_{\text{STP,dry}}$

The volume-based water vapor load has to be converted into the gaseous water volume, which is then related to the dry gas volume plus the volume of vaporous water:

$$H_2O = \frac{\frac{100}{1{,}000} \cdot \frac{22.418}{18}}{\left(1 + \frac{100}{1{,}000} \cdot \frac{22.418}{18}\right)} \cdot 100 = 11.08\ \text{vol\%}$$

Seventh is the sulfuric acid dew point temperature $(\theta_{\text{dp},\,H_2SO_4})$
with eq. (2.16) we get

$$A = \log\left(\frac{300 \cdot 980}{10^9}\right) = -3.532$$

with eq. (2.17) we get

$$B = \log\left(\frac{11.075 \cdot 980}{10^5}\right) = -0.964$$

and with eq. (2.15) we get

$$\theta_{dp,H_2SO_4} = \frac{1{,}000}{1.7842 - 0.0269 \cdot 0.964 + 0.1029 \cdot 3.532 + 0.0239 \cdot 3.532 \cdot 0.964} - 273 = 174.7\ °C$$

Discussion: Now the off-gas specification of Table 1.1 is nearly complete (except the cooling limit and the discussion of the dust specification). We are strongly recommended not to fall below a sulfuric acid dew point temperature of $\theta_{dp,H_2SO_4} = 175$ °C in the dry off-gas purification line. The off-gas temperature of $\theta = 200$ °C is well above the sulfuric acid dew point temperature.

Eighth is the adiabatic cooling limit temperature $\theta_{out,acl}$ plus h_{1+X} and cooling limit temperature $\theta_{out,cl}$ plus specific enthalpy h_{1+X}.

We again make use of the simplifications: $Cp_{N_2,O_2,CO_2,mean} = 1$ kJ kg^{-1} K^{-1} and $Cp_{H_2O,vap,mean} = 1.91$ kJ kg^{-1} K^{-1} and $Cp_{H_2O,l,mean} = 4.19$ kJ kg^{-1} K^{-1}.

By assuming (guessing) a water feed of $m_{H_2O} = 0.062$ kg kg^{-1} gas we get for the adiabatic cooling limit with eqs. (2.14) and (2.13):

$$\theta_{out,acl} = \frac{(1 \cdot 1 + 1 \cdot 0.075 \cdot 1.91) \cdot 200 - 0.062 \cdot 2,501}{(1 \cdot 1 + 1 \cdot 0.075 \cdot 1.91 + 0.062 \cdot 1.91)} = 58.1\ °C$$

$$h_{1+X,200} = \frac{(1 \cdot 1 + 1 \cdot 0.075 \cdot 1.91) \cdot 200 + 1 \cdot 0.075 \cdot 2,501}{1} = 416\ \text{kJ kg}^{-1}\ \text{dry air}$$

$$h_{1+X,58.1} = \frac{(1 \cdot 1 + 0.137 \cdot 1.91) \cdot 58.1 + 1 \cdot 0.137 \cdot 2,501}{1} = 416\ \text{kJ kg}^{-1}\ \text{dry air}$$

Discussion: The specific enthalpy for the feed gas $h_{1+X,200}$ and for the adiabatically water saturated gas $h_{1+X,58.1}$ are same, confirming that the estimated water consumption of $m_{H_2O} = 0.062$ kg kg^{-1} is correct.

With eq. (2.16), $m_{H_2O} = 0.063$ kg kg^{-1} and $\theta_{w,in} = 12$ °C we get for the cooling limit

$$\theta_{out,cl} = \frac{(1 \cdot 1 + 1 \cdot 0.075 \cdot 1.91) \cdot 200 - 0.063 \cdot 2,501 + 0.063 \cdot 4.19 \cdot 12}{(1 \cdot 1 + 1 \cdot 0.075 \cdot 1.91 + 0.063 \cdot 1.91)} = 58.2\ °C$$

and

$$h_{1+X,58.2} = \frac{(1 \cdot 1 + 0.138 \cdot 1.91) \cdot 58.2 + 1 \cdot 0.138 \cdot 2,501}{1} = 418\ \text{kJ kg}^{-1}\ \text{dry air}$$

Discussion: As expected, low temperature of the cooling water does neither much affect the outflow temperature nor affect the water consumption nor the specific enthalpy, but the comparison of results shows a difference, even for low water temperature. Therefore, always have in mind the difference between $\theta_{out,acl}$, $\theta_{out,cl}$ and θ_{dp,H_2SO_4}.

Recommendation: Calculation of the cooling limits and the water consumption is probably performed easier when not done with specific data (per kg of dry air) but the given dry gas mass or mass flow rate, because it produces larger numbers.

Exercise 2.11: First experience I

Due to its great properties acetone (MM = 58.08 g mol^{-1}) is a widely used solvent in (the chemical) industry. Acetone is fully water miscible. According to [13] the lower limit of inflammability in air is 2.55 vol% and the upper limit of inflammability is 12.8 vol%, and the limit for human exposure is 1,000 ppm. The *Handbook of Environmental Data on Organic Chemicals* [14] reports a 100% odor recognition concentration of 300 ppm. According to Chapter 5, General Requirements of "Technical Instructions on Air Quality Control" [15], the emission limit is 50 mg m^{-3} total carbon (TC). And now the question: Is it possible to suffice the legal requirements of 50 mg m^{-3} TC by condensing vaporous acetone from air at ambient pressure P_{tot} = 1,013 hPa?

Solver:

Simply said, we just need to find the temperature at which 50 mg m^{-3} TC load of acetone corresponds to the vapor pressure of acetone. From [7] we may collect vapor pressure data for acetone, as listed in Tab. 2.4.

Tab. 2.4: Vapor pressure of acetone.

θ (°C)	p_0 (hPa)	$\log p_0$
−44.1	5	$\lg p_1 = 0.699$
−10.5	50	$\lg p_2 = 1.699$
55.8	1,000	$\lg p_3 = 3.000$

The Antoine equation (2.20) offers a three-parameter correlation of vapor pressure and temperature:

$$\log p_{\text{constituent},\theta,0} = A - \frac{B}{C+\theta} \tag{2.20}$$

where $p_{H_2O,\theta,0}$ is the vapor pressure of water at temperature θ and θ is the temperature (°C).

If we rearrange eq. (2.20) we may solve for A, B and C according to

$$C = \frac{(\theta_3 - \theta_1)}{\left(1 - \frac{(\lg p_3 - \lg p_2)\cdot(\theta_2-\theta_1)}{(\lg p_2 - \lg p_1)\cdot(\theta_3-\theta_2)} - \theta_3\right)} = 237.459 \tag{2.21}$$

$$B = \frac{(\lg p_3 - \lg p_1)\cdot(\theta_1 + C)\cdot(\theta_3-\theta_1)}{(\theta_3-\theta_1)} = 1,306.0885 \tag{2.22}$$

$$A = \lg p_2 + \frac{B}{(\theta_2 + A)} = 7.4537 \tag{2.23}$$

About 50 mg m^{-3} TC corresponds to $\frac{50}{12}\cdot\frac{1}{3}\cdot 58.08 = 81\,\text{mg acetone m}^{-3}$, and with the density of air at STP (see Exercise 2.1), $\rho_{STP} = 1.286\ \text{kg m}^{-3}{}_{STP}$ we obtain $\frac{50}{12}\cdot\frac{1}{3}\cdot 58.08\cdot\frac{1}{1.286} = 62.73\ \text{mg acetone kg}^{-1}$ air, which is equivalent to $X = 62.73\cdot 10^{-6}$ kg acetone kg^{-1} air.

With eq. (2.12) we get:

$$p_{acetone} = \frac{X\cdot P_{tot}}{\frac{MM_{acetone}}{MM_{air}} + X} = 0.0316\ \text{hPa} \tag{2.24}$$

which, when inserted into eq. (2.8), results in the condensation temperature of

$$\theta = \frac{B}{A - \log(p_{acetone,\theta,0})} - C = -92\ °\text{C} \tag{2.25}$$

Discussion: Seemingly, it is not recommended to propose the condensation technology. It would need evaporation of liquid nitrogen to provide such low temperatures. Besides, the melting point of acetone is just $\theta = -94.7$ °C.

Exercise 2.12: First experience II, the safety challenge.

Besides their toxicity the odor of mercaptans is an effing nuisance. Phenylmercaptan for example has a 100% odor recognition concentration of 0.2 ppb [14], making any pollution abatement technology an economic challenge. Biodiesel has recently been discussed to provide outstanding physical properties for absorbing odorous pollutants from air. But, how about process safety and emission control at ambient pressure?

Solver:

Biodiesel is a mixture of fatty acid methyl esters, suggesting the typical representative methyl oleate for discussing safety and emission issues. To satisfactorily answer the questions, we must find a correlation of temperature and saturation load of methyl oleate in air for emission control purposes, and we have to search for information about the lower limit of inflammability. For convenience we just consider the binary system dry air/methyl oleate.

West [13] provides the vapor pressure data for methyl oleate, as shown in Tab. 2.5.

Tab. 2.5: Vapor pressure of methyl oleate.

θ (°C)	p_0 (hPa)	$\log p_0$
85	0.01	$\lg p_1 = -2$
195.6	10	$\lg p_2 = 1$
340	1,000	$\lg p_3 = 3.000$

With eqs. (2.18)–(2.20) we determine the Antoine constants for methyl oleate with

$A = 8.217$,

$B = 2718.3917$ and

$C = 181.066$.

Table 2.6 shows some relevant physical–chemical data of methyl oleate.

Tab. 2.6: Physical–chemical data of methyl oleate.

Formula	$C_{19}H_{36}O_2$	
Molar mass	296.49	g mol^{-1}
$h_{v,406\ K}$	291.07	kJ kg^{-1}
$Cp_{l,473\ K}$	2	kJ kg^{-1} K^{-1}
$Cp_{v,473\ K}$	1.9	kJ kg^{-1} K^{-1}

Let us start the discussion with the emission control issue. Again Chapter 5 (General Requirements of "Technical Instructions on Air Quality Control") [15] specifies the emission limit of methyl oleate with 50 mg m^{-3} TC.

TC = 50 mg m^{-3} is equivalent to $\frac{50}{12} \cdot \frac{1}{19} \cdot 296.49 = 65$ mg methyl oleate m^{-3}. We may simply transfer from methyl oleate concentration in mg m^{-3} by multiplying with 50/65 = 0.769 into TC concentration in mg TC m^{-3}.

With the Antoine equation (eq. (2.7)) we calculate the vapor pressure of methyl oleate, and with eq. (2.9) we calculate the mass load X of methyl oleate and air. We transfer from mass load X in kg kg^{-1} by multiplying with 10^6 into mass load X in mg kg^{-1}. With the density of dry air at STP we determine the volume-based methyl oleate load. We will not fail much at these load levels when we assume the volume-based methyl oleate load to be $\approx$ the methyl oleate concentration in air, as shown in Tab. 2.7. Finally, we also determine the molar fraction y of methyl oleate in air and convert it into mol%.

Tab. 2.7: Data conversions for methyl oleate.

θ (°C)	$P_{\text{methyl oleate}}$ (hPa)	$X_{\text{methyl oleate}}$ (mg kg^{-1})	TC (mg kg^{-1})	TC (mg m^{-3})	y (mol%)
75	0.004	40.5	31.15	40.07	0.00039

Figure 2.9 shows the outcome for the temperature span 0 °C < θ < 100 °C

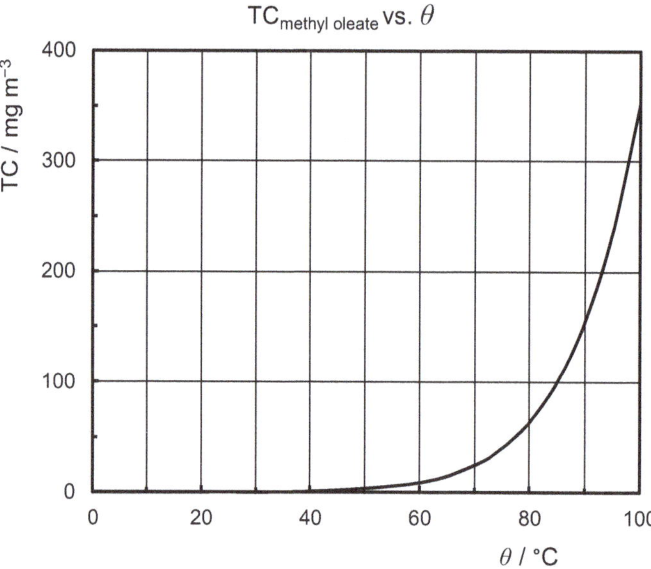

Fig. 2.9: Methyl oleate concentration in dry air at saturation in terms of total carbon (TC).

From Fig. 2.9 we can quickly conclude that the saturation temperature of methyl oleate in dry air must not exceed θ = 77 °C to suffice the emission limit of 50 mg m^{-3} TC. It is a nice outcome, but it doesn't tell us anything about the maximum inflow temperature of dry air into our "adiabatic methyl oleate saturation black box." To quickly answer this question, we make use of eq. (2.26) and the data of Tab. 2.6:

$$h_{1+X}\left[\text{kJ kg}^{-1}\text{ dry air}\right] = \frac{\left(m_{\text{gas}} \cdot Cp_{\text{g}} + X \cdot Cp_{\text{methyl oleate, vap}}\right) \cdot \theta + m_{\text{gas}} \cdot X \cdot h_{\text{vap, methyl oleate, 273.15}}}{m_{\text{gas}}} \tag{2.26}$$

If we insert numbers for θ = 75 °C (but be careful with X, because you must insert X in terms of kg kg^{-1}), we get a result of $h_{1+X} = 75$ kJ kg^{-1}. When comparing this result with h_{air} at θ = 75 °C we would not register a difference because of the low amount of methyl oleate.

What is still left, is the discussion of safety issues. You will probably fail when looking for valid data for the lower inflammability limit and the flash point of methyl oleate in air in literature. So we have to apply rules of thumb. One of these rules of thumbs says that the lower limit of inflammability of combustibles can be approached via half the molar fraction of the substance for stoichiometric combustion with air. Does that help? Let us do it step by step.

Methyl oleate has the formula $C_{19}H_{36}O_2$. For stoichiometric combustion of 1 mol of methyl oleate with oxygen we will need: 19 + 36/4 − 1 = 27 mol of oxygen. Since oxygen has a molar fraction in air of $x = 0.21$ we will need 128.57 mol of air for stoichiometric combustion of methyl oleate. The molar fraction of methyl oleate in air is then $x = 1/(1 + 128.57) = 0.0077$. Half the amount of the molar fraction of methyl oleate in air is then 0.00385. From this outcome we may conclude that the lower inflammability limit of methyl oleate in air is 0.385 $\approx$ 0.4 vol%. The partial pressure of methyl oleate is then $p = 1{,}013 \cdot 0.00385 = 3.9$ hPa. With eq. (2.8) and the Antoine constants for methyl oleate we may now calculate the flash point:

$$\theta = \frac{B}{A - \log(p)} - C = \frac{2{,}718.3917}{8.2170 - \log(3.9)} = 175.4\,°\text{C}$$

Discussion: When comparing the operation limits for emission control with the safety data you will recognize that we are far below operation conditions of safety concerns. However, be careful with rules of thumb.

Finally, we have to discuss your part of the game. You have to check now the validity of the outcome. Perhaps you need a starting point. How about hexane? Table 2.8 shows the physical properties of hexane.

Tab. 2.8: Physical properties of hexane.

Formula	**C_6H_{14}**		**θ (°C)**	**P (hPa)**
Molar mass	86.18	g mol^{-1}	−28.6	10
$h_{v,273\,K}$	334	kJ kg^{-1}	47.7	500
$Cp_{l,473\,K}$	2.38	kJ kg^{-1} K^{-1}	68.4	1,000
$Cp_{v,473\,K}$	1.67	kJ kg^{-1} K^{-1}		

While discussing this example we have addressed several helpful data series. For convenience you may find data in [16]. You may also find validated data when you contact the homepage of the National Institute of Standards and Technology (NIST).

3 Specification of solid/gas and liquid/gas dispersions: properties

Proper design of industrial dust emission control equipment must consider the properties of dispersoids. Qualitative specification of dispersions is mainly derived from Anglo-Saxon standards, which distinguish between dust, smoke, fume, mist and fog.

- Dust, with a mean particle size between 1 and 200 µm, is used for specifying particles emitted from crushing processes.
- Smoke, with a mean particle size between 0.01 and 1 µm, is mainly formed during incineration.
- Fume, with a mean particle size between 0.1 and 1 µm, is formed by sublimation or incineration (e.g., oxidation of metal vapor).
- Mist (fog), with a mean particle size between 5 and 100 µm, is formed by condensation of liquid constituents.

Liquid–gas dispersion, with a mean particle size above 100 µm, is called rain.

The primary characteristic of solid/gas dispersoids is the particle size. The behavior of particles in a fluid depends on its shape. Spherical shape will show the same properties irrespective of its orientation. Off-gas purification rarely has to deal with ideal spherically shaped particles. We, therefore, have to define irregularly shaped particles in terms of the behavior and the size of particles of equivalent sphere. Equivalent spheres may be derived from the sphere

- of the same volume as the particle,
- of the same surface area,
- of the same surface area per unit volume,
- of the same area when projected on to a plane perpendicular to the direction of motion,
- of the same projected area, as viewed from above, when positioned in the maximum stable position,
- which will just pass through a square aperture of same size as the particle and
- with the same settling velocity in a specified fluid.

For dust collection the most important size related property is the dynamic behavior. Large particles of $x > 100$ µm are readily collectible by gravitational methods. The collection of particles with diameters $x < 100$ µm is mainly determined by the resistance to motion which is viscous. Size specification is therefore based on the Stokes settling diameter, which is the diameter of the spherical particle of same density and same terminal settling velocity in viscous flow as the target particle.

The use of the aerodynamic diameter, which is the diameter of a particle of unit density with same terminal settling velocity as the particle in question, is more convenient.

https://doi.org/10.1515/9783110763928-003

The use of the aerodynamic diameter permits the comparison of the dynamic behavior of particles of different size, shape and density.

Decreasing the particle size to the same order of magnitude as the free mean path of gas molecules leads to better settling properties than proposed by Stokes law. Slip flow correction for particles with aerodynamic diameter $x < 1$ µm is considered by the Cunningham correction.

Fine particles, smaller than $x < 0.1$ µm, should not be of much significance due to their small fraction of the total mass. Collection is determined by diffusional deposition. Coagulation and flocculation rate is very high, resulting in a significant growth of particles up to a size of $x < 0.1$ µm. Precipitation of particles between particle size of $x = 0.1$ µm and $x = 2$ µm may cause severe problems because settling is mainly controlled by diffusion.

3.1 Optical properties of dispersions

The consideration of optical properties of solid/gas and liquid/gas dispersions is of decisive importance in off-gas purification due to visibility of off-gas plumes.

Light scattering depends on particle size, the shape of particles and the refraction properties. Uniform description is not possible. Light scattering properties of particles with a size of less than $x < 0.05$ µm is described by Rayleigh's law, which considers scattering to depend on the sixth power of the particle diameter and the fourth power of the wave length. Larger particles cause scattering properties of the sixth and second power of the particle diameter. The color of an off-gas plume indicates the prevalent particle size. In general, the limit of visibility of off-gas plumes is in the range of $C = 10–40$ mg m^{-3}.

Tobacco smoke is a bad example for demonstration of the light scattering phenomenon. Native tobacco smoke forms a bluish plume, which leads to preferred scattering in the ultraviolet region. After inhaling tobacco smoke (a very dangerous action) and exhaling again, the plume will be colored white. What happened? Inside the (heavily poisoned) lungs, the smoke particles are either adsorbed or they may agglomerate or condense water vapor on the particle surface before exhaling. Scattering of the exhaled residue is uniform over the whole region of the visible light due to an increased particle size, resulting in a white plume. The optical density of particle laden off-gas depends on its structural shape, which ordinarily leaves a subjective impression. Light transmission of an off-gas plume correlates with precipitation efficiency via the Lambert–Beer law. Figure 3.1 shows a qualitative sketch of this correlation.

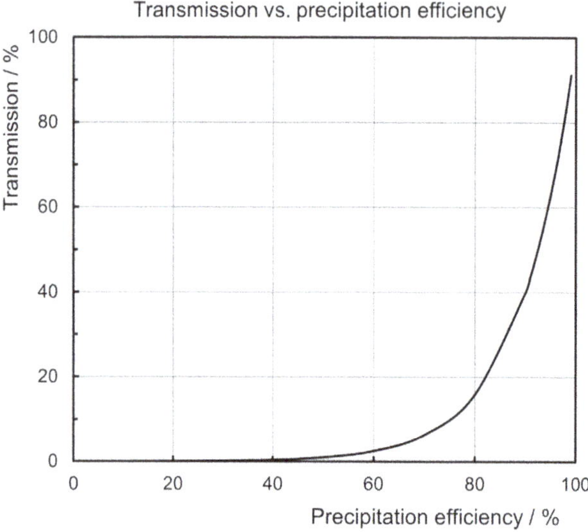

Fig. 3.1: Correlation of light transmission and precipitation efficiency of off-gas plumes.

3.2 Particle size distribution (PSD)

The demand for highly efficient dust and particle precipitation has led to the necessity of specifying the overall precipitation efficiency and the fractional (grade) precipitation efficiency.

Specification by the concentration and by the particle size distribution (PSD) is needed to determine the necessity of measures and the kind of measures to

- establish compliance with legal requirements,
- obtain information for successful precipitator design and
- determine collector performance.

Methods of particle size analysis depend on the expected PSD, as shown by Tab. 3.1.

Tab. 3.1: Particle sizing: applied methods and limits of performance.

Paricle size (μm)	Applied method
>35	Dry sieve
>1.0 to 100	Wet sieve
>0.2 to 20	Cascade impactor
>0.1 to 10	Ultracentrifuge
	Transmission electron microscope
	Scanning electron microscope

Specification of the PSD needs the information about the particle size class and the corresponding quantity. For that purpose, the dust sample is split into several fractions. Particle size analysis is preferably performed by cascade impaction. Figure 3.2 shows a sketch of the measurement setup.

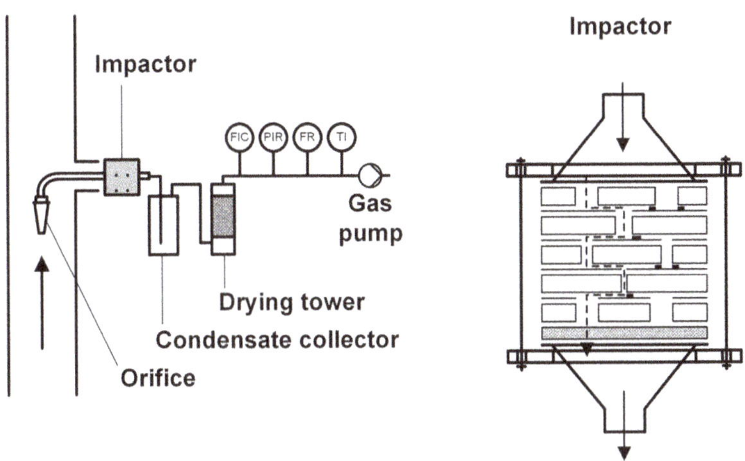

Fig. 3.2: Measurement setup for particle size distribution analysis.

The measurement principle is based on inertia of particles forcing them to collect on an aluminum foil. Within the cascade the gap size decreases and gas velocity increases, forcing particles of different size to collect in different cascade stages. The orifice of the probe head must be adjusted to the gas velocity in the duct (isokinetic operation). As indicated in Fig. 3.2, it is very important to care for correct data logging of the gas flow (FR), the gas temperature (TI) and the pressure drop (PIR), since the gas flow rate FI has to be kept at a fixed value by an FIC to provide isokinetic flow conditions. The quality of data depends on the quality of the measurement setup and on the sophisticated skills of the operator.

The total mass of particulate matter is quantified with a similar setup. As shown in Fig. 3.3 the dust is collected on a filter in the orifice. For correct sampling the gas flow rate, temperature pressure and the correct total volume of the gas must be recorded.

Analysis of samples is based on listing the total amount of dust, the class and the fractional (grade) amount of particles within a class. For our purposes we make use of a "constructed" measurement report, as listed in Tab. 1.1. Table 3.2 shows an outline of this report. The left column of Tab. 3.2 shows the upper limit of the particle size class, the center column shows the cumulative mass and the right column shows the mass concentration of the respective class.

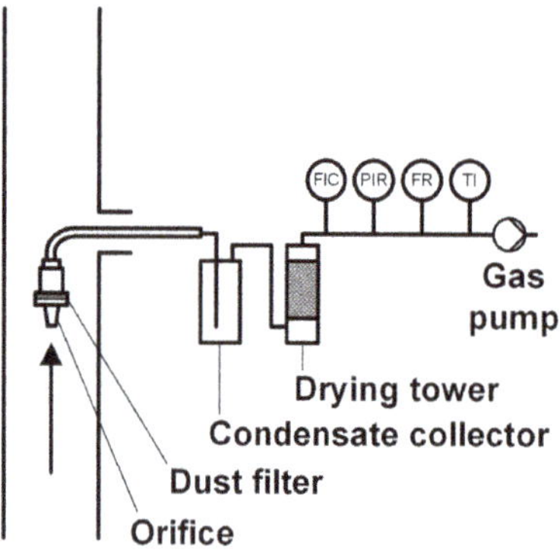

Fig. 3.3: Measurement setup for dust sampling.

Tab. 3.2: Outline of the dust measurement report.

Dust	3,000	mg $m^{-3}_{STP,dry}$
	PSD	
x_{ul} (µm)	Q_3 (mass)	c (mg m^{-3}_{STP})
0.2	0.004	13.3
0.4	0.018	39.6
0.6	0.039	64.8
0.8	0.069	88.3
1.0	0.105	109.6
2.0	0.359	761.0
4.0	0.831	1,416.5
6.0	0.982	452.1
8.0	0.999	52.5
10.0	1.000	2.4

The data of the third column in Tab. 3.2 provide a first impression of the problem we will have to solve when discussing particle precipitation technologies because of the huge amount of fines of as much as C_{dust} = 315.5 mg m^{-3}_{STP} for the particle classes $x \leq 1$ µm; therefore, we have to discuss data interpretation and reporting in more detail.

In general graphical documentation is preferred. ISO 9276–1 (1998–06) recommends the presentation of the sensitivity parameter (particle size x) on the abscissa and the corresponding fractional amount ΔQ_3 on the Y-axis.

We have to distinguish between cumulative presentation $Q_r(x)$, distributive presentation ΔQ_3 and presentation of the frequency $q_r(x)$.

The cumulative amount $Q_r(x_i)$ considers the amount of particles with $x \leq x_i$, $Q_r(x_{min}) = 0$ and $Q_r(x_{max}) \leq 1$, respectively.

The method of (counting) measurement is indicated with the subscript r.

Subscript *r*	Counting method
0	number
1	length
2	area
3	mass, volume

Mass distribution, represented by subscript $r = 3$, is commonly applied in dust analysis.

The cumulative distribution curve $Q(x)$ is either represented by the rate of transmission $D(x)$ (undersize) or by the rate of retention capability $R(x)$ (oversize) according to the following equation:

$$R(x) = 1 - D(x) \tag{3.1}$$

The frequency (probability density) represents the number (mass, amount) of particles ΔQ_r within the class $x + \Delta x$, corresponding to the considered width Δx, as shown in the following equation:

$$q_r(x_1, x_2) = \frac{\Delta Q_r(x_1, x_2)}{\Delta x} = \frac{Q_r(x_2) - Q_r(x_1)}{x_2 - x_1} \tag{3.2}$$

If the particulate distribution Q_r can be presented by a continuous function we will get

$$q_r(x) = \frac{\mathrm{d}Q_r(x)}{\mathrm{d}x} \tag{3.3}$$

With the area Q_r covered by the frequency (weight) curve $q_r(x)$ between x_{min} and x_{max} we obtain

$$\int q_r(x)\mathrm{d}x = 1 = Q_r(x_{\mathrm{max}}) - Q_r(x_{\mathrm{min}}) \tag{3.4}$$

The specification of a distribution is either based on the mean value $x_{50,r}$ or the maximum of distribution, the modal value x_{mod}

$x_{50,r}$ = particle size, corresponding to $Q_r(x_{50,r}) = 0.5$
x_{mod} = particle size, corresponding to $q_r(x) = \max$.

3.3 Particle size distribution (PSD) equations

For further processing, the PSD is specified best by correlating the quantity of particulate matter with particle size. Common approximative functions are based on description with two parameters, considering the sensitivity and the mean. Several functions are recommended standards.

Power distribution function according to Gaudin–Schuhmann (ISO 9276–1 (1998–06)):

$$Q_3(x) = \left(\frac{x}{x_{\max}}\right)^m \tag{3.5}$$

Distribution parameters: $x_{\max}$, m.

The power distribution function is best evaluated by a log–log plot. The distribution parameter m is derived from the slope of the linearized distribution graph.

RRSB distribution function (Rosin–Rammler–Sperling–Bennet) (ISO 9276–1 (1998–06)).

Bennet has defined RRSB distribution according to

$$Q_3(x) = D_3(x) = 1 - e^{-\left(\frac{x}{x'}\right)^n} \tag{3.6}$$

with the distribution parameters: x', n.

The distribution parameter x' represents a statistical mean particle size. It is calculated from the amount $D = 63.2\%$ transmitted (undersize) or from the corresponding amount of retention (oversize) $R = 36.8\%$ deduced from the RRSB function. The parameter x' defines the quality of a sample. The distribution parameter n specifies the smoothness of a distribution. The higher the value of n the smoother a distribution will be. n will usually have values between 0.8 and 2.

By forming double-logarithmic values of a given distribution we obtain a distribution line according to

$$\ln\left(\ln\left(\frac{1}{R(x)}\right)\right) = \ln\left(\ln\left(\frac{1}{1 - D(x)}\right)\right) = n \cdot \ln x + k \tag{3.7}$$

$$k = n \cdot \ln(x') \tag{3.8}$$

Exercise 3.1: Interpretation of dust (particulate) measurement data

In the first step, we complete the PSD specification of Tab. 3.2. The outcome is summarized in Tab. 3.3.

Tab. 3.3: Dust measurement analysis.

x_{ul} µm	x_{mean} µm	$Q_3 = D(x)$	ΔQ_3	Δx	$q_3 = \Delta Q/\Delta x$	$\ln(x_{ul})$	$\ln\left(\ln\frac{1}{1-D(x)}\right)$
0	0	0	0				
0.2	0.1	0.004	0.004	0.2	0.022	−1.61	−5.42
0.4	0.3	0.018	0.013	0.2	0.066	−0.92	−4.03
0.6	0.5	0.039	0.022	0.2	0.108	−0.51	−3.22
0.8	0.7	0.069	0.029	0.2	0.147	−0.22	−2.64
1	0.9	0.105	0.037	0.2	0.183	0.00	−2.20
2	1.5	0.359	0.254	1	0.254	0.69	−0.81
4	3	0.831	0.472	2	0.236	1.39	0.58
6	5	0.982	0.151	2	0.075	1.79	1.39
8	7	0.999	0.017	2	0.009	2.08	1.96
10	9	1.000	0.001	2	0.000	2.30	2.41

From this specification we can graphically determine the mean of the PSD and the maximum of the PSD. Figure 3.4 shows the transmission capability $D(x)$ with the mean, and from the probability chart we can extract the maximum x_{mod}.

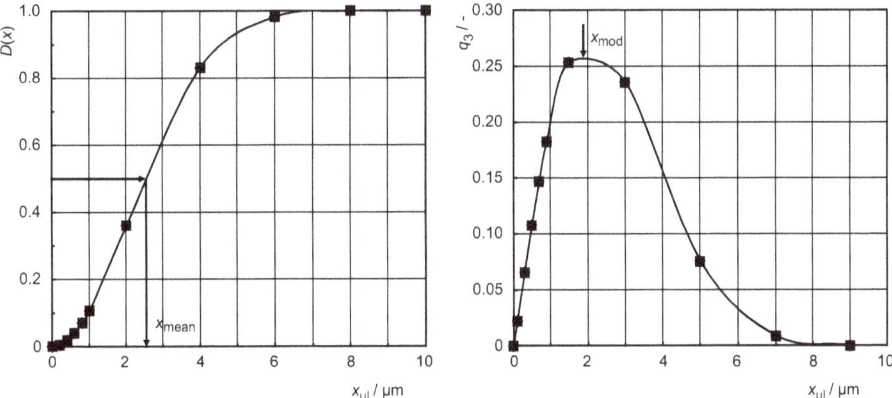

Fig. 3.4: Transmission capability chart and probability chart for the data of Tab. 3.2.

In the second step, we transfer the data of Tab. 3.2 into a chart according to eq. (3.7). The chart is shown in Fig. 3.5.

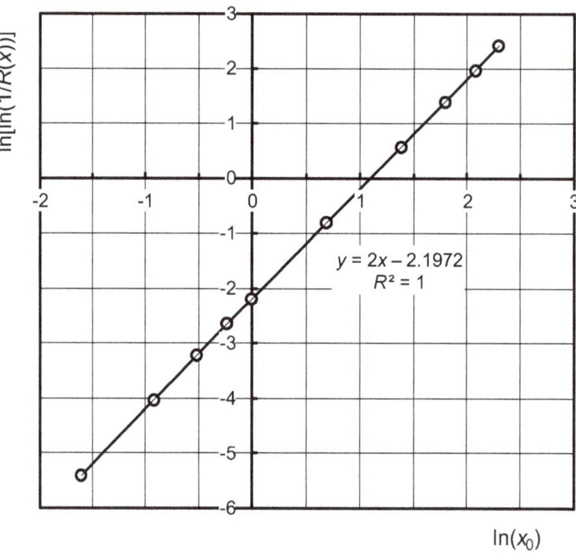

Fig. 3.5: Distribution line of the PSD of Tab. 3.2 (eq. (3.7)).

From the inclination of the distribution line in Fig. 3.5 we obtain the smoothness parameter $n = 2$. From the ordinate distance k, the smoothness parameter n and eq. (3.8) we obtain the statistical mean particle size x':

$$x' = \exp\left(\frac{k}{n}\right) = \exp\left(\frac{-2.1972}{2}\right) = 3\,\mu\text{m}$$

Discussion: The PSD of the particulate matter load of our off-gas, as specified in Tab. 3.2, is well represented by the RRSB distribution function of eq. (3.6). We could also have deduced x' from the left chart in Fig. 3.4. As already mentioned, the particulate matter load of our off-gas has a distinct amount of fines. This specification indicates the need of high-efficiency dust precipitation equipment.

Recommendation: Whenever you have to perform any kind of off-gas measurement you may do it to the best of your knowledge. But, this will be a waste of time if you need to prepare a measurement report, or if you need to gain legal compliance for any activity or installation or operation of installations. Kindly make yourself acquaint with the standards you MUST consider (you may also contact experts); see for example ANNEX VI of Directive 2008/50/EC [17]. Table 3.4 lists off-gas measurement related standards.

Tab. 3.4: Selected off-gas measurement standards.

VDI 2066, Blatt 11 (2018–05)	Particulate matter measurement – dust measurement in flowing gases – measurement of emissions of crystalline silicon dioxide
DIN EN 13284-1 (2018–02)	Stationary source emissions – determination of low range mass concentration of dust – Part 1: manual gravimetric method
DIN EN 14789 (2017–05)	Stationary source emissions – determination of volume concentration of oxygen
DIN EN 14790 (2017–05	Stationary source emissions – determination of vaporous water in ducted systems
DIN EN 14791 (2017–05)	Stationary source emissions – determination of the mass concentration of sulfur dioxide
DIN EN 14792 (2017–05)	Stationary source emissions – determination of the mass concentration of nitrogen oxides
DIN EN 15058 (2017–05)	Stationary source emissions – determination of mass concentration of carbon monoxide
DIN EN 16429 (2021–05)	Stationary source emissions – determination of gaseous HCl concentration in off-gas; standard reference method
DIN EN ISO16911-1 (2013–06)	Stationary source emissions – manual and automatic determination of velocity and volume flow rate in ducts – Part 1: manual reference method
DIN CEN/TS 17286 (2019–07)	Stationary source emissions – monitoring of mercury using sorbent traps
DIN CEN/TS 17337 (2019–08)	Stationary source emissions – determination of mass concentration of several constituents by FTIR

The list is by far not complete. It is not an advertisement but do not hesitate to contact www.beuth.de when you need access to valid standards or even obsolete standards. Latter may sometimes be necessary to understand history files.

4 Legislative framework and dispersion modeling

Directive 2008/50/EC [CELEX 32008L0050] is a mandatory legislative framework for all member states of the EU. The prefix of Directive 2008/50/EC [CELEX 32008L0050] states

> that in order to protect human health and the environment as a whole, it is particularly important to com bat emissions of pollutants at source and to identify and implement the most effective emission reduction measures at local, national and community level. Therefore, emissions of harmful air pollutants should be avoided, prevented or reduced and appropriate objectives set for ambient air quality taking into account relevant World Health Organisation standards, guidelines and programs.

According to Directive 2008/50/EC [CELEX 32008L0050] on ambient air quality and cleaner air for Europe, any negative impact of air pollution on the environment has to be avoided to suffice the basic right of living for both the inanimate and the animate natural life. Annex III of Directive 2008/50/EC [CELEX 32008L0050] summarizes an indicative list of the main polluting substances, as listed further, to be taken into account if they are relevant for fixing emission limit values:

1. Sulfur dioxide and other sulfur compounds
2. Oxides of nitrogen and other nitrogen compounds
3. Carbon monoxide
4. Volatile organic compounds
5. Metals and their compounds
6. Dust
7. Asbestos (suspended particulates and fibers)
8. Chlorine and its compounds
9. Fluorine and its compounds
10. Arsenic and its compounds
11. Cyanides
12. Substances and preparations which have been proved to possess carcinogenic or mutagenic properties or properties which may affect reproduction via the air
13. Polychlorinated dibenzodioxins and polychlorinated dibenzofurans

Referring to Annex VI, Part 6 of Directive 2010/75/EC [CELEX 32010L0075] [6]:

> Sampling and analysis of all polluting substances including dioxins and furans as well as the quality assurance of automated measuring systems and the reference measurement methods to calibrate them shall be carried out according to CEN-standards. If CEN standards are not available, ISO, national or other international standards which ensure the provision of data of an equivalent scientific quality shall apply. Automated measuring systems shall be subject to control by means of parallel measurements with the reference methods at least once per year.

To achieve the goal of improved ambient air quality, air quality standards have been established. Representative current ambient air-quality standards for the protection of health are shown in Tab. 4.1. For comparison the US EPA ambient air quality

https://doi.org/10.1515/9783110763928-004

standards are shown in Tab. 4.1. To avoid confusion, we will briefly discuss the presentation of measurement data:

- The half-hourly mean (HHM) is the mean value of data collected with a monitoring device. When you have to design equipment, you shall have a look on the “maximum half-hourly mean” (MHHM).
- The hourly mean (HM) is derived from two consecutive HHM values.
- The daily average (DA) is calculated from 48 HHM values.
- The annual average (AA) is calculated from 8,760 HM values (or the equivalent of HHMs).

You do not need to fear a bad job now. These data will for sure be prepared by the processor of your measurement device. From practical experience correlations for different immission scales have been deduced. For continuous emission sources and downwind direction [18] recommends the correlations:

$$\frac{c_{\mathrm{DA,\,max}}}{c_{\mathrm{MHHM}}} = 0.45 \qquad \text{and} \qquad \frac{c_{\mathrm{AA}}}{c_{\mathrm{MHHM}}} = 0.1$$

Tab. 4.1: Ambient air-quality standards for the protection of health according to the Directive 2008/50/EC [CELEX 32008L0050] (ANNEX XI) (https://publications.europa.eu/en/publications) and US-(EPA) Standards (U.S. Environmental Protection Agency, 2012a. Air and radiation: National Ambient Air Quality Standards, http://www.epa.gov/air/criteria.html).

Pollutant	Averaging time US-EPA	Primary standards µg m^{-3}	Averaging time ANNEX XI	Primary standards µg m^{-3} [a]
SO_2			1 h	350
SO_2	1 year	80	1 day	125
SO_2			1 year	50
HF			1 year	0.4
CO	1 h	40	8 h	10,000
NO_2	24 h	100	1 h	200
NO_2	1 year	40	1 year	40
O_3	1 h	240 (US)	8 h	120
C_xH_y [b]	3 h	160 (US)		
Benzene			1 year	5
Particulate PM10	1 year	40	1 day	50
			1 year	40
BaP			1 year	0.001
Pb	3 months	1.5	1 year	0.5
As			1 year	0.006
Cd			1 year	0.005
Ni			1 year	0.020
Particulate PM2.5			1 year	25

nn, not specified.
[a] $T = 293.15$ K, $P = 1{,}013$ hPa.
[b] Corrected for methane.

In parallel to the dynamic development of ambient air-quality standards, the legislative framework for emission control has been subject of continuous strengthening within the last decades. Subject of legislative regulations is the definition of standards and limits in pollution control.

The Industrial Emission Directive 2010/75/EC [CELEX 32010L0075] has established a general framework for integrated pollution prevention and control of pollution arising from the activities listed in Annex I (addressing activities of energy industries, production and processing of metals, the mineral industry, the chemical industry, waste management and other activities).

> It should lay down the measures necessary to implement integrated pollution prevention and control in order to achieve a high level of protection for the environment as a whole. Application of the principle of sustainable development should be promoted by an integrated approach to pollution control. The integrated approach means that the permits must take into account the whole environmental performance of the plant, covering e.g., emissions to air, water and land, generation of waste, use of raw materials, energy efficiency, noise, prevention of accidents, and restoration of the site upon closure.

The permit conditions including emission limit values must be based on the "Best Available Techniques" (BAT) as specified in the "BAT Reference Documents" (BREFs). Annex IV of Directive 2008/50/EC [CELEX 32008L0050] and Annex III of Directive 2010/75 [CELEX 32010L0075] specify criteria for determining best available techniques:

1. the use of low-waste technology;
2. the use of less hazardous substances;
3. the furthering of recovery and recycling of substances generated and used in the process and of waste, where appropriate;
4. comparable processes, facilities or methods of operation which have been tried with success on an industrial scale;
5. technological advances and changes in scientific knowledge and understanding;
6. the nature, effects and volume of the emissions concerned;
7. the commissioning dates for new or existing installations;
8. the length of time needed to introduce the best available technique;
9. the consumption and nature of raw materials (including water) used in the process and energy efficiency;
10. the need to prevent or reduce to a minimum the overall impact of the emissions on the environment and the risks to it;
11. the need to prevent accidents and to minimize the consequences for the environment; and
12. the information published by the Commission pursuant to Article 17(2), second subparagraph, or by international organizations.

In harmonized national legislation the German "Technical Instructions on Air Quality Control" (Pursuant to sections 48 and 51 of the German Federal Immission Control Act

[15], also applied outside Germany, "provide instructions to installations requiring a legal permit for operation to protect the general public and the neighborhood against harmful effects of air pollution and to provide precautions against harmful effects of air pollution in order to attain a high level of protection for the environment altogether." The emission limits represent state of the art emission control for several emission sources. Table 4.2 shows selected emission limits according to Chapter 5 of "Technical Instructions on Air Quality Control." Table 4.2 just considers the general requirements except carcinogenic substances, fibers, mutagenic substances, highly toxic substances, odor-intensive substances and soil contaminating substances.

Tab. 4.2: Selected emission limits according to Chapter 5 (General Requirements, of "Technical Instructions on Air Quality Control") [15].

Constituent		**Mass flow rate**	**Emission limit (mg m^{-3})**
Total dust		0.2 kg h^{-1}	20
		>0.4 kg h^{-1}	10
Inorganic particulate matter			
Hg and Tl plus compounds, per substance	Class I	0.25 g h^{-1}	0.05
Pb, Co, Ni, Se, Te (plus compounds, total)	Class II	2.5 g h^{-1}	0.5
Sb, Cr, CN, F, Cu, Mn, V, Sn (plus compounds, total)	Class III	5 g h^{-1}	1
Inorganic gaseous substances			
AsH_3, ClCN, $COCl_2$, PH_3	Class I	2.5 g h^{-1}	0.5
HBr, Cl_2, HCN, HF, H_2S plus compounds, per substance	Class II	15 g h^{-1}	3
NH_3, inorganic chlorine compounds	Class III	0.15	30
SO_x (SO_2, SO_3), NO_x (NO, NO_2) (except combustion off-gas); NO_2 from post combustion and catalytic post combustion	Class IV	1.8	350 200 (at CO <100)
Organic substances (total carbon)		0.5	50
Class I and class I/class II mixtures, pursuant to Annex 4 (acetaldehyde)		0.1	20
Class II (e.g., methyl formate)		0.5	100

Emission of gaseous pollutants is either limited by the mass flow rate or the emission limit of the corresponding class. In case of mixed particular matter the maximum of class II must not be exceeded for class I/class II mixtures, and the class III limit for class I and/or class II and class III mixtures.

Apart from the general requirements the "Technical Instructions on Air Quality Control" [15] also provide "Special Provisions for Certain Types of Installations" as well as guidelines for "Discharge of Waste Gas" and "Measures for the Protection of the Vegetation and of Ecosystems." Latter demands more stringent immission concentration limits than needed for the protection of human health.

To obtain and improve ambient air-quality standards, emission control at emission sources at least has to meet the emission limits. Emission limits are increasingly affected by the local immission level in the vicinity of emission sources. It is therefore necessary to find a link between emission sources, pollutant transmission conditions and immission spots via dispersion modeling.

4.1 Meteorology

Whether an emission source may have a negative impact on the animated and unanimated environment depends on the quality of the emission source (concentration and mass flow) and the atmospheric pollutant transmission conditions. Fig 4.1 shows a simple sketch of the interplay of emission, transmission and immission.

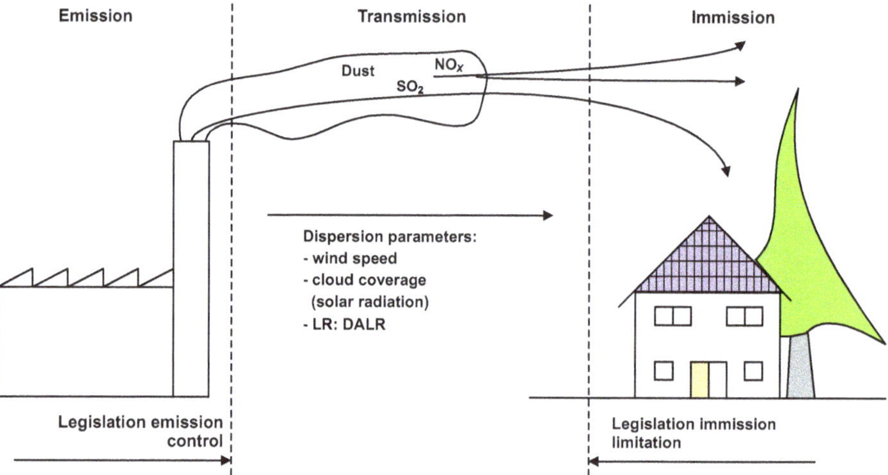

Fig. 4.1: Interplay of emission, transmission and immission.

The problem we are faced with is that we can force the emission source by legislation to comply with appropriate emission standards. We can protect the immission-victim by law against harmful pollutant levels, but we do not have the transmission path under control. While quality of the emission source depends on the design and operation of off-gas purification equipment, transmission conditions cannot be adjusted by technical measures.

For a long time, tall stacks discharging to the atmosphere have been the most common method of disposing of waste gases. The concentration to which animate and unanimated life are exposed at ground level can be reduced significantly by emitting gaseous and particulate pollutants at tall height. Although tall stacks may be effective in lowering the ground-level concentration of pollutants, they do not in themselves reduce the amount of pollutants released to the atmosphere.

To determine the acceptability of a stack as a means of disposing of waste gases, the acceptable ground level concentration of pollutants in the vicinity of an emission source must be estimated.

Awareness of the meteorological conditions prevalent in the area, such as prevailing winds is essential. The orography must be considered so that the stack can be properly located with respect to buildings and hills that might introduce a factor to air turbulence into the operation of the stack.

Wind direction is measured at the height at which the pollutant is released and the mean direction will indicate the direction of travel of pollutants. Wind speed will determine the travel time from a source to a receptor in consideration. It will affect dilution by dispersion in downwind direction. The concentration of air pollutants downwind from a source is inversely proportional to wind speed.

Wind speed has velocity components in all directions so that there are random vertical and horizontal motions. These random motions are essentially responsible for the movement and dispersion of pollutants about the mean downwind path.

If the scale of a turbulent motion is larger than the size of the pollutant plume in its vicinity, the eddy will move that portion of the plume. If an eddy is smaller than the plume, it will diffuse or spread out the plume.

Mechanical turbulence is the induced eddy structure of the atmosphere due to the roughness of the surface over which the air is passing. It increases with increasing wind speed.

Thermal turbulence is mainly induced by solar radiation. Solar radiation will induce a rising component to the lower layer of the atmosphere. Light wind will increase thermal turbulence.

On clear nights with wind, heat is radiated from the Earth's surface, resulting in the cooling of the ground and the air adjacent to it. Then turbulence is at a minimum [19].

The most important factor that influences the degree of turbulence and hence the rate of dispersion in the lower atmosphere is the variation of temperature with the height above ground, referred to as the lapse rate. The dry-adiabatic lapse rate is the temperature change for a rising parcel of dry air. The dry-adiabatic lapse rate (DALR) can be approximated as $\frac{dT}{dz} \approx -1\,^{\circ}\text{C per } 100\text{ m}$

Exercise 4.1: Dry-adiabatic lapse rate

Let us estimate the temperature at a height $h = 100$ m above ground level (at ground level $T_0 = 298.15$ K and $P_0 = 101{,}300$ Pa). Due to adiabatic expansion the temperature changes according to eq. (4.1).

From the first law of thermodynamics we get

$$dh = du + pdv + vdp$$
$$du = dq - pdv$$
$$dq = dh - vdp$$

From the second law of thermodynamics we get

$$dq = Tds$$

With $dh = Cp_0 dT$ and $v = \frac{R \cdot T}{p}$ and $Cp_0 - Cv_0 = R$ we get after inserting and rearranging

$$ds = Cp_0 \cdot \frac{dT}{T} - R \cdot \frac{dp}{p}$$

For isentropic change of state $ds = 0$ and

$$Cp_0 \cdot \frac{dT}{T} = R \cdot \frac{dp}{p},$$

and after rearrangement and integration we obtain

$$\ln \frac{T_2}{T_1} = \frac{R}{Cp_0} \cdot \ln \frac{p_2}{p_1}$$

With $\frac{R}{Cp_0} = \frac{Cp_0 - Cv_0}{Cp_0}$ and $\frac{Cp_0}{Cv_0} = \kappa$ we finally end up with

$$\frac{T_{100}}{T_0} = \left(\frac{P_{100}}{P_0}\right)^{\frac{\kappa-1}{\kappa}} \tag{4.1}$$

with $\kappa_{air} = \frac{Cp_{air}}{Cv_{air}} = \frac{28.98}{20.67} = 1.4$

The pressure of air between ground-level H and any height h above is then $P = \rho \cdot g \cdot (H - h)$.

With $\rho\left[\text{kg m}^{-3}\right] = \frac{P \cdot MM_{air}}{R \cdot T}$ we get for $\frac{dP}{dh} = -\frac{P \cdot MM \cdot g}{R \cdot T}$ and $\frac{dP}{P} = -\frac{MM \cdot g \cdot dh}{R \cdot T}$.

After integration between P_0 and P_{100} we end up with

$$P_{100} = P_0 \cdot e^{-\frac{MM \cdot g \cdot h}{R \cdot T}} = P_0 \cdot e^{-\frac{\rho_0 \cdot g \cdot h}{P_0}} \tag{4.2}$$

If we insert numbers we get:

$$\frac{\kappa - 1}{\kappa} = \frac{1.4 - 1}{1.4} = 0.286$$

$$\frac{MM \cdot g \cdot h}{R \cdot T} = \frac{0.029 \cdot 9.81 \cdot 100}{8.314 \cdot 298.15} = 0.011 \text{ and } P_{100} = P_0 \cdot e^{-\frac{MM \cdot g \cdot h}{R \cdot T}} = 101,300 \cdot e^{-0.011} = 100,144 \text{ Pa}$$

and

$$T_{100} = T_0 \cdot \left(\frac{P_{100}}{P_0}\right)^{\frac{\kappa-1}{\kappa}} = 298.15 \cdot \left(\frac{100,144}{101,300}\right)^{0.286} = 297.17 \text{ K}$$

and we end up with the dry-adiabatic lapse rate (DALR) in the following equation:

$$\frac{dT}{dz} = -(298.15 - 297.17) = -0.98\ {}^{\circ}\text{C per 100 m} \cong -1\ {}^{\circ}\text{C per 100 m} \tag{4.3}$$

The cooling of humid air due to adiabatic expansion will result in an increased relative humidity. Saturation may be reached and further ascent would result in condensation of water vapor. The latent heat released during condensation will reduce the rate of cooling.

Buoyancy of warm (off-)gas is caused by the difference in density to the surrounding air. According to the perfect gas law the temperature and density of the gas flux in consideration are inversely related at fixed pressure.

To explain the interaction of off-gas with the atmospheric ambient we compare a parcel of off-gas, strictly following the dry adiabatic lapse rate, with the surrounding environment [20]. Latter being subject of change by solar radiation, overcast and wind speed.

If the temperature gradient of the atmosphere is the same as the adiabatic lapse rate of the off-gas, as shown in Fig. 4.2, the off-gas will expand or contract in such a manner that its density and temperature remain the same as in the surrounding atmosphere. Buoyancy force will be negligible. The atmosphere is neutrally stable. The parcel of air will remain at the height we place it. Pollutants of the off-gas plume will disperse by coning. We call this dispersion conditions "neutral."

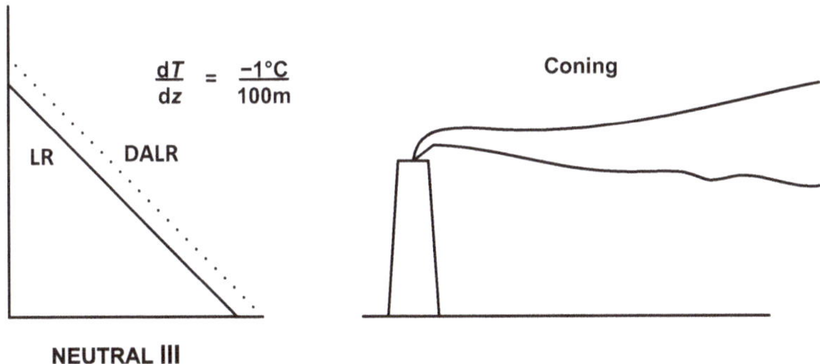

Fig. 4.2: Neutral dispersion conditions.

Slightly unstable conditions or neutral conditions (weak lapse) may occur during both day and night. In cloudy and windy areas it may be the most frequent type of lapse rate characteristic observed.

If the atmospheric temperature (actual lapse rate) decreases faster with increasing altitude than the DALR of our parcel of off-gas, as shown in Fig. 4.3, will have a higher temperature in the parcel than the ambient air. Its density will be lower, resulting in a net upward buoyancy force (unstable atmospheric conditions).

Unstable (strong) lapse rate is usually a fair weather daytime condition, since strong solar heating of the ground is required. Looping may have a negative effect on pollutant dispersion. It may force parcels of undiluted pollutants to ground level.

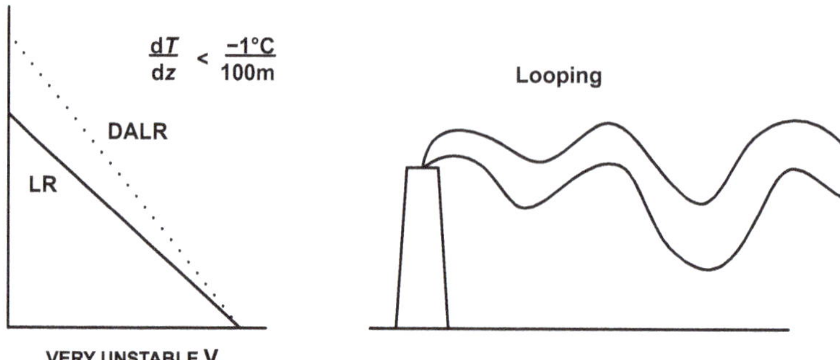

Fig. 4.3: Unstable dispersion conditions.

Transition from unstable conditions to stable conditions, shown in Fig. 4.4, occurs during transition from lapse to inversion. It is observed near sunset, and it may be transitory as well as persist for several hours. Our off-gas pollutant will just dilute between the inversion ceiling and the ground level.

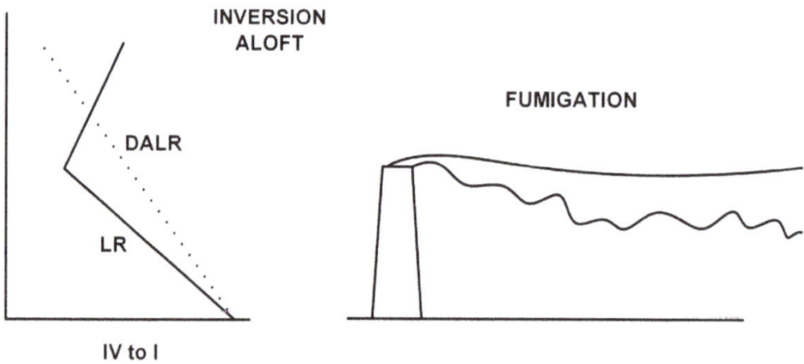

Fig. 4.4: Transition from unstable to stable dispersion conditions.

If the atmospheric temperature decreases more slowly than the DALR, buoyancy forces cause stable atmospheric conditions, shown in Fig. 4.5. The density of air in the parcel will be higher compared with the ambient and as a consequence the parcel will drop to a lower height (bottom).

Stable dispersion (inversion) is principally a nighttime condition. It is favored by light winds, clear skies and snow cover. The condition may persist for several days during winter. Very stable lapse rates are referred to as inversions. Strong stability inhibits mixing across the inversion layer. These conditions extend for a few hundred meters vertically.

The vertical extent is referred to as the inversion depth. We have to distinguish between ground-level inversion, caused by radiative cooling of the ground at night, and inversions aloft, occurring between 500 m and several 1,000 m above ground level.

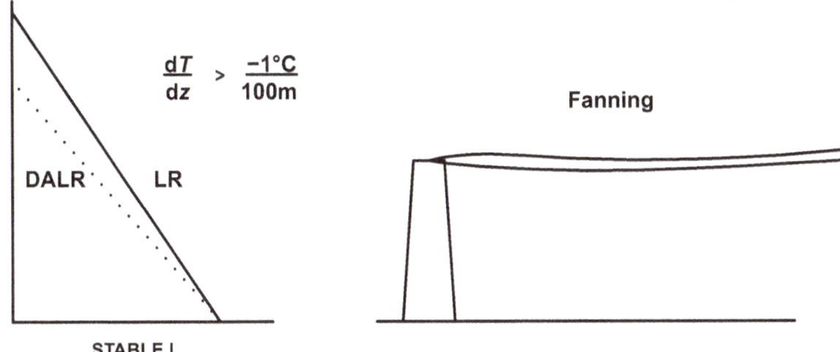

Fig. 4.5: Stable dispersion conditions.

Transition from stable to unstable dispersion conditions, shown in Fig. 4.6, occurs when the nocturnal inversion is dissipated by heat from the morning sun. The lapse rate usually starts at the ground and works its way upward. The inversion layer forms a barrier below the emission height and the ground level. Pollutants emitted above this barrier also dilute in the atmosphere above this barrier. The plume grows upward into a neutral layer of air.

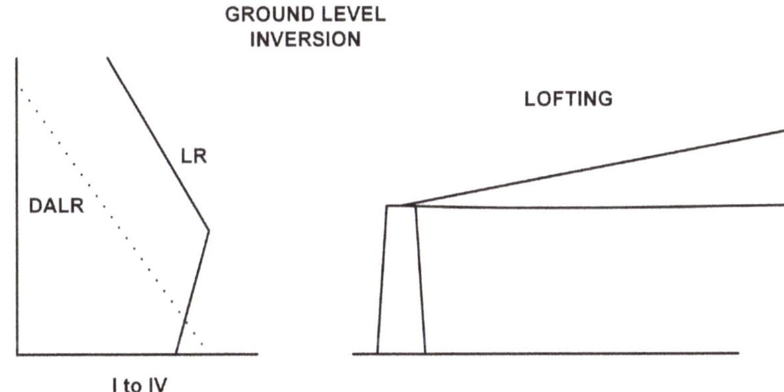

Fig. 4.6: Transition from stable to unstable.

From the viewpoint of air pollution, stable surface layers and low level inversions are undesirable because they minimize the rate of dilution of contaminants in the atmosphere.

Low-level inversion will lead to an accumulation of contaminants. Low-level inversion acts as a barrier to vertical mixing.

As may be concluded from this brief discussion the lapse rate, mainly affected by solar radiation, has a significant effect on dispersion of pollutants in the atmosphere. Figure 4.7 shows a rough correlation of radiation intensity and the dry adiabatic lapse rate.

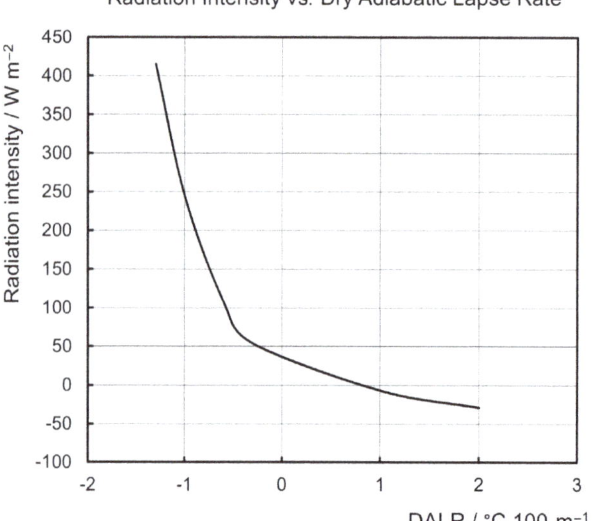

Fig. 4.7: Correlation of solar radiation intensity and the DALR.

Solar radiation intensity varies widely, depending on latitude, season and daytime. As a consequence, the lapse rate is subject of diurnal change and together with prevailing wind and overcast it acts on pollutant dispersion. If you are interested in this topic, you will find a fascinating discussion in [21]. For our needs we just keep in mind that we have to consider the wind speed, solar radiation and overcast in estimation of pollutant dispersion.

4.2 Pollutant dispersion modeling

In Tab. 1.1 we have specified our primary emission source, and in Tab. 4.1 we have specified ambient air-quality standards. We assume that the emission limits of Tab. 4.2 may apply. At this project level we do not have any feedback from off-gas purification technologies. Therefore, we assume that the clean gas will be emitted at cooling limit temperature and emission limits, as precised in Tab. 4.3. The stack opening diameter is recommended to correspond with twice the gas velocity of the mean wind speed. An off-gas velocity of $v_{gas} = 6\ m\ s^{-1}$ at the stack mouth should suffice this recommendation. For discussing pollutant dispersion, we pick out the pollutant SO_2.

We now want to find out with the help of meteorology, whether our off-gas pollutant, as specified in Tab. 4.3, may have a negative impact on the environment. The meteorological factors include wind speed, air temperature, shear of the wind speed with height and atmospheric stability. We have in mind that the maximum ground-level concentration of SO_2 must not exceed 350 μg m^{-3}. The actual ground-level concentration is

Tab. 4.3: Off-gas specification of Tab. 1.1, extended to our needs for dispersion modeling.

Specification	Value	Unit
p_{stat}	980	hPa
θ_{cl}	58.2	°C
Stack diameter D	1	m
N_2	78	vol%
O_2	12	vol%
CO_2	10	vol%
$X_{H_2O,cl}$	138	g $kg^{-1}{}_{STP,dry}$
ROV (according to Tab. 2.1)	3	vol% (dry gas)
$F_{V,STP,dry}$	24,560.6	$m^3{}_{STP,dry}$ h^{-1}
$F_{V,STP,humid}$	30,209	$m^3{}_{STP,humid}$ h^{-1}
$F_{V,actual}$	37,888	am^3 h^{-3}
$\rho_{STP,dry}$	1.342	kg $m^{-3}{}_{STP,dry}$
$\rho_{STP,humid}$	1.241	kg $m^{-3}{}_{STP,humid}$
$\rho_{STP,actual}$	0.990	kg am^{-3}
SO_x (according to Tab. 1.2)	350	mg $m^{-3}{}_{O_2 = 3\ vol\%}$
SO_x (according to Tab. 2.1)	175	mg $m^{-3}{}_{O_2 = 11\ vol\%}$
SO_x	61	ppm
SO_x	4.3	kg h^{-1}

compared with the ambient air-quality standards. But, we must be careful with data interpretation. The reading of Tab. 4.1 says that the immission concentration of SO_2 must not exceed 350 µg m^{-3}. The reading of ANNEX II of Directive 2008/50/EC [CELEX 32008L0050] says that corresponding to the lower assessment threshold 40% of the 24-h immission limit (= 125 µg m^{-3}) must not be exceeded more than three times in any calendar year). Suddenly we talk about an immission limit of 50 µg m^{-3} for human health protection and 8 µg m^{-3} for vegetation protection. Consequently we have to quantify, whether our off-gas concentration of SO_2 of 175 mg $m^{-3}{}_{STP,dry}$ at 11 vol% oxygen, as specified in Tab. 1.1 meets the immission quality standards of Tab. 4.1 under consideration of ANNEX II of Directive 2008/50/EC [CELEX 32008L0050].

Let us start with the estimation of an appropriate stack height. According to Chapter 5, General Requirements, of "Technical Instructions on Air Quality Control" [15], stacks shall have a minimum height of 10 m above ground level (or 3 m at minimum above the ridge of a roof). The height of a stack shall not exceed 250 m. If stack height is beyond 200 m further emission reduction is recommended. For the estimation of the stack height we need the stack diameter, the off-gas temperature at the stack mouth, the volume flow of gas at dry standard conditions and the hourly mass flow of pollutant. With these numbers we may go into a stack height determination nomogram to end up with a stack height of 20 m. You will find this nomogram for example in an App of the German Environmental Agency (Umweltbundesamt). You may also find it in General Requirements, of "Technical Instructions on Air Quality Control" [22].

The effective height of an emission source rarely corresponds with the physical height of a stack due to plume rise. It must be determined empirically. Determination is based on the plume-rise equations of [23–25]. Several source emission characteristics and meteorological factors affect the rise of a plume. Source emission factors include the gas flow rate and the temperature of the effluent at the top of the stack and the diameter of the stack opening. If the plume is caught in the turbulent wake of the stack or of buildings in the vicinity of a stack, the effluent will be rapidly mixed downward toward the ground.

Before we jump into dispersion modeling, we discuss basic approaches for pollutant dispersion. We may assume that the pollutant disperses perpendicular (y-direction of our Cartesian system) to downwind direction x of our point pollutant (pulse) source with pollutant mass m and area A according to the diffusion equation:

$$\frac{\mathrm{d}c}{\mathrm{d}t} = D \cdot \frac{\mathrm{d}^2 c}{\mathrm{d}y^2} \tag{4.4}$$

Equation (4.5) [26] provides a solution of

$$c = \frac{B}{\sqrt{t}} \cdot e^{-\frac{y^2}{4D \cdot t}} \tag{4.5}$$

With the total mass of pollutant m/A corresponding to

$$\frac{m}{A} = \int_{-\infty}^{+\infty} c \cdot \mathrm{d}y \tag{4.6}$$

we will obtain for constant B of eq. (4.5) the correlation of the following equation:

$$B = \frac{m}{A} \cdot \frac{1}{\sqrt{4 \cdot \pi \cdot D}} \tag{4.7}$$

which after resubstituting into eq. (4.5) will result in a handsome expression for the mean concentration of our pollutant for given wind speed u and distance x and therefore $t = (x/u)$, according to the following equation:

$$c_{\text{mean}} = \frac{(m/A)}{\sqrt{4 \cdot \pi \cdot D \cdot t}} \cdot e^{-\frac{y^2}{4 \cdot D \cdot t}} \tag{4.8}$$

with the pulse width $l = \sqrt{4 \cdot D \cdot t}$.

In his famous book *Diffusion*, Cussler [27] performs a quick estimate of the pulse width for a mean diffusion coefficient of gaseous constituents of $D = 1 \cdot 10^{-5}$ m^2 s^{-1}, x = 10 km and u = 15 km h^{-1}, resulting in l = 0.3 m, a disastrous result compared to an experimental result of l = 1,000 m, which he states to be a big error, "even for engineers." However, he wanted to point out the difference between diffusion control and dispersion control. It does not help much, because we need a handsome algorithm for estimating whether our off-gas will cause harm to the environment or not, and we want to do it with our hand calculator.

A Gaussian plume model sounds to be an appropriate atmospheric dispersion formula. "Technical Instructions on Air Quality Control" [28] may offer help (you know, at that time calculations were mainly performed with hand calculators), and eq. (4.9) shows the dispersion formula for dispersion of gaseous pollutants with reflection from the ground and the mixing layer we were looking for

$$C(x,y,z) = \frac{10^6 \cdot F_{\mathrm{m,pollutant}}}{3{,}600 \cdot 2 \cdot \pi \cdot u_{\mathrm{h,\,Stack}} \cdot \sigma_y \cdot \sigma_z} \cdot e^{-\frac{y^2}{2\sigma_y^2}} \cdot \left(e^{-\frac{(z-h)^2}{2\sigma_z^2}} + e^{-\frac{(z+h)^2}{2\sigma_z^2}} \right) \tag{4.9}$$

where x, y, z are Cartesian coordinates of the investigated location (stack coordinates: 0,0,0; ground-level coordinates; x,y,0), $C(x,y,z)$ is the mass concentration of the pollutant in mg m^{-3} at level (x,y,z), z is the height above ground level (m), $F_{\mathrm{m,\,pollutant}}$ is the mass flow of pollutant (kg h^{-1}), h is the effective height of emission (m), σ_y, σ_z is the horizontal and vertical dispersion parameter (m) and $u_{\mathrm{h,Stack}}$ is the wind speed in stack height (m s^{-1}).

Comparison of eq. (4.9) with eq. (4.8) shows how we can overcome the gap between diffusion control and the effect of wind, the clouds and solar radiation, although we can just consider constant wind speed in one direction. The "diffusion coefficient" is hidden in the dispersion parameters.

Equation (4.9) also offers simplifications. If we just want to gain information about the ground-level concentration (z = 0), we may make use of

$$C(x,y,z) = \frac{10^6 \cdot F_{\mathrm{m,\,pollutant}}}{3{,}600 \cdot \pi \cdot u_{\mathrm{h}} \cdot \sigma_y \cdot \sigma_z} \cdot e^{-\frac{y^2}{2\sigma_y^2}} \cdot \left(e^{-\frac{(h)^2}{2\sigma_z^2}} \right) \tag{4.10}$$

If we just want to gain information about the ground-level concentration in the center line (y = 0, z = 0), we may make use of

$$C(x,y,z) = \frac{10^6 \cdot F_{\mathrm{m,\,pollutant}}}{3{,}600 \cdot \pi \cdot u_{\mathrm{h}} \cdot \sigma_y \cdot \sigma_z} \cdot \left(e^{-\frac{(h)^2}{2\sigma_z^2}} \right) \tag{4.11}$$

Before we can perform calculations we need some additional tools in our dispersion modeling toolbox. Let us start with the determination of dispersion classes based on the meteorological phenomena as discussed in Section 4.1, and as suggested in "TA-Luft 86." The local weather conditions, the day time and the wind speed have to be determined according to the classification of Tab. 4.4. u_{r} is the mean wind speed, determined at height z_{a} = 10 m above ground level as per ref. [29].

Then the dispersion parameters σ_y, σ_z, are calculated with the dispersion coefficients F, f and G, g, as listed in Tab. 4.5 and Tab. 4.6. The data of Tabs. 4.5 and 4.6 just differ in the effective height h of emission. The data of Tab 4.5 and 4.6 may be used

Tab. 4.4: Determination of the dispersion class with u_r: mean wind speed at anemometer height [28].

u_r (m s^{-1}) (at h = 10 m)	Nighttime overcast		Daytime overcast		
	0/8–6/8	7/8–8/8	0/8–2/8	3/8–5/8	6/8–8/8
<1.2	I	II	IV	IV	IV
1.3–2.3	I	II	IV	IV	III/2
2.4–3.3	II	III/1	IV	IV	III/2
3.4–4.3	III/1	III/1	IV	III/2	III/2
4.4+	III/1	III/1	III/2	III/1	III/1

for an effective height of emission of $h < 100$ m only. The dispersion parameters are calculated via the power correlations as follows:

$$\sigma_y = F \cdot x^f \tag{4.12}$$

$$\sigma_z = G \cdot x^g \tag{4.13}$$

Tab. 4.5: Dispersion coefficients F, f and G, g, applicable for determining the dispersion parameters σ_y, σ_z, for effective height of emission of $h < 50$ m [28].

Dispersion class	m	s (s^{-2})	Lapse rate (air)	Dispersion parameters			
				F	f	G	g
V	0.09	$5 \cdot 10^{-4}$	Very unstable	1.503	0.833	0.151	1.219
IV	0.2	$5 \cdot 10^{-4}$	Unstable	0.876	0.823	0.127	1.108
III/2	0.22	$1 \cdot 10^{-3}$	Neutral/unstable	0.659	0.807	0.165	0.996
III/1	0.28	$2 \cdot 10^{-3}$	Stable/neutral	0.64	0.784	0.215	0.885
II	0.37	$2 \cdot 10^{-3}$	Stable	0.801	0.754	0.264	0.774
I	0.42	$2 \cdot 10^{-3}$	Very stable	1.294	0.718	0.241	0.662

Tab. 4.6: Dispersion coefficients F, f and G, g, applicable for determining the dispersion parameters σ_y, σ_z, for effective height of emission of 50 m < h < 100 m [28].

Dispersion class	m	s (s^{-2})	Lapse rate (air)	Dispersion parameters			
				F	f	G	g
V	0.09	$5 \cdot 10^{-4}$	Very unstable	0.170	1.296	0.051	1.317
IV	0.2	$5 \cdot 10^{-4}$	Unstable	0.324	1.025	0.070	1.151
III/2	0.22	$1 \cdot 10^{-3}$	Neutral/unstable	0.466	0.866	0.137	0.985
III/1	0.28	$2 \cdot 10^{-3}$	Stable/neutral	0.504	0.818	0.265	0.818
II	0.37	$2 \cdot 10^{-3}$	Stable	0.411	0.882	0.487	0.652
I	0.42	$2 \cdot 10^{-3}$	Very stable	0.253	1.057	0.717	0.486

The off-gas heat energy flux Q of the humid off-gas is calculated with the following eq. (4.14):

$$Q = 1.36 \cdot 10^{-3} \cdot F_{\text{V, gas, humid}} \cdot \left(\theta_{\text{gas}} - \theta_{\text{ambient}}\right) \tag{4.14}$$

where Q is the heat energy flux of the off-gas in MW, $F_{\text{V, gas, humid}}$ is the volumetric flow rate of the humid gas in $m^3\ s^{-1}{}_{\text{STP,humid}}$, θ_{gas} is the off-gas temperature in °C and θ_{ambient} is the temperature of the environment being 10 °C.

If you wonder about eq. (4.14), do not panic. The value of $1.3.6 \cdot 10^{-3}$ considers the mean density $\rho = 1.3$ kg m^{-3} and the mean specific heat of off-gas $Cp \approx 1$ kJ kg^{-1} K^{-1} and vaporous water $Cp_{\text{H}_2\text{O, vapor}} \approx 1.91$ kJ kg^{-1} K^{-1}. You can easily check that with the data of Tab. 4.3.

In the next step we have to focus on plume rise. Plume rise is affected by off-gas buoyancy and the wind speed. Briggs [23] and Carson and Moses [24, 25] and Turner [30] have developed several plume-rise correlations. The sketch in Fig. 4.8 shows the basic approach of Turner [30].

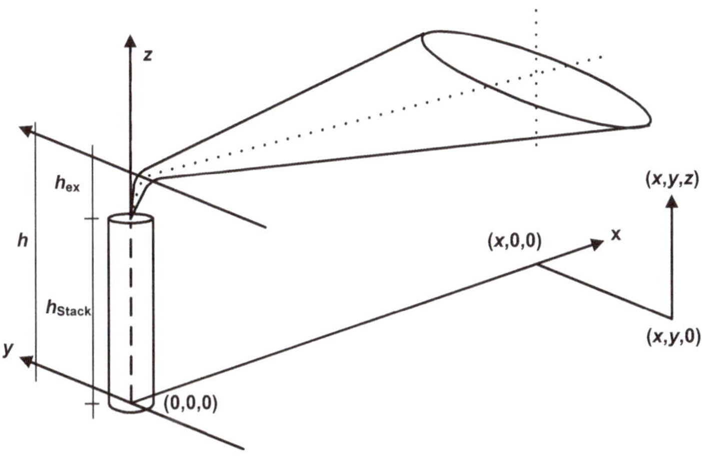

Fig. 4.8: The effect of plume rise on the height of plume dispersion.

According to "TA-Luft 86" [28] the excess height h_{ex} due to thermal plume rise is calculated with eq. (4.15), $u_{\text{h, Stack}}$ with eq. (4.16) and u_{h} with eq. (4.17):

$$h_{\text{ex}} = \frac{K \cdot Q^{\frac{1}{3}} \cdot x^{\frac{2}{3}}}{u_{\text{h, Stack}}} \tag{4.15}$$

where h_{ex} is the excess height (m), K is the constant (see Tab. 4.7), $u_{\text{h, Stack}}$ is the wind speed at stack height (m s^{-1}) and x is the distance in center line (Fig. 4.8) with

$$u_{h,\text{Stack}} = u_r \cdot \left[\frac{h_{\text{Stack}}}{z_a}\right]^m \tag{4.16}$$

where $u_{h,\text{Stack}}$ is the wind speed at stack height (m s^{-1}), u_r is the mean wind speed in m s^{-1}, determined at height z_a (anemometer height = 10 m), h_{Stack} is the stack height (m), z_a is the height of anemometric wind speed determination (m) (standard height: 10 m) and m is the power exponent (see Tab. 4.5 or Tab. 4.6):

$$u_h = u_r \cdot \left[\frac{h}{z_a}\right]^m \tag{4.17}$$

u_h is the wind speed at the effective height of emission (m s^{-1})
with

$$h = h_{\text{Stack}} + h_{ex} \tag{4.18}$$

The power exponent m is linked with the dispersion classes according to the values listed in Tabs. 4.5 and 4.6. Table 4.7 shows the constants K of eq. (4.15).

Tab. 4.7: *K*-values for different dispersion classes for $Q < 6$ MW [28].

Dispersion class	*K*
V	3.34
IV	3.34
III/2	2.84
III/1	2.84
II	3.34
I	3.34

Following the proposal of [23–25], the excess height h_{ex} due to thermal plume rise is calculated with eq. (4.19) or (4.19a), $u_{h,\text{Stack}}$ with eq. (4.20) and u_h with eq. (4.21):

$$h_{ex} = \frac{2.6 \cdot \sqrt{Q}}{u_{h,\text{Stack}}} \tag{4.19}$$

or

$$h_{ex} = 0.6 \cdot \sqrt[3]{\frac{Q}{u_{h,\text{Stack}} \cdot s}} \tag{4.19a}$$

with

$$Q = 3.6 \cdot 10^{-4} \cdot F_{V,\,gas,\,humid} \cdot \left(\theta_{gas} - \theta_{ambient}\right) \tag{4.20}$$

where Q is the heat energy flux of the off-gas (kW), $F_{V,\,gas,\,humid}$ is the volumetric flow rate of the humid gas ($m^3_{STP,\,humid}\,h^{-1}$) θ_{gas} is the off-gas temperature (°C), $\theta_{ambient}$ is the temperature of the ambient environment = 15 °C and s is the stability parameter from Tab. 4.5 or 4.6

$$u_{h,\,Stack} = u_r \cdot \left[\frac{h_{St}}{z_a}\right]^m \tag{4.21}$$

where $u_{h,Stack}$ is the wind speed at stack height (m s^{-1}), u_r is the mean wind speed, determined at height z_a (anemometer height) (m s^{-1}), z_a is the height of speed determination (m, standard height is 10 m) and m is the power exponent (Tab. 4.5 or Tab. 4.6).

It is recommended to decide for the smaller value of h_{ex} from eqs. (4.19) and (4.19a). The power exponent m is correlated with the dispersion class according to the values listed in Tabs. 4.5 and 4.6. The immission concentration $C(x,y,z)$ is determined with eq. (2.22) with the mean wind velocity determined from $u_{h,Stack}$ and u_h with eq. (4.23):

$$C(x,y,z) = \frac{10^6 \cdot F_{m,\,pollutant}}{3,600 \cdot 2 \cdot \pi \cdot u_{mean} \cdot \sigma_y \cdot \sigma_z} \cdot e^{-\frac{y^2}{2\sigma_y^2}} \cdot \left(e^{-\frac{(z-h)^2}{2\sigma_z^2}} + e^{-\frac{(z+h)^2}{2\sigma_z^2}}\right) \tag{4.22}$$

$$u_{mean} = \left(\frac{u_{h,\,stack} + u_h}{2}\right) \tag{4.23}$$

Exercise 4.2: Estimation of pollutant dispersion

We must “convince” SO_2 emission concentration in our off-gas, as specified in Tab. 4.2, to change from inflow concentration of 300 ppm (= 856.5 mg $m^{-3}{}_{STP,dry}$) at actual oxygen content of 11 vol% to 175 mg $m^{-3}{}_{STP,dry}$ to comply with Chapter 5 (General Requirements, of “Technical Instructions on Air Quality Control”) [15]. ANNEX II of Directive 2008/50/EC [CELEX 32008L0050] demands a maximum immission concentration of 50 [μg m^{-3}] for human health protection. Gladly our facility is installed in a plane area with constant wind speed at anemometer height z_a = 10 m of u_r = 1.5 m s^{-1}. From Chapter 5 (General Requirements, of “Technical Instructions on Air Quality Control”) [15] we get a stack height of h_{Stack} = 20 m.

With eq. (4.20) we obtain a heat energy flux of

$$Q = 3.6 \cdot 10^{-4} \cdot F_{V,\,gas,\,humid} \cdot \left(\theta_{gas} - \theta_{ambient}\right) = 3.6 \cdot 10^{-4} \cdot 30,209 \cdot (58.2 - 5) = 578.6\ \text{kJ s}^{-1}$$

(eq. (4.14) would result in 550 kJ s^{-1}. The results are very comparable).

With the wind speed reading at anemometer height u_r (z_a) = 1.5 m s^{-1} we get the wind speed at stack mouth height h_{Stack} = 20 m calculated with eq. (4.21). With eq. (4.19a) we get the excess height, and the height of plume center line according to Fig. 4.8 is then determined via eq. (4.18):

$$u_{h,\,Stack}\left[\text{m s}^{-1}\right] = u_r \cdot \left[\frac{h_{Stack}}{z_a}\right]^m = 20 \cdot \left[\frac{20}{10}\right]^m \tag{4.21}$$

$$h_{ex}[\text{m}] = 0.6 \cdot \sqrt[3]{\frac{Q}{u_{h,\,Stack} \cdot s}} = 0.6 \cdot \sqrt[3]{\frac{578.6}{u_{h,\,Stack} \cdot s}} \tag{4.19a}$$

$$h = h_{Stack} + h_{ex} \tag{4.18}$$

The results for the different dispersion classes of Tab. 4.5, considered in this dispersion modeling approach, are summarized in Tab. 4.8.

Tab. 4.8: Mean velocity u_{mean}, excess height h_{ex} and dispersion height h for different dispersion classes.

Dispersion class	Power law exponent *m*	Stability parameter *s*	u_{mean} (m s^{-1})	h_{ex} (m)	*h* (m)
I	0.42	0.002	2.45	28.1	48.1
II	0.37	0.002	2.32	28.5	48.5
III/2	0.22	0.001	1.95	31.6	51.6
IV	0.2	0.0005	1.91	32.0	52

Referring to Tab. 4.5 dispersion class III/1 is outside the range of u_r. In the next step we determine the ground-level concentration of SO_2 ($C_{x,0,0}$) in downwind direction center line with eq. (4.22) after having prepared the dispersion parameters σ_y and σ_z for the independent variable x. Table 4.9 shows the results for dispersion class III/2.

Tab. 4.9: Results for dispersion class III/2.

x (m)	σ_y (m)	σ_z (m)	*C*(*x*,0,0) (mg m^{-3})
100	27.09	16.20	0.002
200	47.40	32.31	0.034
245	55.84	39.55	0.037
300	65.75	48.38	0.034
400	82.94	64.44	0.026
500	99.30	80.47	0.020
600	115.04	96.50	0.015
700	130.28	112.51	0.012
800	145.10	128.52	0.010
900	159.57	144.51	0.008
1,000	173.73	160.50	0.007
1,100	187.62	176.49	0.006
1,200	201.27	192.46	0.005
1,300	214.70	208.44	0.004
1,400	227.93	224.40	0.004
1,500	240.99	240.36	0.003

The maximum ground-level concentration for dispersion class III/2 in the downwind center line is determined via $\sigma_{z,max}$ in m as follows:

$$\sigma_{z,\,max} = \frac{h}{\sqrt{2}} \tag{4.24}$$

Figure 4.9 shows the outcome for the considered dispersion classes and the immission maximum for dispersion class III/2 at $x = 245$ m.

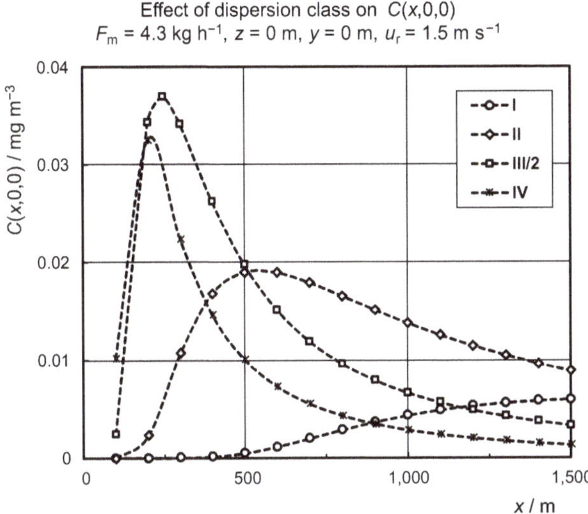

Fig. 4.9: Comparison of the ground-level concentration for different dispersion classes. SO_2 emission of $c_{SO_2} = 175$ mg m^{-3}, stack height $h_{Stack} = 20$ m, off-gas temperature $\theta_{gas} = 58.2$ °C, temperature of ambient air $\theta_{air} = 15$ °C, off-gas flow $F_{V,actual} = 37{,}888$ am^3 h^{-1}.

With this algorithm the immission concentration profiles perpendicular to the direction of prevailing wind y, and the dispersion height z in the downwind center line ($y = 0$) can be constructed. The concentration profiles are shown in Fig. 4.10.

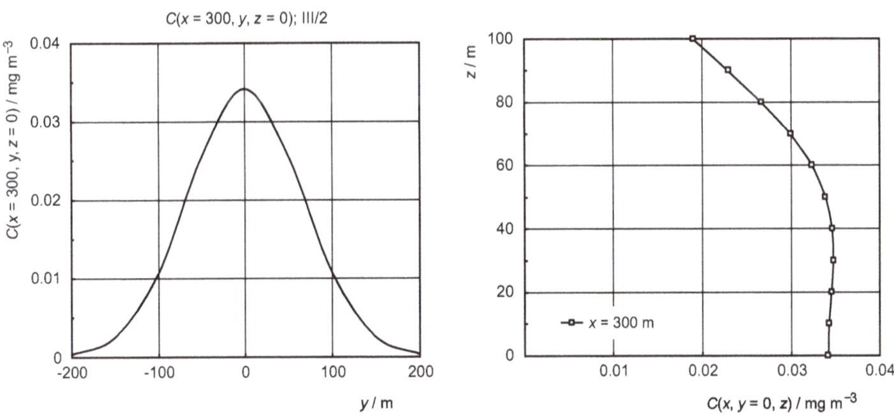

Fig. 4.10: Immission concentration profile perpendicular to the dispersion direction, and over the height z in the centerline of dispersion ($y = 0$) at $x = 300$ m.

! Discussion: Immission concentration passes a maximum at different distance x for stable (class II), neutral (class III/2) and unstable (class IV) dispersion conditions, to drop afterward with increasing distance from the emission source. The lag of dispersion class I (very stable dispersion conditions) close to the emission source is expected. From Fig. 4.9 we may conclude that a minimum precipitation efficiency for SO_2 of 80%, calculated from the off-gas SO_2 concentration (c_{SO_2} = 856.5 mg $m^{-3}{}_{STP,dry}$) and the lower assessment threshold (c_{SO_2} = 175 mg $m^{-3}{}_{STP,dry}$), as specified in ANNEX II of Directive 2008/50/EC [CELEX 32008L0050]), would meet the immission standards for human health protection.

However, we are not recommended to base the design of SO_2 precipitation equipment on the minimum needs, we have to perform better, as will be discussed in detail in the technology part of the book.

It has to be outlined that this paragraph on dispersion modeling is just a first introduction. If you are interested in this important subject in more detail you may for example collect more information in [31]. You may also visit Appendix W to part 51 of number 40 of the Code of Federal Regulations (but do not get lost). Anyhow, this discussion of pollutant dispersion hopefully also suggests the recommendation of consulting an expert in case of necessity. You may have registered that dispersion of particulate matter has not been addressed. We will not ignore particulate matter, but have to make a quick excursion prior to that to elucidate the smog problem.

4.3 Smog: photochemical smog

Perhaps you may have been in touch with the blue haze phenomenon of the Great Smoky Mountains. Cherokee mythology explained it with "a great smoke Manitou had, while having a rest" in this beautiful area of Tennessee/North Carolina. The environmental chemists and the biologists explain this phenomenon with huge terpene evaporation from the coniferous woods in this area under high solar irradiance and environmental pollution. We have already addressed the phenomenon in Section 3.1, and we fear it in pollution control. You can easily slip into a blue haze disaster. Ply wood veneer driers may suffer from blue haze emissions when elevated temperature in the drier induces formation of heterogeneous azeotropic mixtures of terpenes and water. At limited process efficiency sulfuric acid plants had severe problems with SO_3 loss of the SO_3 absorption tower, also forming a bluish plume due to submicronic aerosol size. Fumes from metallurgical kilns may also contribute to man-made blue haze (they can also form a brown haze when dedusting of fines fails). Blue haze of automobile exhaust indicates damaged oil control rings in the engine. In combination with air humidity and solar irradiation and low wind these pollutants may accumulate to an ugly photochemical smog problem (smog is originating in the ingredients smoke and fog) many urban areas suffer from. Photochemical smog shows up in a more or less dense brownish plume covering large areas. The brownish color is mainly caused by the NO_2 load, but light scattering of particulate matter in the particle size range of 500 nm may also contribute to the brownish color of smog. The "nice" bluish haze of the Smokies may also change color during the day due to solar interaction (plus man-made pollutants such as NO_x).

The background story of the blue haze in the Smokies: Have you ever met a spruce tree in the health care department of a super market ask for sunscreen? Rather not. When you expose your skin to sun you are very recommended to protect it with an effective sunscreen to avoid bad sunburns. Since spruce trees and pines cannot protect their treetop with a sunscreen, they have developed an alternative by emitting terpenes such as *ß*-pinene or γ-terpinene. These monoterpenes have a fantastic UV-light absorbance below $\lambda = 300$ nm. While we get the sunscreen removed in the shower, trees again need an alternative. With suns help, oxygen and even man-made pollutants they let the terpene protection shield decompose in the afternoon to get rid of it; simple and effective.

We have to distinguish between primary smog and secondary smog. Primary smog is originating in pollutant emissions from point source or diffuse source, while secondary smog is formed either by chemical reaction of pollutants, or by photochemical conversion. Particulate matter of ammonium sulfate for example is formed by precipitation of airborne ammonia with sulfuric acid formed from airborne sulfur dioxide, humidity, oxygen with and without airborne solid catalysts, and sometimes with the help of sun. Ammonia may travel hundreds of kilometers from the source to the location where it finally forms ammonia sulfate. Photosmog is a secondary smog, formed from a poisonous cocktail of many man-made pollutants and the sun.

The Sun, fueling life on the Earth, provides as much as 1,368 W m^{-2} of irradiance to the edge of Earth's atmosphere over the whole span of UV radiation, visible light and infrared radiation. Figure 4.11 shows the cumulative irradiance for $\lambda < 800$ nm [13], and the derivative provides the specific irradiance within the selected wavelength span.

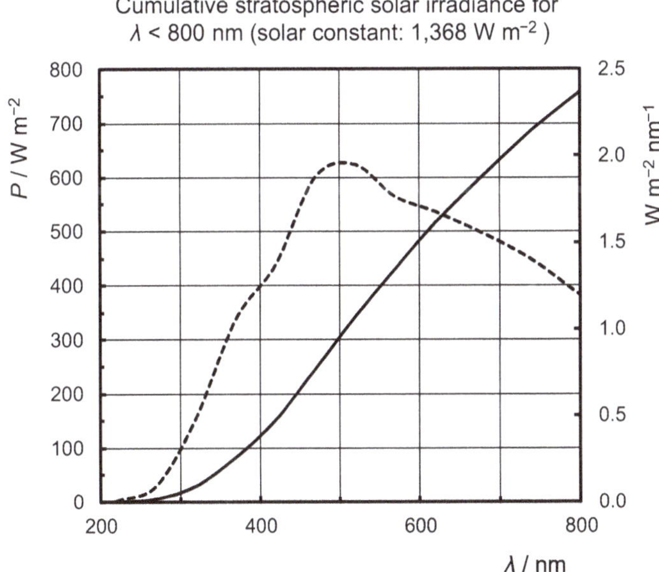

Fig. 4.11: Cumulative solar irradiance and specific solar irradiance for $\lambda < 800$ nm.

Depending on the latitude, the season, the atmosphere and backscattering up to 50% of the solar radiation may touch ground level. Figure 4.12 shows a representative seasonal energy trend of solar energy for the northern hemisphere and the irradiance [32].

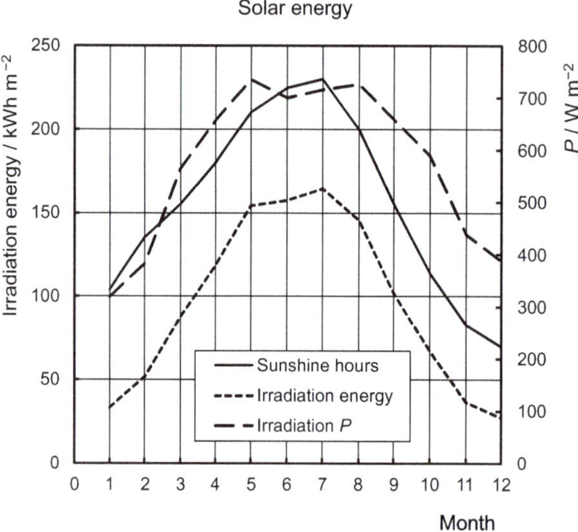

Fig. 4.12: Seasonal trend of the solar irradiation energy for Central Europe.

Radiation energy of photons depends on the wavelength. According to the Planck equation

$$E\,[\mathrm{J\,mol^{-1}\,photons}] = N_\mathrm{L} \cdot h \cdot \nu = N_\mathrm{L} \cdot \frac{h \cdot c}{\lambda} \tag{4.25}$$

with $h = 6.6265 \cdot 10^{-34}$ J s^{-1} and $c = 2.998 \cdot 10^{8}$ m s^{-1} and the Loschmidt number $N_\mathrm{L} = 6.023 \cdot 10^{23}$ molecules per mol, we obtain the energy per mole of photons provided by the sun within the chosen wavelength span. Fig 4.13. shows the photon energy correlation for 200 nm < λ < 800 nm.

This energy may act on anything, the terpenes evaporated from coniferous woods in the Smokies as well as man-made pollutants. Table 4.10 provides data for the bond strength of selected chemical bonds and the energy of photons of a specific wavelength λ needed to cleave these bonds.

From comparison of Tab. 4.10 with Fig. 4.13 we may conclude that within the given wavelength span sun has the power to easily attack and cleave any of the mentioned constituents to synthesize a variety of new constituents with very different physical and toxicological properties. If we provide the whole variety of precursors and let sun do her job we finally end up with a smog problem. Let us pick out traffic in urban areas. We may assume that folks hurry to work in the morning and move

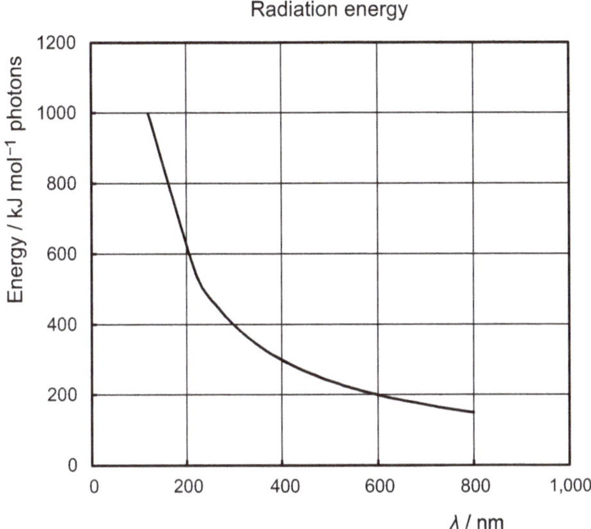

Fig. 4.13: Photon energy for 200 nm < λ < 800 nm.

Tab. 4.10: Bond strength of selected bonds [13] and the corresponding wavelength λ.

Bond type	Substance	Bond strength (kJ mol^{-1})	λ (nm)
H-CH_2OH	Methanol	360	332
H-OCH_3	Methanol	434	276
H-CH_3	Methane	435	275
H-OH	Water	498	240
H-OC_2H_5	Ethanol	435	275
H-SH	Dihydrogen sulfide	377	318
HO-OH	Hydrogen peroxide	213	560
O-NO	Nitrogen dioxide	306	392

back home in the afternoon. Hundreds and thousands of cars may emit some unburned fuel and NO_x within a few hours. The records of environmental monitoring systems confirm that. Figure 4.14 shows the NO_x trend of a monitoring station close to a highly frequented road with a NO_x peak in the morning and a second peak in the late afternoon [33]. Figure 4.14 also shows a second constituent, ozone, for sure not emitted through the exhaust of vehicles. Ozone is a secondary pollutant, synthesized by the sun via man-made precursors, with the yield passing a maximum in the afternoon after sun has provided a maximum of irradiation energy.

Man-made NO_x and the secondary pollutant ozone together with solar irradiation provide a powerful self-sustaining oxidation potential with a tremendous poisonous

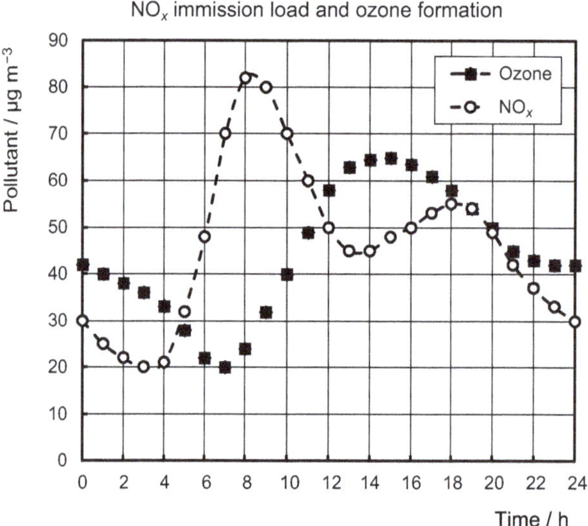

Fig. 4.14: Diurnal NO$_x$ immission and diurnal ozone formation in urban areas, mainly affected by traffic; Graz Griesplatz, 2 February 2020.

effect on the animate and the unanimated environment. Below solar irradiation of λ= 310 nm ozone may interact with vaporous water to form the probably most powerful oxidizer, the hydroxyl radical:

$$O_3 + h\nu \rightleftharpoons O_2 + O^* \qquad \lambda < 310\,\text{nm}$$

$$O^* + H_2O \rightleftharpoons 2\,OH$$

Nitrogen dioxide may cleave into nitrogen oxide and oxygen radicals below λ < 430 nm, to be reoxidized with ozone:

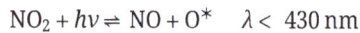

$$NO_2 + h\nu \rightleftharpoons NO + O^* \qquad \lambda < 430\,\text{nm}$$

$$O^* + O_2 \rightleftharpoons O_3$$

$$O_3 + NO \rightleftharpoons NO_2 + O_2$$

$$O_3 + NO_2 \rightleftharpoons NO_3 + O_2$$

Ozone may directly attack non saturated hydrocarbons and form very reactive ozonides and in consecutive steps very undesirable oxygenated hydrocarbons such as aldehydes and free hydrocarbon-based radicals:

$$RHC{=}CHR + O_3 \rightarrow RHC\underline{\text{-O-O-O-}}CHR \rightarrow RCHO + \dot{R}O + HC\dot{O}.$$

$$\mathrm{RCHO} \rightarrow \dot{\mathrm{R}} + \mathrm{HC}\dot{\mathrm{O}}.$$

Latter, very reactive too, may pick up oxygen to form peroxy radicals. Peroxy radicals may interact with nitrogen oxide to form nitrogen dioxide and alkoxy radicals:

$$\dot{\mathrm{R}} + \mathrm{O_2} \rightarrow \mathrm{RO}\dot{\mathrm{O}}$$

We have not mentioned another famous member of this orchestra, SO_2. SO_2 may undergo direct interaction with the sun and humidity to form sulfuric acid aerosols, but also with ozone and nitrogen oxides. Statistics [34] report impressive correlations of smog and the death toll. To sum up, the synthesis of poisonous species from man-made precursors (pollutants) and the sun does not offer any way out of this diabolic loop, except pollutant deposition and except minimizing emission of precursors.

4.4 Particulate matter: dispersion and deposition

When discussing the optical properties of solid and liquid dispersion in Section 3.1 and the smog problem in Section 4.3 we developed a link between visibility of dispersion plumes and the interaction with sunlight, also raising the questions, what makes these smog plumes of submicronic particulate matter become a persisting annoyance, and why does gravity seemingly fail? The main physical property of particulate matter, the density, covers a wide range between $\rho_P = 800$ kg m^{-3} and several thousand kg m^{-3}, depending on the chemical nature of the constituents, whether we talk about hydrocarbon-based particulate matter or metal oxides. To gain an idea of the settling behavior of submicronic particulate matter we do some calculations in Exercise 4.3.

Exercise 4.3: Settling properties of submicronic particulate matter

Solver:

We place a single particle at a given height h above ground level, referring to Tab. 4.8 for example at h = 48 m. Then we let act a force balance on our particle, as shown in Fig. 4.15.

Without detailed discussion we may assume "Stokes" sedimentation conditions, resulting in the drag coefficient C_D according to the following equation:

$$C_D = \frac{24}{\mathrm{Re_P}} \text{ with } \mathrm{Re_P} = \frac{w_P \cdot d_P \cdot \rho_{gas}}{\mu_{gas}} \tag{4.26}$$

Now we may solve the force balance for the settling velocity w_P to obtain

$$w_P = \frac{\left(\rho_P - \rho_{gas}\right) \cdot g \cdot d_P^2}{18 \cdot \mu_{gas}} \tag{4.27}$$

Then we may make an estimate of the settling velocity w_P, the settling time $t_{settling}$ in (h) and the travelling distance s in km for a horizontal gas velocity of $u_{mean} = 1$ m s^{-1} for an assumed air temperature of θ = 10 °C,

$P = 1{,}013$ hPa, $\rho_P = 1{,}000$ kg m^{-3}, $\rho_{gas} = 1.24$ kg m^{-3} and $\mu_{gas} = 2 \cdot 10^{-5}$ Pa s^{-1}, as listed in Tab. 4.11. Table 4.11 also shows the particle size dependent deposition rate w_{dp} according to "TA-Luft 21" [15].

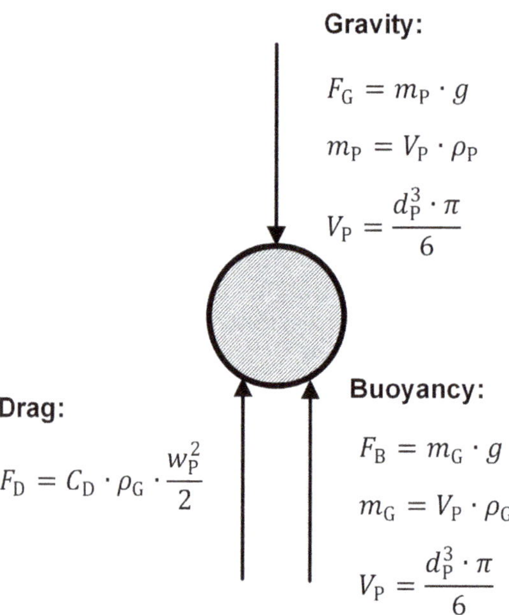

Gravity:
$F_G = m_P \cdot g$
$m_P = V_P \cdot \rho_P$
$V_P = \frac{d_P^3 \cdot \pi}{6}$
Drag:
$F_D = C_D \cdot \rho_G \cdot \frac{w_P^2}{2}$
Buoyancy:
$F_B = m_G \cdot g$
$m_G = V_P \cdot \rho_G$
$V_P = \frac{d_P^3 \cdot \pi}{6}$

Fig. 4.15: Force balance.

Tab. 4.11: Settling velocity of particles w_P, settling time $t_{Settling}$ and traveling distance s for $u_{mean} = 1$ m s^{-1} versus particle size d_P, plus deposition rate data from "TA-Luft 21" [15].

Particle size d_P (µm)	Settling velocity w_P (m s^{-1})	Settling time $t_{settling}$ (h)	Traveling distance s (km)	Class TA-Luft 86	Particle size d_P (µm)	Deposition rate w_{dp} (m s^{-1})
0.2	$1.21 \cdot 10^{-6}$	11,022.9	39,682			
0.4	$4.84 \cdot 10^{-6}$	2,755.7	9,921			
0.6	$1.09 \cdot 10^{-5}$	1,224.8	4,409			
0.8	$1.94 \cdot 10^{-5}$	688.9	2,480			
1.0	$3.02 \cdot 10^{-5}$	440.9	1,587			
2.0	$1.21 \cdot 10^{-4}$	110.2	397	1	<2.5	0.001
4.0	$4.84 \cdot 10^{-4}$	27.6	99			
5.0	$7.65 \cdot 10^{-4}$	17.6	63			
6.0	$1.09 \cdot 10^{-3}$	12.2	44			
8.0	$1.94 \cdot 10^{-3}$	6.9	25			
10.0	$3.02 \cdot 10^{-3}$	4.4	16	2	2.5–10	0.01
20	$1.21 \cdot 10^{-2}$	1.1	4			
50	$7.56 \cdot 10^{-2}$	0.2	1	3	10–50	0.05
>50				4	>50	0.2

Discussion: Following the readings of Tab. 4.11 the poor 0.2 μm particle would need to travel about once Earth's equatorial circumference to touch ground. These findings may explain why pollutant dispersion modeling considers particulate matter of $d_P < 2.5$ μm as "quasi-gaseous." By the way, the particle diameter in Tab. 4.11 has to be specified as aerodynamic diameter (per definition: The aerodynamic diameter of a particle is the equivalent diameter of a spherical particle of density $\rho_P = 1{,}000$ kg m^{-3} to a specific particle of same settling velocity w_P.

Dispersion modeling of particulate matter does not consider submicronic particles. Their dispersion properties are similar to gaseous pollutants. TA-Luft 86 [28] recommends four particle classes, shown in Tab. 4.11. For modeling dispersion of particulate matter eq. (4.9) has to be extended to the following equation:

$$C(x,y,z) = \frac{10^6 \cdot F_{m,i}}{3{,}600 \cdot 2 \cdot \pi \cdot u_{h,Stack} \cdot \sigma_y \cdot \sigma_z} \cdot e^{-\frac{y^2}{2\sigma_y^2}} \cdot \left(e^{-\frac{(z-h)^2}{2\sigma_z^2}} + e^{-\frac{(z+h)^2}{2\sigma_z^2}} \right) \cdot e^{\left[-\sqrt{\frac{2}{\pi}} \cdot \frac{w_{dp}}{u_h} \cdot \int_0^x \frac{1}{\sigma_{z,\tau}} \cdot e^{-\frac{(h)^2}{2\sigma_{z,\tau}^2}} d\tau \right]} \tag{4.9a}$$

with $\tau = \frac{x}{u_h}$

$C(x,y,z)$ has to be determined for every particle class, as shown in Tab. 4.11. For estimation of particle deposition we have to determine $C(x,y,0)$ and add up the deposition results $d(x,y,z)$ for each particle class i according to the following equation:

$$d(x,0,0)\ \left[\text{mg m}^{-2}\ \text{day}^{-1}\right] = 86{,}400 \cdot \sum_{i=1}^{i=4} w_{dp} \cdot C_{i(x,y,0)} \tag{4.9b}$$

The maximum dust deposition rate of nonhazardous dust must not exceed 350 μg m^{-2} day^{-1}. You may find deposition rate data for dry deposition as well as wet deposition of specific particulate pollutants and gaseous pollutants in [15]. To make you familiar with the procedure we will simulate an emergency case.

Exercise 4.4: Off-gas emergency

In Tab. 1.1 we have specified our basic off-gas with a temperature of $T = 200$ °C, a SO_x concentration of 300 ppm and a dust concentration of 3,000 mg m^{-3} plus particle size distribution. The off-gas will be subject of several purification technologies in part II of this book, including dedusting, $DeSO_x$ measures and $DeNO_x$ measures. To avoid breakdowns in case of emergency we have to install a bypass duct from the incineration plant to the stack. Our job is now to simulate an emergency situation. We have to check for the SO_x immission concentration and the particulate matter immission concentration as well as particle deposition. We will fail the emission limits of Tab. 4.2. We have to compare the results of pollutant dispersion modeling with the ambient air-quality standards of Tab. 4.1. We perform modeling with dispersion class III/2.

Solver:

We start with investigating the effect of the emergency bypass on SO_x immission. After a detailed discussion in Exercise 4.2 we are already familiar with that and will therefore discuss the results, shown in Fig. 4.16, without repeating the procedure. When we emit the full SO_x load of 300 ppm, we have to expect an elevated immission level. Higher off-gas temperature will act on the plume rise and therefore have a damping effect on the immission level.

Effect of emergency dispersion on SO2 GLC C(x,0,0)

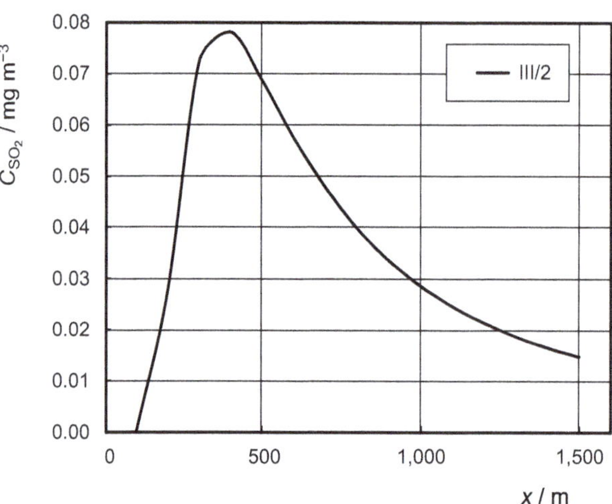

Fig. 4.16: SO_2 immission trend for 300 ppm off-gas concentration and dispersion class III/2.

Compared to the outcome of Exercise 4.2 the immission concentration of SO_x is more than twice as high. (You may wonder about that, because emission rate of the "emergency case" is nearly fivefold the emission rate discussed in Exercise 4.2. The reason is very simple. It is the different plume rise. Off-gas temperature in Exercise 4.2 was $\theta = 58$ °C. Off-gas temperature in the emergency case is $\theta = 200$ °C.) The SO_2 immission concentration will not exceed the ambient air-quality standards of Tab. 4.1.

In the second step we investigate the effect of elevated dust emission on dust immission in downwind direction. The off-gas dust concentration is 3,000 mg m^{-3}. According to the particle size distribution of our off-gas in Tab. 4.2 we have to discuss the dispersion of two classes. According to the data of Tab. 4.11 50% of the particulate matter of our off-gas is part of class 1 with a particle size of $d_P < 2.5$ µm and 50% is part of class 2 with a particle size of 2.5 µm $< d_P <$ 10 µm. For modeling we have to solve the integral of the particle related exponential of eq. (4.9a) by applying the Simpson rule, as shown in eq. (4.9c), before we can complete dispersion modeling:

$$\int_0^x \frac{1}{\sigma_{z,\tau}} \cdot e^{-\frac{(h)^2}{2\sigma_{z,\tau}^2}} d\tau = \frac{\Delta\tau}{3} \cdot \left[\begin{array}{l} 4 \cdot \frac{1}{\sigma_z(\tau_1)} \cdot e^{\left(\frac{h^2}{2\sigma_z^2(\tau_1)}\right)} + 2 \cdot \frac{1}{\sigma_z(\tau_2)} \cdot e^{\left(\frac{h^2}{2\sigma_z^2(\tau_2)}\right)} + 4 \cdot \frac{1}{\sigma_z(\tau_3)} \cdot e^{\left(\frac{h^2}{2\sigma_z^2(\tau_3)}\right)} + \cdots \\ \cdots + 4 \cdot \frac{1}{\sigma_z(\tau_{n-1})} \cdot e^{\left(\frac{h^2}{2\sigma_z^2(\tau_{n-1})}\right)} + \frac{1}{\sigma_z(\tau_n)} \cdot e^{\left(\frac{h^2}{2\sigma_z^2(\tau_n)}\right)} \end{array} \right] \quad (4.9c)$$

The residence time τ in eq. (4.9c) is derived from the distance x and the mean wind speed u_{mean} from eq. (4.23).

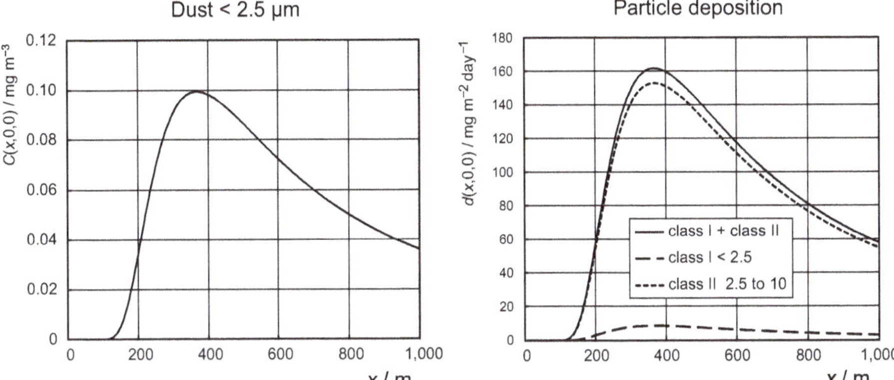

Fig. 4.17: Dust immission concentration of the class $d_P < 2.5$ µm, and deposition rate for both classes.

We determine the ground-level dust immission concentration in downwind direction C(x,0,0) according to eq. (4.9a) and the deposition rate d(x,0,0) according to eq. (4.9b). The left-hand side graph of Fig. 4.17 shows the dust immission concentration of class 1 particulate matter, $d_P < 2.5$ µm, and the right-hand side graph shows the deposition rate for both classes.

Discussion: As expected, the immission concentration exceeds the primary air quality standards for PM2.5 of Tab. 4.1 by nearly 400%. The particle deposition rate would by far exceed the limits for non-hazardous dust of 350 µg m^{-2} day^{-1}.

Comment: Assumingly you may not be able or willing to solve eq. (9c) with your hand calculator. You may also wonder about the four particle size classes in Tab. 4.11, since it is not a huge job to determine the settling velocity w_P of particles. When solving eq. (9c) you will end up with values close to 1. The error will not be significant when we ignore it. You will obtain a serious approach when you insert eq. (9a) without extension into eq. (9b). It is rather recommended to make use of real settling velocity data for several particle classes, as shown in the left two data columns of Tab. 9.11. In the worst case you end up with higher deposition rate data, which will not be disadvantageous.

4.5 Pollutant accumulation

Tab 4.11 tells us that submicronic particulate matter as well as gaseous pollutants will accumulate locally when not removed by the wind, with tremendous impact on animate and unanimated life. Just have the London smog problem of 1952 [34] in mind. Exercise 4.5 may provide an estimate for pollutant accumulation.

Exercise 4.5: The smog pot

The background story:

Graz city, a nice place to go, with an area of 128.6 km^2 and a population of 300,000 is located in the southeastern part of the Alps. The elevation of the city is 316 m upon sea level. It is surrounded by smooth mountains with an elevation of 600 m. Mur river enters the valley through the north-northwest gate with an approximate width of 4 km times 247 m, and it leaves the city through an open gate in the South. With about 130,000 cars per day entering and leaving the city limit in any direction, Graz, like many other crowded areas, suffers from a traffic problem with plenty of particulate and gaseous pollutants being left in the city. Fig 4.14 may give you an idea of the problem we have to face. The traffic provided NO_x load is about 1,430 t a^{-1} [35] equivalent to an approximate mean of 160 kg h^{-1}. The wind rose [36] shown in Fig. 4.18, well reflects the topography of the area with a mean share of calms of about 11% and a main velocity range of 39% for 0.1 m $s^{-1} < v_{Wind} <$ 0.8 m s^{-1}. These boundary conditions suggest a tendency to formation of inversion layers, as confirmed by wind profile measurements [37] and detailed analysis of the smog problem [38]. With these boundary conditions in mind we construct a smog pot. We have to find out, whether we run into a smog problem at low wind speed.

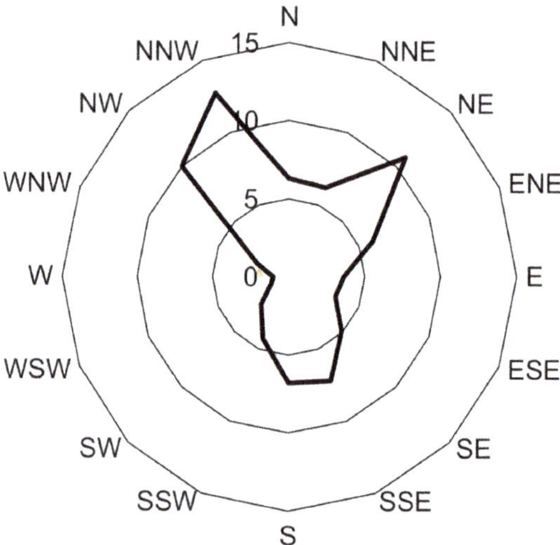

Fig. 4.18: Wind rose all classes in %, monitoring station Graz University.

Assumptions:

The topography of the Graz area suggests the assumption of a well-mixed smog pot, shown in Fig. 4.19.

The NNW-wind enters the smog pot without any preload. The NE-wind and the S-wind is "ignored." We investigate the effect of wind breakdown from NNW, with wind speed as low as v_{Wind} = 0.3 m s^{-1} and v_{Wind} = 0.1 m s^{-1}. We only consider constant NO_x source from traffic of F_{m,NO_x} = 160 kg h^{-1}. We assume $c_{NO_x,\text{smog pot}}$ = 0 when we start the smog accumulation experiment. NO_x is not consumed in the smog pot. $F_{V,in} = F_{V,out}$ (all the air entering the smog pot through the NNW-gate leaves through the S-gate. Due to well mixing the NO_x concentration is same in the smog pot and in the outflow $F_{V,out}$.

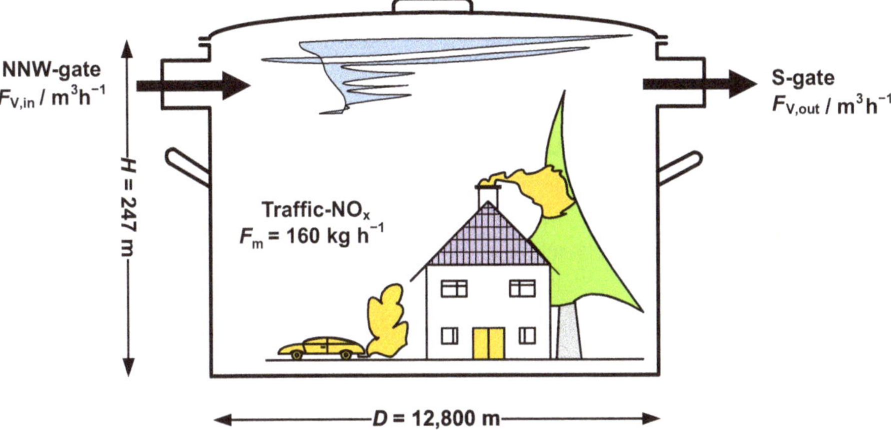

Fig. 4.19: The well-mixed smog pot.

Solver:
We start with a mass balance: All mass of NO_x entering the smog pot will either leave the smog pot or accumulate in the smog pot. In symbols we finally get eq. (4.28).

We start with the balance,

$$F_{V,\,in} \cdot c_{NO_x,\,in} + F_{m,\,traffic} = F_{V,\,out} \cdot c_{NO_x,\,out} + V_{smog\,pot} \cdot \frac{dc_{NO_x}}{dt}$$

$$c_{NO_x,\,in} = 0$$

rearrange it (it is not a magic when we multiply and divide the first term with F_V),

$$F_{m,\,traffic} - F_{V,\,out} \cdot c_{NO_x,\,out} = V_{smog\,pot} \cdot \frac{dc_{NO_x}}{dt}$$

$$F_{m,\,traffic} \cdot \frac{F_{V,\,out}}{F_{V,\,out}} - F_{V,\,out} \cdot c_{NO_x,\,out} = V_{smog\,pot} \cdot \frac{dc_{NO_x}}{dt}$$

$$\frac{F_{m,\,traffic}}{F_{V,\,out}} - c_{NO_x,\,out} = \frac{V_{smog\,pot}}{F_{V,\,out}} \cdot \frac{dc_{NO_x}}{dt} = \tau \cdot \frac{dc_{NO_x}}{dt}$$

finish the job,

$$\frac{F_{m,\,traffic}}{F_{V,\,out}} = a \text{ and } \frac{dc_{NO_x}}{a - c_{NO_x,\,out}} = \frac{dt}{\tau}$$

$$\int_{c_{NO_x,\,out}=0}^{c_{NO_x,\,out}} \frac{dc_{NO_x}}{a - c_{NO_x,\,out}} = \frac{t}{\tau} \text{ and } -\ln\left(\frac{a - c_{NO_x,\,out}}{a}\right) = \frac{t}{\tau}$$

$$\left(\frac{a - c_{NO_x,\,out}}{a}\right) = e^{-\frac{t}{\tau}}$$

And rearrange for $c_{NO_x,\,out}$, to finally obtain

$$c_{NO_x,\,out} = a \cdot \left(1 - e^{-\frac{t}{\tau}}\right) \tag{4.28}$$

If we now insert numbers, we will get for $v_{Wind} = 0.3\ m\ s^{-1}$ and $t = 50$ h:

Inflow area $A = 4{,}000 \cdot 247 = 98{,}800\ m^2$.

$F_{V,in} = F_{V,out} = F_V = 3{,}600 \cdot v_{Wind} \cdot A = 3{,}600 \cdot 0.3 \cdot 98{,}800 = 1.07 \cdot 10^9\ m^3\ h^{-1}$

$$a = \frac{F_{m,traffic}}{F_V} = \frac{160 \cdot 10^9}{1.07 \cdot 10^9} = 149.5\ \mu g\ m^{-3}$$

$$V = \frac{D^2 \cdot \pi}{4} \cdot H = \frac{12{,}800^2 \cdot \pi}{4} \cdot 247 = 3.18 \cdot 10^{10}\ m^3$$

$$\tau = \frac{V}{F_V} = \frac{3.18 \cdot 10^{10}}{1.07 \cdot 10^9} = 29.8\ h$$

$$c_{NO_x,out,t} = a \cdot \left(1 - e^{-\frac{t}{\tau}}\right) = 149.5 \cdot \left(1 - \exp^{-\frac{50}{29.8}}\right) = 122\ \mu g\ m^{-3}$$

At steady state $c_{NO_x,out} = a = 149.5\ \mu g\ m^{-3}$. The NO_x immission limit alarm would sound at $c_{NO_x,out} = 400\ \mu g\ m^{-3}$, but the alarm level will not be obtained. We do not need to panic.

Now we investigate NO_x accumulation in the smog pot for $v_{Wind} = 0.1\ m\ s^{-1}$. We compare it with NO_x accumulation for $v_{Wind} = 0.3\ m\ s^{-1}$. Figure 4.20 shows the outcome.

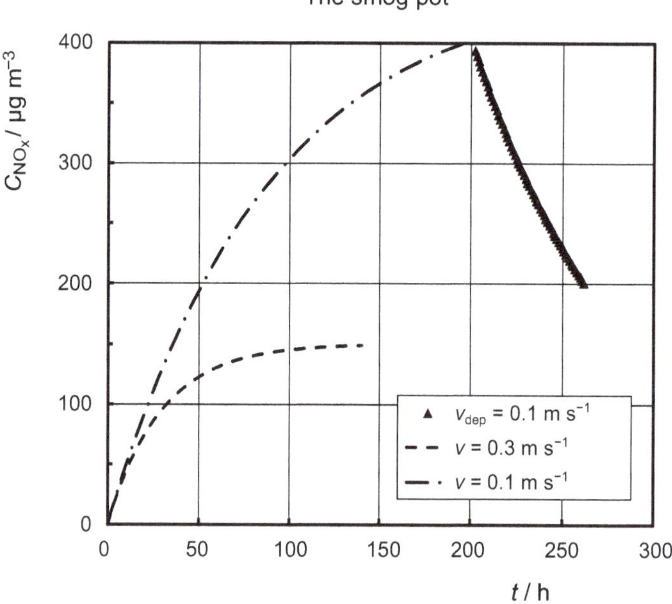

Fig. 4.20: Comparison of NO_x accumulation for $v_{Wind} = 0.3\ m\ s^{-1}$ and for $v_{Wind} = 0.1\ m\ s^{-1}$. At $t = 200$ h, F_{NO_x} is set to $F_{NO_x} = 0$ for $v_{Wind} = 0.1\ m\ s^{-1}$.

A quick analysis of NO_x accumulation for $v_{Wind} = 0.1\ m\ s^{-1}$ tells us that we slip into a NO_x immission problem because of the huge NO_x concentration at steady state of

$$a = \frac{F_{m,traffic}}{F_V} = \frac{160 \cdot 10^9}{3.56 \cdot 10^8} = 449.8\ \mu g m^{-3}$$

Figure 4.20 shows that we would exceed the NO_x immission alarm level of $C_{NO_x,\,out} = 400\ \mu g\ m^{-3}$ after 198 h. In fully developed panic we stop the traffic to get the NO_x immission level under control and adjust the mass balance to the new situation without traffic-based NO_x source:

$$0 = F_{V,\,out} \cdot C_{NO_x,\,out} + V_{smog\ pot} \cdot \frac{dC_{NO_x}}{dt}, \text{ and consequently}$$

$$C_{NO_x,\,out} = C_{NO_x,\,198} \cdot \left(e^{-\frac{t}{\tau}}\right) = 400 \cdot e^{-\frac{t}{29.8}} \tag{4.29}$$

(But be careful. For calculation of NO_x depletion, you have to start with $t = 0$; for presentation you of course have to add the time of depletion to the final time of accumulation, which was 198 h.) Figure 4.20 shows the outcome of NO_x depletion for $v_{Wind} = 0.1\ m\ s^{-1}$. Even when you completely stop traffic it will need about 50 h to bring down the NO_x level to less than 200 µg m^{-3}.

Discussion: It is a very simple approach. Therefore, handle the outcome with care, because we have ignored the baseline level, we have ignored other pollutant sources, and we have assumed constant mass inflow of NO_x by traffic, although Fig. 4.14 suggests fluctuating immission concentration. However, we just wanted to discuss the principle. You are not limited in improving this approach. Referring to Fig. 4.14 you may for example consider fluctuating NO_x source with a frequency of 12 h. We will discuss it in Exercise 4.6.

Exercise 4.6: The improved smog pot

We nearly keep everything same as in Exercise 4.5.

Assumptions:

We investigate the effect of wind breakdown from NNW, with wind speed as low as $v_{wind} = 0.1\ m\ s^{-1}$. We split the NO_x source of $F_{m,\,NO_x} = 160\ kg\ h^{-1}$ into a static amount of $F_{m,\,NO_x} = 80\ kg\ h^{-1}$ and a dynamic amount of $F_{m,\,NO_x} = 80\ kg\ h^{-1}$ ($f_{stat} = 0.5$). Frequency of the dynamic amount is $f_{frequency} = 2$ per day (every 12 h), and we assume sinoidal fluctuation. We assume $C_{NO_x,\,smog\ pot} = 0$ when we start the smog accumulation experiment. NO_x is not consumed in the smog pot. $F_{V,in} = F_{V,out}$ (all the air entering the smog pot through the NNW-gate leaves through the S-gate. Due to well mixing the NO_x concentration is same in the smog pot and in the outflow $F_{V,out}$.

Solver:

We construct the time dependent source $F_{m,t}$. It is made up of the static amount $F_{m,\,static}$ and the time dependent dynamic amount $F_{m,\,dynamic} \cdot f(t^*)$, as follows:

$$F_{m,\,t} = F_{m,\,static} + F_{m,\,dynamic} \cdot f(t^*) \tag{4.30}$$

The dynamic time function $f(t^*)$ considers the static fraction $f_{stat} = 0.5$ and the dynamic fraction $f_{dyn} = (1 - f_{stat}) \cdot \sin\left(f_{frequency} \cdot \pi \cdot \frac{(t + t_{delay})}{t_{fluctuation}}\right)$ with the frequency factor $f_{frequency} = 2$, the actual time t of our experiment, the delay time $t_{delay} = 8$ h to correspond with the NO_x immission trend of Fig. 4.14, and the time span of fluctuation $t_{fluctuation} = 12$ h, shown in the following eq. (4.31):

$$f(t^*) = \left[f_{stat} + (1 - f_{stat}) \cdot \sin\left(f_{frequency} \cdot \pi \cdot \frac{(t + t_{delay})}{t_{fluctuation}}\right)\right] \tag{4.31}$$

Then we apply the same algorithm as discussed in Exercise 4.5:

$$a\left[\mu g\, m^{-3}\right] = \frac{F_{m,t}}{F_V}$$

$$V = \frac{D^2 \cdot \pi}{4} \cdot H = \frac{12,800^2 \cdot \pi}{4} \cdot 247 = 3.18 \cdot 10^{10}\ m^3$$

$$\tau = \frac{V}{F_V} = \frac{3.18 \cdot 10^{10}}{3.56 \cdot 10^8} = 89.3\ h$$

$$c_{NO_x,\,out}\left[\mu g\, m^{-3}\right] = a \cdot \left(1 - e^{-\frac{t}{\tau}}\right)$$

Let us check the algorithm of eq. (4.31) for (arbitrarily chosen) $t = 32$ h, and insert numbers:

$$f(t^*) = \left[f_{stat} + (1 - f_{stat}) \cdot \sin\left(f_{frequency} \cdot \pi \cdot \frac{(t + t_{delay})}{t_{fluctuation}}\right)\right] = \left[0.5 + 0.5 \cdot \sin\left(2 \cdot \pi \cdot \frac{(32 + 8)}{12}\right)\right] = 0.931$$

$$F_{m,32} = F_{m,static} + F_{m,dynamic} \cdot f(t^*) = 80 + 80 \cdot 0.931 = 154\ kg\ h^{-1}$$

$$a = \frac{F_{m,t}}{F_V} = \frac{154 \cdot 10^9}{3.56 \cdot 10^8} = 434\ \mu g\, m^{-3}$$

$$c_{NO_x,\,out,\,t=32} = a \cdot \left(1 - e^{-\frac{t}{\tau}}\right) = 434 \cdot \left(1 - e^{-\frac{32}{89.8}}\right) = 130.8\ \mu g\, m^{-3}$$

Figure 4.21 shows the NO_x trend for the first 200 h. In comparison with Exercise 4.4, the NO_x immission limit alarm level of $c_{NO_x} = 400\ \mu g\ m^{-3}$ will not be obtained.

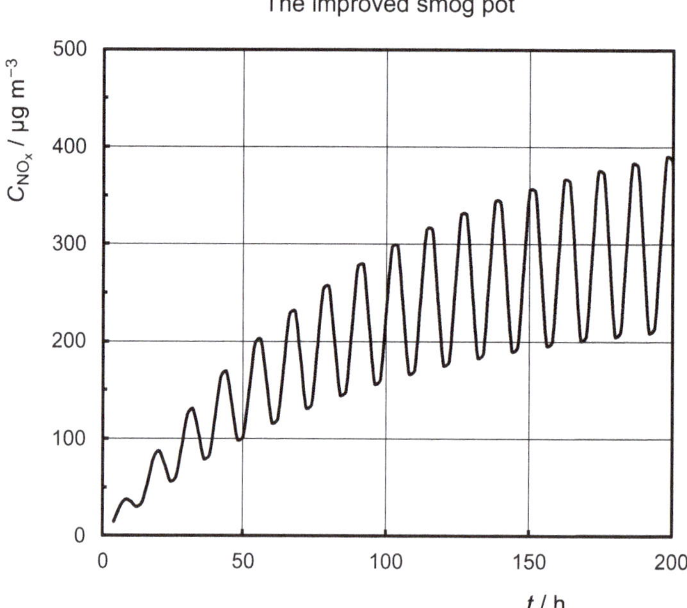

Fig. 4.21: NO_x immission trend for fluctuation time span $t_{fluctuation} = 12$ h, static NO_x source of $F_{m,NO_x} = 80\ kg\ h^{-1}$ and a dynamic NO_x source of $F_{m,NO_x} = 80\ kg\ h^{-1}$.

If you are not happy with the amplitude you just have to adapt eq. (4.31) to your needs, for example:

$$f(t^*) = \left[f_{stat} + (1 - f_{stat}) \cdot \left[\sin\left(f_{frequency} \cdot \pi \cdot \frac{(t + t_{delay})}{t_{fluctuation}} \right) \right]^2 \right]$$

!

Discussion: Smog problems are much more complex than suggested by these examples. The basic idea behind the examples was, to demonstrate how we might approach the problem, and to underline the pool of information you need for approaching smog problems.

If you want to do this calculation with airborne particulate matter from traffic you may make use of the correlation of traffic-based NO_x emission and traffic-based PM10-emission for either NO_x source in kg h^{-1} or NO_x concentration in μg m^{-3} according to the following equation:

$$PM10 = 0.0141 + 0.0316 \cdot NO_x \tag{4.32}$$

This PM10 correlation considers airborne particulate matter from the exhaust as well as the tire wear, break wear, road wear and resuspended dust. Detailed data on this subject are published in [39].

4.6 Odor and odor dispersion

Let us start with the great odor of a perfume you may have in mind. Perfumes consist of a liquid carrier and highly diluted fragrances, the mixture of which is responsible for (hopefully) pleasant hedonics. Amber (do you know, where it comes from?) for example is an ingredient of very expensive perfumes, which in high concentration may facilitate ugly hedonics.

In off-gas purification we are rarely faced with nice odors, but rather with odor nuisance we hardly can specify by its specific composition. Intensive mass animal farming is still a huge problem. Therefore, we will preferably discuss this topic with the example of industrial pig farming. The poor hog of a piggery with a life span of four months from birth to market for example produces about 0.6 m^3 of urine, which together with the excrements is collected in manure tanks to be finally dispersed on soil for fertilizing purposes. Off-gas from the stables and manure storage tanks is still subject of severe complaints by the neighborhood. Besides water and more than 200 identified ingredients, pig manure mainly contains up to 2,000 mg L^{-1} of ammonia, pretty fast converted from urea with the enzyme ureases, and the main odor determining ingredients acetic acid with an odor detection threshold of ODT = 0.06 μg L^{-1}, butanoic acid with an ODT of ODT = 50 μg L^{-1}, 3-methylbutanoic acid (ODT 0.0015 μg L^{-1}), dimethyl sulfide (ODT 0.0005 μg L^{-1}), *p*-cresol (ODT 0.0007 μg L^{-1}) and 4-ethylphenol (ODT 0.001 μg L^{-1}) [40]. These constituents are also emitted through the exhaust, driving the neighborhood crazy.

The pH value of pristine manure is about pH = 7.5, the optimum pH range for fast enzymatic conversion of urea to ammonia. From an engineer's point of view, it should not be a big issue to get the odor problem of pig manure under control. Referring to

[41], conversion of urea with urease is strongly pH dependent. At low pH and at high pH conversion is negligible. When pH is elevated above pH = 10, all acid-based odorants are neutralized and run out of odor. The phenols form phenolates and the sulfur carriers form sulfides and get lost of odor too. The farmer would just have to add lime to pristine urine continuously (we talk about 200 L h^{-1} of urine per 1,000 pigs in the stable), to raise the pH value beyond pH = 10.

Let us focus on some basic aspects of odors and odor dispersion before we come back to the hog. The odorous properties of substances cannot be quantified with a balance. Whether we smell traces of a vaporous substance or not depends on the individual interaction with the mucus of our nose, an absorption process, and the transfer to the olfactory receptor [42]. Individuals of a population collective differ in their "odor absorption" ability and the receptor sensitivity. For optimum absorption/reception properties the odorous substance shall have balanced hydrophilic/lipophilic properties. In some cases, receptor sensitivity may be orders of magnitudes better than the best resolution analytics available on the market. This uncertainty in registering odor suggests the need of panels for classifying intensity and hedonic of odors. Table 4.12 shows the wide span of olfactometric sensitivity [14, 43]. The question is, how to quantify properties you seemingly cannot quantify?

The European Union has compromised on the specification of the European odor unit OUE. One OUE is the reproducible mass equivalent of 123 μg of *n*-butanol per m^3 of clean air (1 OUE = 123 μg m^{-3} *n*-butanol = 1 EROM). This quantity is the odor threshold of n-butanol, registered by 50% of a panel, and it is the European Reference Odor Mass (EROM)[DIN EN 13725:2003–07]. We may now ask the qualified olfactometry panel to perform experiments with different substances and will get the feedback of the odor detection threshold (ODT).

The quantity of 1 OU just identifies an odorous load at ODT. At least five OUs are needed to identify the substance hidden behind an odor. As may be deduced from

Tab. 4.12: Span of the olfactometric sensitivity: (a) Nagata [43] and (b) Verschueren [14].

Odorant	Molar mass (g mol^{-1})	ODT (a) (μg m^{-3})	ODT (μg m^{-3})	100% recognition (b) (mg m^{-3})
NH_3	17	1,137.48	66.911	41.71
Acetic acid	60	16.06	0.268	5.35
Butanoic acid	88	0.75	0.008	0.08
Dimethyl sulfide	62	8.30	0.134	0.28
p-Cresol	108	0.26	0.002	0.96
H_2S	34	0.62	0.018	1.52
n-Butanol	74	123	1.695	16,504

Tab. 4.12, the actual concentration (number of molecules per m^3) of different substances at ODT is very different. To underline the problem, Fig. 4.22 shows the ODT of several homologous series [43].

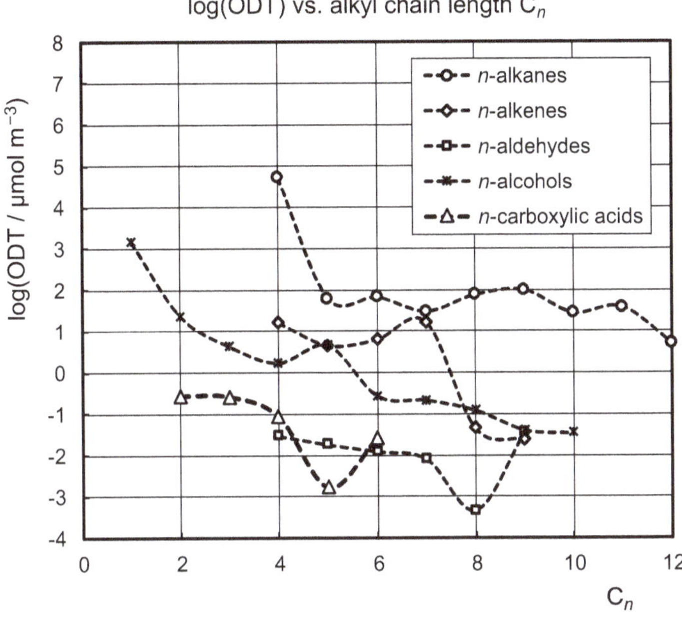

Fig. 4.22: Log(ODT) versus chain length of several homologous series. For better comparability, the ODT is expressed in terms of µmol m^{-3}.

The physical properties of homologous series of hydrocarbons and substituted hydrocarbons, as shown in Fig. 4.22, do in principle not show a clear trend, whether it is the boiling point, the density or any physical property or even the ODT.

Our nose does not correlate odor concentration and odor sensibility linearly but on a logarithmic scale according to the Weber–Fechner law:

$$P = k \cdot \log \frac{C}{C_0} \tag{4.33}$$

with the perception P, the sense specific constant k and the concentration C (e.g., $C_0 = 1$ OU). The intensity of perception IP roughly correlates with the concentration (in terms of OU).

It is:

- zero for $C = 1$ OU,
- very faint up to $C = 10$ OU (1),
- faint up to $C = 100$ OU (2),
- distinct up to $C = 500$ OU (3),

- strong up to $C = 1{,}000$ OU (4) and
- very strong beyond (5).

The numbers in brackets specify the approximate intensity level IP. Referring to [44], the intensity of perception IP may be calibrated to the odor intensity of aqueous mixtures of *n*-butanol. The correlation is shown in Fig. 4.23.

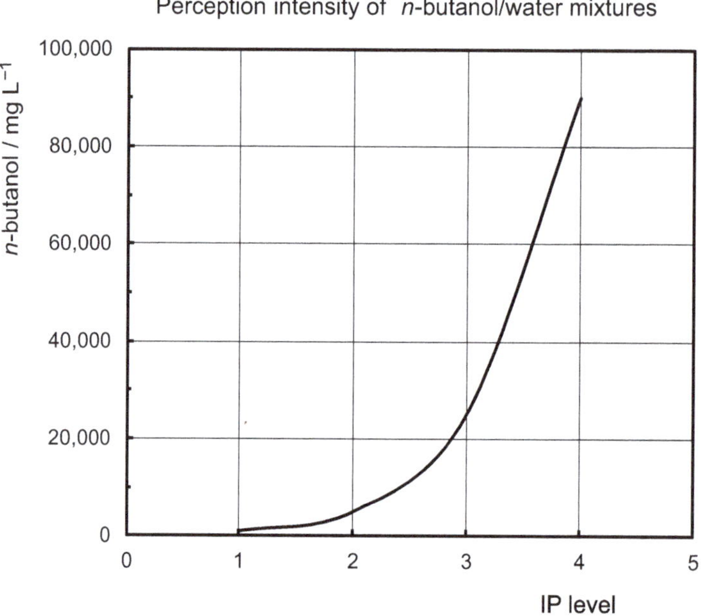

Fig. 4.23: Intensity of perception level (IP) of water/*n*-butanol mixtures at $T = 293$ K.

The hedonic also depends on the individual perception, whether an odor is classified as "very pleasant" or "inacceptably annoying," and it does not linearly correlate with the amount of an odorant in the mixture. To provide a link to practical application, odor emission of industrial pig farming may obtain 50 OU s^{-1} and livestock unit (in case of rearing pig a livestock unit is equivalent 500 kg of animals or about 8 pigs) [45]. In her master's thesis Spernbauer B. [46] concludes from experiments and simulation that odor emissions of 40 [OU s^{-1}] and livestock unit would well reflect the real situation. She points out that seemingly rate and fodder frequency affect odor emissions. In [47] up to 40% less ammonia emission from pig farming is reported, when multifrequency fodder application is practiced. Late investigations report odor emissions of 140 OU s^{-1} and livestock unit [48]. If you are specifically interested in odor immission aspects of animal farming you may find a very comprehensive study in [49]. If you are interested in medical facts of odor immission, the report [50] will provide very detailed insight into aspects you cannot weigh with a balance.

According to [51], the odor immission level in residential areas shall not exceed:
- 15 OU m^{-3} for low nuisance level,
- 5 OU m^{-3} for medium nuisance level and
- 4 OU m^{-3} for high nuisance level.

The nuisance level of off-gas from pig farming for example is classified "medium." An OU level of 5 [OU m^{-3}] should therefore be an acceptable immission level in residential areas. Duration of nuisance from animal farming in residential areas shall be less than 10% per year [52].

To avoid odor nuisance in the vicinity of industrial pig farming stables the stack mouth should at least be installed 10 m above ground level. The stack diameter shall suffice an off-gas velocity of $v > 7$ m s^{-1} at minimum. The off-gas flow rate differs widely, depending on the applied standard. An air exchange rate of 120 m^3 h^{-1} per animal is recommended in [53]. The "Bayerische Landesanstalt für Landwirtschaft Institut für Landtechnik und Tierhaltung" recommends a mean gas velocity of 0.2–0.6 m s^{-1} in summer, resulting in 864 m^3 h^{-1} per animal for 1.2 m^2 per animal. The Austrian Directive [54] recommends an animal specific area for rearing pig farms up to 1 m^2 per animal, depending on the mass per animal. Referring to DIN 18910, part 1 [55], the mean gas velocity of 0.2–0.6 m s^{-1} in summer is confirmed. Referring to DIN 18910 [56] a specific winter air exchange rate of 16.1 m^3 h^{-1} for pigs with a mass of 100 kg shall be applied. He figures out that in principle the fresh air demand and the off-gas flow rate depend on the season and the age (mass) of rearing pig, and may vary by a factor of 11.

Exercise 4.7: The hog

We have to investigate odor dispersion from rearing pig farming. Since we try to be careful, we assume conservative boundaries.

Data:
Number of pigs in the stable: 500
Mass per pig: 75 kg
Specific area/pig a: 1.5 m^2
Specific air exchange rate SAER: 120 m^3 h^{-1} per animal
Odor emission rate OER_{sp}: 140 OU s^{-1} per livestock unit
The off-gas is collected in 1 stack, located in the center of the stable
Wind speed is 2 m s^{-1}

Our job:
We have to estimate the minimum distance from the stable in downwind direction to obtain 1 OU m^{-3}. We just estimate the ground-level concentration for dispersion class III/2. We assume same off-gas temperature as ambient $\theta = 20$ °C.

Solver:
In the first step, we construct the size of the stable. The area A is calculated from the number of animals n and the specific area a:
$A = n \cdot a = 500$ animals $\cdot$ 1.5 m^2/animal = 750 m^2
The width of the stable is 20 m. Then the length is 37.5 m.

Height to roof top: 7 m
Stack height is chosen to be 10 m.
Off-gas velocity at stack mouth is fixed with 7 m s^{-1}.
Then we have to estimate the odor emission rate. About 500 pigs of 75 kg mass (recommended conversion factor: 0.15) correspond to a livestock unit LSU of
$LSU = n \cdot f = 500 \cdot 0.15 = 75$ LSU.
The OER is calculated from the number of pigs n in terms of LSU and the specific OER_{sp}:
$OER = n \cdot OER_{sp} = 75 \cdot 140 = 10{,}500$ OU $s^{-1} = 37.8$ MOU h^{-1}.
The total air exchange rate AER is calculated from the number of animals and the specific air exchange rate SAER with $AER = 500 \cdot 120 = 60{,}000$ m^3 h^{-1}.
The off-gas-concentration is $C = 37.8 \cdot 10^6/60{,}000 = 630$ OU m^{-3}.

We trust in a correct estimate of our Gaussian plume model as discussed in Section 4.2. We investigate dispersion class III/2. The results for dispersion class III/2 from Tab. 4.5 is given in Tab. 4.13.

Tab. 4.13: Mean velocity u_{mean} excess height h_{ex} and dispersion height h for dispersion class III/2.

Dispersion class	u_{mean} (m s^{-1})	h_{ex} (m)	h (m)
III/2	2.0	0	10

In the next step, we determine the ground-level concentration of OU ($C(x,0,0)$) in downwind direction center line with eq. (4.22) after having prepared the dispersion parameters σ_y, σ_z for the independent variable x. Table 4.14 shows the results for dispersion class III/2.

Tab. 4.14: Ground-level concentration for dispersion class III/2.

x (m)	σ_y (m)	σ_z (m)	$C(x,0,0)$ (OU m^{-3})
100	27.09	16.20	3.15
200	47.40	32.31	1.04
300	65.75	48.38	0.51
400	82.94	64.44	0.31
500	99.30	80.47	0.21
600	115.04	96.50	0.15
700	130.28	112.51	0.11
800	145.10	128.52	0.09
900	159.57	144.51	0.07
1,000	173.73	160.50	0.06
1,100	187.62	176.49	0.05
1,200	201.27	192.46	0.04
1,300	214.70	208.44	0.04
1,400	227.93	224.40	0.03
1,500	240.99	240.36	0.03

Figure 4.24 shows the ground-level odor concentration IC for dispersion class III/2.

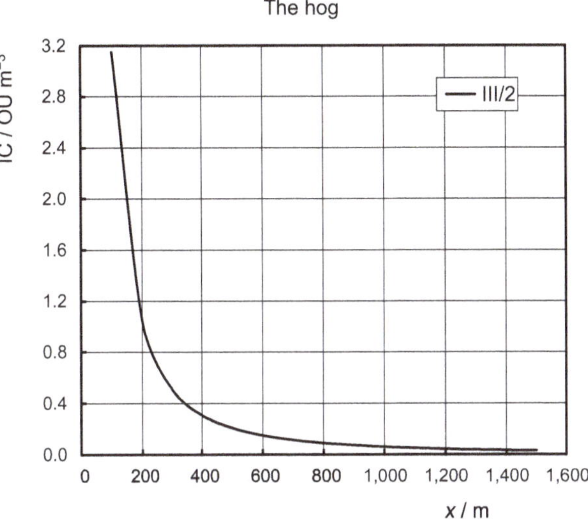

Fig. 4.24: Ground-level odor immission concentration IC in OU m^{-3} for emission of $C_0 = 630$ OU m^{-3}, stack height $h_{Stack} = 10$ m, off-gas temperature $\theta_{gas} = 20$ °C, temperature of ambient air $\theta_{air} = 20$ °C, off-gas flow rate $F_{V,actual} = 60{,}000$ am^3 h^{-1}.

Discussion: Our estimate considers constant odor emission rate and constant off-gas flow rate. The reading of Fig. 4.23 says that we fall below 1 OU (= ODT) in a distance of 200 m from our point source. We would fall below the level of irrelevance (0.4 OU) in a distance of 350 m. Following experience-based recommendations [51] we should rather consider a distance of 4,500 m, just a factor of 13, seemingly a little bit much, suggesting the question, how to find a way out?

- You might make use of the freeware ADAS – Austrian database for air quality assessment near small sources [ADAS] (http://www.umwelt.steiermark.at/cms/beitrag/11257761/2222407/), if applicable.
- You might make use of the freeware AUSTAL [AUSTAL] (http://www.austal2000.de).
- You might make use of GRAL [GRAL] (https://github.com/Gral Dispersion Model).
- You might consider to contact an odor pollutant dispersion modeling expert.

Tab. 4.15 lists selected standards you are recommended to consider in dispersion modeling.

Tab. 4.15: Selected pollutant dispersion standards.

VDI 3782 Blatt 3 (2019–12)	Atmospheric dispersion – plume rise
VDI 3782 Blatt 5 (2006–04)	Environmental meteorology – atmospheric dispersion models – deposition parameters
VDI 3782 Blatt 6 (2017–04)	Environmental meteorology – atmospheric dispersion models – determination of Klug/Manier dispersion categories

Tab. 4.15 (continued)

VDI 3783 Blatt 7 (2017–05)	Environmental meteorology – prognostic microscale wind field models – evaluation for flow around buildings and obstacles
VDI 3783 Blatt 10 (2010–03)	Environmental meteorology – diagnostic microscale wind field models – air-flow around buildings and obstacles
VDI 3786 Blatt 1 (2013–08)	Environmental meteorology – meteorological measurements – fundamentals
VDI 3786 Blatt 2 (2018–05)	Environmental meteorology – meteorological measurements – wind
VDI 3788 Blatt 1 (2000–07)	Environmental meteorology – dispersion of odorants in the atmosphere – fundamentals
VDI 3790 Blatt 2 (2017–06)	Environmental meteorology – emissions of gases, odors and dusts from diffuse sources – fundamentals
VDI 3790 Blatt 3 (2010–01)	Environmental meteorology – emission of gases, odors and dusts from diffuse sources – storage, transshipment and transportation of bulk materials
VDI 3882 Blatt 2 (2021–11)	Olfactometry – determination of odor intensity
VDI 3884 Blatt 1 (2015–02)	Olfactometry – determination of odor concentration by dynamic olfactometry – supplementary instructions for application of DIN EN 13725
VDI 3945 Blatt 3 (2020–04)	Environmental meteorology – atmospheric dispersion models – particle model
BS ISO 28902-3 (2018-11-15)	Air quality – environmental meteorology. Ground-based remote sensing of wind by continuous-wave Doppler lidar
DIN ISO 28902-1 (2012–06)	Air quality – environmental meteorology – part 1: ground-based remote sensing of visual range by lidar (ISO 28902-1:2012)
DIN ISO 28902-3 (2019–04)	Air quality – environmental meteorology – part 3: ground-based remote sensing of wind by continuous-wave Doppler lidar (ISO 28,902-3:2018)
DIN ISO 4225 (2020–10)	Air quality – general aspects – vocabulary (ISO 4225:2020)
DIN EN 15483 (2009–02)	Ambient air quality – atmospheric measurements near ground with FTIR spectroscopy; German version EN 15483:2008
DIN EN 13725 (2022–06)	Stationary source emissions – determination of odor concentration by dynamic olfactometry and odor emission rate

The list is not complete. A comprehensive list of standards is published in attachment 5 of BImSchG 2021 [15]. It is not an advertisement, but do not hesitate to contact www.beuth.de when you need access to valid standards or even obsolete standards. Latter may sometimes be necessary to understand history files.

5 Summary

In the latest biennial report on air quality in Styria [57] the federal government of Styria published several time series of the immission concentration level in northern Graz. These statistics reflect very well the huge progress in air quality achieved in the last decades. The left-hand side graph of Fig. 5.1 shows the impressive time series of the SO_2 immission level, which compares well with an exponential decay, and the right-hand side graph shows the time series of NO_x immission, PM2.5-immission and PM10-immission.

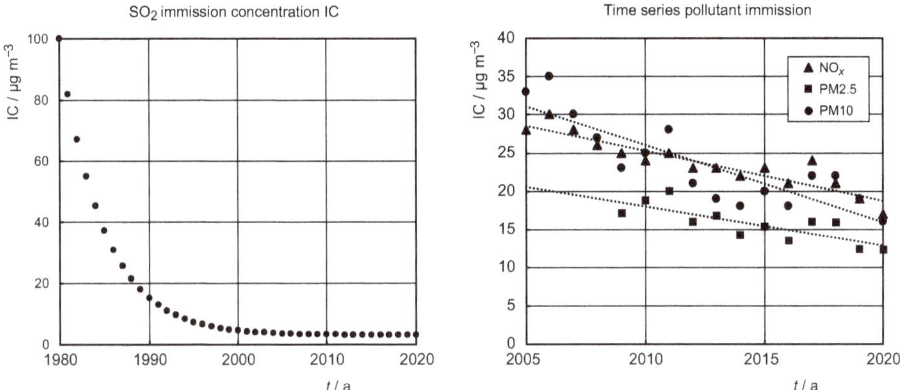

Fig. 5.1: Time series of SO_2 immission, NO_x immission, PM2.5- and PM10-immission in northern Graz.

The SO_2 immission level has dropped by 97% within the last four decades, reflecting the great progress in primary and secondary measures. For different reasons NO_x immission and PM-immission do not by far show comparable progress. Deep desulfurization of fuels and progress in pollution control may partially explain the SO_2 trend, while NO_x and PM emissions, especially from traffic, hardly can be reduced by primary and secondary measures, suggesting plenty of space for new approaches. However, when we have improvements in mind we also need to consider the (Boltzmann) cost law of pollution control, shown in Fig. 5.2:

$$\text{Cost} \approx -\ln\left(\frac{C}{C_0}\right)$$

When we think about technical measures in pollution control we shall keep the correlation of efficiency and cost in Fig. 5.2 in mind.

The correlation of efficiency and cost is not linear. Pollution control processes become the more expensive the higher the expected precipitation efficiency is. We shall not exceed the feasibility limit if we want to avoid waste of resources (= the outcome does not compensate the expenditure anymore).

https://doi.org/10.1515/9783110763928-005

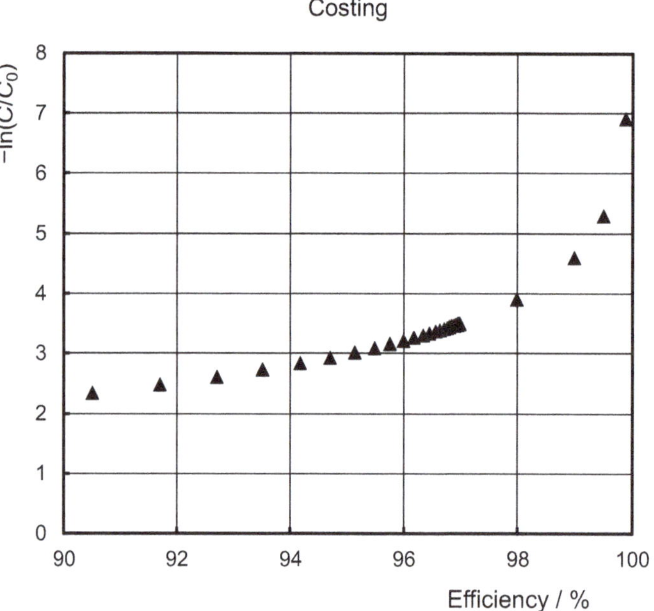

Fig. 5.2: Correlation of precipitation efficiency and cost. The ordinate represents multiples of cost.

However, we have to keep in mind the tremendous numbers of fatalities [EEA report no. 19]. An annual death toll of more than 300,000 must fuel our phantasy and skills to improve (technical) measures.

The intention behind discussing the basics the way we did is to prepare the major aspects we have to consider in off-gas purification. Our focus must not be limited to discussing technical measures; we must be ready to develop a global view on a specific problem. We have to collect all information, including the process data, the interaction of specific emission sources with the environment, the topographic and the meteorological situation, the background immission level and, of course, the legislative framework, before we start with thinking about any appropriate technical measures in terms of “best technologies.”

The second part of our project has a distinct focus on technical measures, on the design of equipment and on the estimation of process efficiency. Referring to our off-gas specification in Tab. 1.1 we will start with “dry” off-gas purification technologies, specifically dedusting. With nearly 50% of the particulate matter being classified as PM2.5 the PSD of the dust load of our off-gas is really bad. We will need to focus on fine dust precipitation technologies. Then we have to keep an eye on constituents we can get under control by absorption/adsorption technologies, before we make use of the support of chemical reaction engineering. The off-gas specification of Tab. 1.1 will be our guide. When needed, we will discuss additional examples. Whenever possible

Tab. 5.1: Pollutants, applied force and off-gas purification equipment.

Pollutant	Force	Equipment
Particulate matter	Centrifugal force	Cyclones
	Electrostatic discharge	Electrostatic precipitators
	Inertia/flow line interception	Filter bag house
	Inertia/flow line interception	Venturi scrubber
SO_2, HF, HCl, NO_x, BaPs, PCDDs, VOCs	Chemical potential	Absorbers
	Chemical potential	Adsorbers
	Chemical conversion	Reactors

we will offer shortcut algorithms as well as detailed design procedures. Table 5.1 provides the link to the technology part.

Depending on the off-gas specification we will rarely be able to solve the problem with a single technology. Our job will be to design the optimum combination of technologies in terms of efficiency, cost of investment and cost of operation. We also have to care for minimum waste production. Inspired by the healing effect of rain on air quality we tend to solve emission problems by washing, but shall keep in mind that washing just transforms the off-gas problem into a wastewater problem. If we cannot solve the wastewater problem, we simply fail.

Part II: **Technologies**

6 Particulate matter precipitation

The essential condition for particulate matter (PM) precipitation is the impact of a force acting on the particles in the off-gas, which gives the particles a different direction of movement to the gas path. The retention time of the particles must be high enough to enable migration from the gas path to the separation site (collecting surface).

The basic steps of particle precipitation by any mechanism and device are:

- the separation of the gas-borne particles by collecting on a surface,
- the retention of the collected particles and
- the removal of the particles from the separation device for recovery or disposal.

Particle precipitation is controlled by physical principles. The equipment to remove PM makes use of several physical mechanisms:

- Gravitational settling
- Centrifugal force
- Inertial impaction and interception
- Electrostatic precipitation
- Diffusion (Brownian motion)
- Thermophoresis and diffusiophoresis

Thermophoresis and diffusiophoresis play a minor role in particle separation but should be mentioned anyway. Several collection mechanisms may act and be observed in dust precipitation simultaneously. The collection mechanism acting on the particle strongly depends on the particle size. The individual contribution of a mechanism depends on the gas and particle properties, the geometry of the precipitation device and the fluid-flow pattern. Due to the difficult theoretical treatment of the complex multiphase flow pattern, simplified assumptions govern the equipment design. The design of industrial-scale separation devices is mainly based on empirical and semiempirical methods.

6.1 Performance of particle separation devices

The performance of particle separation devices (PSDs) is defined by the overall degree of separation E. It is given by the mass flow rate of the precipitated particles $\dot{m}_{\mathrm{Prec}}$ to the mass flow rate of PM in the off-gas entering the separation device $\dot{m}_{\mathrm{PM,\,Off\text{-}gas}}$:

$$E = \frac{\dot{m}_{\mathrm{Prec}}}{\dot{m}_{\mathrm{PM,\,Off\text{-}gas}}} \tag{6.1}$$

https://doi.org/10.1515/9783110763928-006

With the particle concentration of the off-gas $c_{\text{Off-gas}}$ and the particle concentration of the cleaned gas leaving the separation device $c_{\text{Clean gas}}$, the overall degree of separation E is

$$E = 1 - \frac{c_{\text{Clean gas}}}{c_{\text{Off-gas}}} \tag{6.2}$$

The particle diameter plays a major role in particle precipitation. An adequate evaluation of the performance of a PSD can only be made if the separation efficiency is determined as a function of the particle diameter. Therefore, the separation efficiency $T(x)$ (or the grade separation efficiency) is used, which gives the probability for a particle with the particle size x to be separated. For every particle size x, the ratio of the precipitated particle mass to the particle mass in the off-gas is

$$T(x) = \frac{m_{\text{Prec}}}{m_{\text{PM, Off-gas}}} \cdot \frac{q_{3,\text{ Prec}}(x)\mathrm{d}x}{q_{3,\text{ Off-gas}}(x)\mathrm{d}x} \tag{6.3}$$

and accordingly

$$T(x) = E \cdot \frac{q_{3,\text{ Prec}}(x)}{q_{3,\text{ Off-gas}}(x)} \tag{6.4}$$

The overall degree of separation E may thus be calculated by integration or summation of the separation efficiency over the entire particle range:

$$E = \int T(x) \cdot q_{\text{i}}(x)\mathrm{d}x = \sum T(x) \cdot \Delta Q_{\text{i}}(x) \tag{6.5}$$

For $T(x) = 0$, the total amount of particles of a specific size x in the off-gas remains in the clean gas (no particle separation) and for $T(x) = 1$, every particle of the off-gas gets separated.

The cutoff diameter x_{T} is an important value for the design of separation devices. It is the particle diameter that corresponds to 50% separation efficiency:

$$T(x_{\text{T}}) = 0.5 \tag{6.6}$$

For application in the high-efficiency separation range, the penetration factor P offers more significant information about the performance of the precipitation device. It is defined by the ratio of the amount of passing particles to the amount of particles in the off-gas:

$$P = \frac{\dot{m}_{\text{Passed}}}{\dot{m}_{\text{PM, Off-gas}}} = 1 - E \tag{6.7}$$

In processing radioactive dust or in cleanroom technology, the decontamination factor (DF), which is the reciprocal of the penetration factor P, is used for specifying the performance. Its logarithmic value is called decontamination index. Based on logarithmus naturalis, it represents the number of transfer units in dust collection. It is suitable for correlating the collector performance data.

6.2 Outlook

In this chapter we discuss different devices for particle separation. You will learn how to design precipitation equipment and how to interpret separation performance. The calculations can easily be carried out with paper and pen and the hand calculator or with the help of a spreadsheet tool and serve as an estimate for apparatus design, e.g., needed on site. All calculations are based on our off-gas report listed in Tabs. 1.1 and 3.2. For convenience, the data are summarized in Tab. 6.1. The volumetric flow rate of our off-gas is $F_{V,g}$ = 10,000 $m^3{}_{STP,dry}$ h^{-1} for all examples. At the end we will find the best suitable separation equipment (or arrangement of requirements) for the treatment of our off-gas to meet the given emission limits. To remember, the total concentration of dust in our clean gas must not exceed $c_{Clean\ gas}$ = 20 mg m^{-3} at particle flow rates of 0.2 kg h^{-1} and $c_{Clean\ gas}$ = 10 mg m^{-3} at particle flow rates >0.4 kg h^{-1}.

Tab. 6.1: Off-gas specification.

Off-gas specification			**Dust, PSD**		
p_{stat}	980	hPa	x_{ul} **(μm)**	Q_3	c **(mg** $m^{-3}{}_{STP,dry}$**)**
p_{dyn}	981.095	hPa	0.2	0.004	13.30
T	200	°C	0.4	0.018	39.56
			0.6	0.039	64.77
N_2	78	%	0.8	0.069	88.29
O_2	12	%	1.0	0.105	109.56
CO_2	10	%	2.0	0.359	760.98
H_2O	100	g Nm^{-3}	4.0	0.831	1,416.50
O_2-ref.	3	%	6.0	0.982	452.09
NO_x	300	ppm	8.0	0.999	52.50
SO_x	300	ppm	10.0	1.000	2.40
$F_{V,g}$	10,000	$m^3{}_{STP,dry}$ h^{-1}			
$F_{V,g}$	11,244	$m^3{}_{STP,humid}$ h^{-1}	c	3,000	mg $m^{-3}{}_{STP,dry}$
$F_{V,g}$	20,111	am^3 h^{-1}			
$\rho_{g,dry}$	1.34	kg $m^{-3}{}_{STP,dry}$			
$\rho_{g,humid}$	1.28	kg $m^{-3}{}_{STP,humid}$			
$\rho_{g,actual}$	0.72	kg am^{-3}			
μ_g	$2 \cdot 10^{-5}$	Pa s			
ρ_P	2,000	kg m^{-3}			

6.3 Equipment

6.3.1 Gravity settler

The gravity settler is a very simple separation device. It consists of an expansion chamber, in which the gas velocity of the particle-laden gas drops to let the particles settle under the action of gravity. Gravity settlers were the first devices used for precipitation of particles to control particulate emissions. However, the use of gravity settler is limited to precipitation of particles with particle size >40 µm in diameter, and the separation efficiency is limited to the efficiency of less than 50%. Today's emission standards have contributed to replace the settling chamber by more efficient separation devices. This is why gravity settlers are mainly used as a preseparator for preconditioning particle-laden off-gas or may be used in research for investigations on the particle settling behavior.

Figure 6.1 depicts a horizontal gravity settler. The gravity chamber is constructed as a long horizontal box with dust collection hopper at the bottom of the box. The particle-laden gas stream enters the device at the gas inlet which is followed by an expansion section. The expansion of the gas reduces the gas velocity and hence the particles are subject to the force of gravity. All particles that cover at least the vertical path H on the horizontal path L due to their settling velocity w_P are separated from the gas flow. This fact immediately gives a relationship between the dimensions of the separation device (H, L), the volume flow rate $F_{V,g}$ (or the relating gas velocity v_g) and the particle size (or the relating settling velocity w_P), shown as follows:

$$\frac{H}{L} = \frac{w_P}{v_g} = v_s \cdot \frac{A}{F_{V,g}} \tag{6.8}$$

The velocity of the gas flow v_g is calculated with volumetric flow rate $F_{V,g}$ divided by the cross-sectional area A of the device. Each particle will be separated if its settling velocity w_P is

$$\frac{H}{w_P} < \frac{L}{v_g} \tag{6.9}$$

Based on this fact, it is uncommon to use the "classical" cutoff diameter at 50% separation efficiency, for the design of a gravity settler. Instead, it is more significant to define the cutoff diameter $x_{T,100}$ at 100% separation efficiency.

Let us discuss this statement with an example.

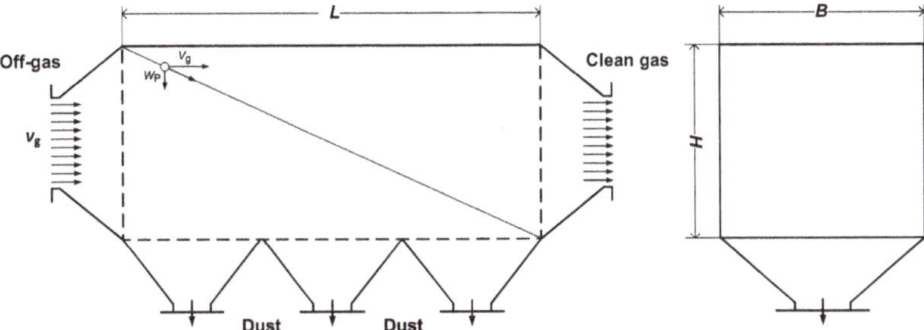

Fig. 6.1: Schematic drawing of a gravity settler with draw-in geometric dimensions.

Exercise 6.1: Design of a gravity settler

Although we already learned that gravity settlers are not suitable for the separation of particles of particle size ≤40 μm (which is the case for our off-gas composition), we take a closer look on their design basics. The separation efficiency of a gravity settler with the length L = 6 m, the height H = 1 m and the width B = 3 m shall therefore be determined for the operation conditions summarized in Tab. 6.1. We will calculate the cutoff diameter $x_{T,100}$ for the given operation conditions.

For the calculations we assume:

- constant gas velocity v_g across the cross-sectional area of the device,
- homogeneously distributed particles in the gas stream and
- equal gas velocity v_g and particle velocity w_P.

Solver:

From the force balance shown in Fig. 4.15 (Exercise 4.3) and the assumption of Stokes sedimentation conditions, we get the settling velocity w_P of our related particles:

$$w_P = \frac{\left(\rho_P - \rho_g\right) \cdot g \cdot x^2}{18 \cdot \mu_g} \tag{6.10}$$

Considering equilibrium of the residence time of the gas t ($t = L/v_g$), and the time needed for particle settling t_P ($t_P = H/w_P$)

$$L \cdot w_P = H \cdot v_g \tag{6.11}$$

the settling velocity is then

$$w_P = \frac{H}{L} \cdot v_g \tag{6.12}$$

By rearranging eq. (6.10) and with the settling velocity w_P as defined in eq. (6.12) we get the cutoff diameter $x_{T,100}$

$$x_T = \sqrt{\frac{18 \cdot \mu_g \cdot H \cdot v_g}{\left(\rho_P - \rho_g\right) \cdot g \cdot L}} = \sqrt{\frac{18 \cdot 2 \cdot 10^{-5} \cdot 1 \cdot 1.86}{(2{,}000 - 0.72) \cdot 9.81 \cdot 6}} = 7.5 \cdot 10^{-5}\ \text{m} \tag{6.13}$$

with the gas velocity v_g

$$v_g = \frac{F_{V,g}}{A} = \frac{F_{V,g}}{H \cdot B} = \frac{\frac{20,111}{3,600}}{1 \cdot 3} = 1.86\ \mathrm{m\ s^{-1}}$$

The separation efficiency $T(x)$ of the gravity settler can be calculated with the following equation:

$$T(x) = \frac{L \cdot w_P}{H \cdot v_g} \tag{6.14}$$

Discussion: The calculation of the cutoff diameter ($x_{T,100}$ = 75.5 μm) confirms that this equipment cannot fulfill our separation task. Particles with diameters ≥75.5 μm will reach the collection hoppers, and particles with lower diameters will escape the gravity settler at its outlet. Referring to our off-gas specification, basically no particles will be separated. The settling velocity w_P and the resulting separation efficiency for the PSD of our off-gas are summarized in Tab. 6.2.

Tab. 6.2: Settling velocity w_p and separation efficiency $T(x)$ for the gravity settler.

PSD (μm)	x_{mean} (μm)	Q_3 (off-gas)	ΔQ_3	Δm (kg)	w_P (m s^{-1})	T(x)	$\dot{m}_{Prec}$ (kg h^{-1})	$c_{Off\text{-}gas}$ (mg m^{-3})	c_{Prec} (mg m^{-3})	$c_{Clean\ gas}$ (mg m^{-3})
0.0	0.0	0.00	0.00	0.00	0.00	0.00	0.00	0.00	0.00	0.00
0.2	0.1	0.00	0.00	0.13	0.00	0.00	0.00	13.30	0.00	13.30
0.4	0.3	0.02	0.01	0.40	0.00	0.00	0.00	39.56	0.00	39.56
0.6	0.5	0.04	0.02	0.65	0.00	0.00	0.00	64.77	0.00	64.77
0.8	0.7	0.07	0.03	0.88	0.00	0.01	0.00	88.29	0.01	88.29
1.0	0.9	0.11	0.04	1.10	0.00	0.01	0.00	109.56	0.02	109.54
2.0	1.5	0.36	0.25	7.61	0.0001	0.04	0.02	760.98	0.30	760.68
4.0	3.0	0.83	0.47	14.17	0.0005	0.16	0.02	1,416.50	2.24	1,414.26
6.0	5.0	0.98	0.15	4.52	0.0014	0.44	0.02	452.09	1.98	450.11
8.0	7.0	1.00	0.02	0.52	0.0027	0.86	0.00	52.50	0.45	52.05
10.0	9.0	1.00	0.00	0.02	0.0044	1.42	0.00	2.40	0.03	2.37
							Σ 0.05			Σ 2,994.92

Although this equipment is not suitable for our separation task, let us have a look at the parameters affecting the separation efficiency for the case you ever need to design a gravity settler.

Exercise 6.2: The effect of settler dimensions

To what extent does $x_{T,100}$ increase/decrease if the length of the gravity settler is changed to L = 3 m and L = 9 m. What do you expect? How does the change of the settler height affect $x_{T,100}$ if the gas velocity v_g is kept constant?

Solver:

In Fig. 6.2a, the separation efficiency curve $T(x)$ for the chamber length of L = 3, 6 and 9 m is depicted. Figure 6.2b shows $T(x)$ for the chamber height of H = 1, 3 and 6 m. We have expected that with increasing length of the separation chamber the separation efficiency increases, resulting in a decrease of the cutoff diameter. A settling chamber of infinite length can theoretically separate all particles. From Fig. 6.2b, we see that increasing the chamber height H decreases the separation efficiency. This is because the settling path of the particles increases (and thus the settling time of the particles t_P), whereas the gas velocity v_g remains constant.

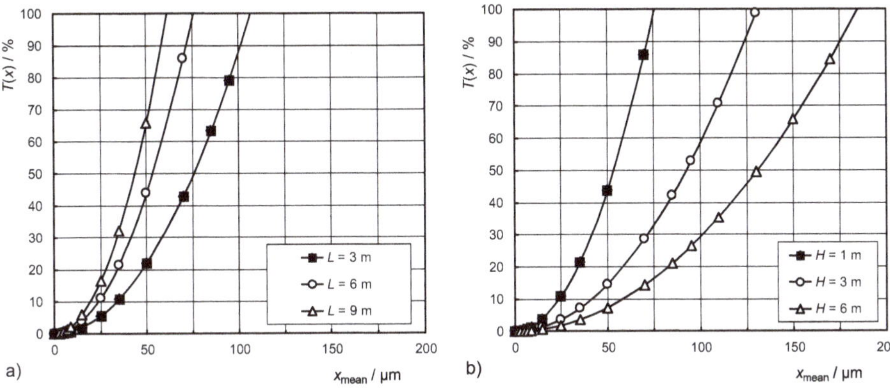

Fig. 6.2: (a) Separation efficiency curve evaluated for the chamber length L = 3, 6 and 9 m; (b) separation efficiency curve evaluated for the chamber height H = 1, 3 and 6 m. The calculations are based on the off-gas specifications and operation conditions listed in Tab. 6.1.

Since we just talked about the effect of the gas velocity v_g, the question arises to what extent the gas velocity impacts the separation efficiency? Let us have a look at Exercise 6.3 to answer this question.

Exercise 6.3: Effect of gas velocity

With the same dimension of the gravity settler (L = 6 m, H = 1 m and B = 3 m) we determine the separation efficiency for three different gas velocities v_g = 0.5, 1.68 and 5 m s^{-1}. We need to adjust the volumetric flow rate as follows:

$v_g = 0.5\ \text{m s}^{-1} \rightarrow F_{V,g} = 1{,}750\ \text{m}^3_{\text{STP,dry}}\ \text{h}^{-1}$
$v_g = 1.68\ \text{m s}^{-1} \rightarrow F_{V,g} = 10{,}000\ \text{m}^3_{\text{STP,dry}}\ \text{h}^{-1}$
$v_g = 5\ \text{m s}^{-1} \rightarrow F_{V,g} = 29{,}000\ \text{m}^3_{\text{STP,dry}}\ \text{h}^{-1}$
if we keep the cross-sectional area A ($A = H \cdot B = 3\ \text{m}^2$) of the gravity settler constant.

Discussion: The separation efficiency $T(x)$ for the three different gas velocities v_g is shown in Fig. 6.3. What we see is that with increasing gas velocity (= increasing volumetric flow rate), the separation efficiency decreases. The cutoff diameter at v_g = 5 m s^{-1} is $x_T \approx 125$ µm, as we can read from the diagram. At low gas velocities of 0.5 m s^{-1} we achieve a cutoff diameter of x_T = 45 µm, which is an appreciable separation performance (for a gravity settler).

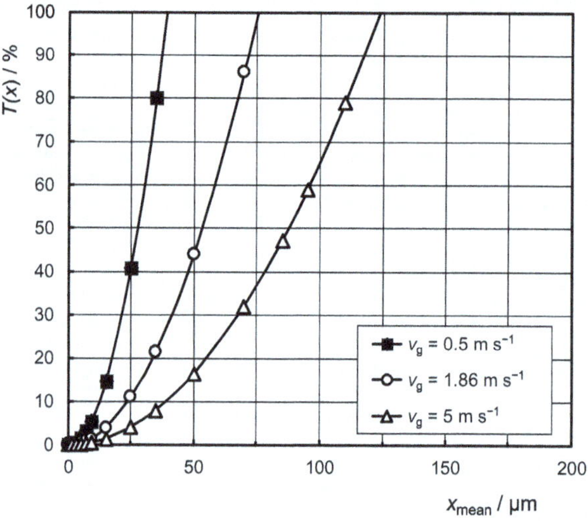

Fig. 6.3: Effect of the gas velocity v_g on the separation efficiency of a gravity settler with the geometric dimensions of L = 6 m, H = 1 m and B = 3 m for the off-gas specification shown in Tab. 6.1.

6.3.1.1 Summary

The outcome of our discussion tells us that the gravity settler is not a suitable equipment for our separation task. The gravity settler may rather be used as a coarse grain preseparator or as a coarse grain classifier.

6.3.2 Cyclone separators

Cyclone separators can be compared with gravity settlers, in which gravitational acceleration is replaced by centrifugal acceleration. As early as in the year 1885, the American inventor J. M. Finch from the Knickerbocker Company received the patent for a (the first) cyclone separator [58]. Although it has a very complex design and bears little resemblance with nowadays standard cyclones, as shown in Fig. 6.4, it already works with the main mode of action by using the principle of centrifugal acceleration for particle separation. Improvements in the design of the separation chamber rapidly followed, as for example by the American inventor O. M. Morse, in the year 1889 [59]. The design of his "dust collector" looks more similar to today's state-of-the-art cyclone separators, as shown in Fig. 6.5. Since then many researchers have been working on the development and the design of cyclone separators, such as the well-known scientist L. Prandtl in the year 1901 [60]. Due to simple construction of cyclone separators resulting in low manufacturing cost, compactness and lack of moving parts, they continued to grow in

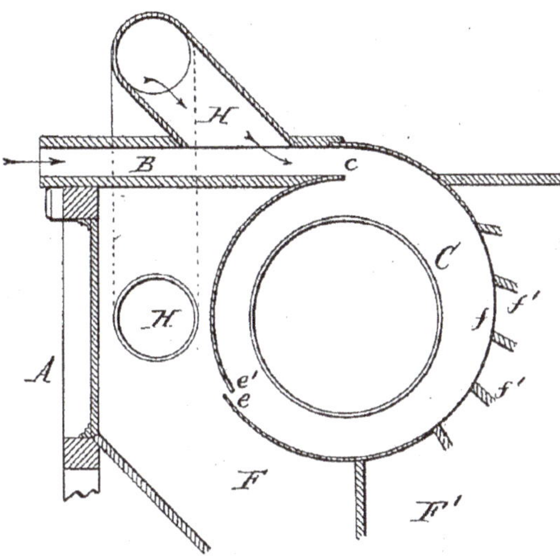

Fig. 6.4: Excerpt of the J. M. Finch patent of a dust collector [58].

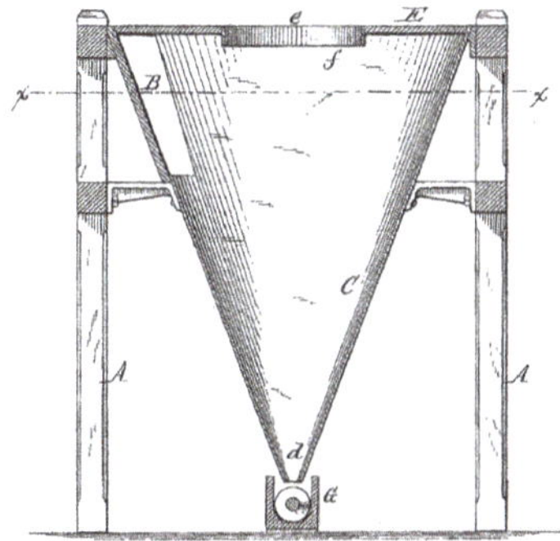

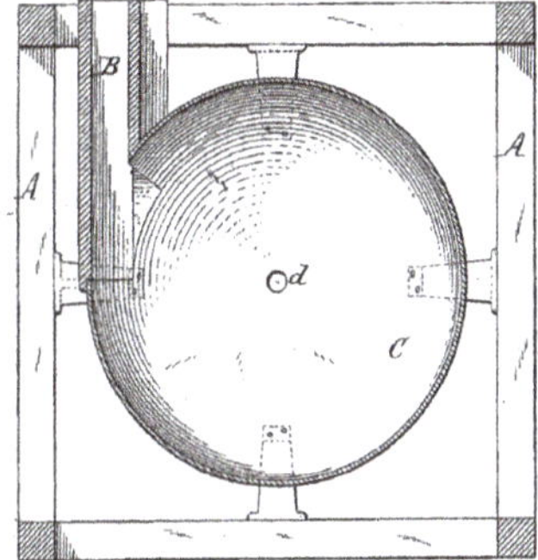

Fig. 6.5: Excerpt of O.M. Morse's patent of a dust collector [59].

popularity and were improved in construction and operation [61]. E. Rammler, P. Rosin, W. Intelmann, and E. Feifel established the basis for the scientific calculations, and contributions of pioneers such as L. Leineweber, G.B. Shepherd, C.E. Lapple, A.J. Linden, W. Barth and E. Muschelknautz have further increased the development and understanding of cyclones [61]. However, with his publication in the year 1956, Barth W. laid the foundation for the state-of-the-art calculation and design of cyclone separators [62]. Based on the publication of Barth W., Muschelknautz et al. [63] developed the phenomenological interfacial model which is widely used (up to now) as a standard model for the industrial design of cyclone separators [64]. This empirical standard model has been verified by numerous experimental investigations and can be found in the VDI Wärmeatlas [65].

Nowadays, the most common type of cyclone separator is the reverse flow cyclone, as shown in Fig. 6.6. In this type, the particle-laden gas is fed tangentially, or axially with swirl vanes, at the inlet channel of the cylindrical part of the cyclone. This inlet geometry induces a three-dimensional highly turbulent swirling flow. Due to the subsequent conical part of the cyclone, the rotational speed of the swirling flow further increases, using the centrifugal force to separate particles from the off-gas. The velocity of the swirling flow can achieve values several times the average feed velocity. The particles are thus separated very fast by centrifugal force. The gas involves a double vortex by spiraling downward along the wall of the cyclone toward the apex of the conical section and then reverses its axial direction upward to the center. The clean gas exits via the so-called vortex finder at the top of the cyclone. At the bottom of the conical section, the separated particles are collected in a hopper. In order to prevent the vortex flow to carry off already separated particles, the transition from the cone to the hopper is equipped by a vortex shield, the so-called apex cone.

The flow inside the cyclone can be divided into an outer, downward-spiraling vortex and an inner, upward-spiraling one [66][61], as illustrated in Fig. 6.6. The downward flow is critically important as it is the prevalent mechanism for transporting the collected particles from the cone wall to the bottom of the cyclone. The flow field inside the cyclone is affected by the inlet flow, the main flow, and the boundary layer flow.

Basically, cyclone separators are specified by their inlet configuration, the body shape, and the flow inlet and outlet directions. The basic inlet configuration of the reverse flow cyclone is the tangential, spiral or axial inlet configuration, as shown in Fig. 6.7. Spiral inlet is preferred over tangential inlet due to the asymmetric flow pattern of latter construction. The axial inlet slot needs swirl vanes to generate the vortex flow. This type of cyclone is therefore also called swirl tube [67]. The flow pattern in the separation chamber of a swirl tube is qualitatively similar to that in conventional cyclones [66]. Although the axial inlet design is more complex, it is often used in multicyclones because of the minimum space requirement.

6.3.2.1 Applications

Cyclone separators are widely used in industry since many decades. Within the limits of performance, they are the least expensive precipitation device regarding investment and operation cost. This is mainly due to the simple and robust design working without any moving parts at a generally low pressure drop. Cyclone separators are especially suitable for the application at high temperatures and pressures. They can be operated under extreme conditions, e.g., at temperatures up to 1,200 °C and pressures up to 100 bar. Furthermore, cyclone separators can process high volume flow rates with high solid concentrations (X_{Dust} = 30 kg_{Dust} kg_{Gas}^{-1}).

In general, cyclone separators are either used for product recovery during the production state or for cleaning the inlet or exhaust gas stream. Early cyclones were

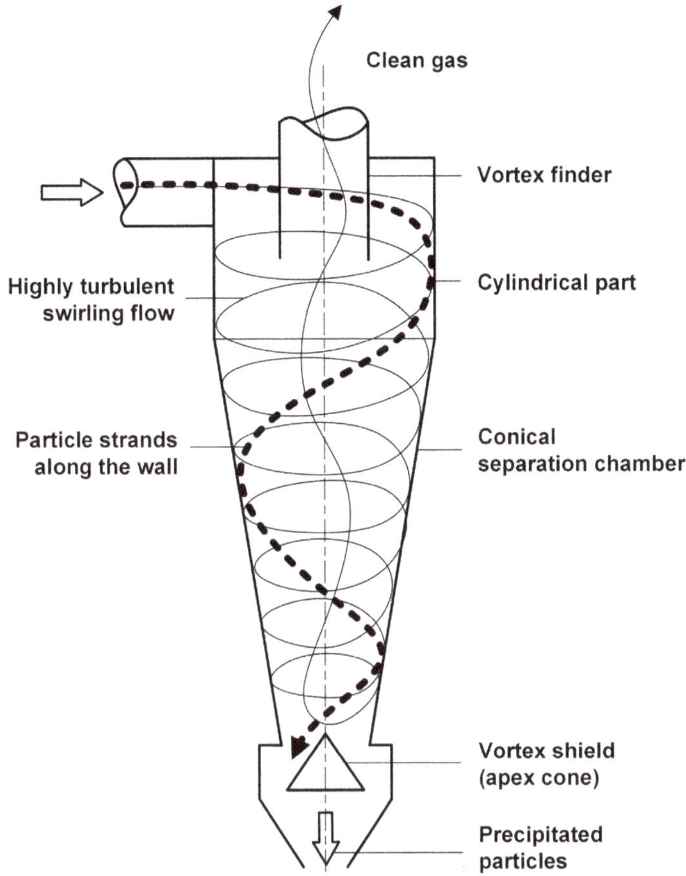

Fig. 6.6: Schematic illustration of a reverse flow cyclone with flow conditions inside the cyclone.

applied to gain dust created from mills processing grains and wood products. However, applications have increased in almost every part of industry where particles (or droplets) need to be removed from a gas stream. This is for example in the

- iron and steel industry (e.g., blast furnace),
- oil and gas industry,
- for power generation,
- sand, cement and coking industries,
- food industry,
- chemical industry,
- wood industry

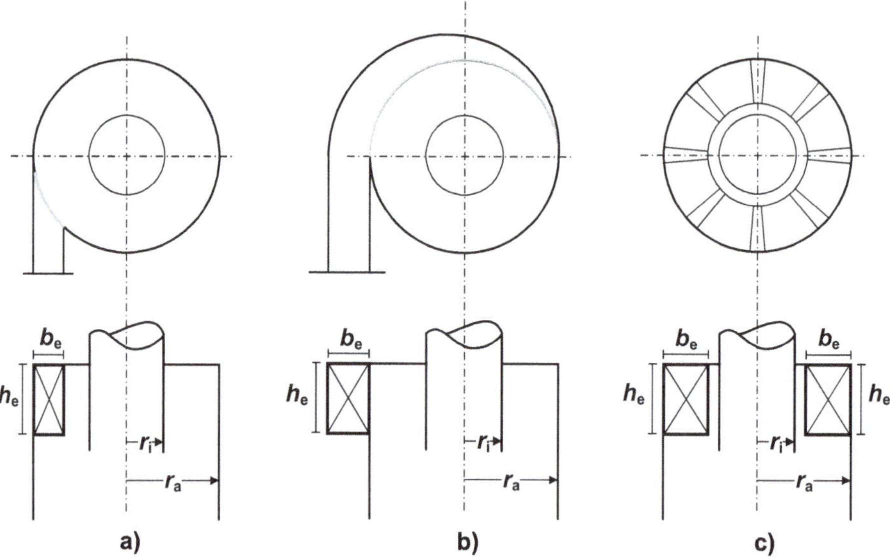

Fig. 6.7: Basic inlet configurations of a cyclone: (a) tangential inlet, (b) spiral inlet and (c) axial inlet (h_e, inlet height; b_e, inlet width; r_i, radius of the vortex finder; r_a, radius of the cyclone).

Despite many advantages, cyclone separators fail when it comes to separation of fine particulates. Particles of less than 2.5 µm cannot be readily separated with commercial cyclones. Therefore, they are often used as a preseparation step to protect downstream bag filters, wet separators or electrostatic precipitators (ESPs) [68]. Since the requirements for the separation efficiency have increased in recent years, which is mainly due to environmental constrains, many attempts have been made to improve the design of cyclone separators and thus the separation efficiency. A promising modification, for example, is the hybrid filter cyclone, which is a combination of a cyclone with internal filter elements.

The application of computational fluid dynamics (CFD) is increasingly applied to design and optimize cyclone separators. Initially, two-dimensional analysis has been performed using radial symmetry of the flow in the cyclone. The two-dimensional approach was necessary in previous times due to lower computer performance but led to inadmissible simplifications [69]. Consequently, studies with three-dimensional predictions followed, most of them investigating the modification of one or more geometries of the cyclone. The effect of vortex finder modifications on precipitation efficiency was, for example, investigated with numerical simulations [70–73]. The modification of the inlet and its impact on separation performance was studied as well [74, 75]. The impact of the cyclone size, the cyclone diameter and cyclone length on the separation performance were also investigated [76–78].

The enormous number of papers (the mentioned papers are just a small number of existing publications) show that CFD (or numerical simulations) is a good and fast method to study the influence of different parameters on the separation performance. However, never forget that every simulation needs to be validated!

The advantages and disadvantages of the cyclone separator are summarized in Tab. 6.3.

Tab. 6.3: Advantages and disadvantages of the cyclone separator.

Advantages	Disadvantages
Simple construction principle	Limited precipitation efficiency
No installations	Dislikes fluctuating gas flow rate
High reliability	Low collection efficiency of fine dust
Good performance at high temperature	
Application at high pressure	
Low temperature loss	

6.3.2.2 Flow pattern and mode of action

The actual flow conditions in cyclone separators are complex and hard to capture quantitatively [79]. Basically, the gas flows along a spiral path and is swirling strongly induced by the tangential gas inlet. Hence, the tangential velocity is more crucial for the separation efficiency than the axial and radial velocities.

The tangential velocity of the gas increases, according to the principle of angular momentum, radially inward and drops back to zero near the cyclone axis [79], as shown in Fig. 6.8. It reaches its maximum at the edge of the vortex finder (r_i). The tangential velocity profile is similar to a Rankine vortex. It can be divided into two types of ideal swirling flow regions: a quasi-free vortex flow region which is surrounding an inner region of quasi-forced vortex flow [66]. In the quasi-free vortex flow region, the fluid is regarded as a frictionless fluid (infinite viscosity); thus, the fluid elements in the swirl have the same angular velocity Ω at all radial positions. In the quasi-forced vortex flow region, the fluid is regarded as a fluid with no viscosity resulting in a swirl with the same tangential velocity distribution as a rotating solid body. In the real swirling flow, the tangential velocity is intermediate between these two extremes. The tangential component of the velocity increases with decreasing radius of the vortex finder.

Figure 6.8 also pictures the axial velocity profile, with the outer region of the downwardly directed axial flow and the inner region of upwardly directed axial flow in the cyclone. The axial velocity profile often drops around the cyclone axis, which can lead to reversed downwardly directed flow at this area [80]. The change of the radial flow direction is located about two-thirds of the cyclone radius.

The radial velocity profile is directed from the outer vortex to the inner one and is distributed over the length of the cyclone. It generally starts below the vortex finder and is not uniformly distributed along the cyclone height, as shown in Fig. 6.8. The radial velocity is directed toward the center. From the center to the projected cylindrical surface area of the vortex finder, the radial flow is directed outward. For basic calculations, the radial velocity can either be neglected since it is the smallest velocity component [66], or it is assumed to be constant over the area of zero axial velocity.

The circumferential velocity $v_{\varphi i}$ reaches its maximum at the radius of the vortex finder r_i, as shown in Fig. 6.8.

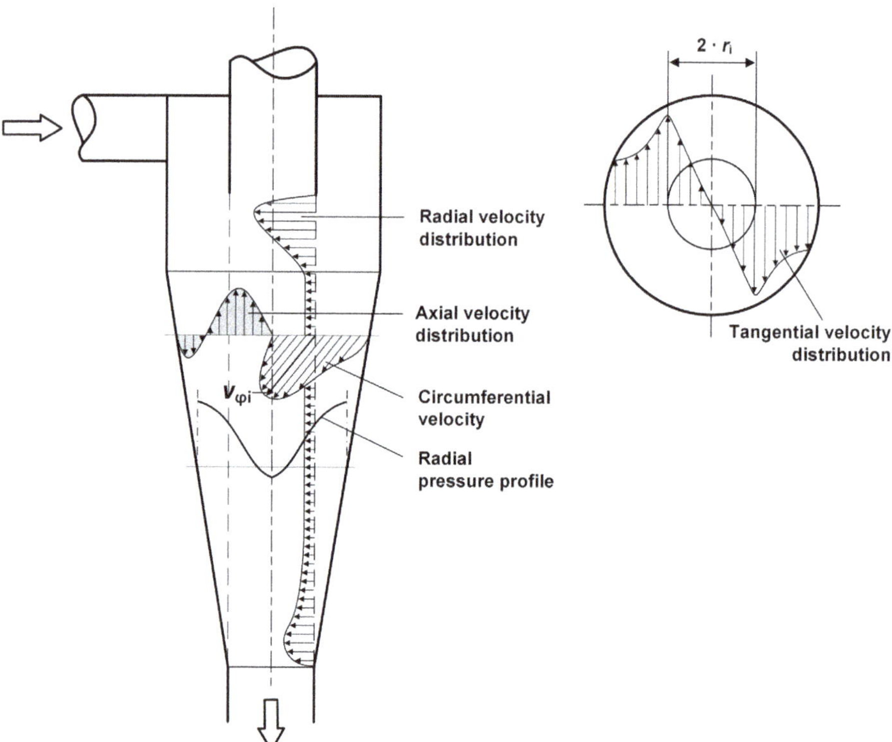

Fig. 6.8: Radial, axial, tangential and circumferential velocity distributions as well as the pressure profile indicated in a cyclone.

There are several parameters influencing the separation performance of the cyclone. The separation efficiency exemplarily increases with increasing gas velocity due to the increasing tangential velocity. High dust load can act by disintegrating as well as agglomerating. Flow pattern and dust properties cause one or the other effect to dominate. Agglomeration, supported by elevated dust load, may lead to an improvement of the separation efficiency and a decrease of the pressure drop. The design of the vortex finder reveals the centrifugal force (about 90% of the dissipated energy), the eccentricity of the flow pattern and the pressure drop. The inlet and outlet area also influences the separation efficiency. Decreasing the inlet area will increase the collection efficiency. In case of a rectangular inlet, the ratio of the inlet width and the inlet height should preferably have a value of approximately 0.15. Large cyclone height and large angle of the cone result in a lower pressure drop.

Despite many years of research and the application of CFD simulations, the flow inside the cyclone is so complex that the real flow conditions cannot be reproduced until now.

6.3.2.3 Pressure drop

The static pressure inside the cyclone correlates with the fluid velocities, which can be explained by the law of energy conservation as well as by Bernoulli's equation. The Bernoulli equation for steady flow of a frictionless fluid with constant density is given as follows:

$$\frac{p}{\rho} + g \cdot h + \frac{1}{2} \cdot v^2 = \text{constant along a streamline} \tag{6.15}$$

In cyclones, the gravitational force (second term in the Bernoulli's equation $g{\cdot}h$) is small compared to the other forces and can be neglected. As it can be seen via Bernoulli's equation, higher fluid velocities are accompanied with lower pressures [79]. Thus, the static pressure and the total pressure strongly depend on the radial position inside the cyclone. The pressure reaches its maximum at the wall of the cyclone and drops significantly in the center. The radial pressure profile inside a cyclone is schematically shown in Fig. 6.8.

In the actual flow, the fluid is not frictionless and mechanical dissipation will cause Bernoulli's trinomial to decrease along a streamline. However, the frictionless approach delivers reasonable approximation for the determination of the pressure drop.

The pressure drop of a cyclone can be subdivided into three sections [61]:

- pressure loss at the entry,
- pressure loss in the separation chamber (main body) and
- pressure loss in the vortex finder.

Flow pattern, turbulence and velocity distribution as well as the geometry of the cyclone correlate and determine the pressure drop, and thus the collection efficiency. The flow pattern is controlled by the inlet velocity, the main gas flow, the gas flow in the region of the boundary layer and the gas outlet. At the boundary layer of the wall, the gas velocity is significantly lower due to friction forces that cause lower centrifugal forces of the vortex. However, the pressure gradient of the prevailing main flow is imposed on the conical wall of the separation chamber. As a result, the compressive force is significantly greater than the centrifugal force. This imbalance creates a strong secondary flow along the conical wall, visible as particle strands. This characteristic strand flow helps to carry particles attached to the wall downward. The dust adhering to the wall contributes to a higher separation efficiency, which means that high dust load improves the degree of separation [68].

6.3.2.4 Apparatus design and prediction of the separation efficiency

When designing a cyclone, high separation efficiency needs to be combined with low pressure drop, resulting in a low cutoff particle diameter. Both parameters usually increase with increasing volumetric flow rate and particle concentration so that a compromise must be found. The ideal cyclone is characterized by a small cutoff diameter at low pressure loss [79]. The flow in the cyclone can only carry a limited amount of particles [80]. If the particle concentration at the inlet is higher than the so-called critical load, the excess particles are separated directly after entering the cyclone [63, 82]. Only the mass fraction that is limited by the "critical load" is separated by the actual centrifugal force. According to Muschelknautz [83], two different separation mechanisms occur in a cyclone:

- Precipitation in the vortex (equilibrium of the cutoff particle diameter according to Barth)
- Precipitation due to particle strands at the inlet area according to Muschelknautz

By Exercise 6.4, the design of a cyclone according to Barth/Muschelknautz algorithm is shown. The denotation of the geometrical dimensions of the cyclone are illustrated in Fig. 6.9.

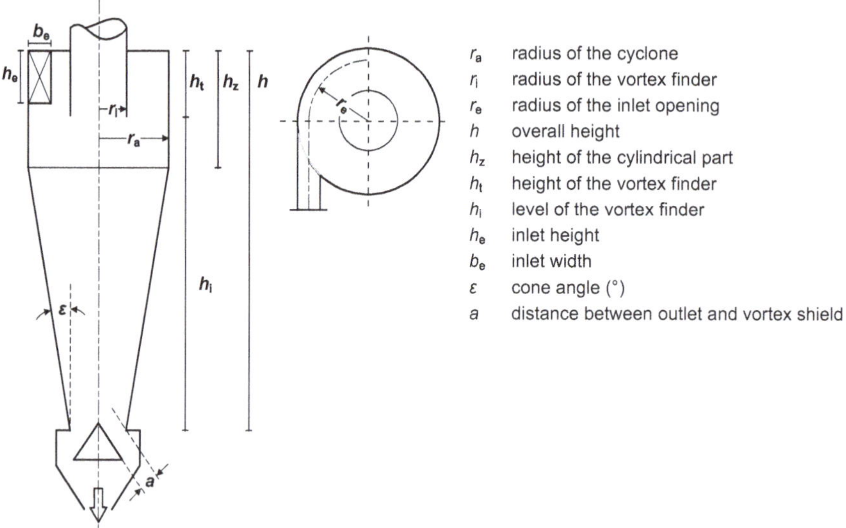

Fig. 6.9: Denotations of the geometrical dimensions of the cyclone.

Exercise 6.4: Cyclone design

For our off-gas specification of Tab. 6.1, we want to design a cyclone with a tangential squared inlet. With the determined geometric cyclone dimensions, we will then calculate the separation efficiency and the cutoff diameter.

Solver:

At the level of the vortex finder, the circumferential velocity $v_{\varphi i}$ reaches its maximum. The balance of centrifugal force and drag force against the radial velocity component at the level of the vortex finder is decisive for the achievable particle size to be separated. This particle size can be defined with the cutoff diameter x_T. Our aim is therefore to determine x_T, which is according to Barth [62]:

$$x_T = \sqrt{\frac{18 \cdot \mu_g \cdot v_r(r_i) \cdot r_i}{\left(\rho_p - \rho_g\right) \cdot v_{\varphi i}^2}} \tag{6.16}$$

For the design of the cyclone, we can follow a simple design guideline to calculate the radial velocity of the gas at the radius of the vortex finder $v_r(r_i)$ and the circumferential gas velocity at the radius of the vortex finder $v_{\varphi i}$.

First step:

First, we need to estimate the gas velocity v_i at the cross section of the vortex finder. Table 6.4 lists the recommended values for v_i as well as geometric ratios for cyclones with the tangential inlet design. For our example let us first choose $v_i = 15\ \text{m s}^{-1}$ (later we will have a look on the effect of lower gas velocities v_i).

Tab 6.4: Recommended values and geometric ratios for cyclones with tangential inlet [84].

Recommended geometric ratios			
v_i	5–15 m s^{-1}	h/r_i	10–13
r_a/r_i	3–4	h_i/r_i	7.5–10
b_e/r_a	0.19–0.27	F_e/F_i	0.44–0.9

Second step:

With $v_i = 15\ \text{m s}^{-1}$ and the actual gas flow rate $F_{V,g} = 20{,}111\ \text{am}^3\ \text{h}^{-1}$, we can calculate the radius of the vortex finder r_i which is after rearranging:

$$r_i = \left(\frac{F_{V,g}}{\pi \cdot v_i}\right)^{0.5} = \left(\frac{\frac{20{,}111}{3{,}600}}{3.14 \cdot 15}\right)^{0.5} = 0.34\ \text{m}$$

Third step:

The radius of the cyclone r_a can easily be calculated using the recommended geometric ratios in Tab. 6.4. If we decide for the radius ratio $r_a/r_i = 4$, r_a is then

$$r_a = 4 \cdot r_i = 4 \cdot 0.34 = 1.38\ \text{m}$$

Fourth step:
For the calculation of the overall cyclone height h, we decide for the ratio $h/r_i = 12$:

$$h = 12 \cdot r_i = 12 \cdot 0.34 = 4.13 \text{ m}$$

Fifth step:
We calculate the level height of the vortex finder h_i with the ratio $h_i/r_i = 8$:

$$h_i = 8 \cdot r_i = 8 \cdot 0.34 = 2.76 \text{ m}$$

Sixth step:
For the calculation of the inlet width b_e, we choose the ratio $b_e/r_a = 0.2$:

$$b_e = 0.2 \cdot r_a = 0.2 \cdot 1.38 = 0.28 \text{ m}$$

Seventh step:
The cross-sectional area of the inlet tube F_e is exemplarily calculated with the ratio value $F_e/F_i = 0.45$ with F_i being the cross-sectional area of the vortex finder and thus $F_i = r_i^2 \cdot \pi = 0.37 \text{ m}^2$:

$$F_e = 0.45 \cdot F_i = 0.45 \cdot 0.37 = 0.17 \text{ m}^2$$

Eighth step and ninth step:
The inlet height h_e and the height of the vortex finder h_t can easily be calculated with the geometric dimensions already determined:

$$h_e = \frac{F_e}{b_e} = \frac{0.17}{0.28} = 0.61 \text{ m}$$

$$h_t = h - h_i = 4.13 - 2.76 = 1.38 \text{ m}$$

Tenth step:
Now we can calculate the inlet gas velocity v_e which results directly from the actual volumetric gas flow rate and the cross-sectional area of the inlet:

$$v_e = \frac{F_{V,g}}{F_e} = \frac{\frac{20{,}111}{3{,}600}}{0.17} = 33.33 \text{ m s}^{-1}$$

Eleventh step:
Next, we need to calculate the wall friction factor λ which depends on the dust load B, either with eq. (6.16) or (6.17).

For $B > 1$:
$$\lambda = 0.005 \cdot \left(1 + 3 \cdot \sqrt{B}\right) \tag{6.16}$$

For $B < 1$:
$$\lambda = 0.005 \cdot \left(1 + 2 \cdot \sqrt{B}\right) \tag{6.17}$$

In our case, the dust load B is

$$B = \frac{\dot{m}_{Dust}}{\dot{m}_{Gas}} = \frac{300\left[\text{mg m}^{-3}{}_{\text{STP, dry}}\right] \cdot (10^{-6}) \cdot 10{,}000\left[\text{m}^3{}_{\text{STP, dry}}\text{ h}^{-1}\right]}{10{,}000\left[\text{m}^3{}_{\text{STP, dry}}\text{ h}^{-1}\right] \cdot 1.341\left[\text{kg m}^{-3}{}_{\text{STP, dry}}\right]} = 0.002$$

and thus, the wall friction factor λ is

$$\lambda = 0.005 \cdot \left(1 + 2 \cdot \sqrt{B}\right) = 0.005 \cdot \left(1 + 2 \cdot \sqrt{0.002}\right) = 0.0055$$

We will need the wall friction factor for the latter calculation of the circumferential gas velocity $v_{\varphi;i}$.

Twelth step:
The radius of the inlet opening r_e is

$$r_e = r_a - \frac{b_e}{2} = 1.38 - \frac{0.28}{2} = 1.24\,\text{m}$$

Thirteenth step:
Barth [62] characterized the inlet gas stream with the aid of the inlet obstruction factor α. The inlet obstruction factor α is the ratio of the inlet angular momentum to the inflow angular momentum:

$$\alpha = \frac{\text{inlet angular momentum}}{\text{input angular momentum}} = \frac{v_e \cdot r_e}{v_{\varphi a} \cdot r_a} \tag{6.18}$$

For slit inlets, Kimura [85] developed the following correlation for the prediction of α:

$$\alpha = 1 - 0.36 \cdot \sqrt{\frac{F_e}{F_i} \cdot \left(\frac{b_e}{r_a}\right)^{0.45}} = 0.883 \tag{6.19}$$

The circumferential velocity at the cyclone radius $v_{\varphi a}$ is thus

$$v_{\varphi a} = \frac{v_e \cdot r_e}{r_a \cdot \alpha} = \frac{33.33 \cdot 1.24}{1.38 \cdot 0.883} = 33.98\,\text{m}\,\text{s}^{-1}$$

Fourteenth step:
Now we need to calculate the circumferential gas velocity at the radius of the vortex finder $v_{\varphi i}$ and the radial velocity $v_r(r_i)$ at the radius of the vortex finder.

Barth and Muschelknautz refrained from approximating the circumferential gas velocity $v_{\varphi i}$ and calculated it with the aid of an alternative flow model instead. The flow model and the principal of angular momentum gives the circumferential gas velocity $v_{\varphi i}$ based on the gas velocity of the vortex finder v_i, as follows:

$$v_{\varphi i} = \frac{v_i}{\frac{F_e \cdot r_i \cdot \alpha}{F_i \cdot r_e} + \frac{\lambda \cdot h}{r_i}} \tag{6.20}$$

$$v_{\varphi i} = \frac{15}{\frac{0.17 \cdot 0.34 \cdot 0.883}{0.37 \cdot 1.24} + \frac{0.0055 \cdot 4.13}{0.34}} = 85.21\ \text{m}\,\text{s}^{-1}$$

The radial velocity at the radius of the vortex finder $v_r(r_i)$ is

$$v_r(r_i) = \frac{F_{V,g}}{2 \cdot \pi \cdot r_i \cdot (h - h_t)} = \frac{\frac{20{,}111}{3{,}600}}{2 \cdot 3.14 \cdot 0.34 \cdot (4.13 - 1.38)} = 0.94\ \text{m}\,\text{s}^{-1}$$

Now let us move on to business.

Fifteenth step:
Finally, we can calculate the cutoff particle diameter x_T which is

$$x_T = \sqrt{\frac{18 \cdot \mu_g \cdot v_r(r_i) \cdot r_i}{\left(\rho_p - \rho_g\right) \cdot v_{\varphi i}^2}} = \sqrt{\frac{18 \cdot 20 \cdot 10^{-6} \cdot 0.94 \cdot 0.34}{(2{,}000 - 0.72) \cdot 85.21^2}} = 2.83 \cdot 10^{-6}\ \text{m}$$

The determination of the cutoff particle diameter is the prerequisite for calculations of the separation efficiency *T*(*x*) and the overall separation efficiency *E* of the cyclone.

Sixteenth and final step:
For convenience, we suggest the following approach for the calculation of separation efficiency *T*(*x*):

$$T(x) = \frac{1}{\left(1 + \frac{x}{x_T}\right)^{-x_T}} \tag{6.21}$$

The resulting separation curve *T*(*x*) for our off-gas specification is shown in Fig. 6.10a.

For the overall separation efficiency *E* we get

$$E = \sum dQ \cdot T(x) = 43.6\%$$

The overall separation efficiency can also be calculated with the particle mass at the inlet $\dot{m}_{PM,Off\text{-}gas}$ and the mass of precipitated particles $\dot{m}_{Prec}$:

$$E = \frac{\dot{m}_{Prec}}{\dot{m}_{PM,Off\text{-}gas}} = \frac{13.1\left[\text{kg h}^{-1}\right]}{30\left[\text{kg h}^{-1}\right]} \cdot 100 = 43.6\%$$

! Discussion: The curve of the separation efficiency *T*(*x*) and the PSD $Q_3(x)$ of the off-gas and the clean gas is shown in Fig. 6.10a. Figure 6.10b depicts the probability distribution function Δq_3 of the off-gas and the clean gas. The calculation results are summarized in Tab. 6.5.

As we see from the separation efficiency and from Tab. 6.5, particles were separated in a magnitude of 13.1 kg h^{-1}. The particle mass flow of our off-gas is reduced from 30 to 16.9 kg h^{-1}.

If we take a closer look on the concentrations listed in Tab. 6.5, we see that the PM1 particle concentration (the sum of smaller particle classes ≤1 μm) in untreated and cleaned off-gas is 316 mg m$^{-3}_{STP}$ and 309 mg m$^{-3}_{STP}$, respectively. As an effect, PM1 concentration was reduced by only 2%. However, the reduction of PM2.5 (≤2.5 μm) and PM10 concentration (≤10 μm) is increasing to 11% and 44%, respectively. The total concentration of the clean gas is $c_{Clean\ gas}$ = 1,690.8 mg m$^{-3}_{STP}$ and thus we are far off from meeting the emission limits. We can conclude that the efficiency of cyclone separation is not significant for fine particles ≤2.5 μm but becomes more important with increasing particle size for coarse particles.

Tab. 6.5: Calculation results gained with the design guideline based on Barth/Muschelknautz algorithm for the off-gas specification from Tab. 6.1.

PSD (µm)	x_{mean} (µm)	Q_3 (off-gas)	ΔQ_3	Δm (kg)	$T(x)$	$\dot{m}_{Prec}$ (kg h^{-1})	$\dot{m}_{Passed}$ (kg h^{-1})	Q_3 (clean gas)	$c_{Off\text{-}gas}$ (mg m^{-3})	c_{Prec} (mg m^{-3})	$c_{Clean\ gas}$ (mg m^{-3})
0.0	0.0	0.00	0.00	0.00	0.00	0.00	0.00	0.00	0.00	0.00	0.00
0.2	0.1	0.00	0.00	0.13	0.00	0.00	0.13	0.01	13.30	0.00	13.30
0.4	0.3	0.02	0.01	0.40	0.00	0.00	0.39	0.03	39.56	0.07	39.49
0.6	0.5	0.04	0.02	0.65	0.01	0.00	0.64	0.07	64.77	0.48	64.29
0.8	0.7	0.07	0.03	0.88	0.02	0.02	0.87	0.12	88.29	1.66	86.63
1.0	0.9	0.11	0.04	1.10	0.04	0.04	1.05	0.18	109.56	4.12	105.43
2.0	1.5	0.36	0.25	7.61	0.14	1.08	6.53	0.57	760.98	108.30	652.67
4.0	3.0	0.83	0.47	14.17	0.54	7.67	6.50	0.95	1,416.50	766.66	649.84
6.0	5.0	0.98	0.15	4.52	0.83	3.77	0.75	1.00	452.09	376.83	75.26
8.0	7.0	1.00	0.02	0.52	0.93	0.49	0.04	1.00	52.50	48.74	3.76
10.0	9.0	1.00	0.00	0.02	0.96	0.02	0.00	1.00	2.40	2.32	0.09
				Σ30		Σ13.09	Σ16.91		Σ3,000	Σ1,309.2	Σ1,690.8

When comparing the cumulative function Q_3 of the PSD of the off-gas and the clean gas, as shown in Fig. 6.10a, we will recognize that the PSD changes when separating particles with a cyclone. Q_3 of the clean gas slightly shifts to smaller particle diameters and becomes slightly steeper. This effect is also shown in the density distribution functions q_3, where the maximum of the curve shifts to lower diameter. The maximum values shift from 2 µm from the off-gas to 1.4 µm of the clean gas. The respective q_3 value indicates that the relative amount of fines in our clean gas increased from 0.25 to 0.39.

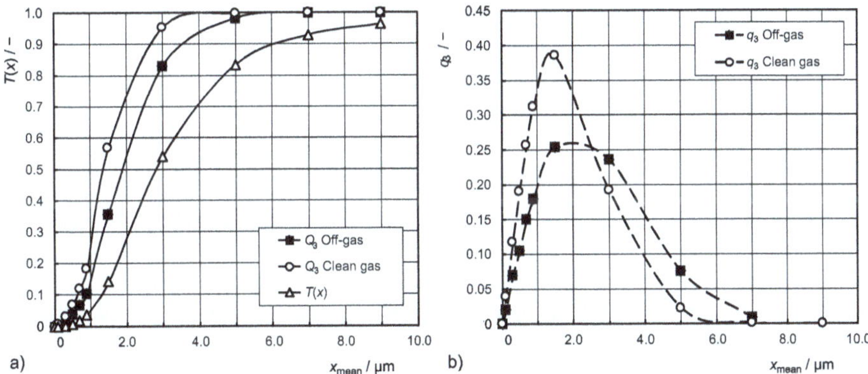

Fig. 6.10: (a) Grade separation efficiency $T(x)$ and cumulative distribution function Q_3 of the off-gas and the clean gas. (b) Density distribution function q_3 of the off-gas and the clean gas. The results correspond to Exercise 6.4.

?

We designed the cyclone based on the recommended values for the geometric dimension and recommended velocities at the vortex finder v_i, as listed in Tab. 6.4. What will be the outcome when we vary the design values within the recommended boundaries? What is the effect of the vortex finder velocity v_i on the separation efficiency? Can we do a better job by changing it? Let us consider this question with Exercise 6.5.

Exercise 6.5: What if . . .?

a) . . . we change the velocity of the vortex finder v_i?

Let us start changing the velocity at the vortex finder radius v_i and see what happens to the separation efficiency. The calculation steps remain the same as already shown in Exercise 6.3. We recommend creating a plot of the separation curve over the mean particle size so that the effect of changing parameters on the separation curve can immediately be seen and interpreted. For convenience, we already did this job, as you can see in Fig. 6.11. Figure 6.11a depicts the effect of v_i. Lower velocities lead to poor separation efficiency curves. With increasing velocity, the separation efficiency increases too. If we increase the velocity by a factor of 3 from $v_i = 5\ \mathrm{m\ s^{-1}}$ to $15\ \mathrm{m\ s^{-1}}$, we can decrease the cutoff diameter (at $T(x) = 0.5$) from $x_T = 6.45$ to $2.83\ \mu\mathrm{m}$.

b) . . . we change the geometric dimensions of the cyclone?

The effect of varying geometric dimensions is shown in Fig. 6.11(b–f). Decreasing the ratio of the cross-sectional area of the inlet to the cross-sectional area of the vortex finder F_e/F_i, as shown in (b), increases the separation efficiency. The ratio is within the recommended values. Increasing the ratio of the radius of the cyclone to the radius of the vortex finder r_a/r_i, as shown in (c), increases the separation efficiency slightly. The impact of the ratio h/r_i and h_i/r_i is rather small, as you can see in Fig. 6.11(d and e). Finally, the ratio of the inlet width to the cyclone ratio b_e/r_a does not affect the separation efficiency.

In practice, product improvements or production adjustments require frequent changes of operation conditions. The cutoff diameter x_T correlates with the square root of the inverse gas flow rate. For a quick estimation of the changes on the separation efficiency, the following approaches can thus be used (the derivation of the equation can be found in [86]):

$$x_T \sim (r_i)^n$$

$$x_T \sim \left(\frac{1}{F_{V,g}}\right)^{0.5}$$

!

Discussion: The best separation performance can be achieved at high velocities at the radius of the vortex finder, if the ratio of F_e/F_i is kept low ($F_e/F_i = 0.45$) and the ratio of r_a/r_i is kept at the upper range ($r_a/r_i = 4$). The velocity at the radius of the vortex finder can either be increased with a decrease of the radius of the vortex finder r_i or an increase of the volumetric flow rate $F_{V,g}$. Even if the effect of the geometric ratios h/r_i, h_i/r_i and b_e/r_a is marginal, stay within the recommended limits as given by the best practice experience.

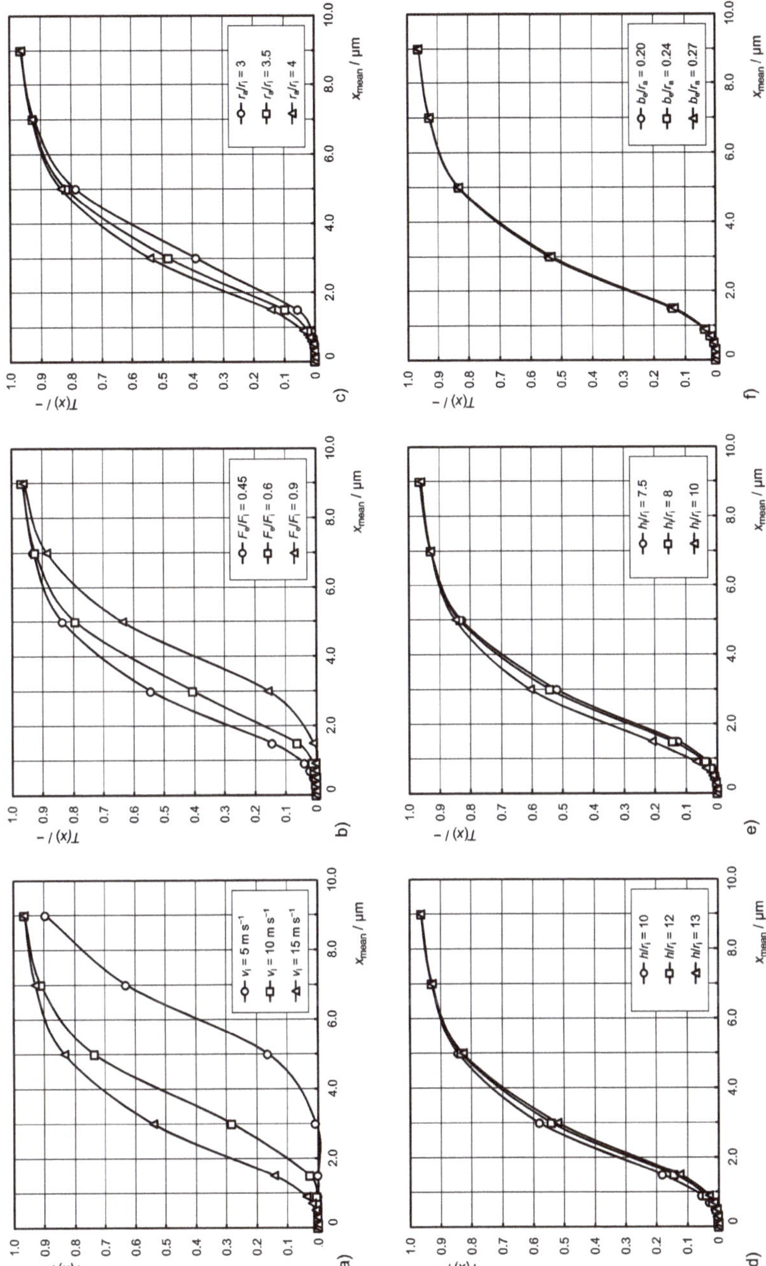

Fig. 6.11: Separation efficiency for (a) varying velocities at the vortex finder radius v_i; (b) varying ratio of the cross-sectional area of the inlet and the outlet at the vortex finder F_e/F_i; (c) varying radius ratios r_a/r_i; (d) varying ratios of h/r_i; (e) varying ratios of h_i/r_i; and (f) varying ratios of b_e/r_a. The calculations are based on the operation conditions and specification of the off-gas as listed in Tab. 6.1.

Exercise 6.6: A matter of power

From Exercise 6.5, we learned that the separation efficiency of a single cyclone is mostly affected by the velocity at the vortex finder v_i. This in turn means that the separation efficiency can be increased by decreasing the radius of the vortex finder or by increasing the volumetric flow rate.

To what extent does an increase of the volumetric flow rate affect the cutoff diameter, the resulting separation efficiency and the theoretically required power P_{theor} of the cyclone? What do you expect? Is it economically feasible to increase the volumetric flow rate?

To answer this question, we will compare P_{theor} for the volumetric flow rates $F_{V,g}$ = 5,000, 10,000 and 20,000 $m^3{}_{STP,dry}\ h^{-1}$. The radius of the vortex finder for all cases is r_i = 0.34 m. For the calculation, the geometric ratios needed should be the same as used in Exercise 6.4.

Solver:

P_{theor} can be calculated from the product of the pressure drop and the actual volumetric flow rate, as follows:

$$P_{theor}[W] = \Delta p \cdot F_{V,g} \tag{6.22}$$

The pressure drop can be estimated on the basis of the Bernoulli approach

$$\Delta p\left[\mathrm{N\,m^{-2}}\right] = \frac{\rho_a \cdot v_i^2 \cdot (\xi_e + \xi_a + \xi_i)}{2} \tag{6.23}$$

with the pressure loss coefficient of the inlet ξ_e = 0 (for slit inlet), the pressure loss coefficient of the main flow ξ_a = 2.15 and the pressure loss coefficient of the vortex finder ξ_i = 23.2 [84].

The velocity of the vortex finder v_i is calculated with the actual volumetric flow rate divided by the vortex finder area:

$$v_i = \frac{F_{V,g}}{\pi \cdot r_i^2}$$

Discussion: The calculation results are summarized in Tab. 6.6. Let us interpret them!

If we increase the volumetric flow rate from 5,000 to 10,000 $m^3{}_{STP,dry}\ h^{-1}$, the cutoff diameter decreases by 28% (from x_T = 3.9 to 2.8 µm), and the total separation efficiency E increases by 76%. On the other hand, the required theoretical power demand increases by a factor of 8.

If we have a look on the fourfold increase of the volumetric flow rate ($F_{V,g}$ = 20,000 $m^3{}_{STP,dry}\ h^{-1}$), we see that we are able to decrease the cutoff diameter by half and thus increase the separation efficiency by 130%. Nevertheless, the required power demand increases by a factor of 64! Furthermore, the velocity at the vortex finder v_i exceeds the suggested limits.

Tab. 6.6: Calculation results of Exercise 6.6.

	Unit	$F_{V,g}$ = 5,000 $m^3{}_{STP,dry}\ h^{-1}$	$F_{V,g}$ = 10,000 $m^3{}_{STP,dry}\ h^{-1}$	$F_{V,g}$ = 20,000 $m^3{}_{STP,dry}\ h^{-1}$
$F_{V,g}$	$am^3\ h^{-1}$	10,055.5	20,111.0	40,222.0
v_i	$m\ s^{-1}$	7.7	15.4	30.8
Δp	$N\ m^{-2}$	537.8	2,151.2	8,604.9
P_{theor}	W	1,502.2	12,017.6	96,141.1
P_{theor}	kW	1.5	12.0	9.1
$P_{\eta=0.75}$	kW	2.0	16.0	128.2
x_T	µm	3.9	2.8	2.0
E	%	25.3	44.6	58.2

When considering whether the high required power is economically acceptable or not, we should keep other technological options of increasing the separation efficiency in mind. How about changing the arrangement, dimensions, and number of cyclones? Can we do a better job though?

Exercise 6.7: Multicyclone

We will now compare the cutoff diameter and the separation efficiency for a single cyclone $n = 1$ with $n = 10$ and $n = 30$ cyclones that are arranged parallel. For the calculations we use the volumetric flow rate $F_{V,g}$ = 10,000 $m^3_{STP,dry}\ h^{-1}$ at the same operation conditions summarized in Tab. 6.1. The velocity of the vortex finder for all cases should be $v_i = 15\ m\ s^{-1}$. Again, the geometric ratios needed for the calculation should be the same as used in Exercise 6.4.

How do the main dimensions of the cyclone change with increasing number of n? What is the theoretical power P_{theor} needed? Does a series arrangement of cyclones make sense?

Solver:

For the calculations, we need to split the volumetric flow rate to the number of cyclones used. If we keep v_i constant at 15 $m\ s^{-1}$, the radius of the vortex finder will decrease, as shown in Tab. 6.7. Following the calculation guideline, we get the geometric dimensions of the cyclones as well as the separation performance. The results are summarized in Tabs. 6.7 and 6.8. The curves of the separation efficiency for all cases are shown in Fig. 6.12.

Tab. 6.7: Results of Exercise 6.7.

	Unit	$n = 1$	$n = 10$	$n = 30$
$F_{V,g}$	$m^3_{STP,dry}\ h^{-1}$	10,000	10,000	10,000
v_i	$m\ s^{-1}$	15.0	15.0	15.0
r_i	m	0.34	0.11	0.06
h	m	4.13	1.31	0.75
r_a	m	1.38	0.44	0.25
Δp	$N\ m^{-2}$	2,043.5	2,043.5	2,043.5
$P_{theor,Single}$	W	11,416.0	1,141.6	380.5
	kW	11.4	1.1	0.38
ΣP_{theor}	kW	11.4	11.4	11.4
$P_{\eta=0.75,Single}$	kW	15.2	1.5	0.5
$\Sigma P_{\eta=0.75}$	kW	15.2	15.2	15.2
x_T	µm	2.8	1.6	1.2
E	%	43.6	63.4	67.4

Tab. 6.8: Concentrations for the PSD for $n = 1$, $n = 10$ and $n = 30$ cyclones arranged parallell.

		$n = 1$		$n = 10$		$n = 30$	
x_{mean} (μm)	$c_{Off\text{-}gas}$ ($mg\ m^{-3}$)	c_{Prec} ($mg\ m^{-3}$)	$c_{Clean\ gas}$ ($mg\ m^{-3}$)	c_{Prec} ($mg\ m^{-3}$)	$c_{Clean\ gas}$ ($mg\ m^{-3}$)	c_{Prec} ($mg\ m^{-3}$)	$c_{Clean\ gas}$ ($mg\ m^{-3}$)
0.0	0.00	0.00	0.00	0.00	0.00	0.00	0.00
0.1	13.30	0.00	13.30	0.16	13.14	0.62	12.68
0.3	39.56	0.07	39.49	2.60	36.96	6.19	33.37
0.5	64.77	0,48	64.29	8.86	55.91	16.57	48.20
0.7	88.29	1.66	86.63	18.81	69.49	30.07	58.23
0.9	109.56	4.12	105.44	31.51	78.05	45.10	64.45
1.5	760.98	108.30	652.68	362.61	398.37	429.79	331.18
3.0	1,416.50	766.64	649.86	1,038.04	378.46	1,062.41	354.09
5.0	452.09	376.83	75.26	389.16	62.94	383.22	68.87
7.0	52.50	48.74	3.76	47.96	4.54	46.89	5.61
9.0	2.40	2.32	0.09	2.26	0.14	2.21	0.19
	Σ 3,000.0	Σ 1,309.2	Σ 1,690.8	Σ 1,902.0	Σ 1,098.0	Σ 2,023.1	Σ 976.9

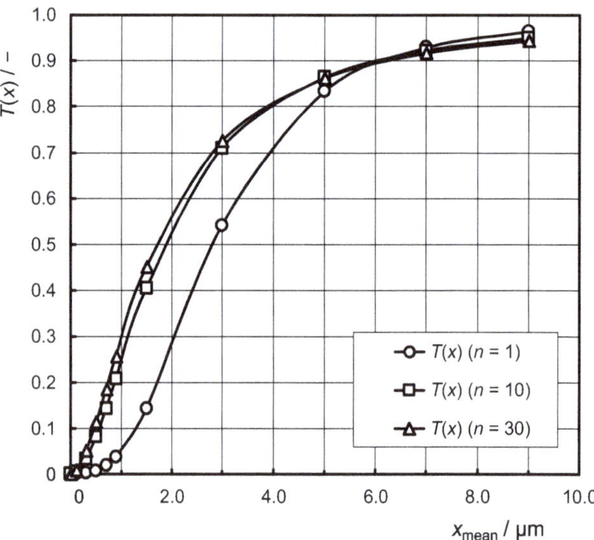

Fig. 6.12: Separation efficiency curve for $n = 1$, $n = 10$ and $n = 30$ cyclones that are arranged parallell.

Discussion: With an increasing number of cyclones n, the particle separation efficiency can be improved from 43.6% for a single cyclone to 63.4% for $n = 10$ cyclones and 67.4% for $n = 30$ cyclones. For all cases, the pressure loss remains the same. Thus, the power required for $n = 30$ cyclones is the same as for a single cyclone. The effect on the separation efficiency of the increase from a single cyclone to $n = 10$ is more distinct than the increase from $n = 10$ to $n = 30$ cyclones. "The more the better" does not work in this case. At least it is an economical optimization problem, whether more cyclones are worth additional cost, and decisions depend on the operation mode. Applications with highly fluctuating gas flow are more likely to benefit from multiple cyclones by avoiding efficiency loss of the single apparatus at lower gas loads.

From the results listed in Tab. 6.7, we see that with a multicyclone system with n cyclones, the linear dimensions of the cyclone decrease with the factor of $n^{0.5}$. This originates from the theory of similarity for scale-down. Thus, the correlation between a single cyclone and a multicyclone can be given as follows:

$$r_{i_{Multi}} = \frac{r_{i_{Single}}}{n^{\frac{1}{2}}}$$

$$\frac{x_{Multi}}{x_{Single}} = \left(\frac{r_{i_{Multi}}}{r_{i_{Single}}}\right)^{\frac{1}{2}} = \left(\frac{1}{n}\right)^{\frac{1}{4}}$$

The use of a multicyclone can increase the separation efficiency to a certain extent. Although with the multicyclone a reduction of fines (PM1) from 11% to 20% for $n = 10$ and to 31% for $n = 30$ could be achieved, the overall amount in the clean gas is still too high to meet the emission limits, as shown in Tab. 6.8.

A series connection of exemplarily two cyclones would slightly increase the separation performance to the double of pressure loss. Therefore, series arrangements are less qualified for industrial use.

6.3.2.5 Summary

Based on the exercises, we were able to confirm that the cyclone is a coarse dust separator with limited separation efficiency. It can be applied to pretreat off-gas or to act as a safety device (e.g., to protect downstream filters), but fails when it comes to the precipitation of fines.

The separation efficiency of cyclones is a function of the vortex finder velocity v_i. With higher v_i the separation efficiency could be increased. Besides, the separation efficiency is a function of the geometric dimension of the cyclone. The ratio of the cross-sectional area of the inlet and the cross-sectional area of the vortex finder Fe/F_i has the greatest impact on the separation efficiency. The parallel arrangement of multiple cyclones can increase the separation performance. Exemplarily, with a multicyclone of 30 cyclones, the separation efficiency for our off-gas was raised from 43.6% to 67.4% at constant pressure drop.

Although the cyclone is not suitable for the treatment of our off-gas, as specified in Tab. 1.1, it should not be underestimated. Its simple construction principle, the absence of installations and particularly the reliable performance at high temperatures and pressure make the cyclone still to a commonly used separation device.

6.3.3 Electrostatic precipitators

The basic operation principle of an ESP is the precipitation of particles by charging them to induce a perpendicular migration of PM to flow direction to the collector. The gas-borne particles pass through an electric field after being initially charged by means of corona discharge. The charged particles migrate through the electric field and deposit on collecting electrodes having the opposite charge.

The precipitation process may be divided into the following process steps:

- Generation of an electric field at the discharge electrode for corona discharge to create charge carriers
- Particle charging with charge carriers
- Migration of the charged particle through the electric field to the collecting electrode
- Removal of the particle from the collecting electrode

A significant progress in industrial off-gas purification was made with the invention of dry electrostatic precipitation in the early twentieth century. The treatment of large gas flow rates with high particle concentration as well as the treatment of hot gas streams or the precipitation of corrosive particles has been possible since then. ESPs are capable of achieving collection efficiencies greater than 99%.

Cottrell's pioneering work, which was significantly supported by Schmidt, laid the foundation for applications of ESPs. Although the precipitation phenomenon was already demonstrated in 1883 by Sir Oliver Lodge [87], it soon became apparent that more electrical power was necessary to ensure continuous precipitation of particles [88]. Thus, Cottrell's success with the implementation of ESPs in the early twentieth century can be attributed to the special attention he paid to high-voltage generation.

ESPs have a very long history and have a significant technical importance in off-gas purification. Hence, a vast number of literatures can be found. Reference is made to White's book [89], which was already published in 1969, and is often denoted as "the" basic work. Recent publications consider the numerical simulation of particle precipitation, corona discharge or flow mechanism, exemplarily mentioned in [90–96]. An interesting recent publication deals with the use of ESPs to reduce PM in urban ambient air [97].

6.3.3.1 Electrical properties of gases and specification of corona discharge

Gases have nearly ideal electrical insulation properties. At low potential difference, it is impossible to monitor any conductivity or current. However, when the potential difference (voltage) is increased above a critical level, a transition from the insulating to the conducting state takes place. This transitional state is called electrical breakdown or gas discharge. Electrical discharge in gases may be of different occurrence. It is distinguished between spark, electric arc, glow and corona discharge. Electrical conductivity in gases must be induced by ionization.

Corona discharge is induced by impact ionization. It is a self-sustaining type of electrical discharge in the gas at ambient pressure, induced between the discharge electrode and the collecting electrode. Free electrons in the gas phase absorb the energy of the electrical field and are accelerated while traveling through it. The subsequent collision of charged electrons with gas molecules results in emission of further electrons by "knocking out" electrons of the struck gas molecules. The energy of the impacting electron must be higher than the ionization energy of the struck molecules. This process creates more electrons and positive gas ions, which in turn can collide again with gas molecules, as schematically and experimentally shown in Fig. 6.13. This mechanism creates the so-called charge carriers, which are responsible for charging particles in the gas stream.

In the ESP, precipitation of PM by corona discharge is performed by creating an electrical field between thin wires (= discharge electrode) and grounded collection plates or tubes. The electrical field is maintained by high voltage applied at the discharge electrode. Under operational geometric conditions for the discharge electrode and the collecting electrode, zones of high field strength are visible by forming regularly shaped (positive corona) or irregularly shaped (negative or alternating corona) bluish plume, as shown in Fig. 6.13. This independent gas discharge, the corona discharge, occurs within the range of the corona onset voltage U_0 and the breakdown voltage U_b. Corona discharge at ambient pressure is limited to a geometric minimum ratio of the distance between the discharge and collecting electrodes b and the radius of the discharge electrode r of at least $b/r = 2.71$. Why this value corresponds to the Euler number has not yet been clarified. Besides the geometric dimensions, the gas composition, the temperature, the pressure, the type of discharge electrodes, the type of discharge current and the polarity of the current–voltage characteristics specify the performance of an ESP. Within the range of corona discharge, a vast number of positively and negatively charged ions is formed. Any particle passing this field will be charged within a few milliseconds due to intensive ion impact. One cubic centimeter of gas in an ESP contains approximately 10^8 negative ions. For comparison, a dense aerosol is a carrier of approximately 10^5 particles per cubic centimeter (100 µg m^{-3}; $x_T = 1$ µm) [98].

Particles distributed in a gas stream may have a basic electrical charge, originating exemplarily from frictional processes. However, this basic electrical charge does not suffice the needs for efficient precipitation; therefore, artificial charging of particles is required.

Particle motion of the charged particle is caused by the Coulomb force. The Coulomb force is proportional to the product of particle charge and the electrical field strength. This results in acceleration of nearly 3,000 times gravitational acceleration for particles with 1 µm of diameter and still 300 times the gravitational acceleration for 10 µm particle diameter [98]. Frictional forces and inertia forces counteract Coulomb force. While inertia may be neglected, the frictional force according to the

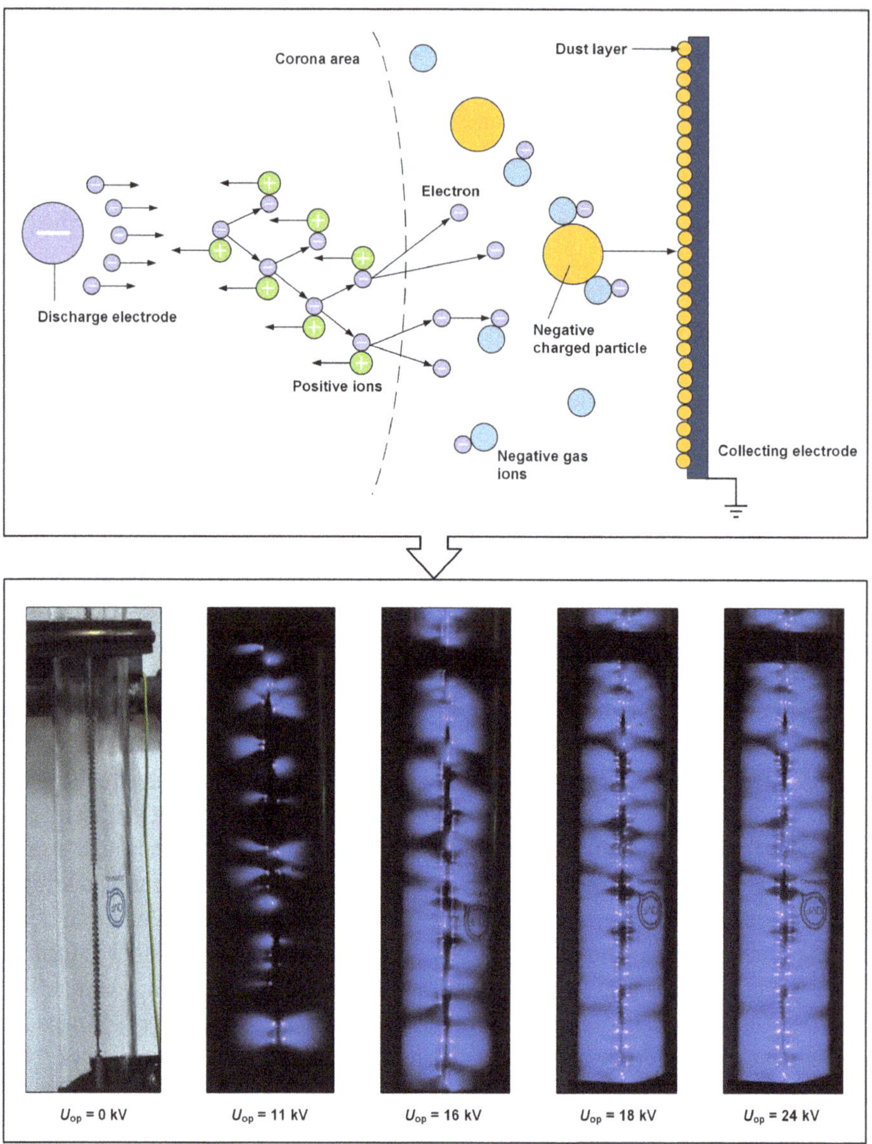

Fig. 6.13: The principle of corona discharge and the experimental demonstration in a lab-scale wet electrostatic precipitator (WESP) at 298 K. The top figure shows schematically the principle of corona discharge, and the five pictures below show different operation states of negative corona discharge, starting with the experimental setup, the onset (11 kV), 16 kV, 18 kV and 24 kV (close to breakthrough) operation voltage.

Stokes law is decisive. Particle migration in ESPs is thus controlled by the Coulomb and Stokes force. The velocity of particles migrating to the collecting electrode is called the rate of particle migration $w(x)$. The rate of particle migration is in a magnitude of about 3–20 cm s^{-1}. Process performance of EPSs is based on the ratio of the distance between the discharge and collecting electrodes b to the rate of particle migration $w(x)$, which needs to be less than the ratio of the length of the ESP L to the gas velocity v_g, as shown in the following equation:

$$\frac{b}{w(x)} < \frac{L}{v_g} \tag{6.24}$$

6.3.3.2 Industrial ESPs and requirements for successful particle precipitation

There are two basic designs of ESPs found in practice, as illustrated in Fig. 6.14. In the single-stage design, charging and particle migration occur simultaneously. In the two-stage design, particle charging is performed in the first step, while the charged particles subsequently enter the precipitation zone with the collecting electrodes. However, the two-stage ESPs have not been established in the industry and are rarely used nowadays.

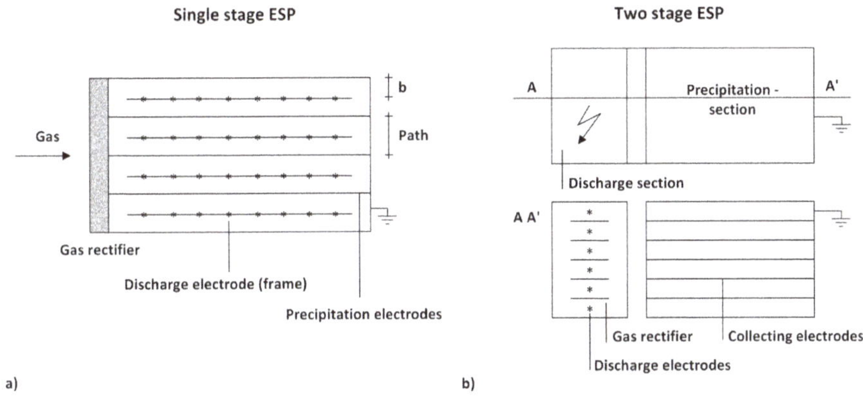

Fig. 6.14: Schematic drawing of (a) a single-stage and (b) a two-stage ESP with main geometric dimensions.

For industrial application, the plate-type ESP and the tube-type ESP are utilized, and both types are designed as a single-stage design. In the past, the plate-type ESP was more common. A schematic drawing of the plate-type ESP is shown in Fig. 6.15. The discharge electrodes are arranged within a conducting frame at a distance of at least $2 \cdot b$, while b is the distance between the discharge electrode and the collecting electrode. The mode of operation of dry ESPs is preferably dry, although the wet operation was demonstrated in application scale [99]. In dry operation, periodic dedusting of the collecting electrode and the discharge electrode is performed mechanically by striking hammers. During this time, the gas flow is not interrupted and can thus absorb precipitated particles. Therefore, the ESP is divided lengthwise into chambers to reduce these emissions, as shown in Fig. 6.19. As a consequence, the applied electrical voltage can also be adjusted lengthwise, which is necessary due to the decreasing particle concentration along the ESP. In the case of wet operation (= complete water saturation of the off-gas), abbreviated as WESP, the precipitated particles drain off at the collecting plates.

The discharge electrode of a tube-type ESP is located in the center of the tubular collecting electrode. For industrial applications, the collecting electrodes are arranged parallelly, as shown in Fig. 6.16. The honeycomb structure is also used frequently in the industry. It is necessary for both ESP designs that the gas velocity is constant; therefore, a gas rectifier at the inlet of plate-type ESPs is inevitable. The gas distribution across the inlet area of the precipitator must be as even as possible.

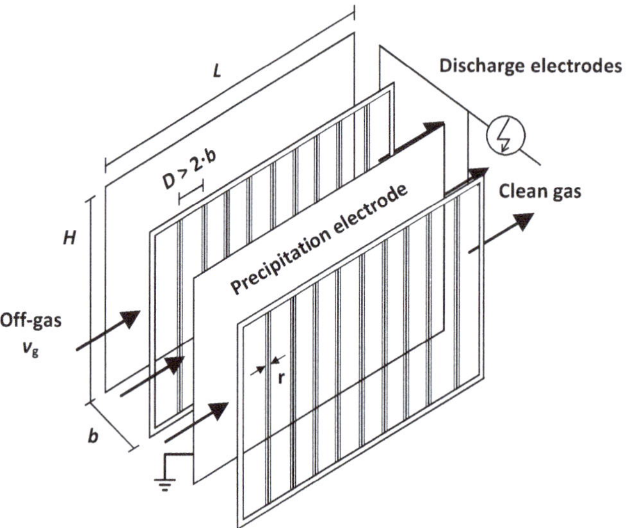

Fig. 6.15: Schematic drawing of a plate-type ESP.

Electrostatic precipitation always has to consider several specific operation constraints affecting the separation performance negatively. In dry ESP, gas temperature, moisture of the gas, and the specific electrical resistance of PM must not differ significantly from (experimentally evaluated) design specification [100]. Operating the ESP out of specification may result in a rapid loss of the separation efficiency. Deviations of the geometric assembly of the discharge and collecting electrodes may also lead to heavy losses of separation performance, and meticulous construction is therefore necessary. The particles to be precipitated must nearly have ideal adhesion properties, since they must be collected on the collecting electrode, on the one hand, and removed by rapping without redistributing in the off-gas, on the other hand. Therefore, the present application focuses on WESP. WESP does not have to consider the specific

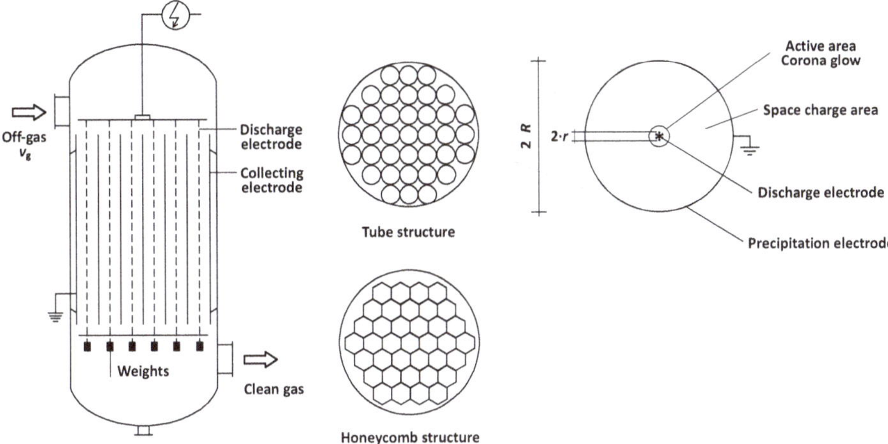

Fig. 6.16: Schematic drawing of a tube-type ESP.

electrical resistance of the particles to be precipitated. The adhesive properties of the dust do not limit precipitation as the collecting electrode is formed by an aqueous film which is renewed continuously.

6.3.3.3 Corona discharge and design basics

Corona discharge of an electrostatic field is initiated by exceeding the onset field intensity E_0. For a given radius of the discharge electrode r and the relative gas density δ, the corona onset field intensity can be estimated according to the empirical equation developed by Peek [101]:

$$E_0\left[\mathrm{kV\,m^{-1}}\right] = 3{,}000 \cdot \delta + 90 \cdot \sqrt{\frac{\delta}{r}} \tag{6.25}$$

With the relative gas density δ, considering the temperature and pressure related to T_0 = 293.15 K and P_0 = 1,013 bar according to

$$\delta = \frac{T_0 \cdot P}{P_0 \cdot T} \tag{6.26}$$

The empirical equation for the corona onset field intensity E_0 does neither consider the gas composition, nor the gas velocity v_g. Experience has led to an empirically obtained correction of E_0 for v_g according to Siebenhofer [99] as follows:

$$E_{0,\,\mathrm{corr}} = E_{0,\,\mathrm{calc}} \cdot \left(3 + 3 \cdot \ln v_g - \frac{0.23}{\ln v_g}\right) \tag{6.25a}$$

Equation (6.25a) considers the change of E_0 for gas velocity within $0.8\ m\ s^{-1} < v_g < 2.5\ m\ s^{-1}$. The dependency of the corona onset field strength E_0 on the gas velocity v_g shows a maximum for plate ESPs in the range of the gas velocity of $1.8\ m\ s^{-1} < v_g < 2\ m\ s^{-1}$.

The calculation algorithm requires the definition of the electrostatic charge unit according to the Giorgi system and thus derived the electrostatic SI unit for 1 A. The following conversion factors can be applied:

$$1\,\mathrm{A} = 9.49 \cdot 10^4\ \mathrm{kg}^{0.5}\ \mathrm{m}^{1.5}\ \mathrm{s}^{-2}$$

The second basis unit results from $1\ W = 1\ J\ s^{-1}$:

$$1\ \mathrm{kV} = 1.05 \cdot 10^{-2}\ \mathrm{kg}^{0.5}\ \mathrm{m}^{0.5}\ \mathrm{s}^{-1}$$

The conversion factors are based on the convention (ESU = electrostatic unit):

$$\begin{aligned} 1\,\mathrm{ESU} &= 1 \cdot \sqrt{\mathrm{dyn}} \cdot \mathrm{cm} \\ 1\,\mathrm{Cb} &= 3 \cdot 10^9 \cdot \mathrm{ESU} \end{aligned} \tag{6.27}$$

After estimation of the corona onset field intensity E_0, the corona onset voltage U_0 of plate- and frame-type precipitators is calculated with eq. (6.28), considering the distance b of the discharge electrode and the collecting electrode [89]:

$$U_0\,[\mathrm{kV}] = E_0 \cdot r \cdot \ln\left(\frac{2 \cdot b}{r \cdot \pi}\right) \tag{6.28}$$

The corona onset voltage U_0 for tube-type precipitators can be calculated with eq. (6.29) with R as the radius of the tube:

$$U_0\,[\mathrm{kV}] = E_0 \cdot r \cdot \ln\left(\frac{R}{r}\right) \tag{6.29}$$

Exemplarily, the curve of the onset field intensity E_0 over the radius of the discharge electrode r ($\delta = 0.6$) is plotted in Fig. 6.17a. Smaller radius of the discharge electrode results in higher onset field intensity. The corona onset voltage U_0 as a function of the discharge electrode radius r is exemplarily shown in Fig. 6.17b for a plate-type ESP for the electrode distance $b = 0.05$ m, $b = 0.10$ m and $b = 0.4$ m. With increasing distance b, the onset voltage increases too.

After transformation into electrostatic SI units, the corona onset voltage U_0 has the unit $kg^{0.5}\ m^{0.5}\ s^{-1}$. The maximum operation voltage U_{max} or the breakdown voltage, respectively, is about $U_{max} = 6–8\ kV\ cm^{-1}$ (mainly depending on the gas composition, temperature and pressure) of the electrode distance between discharge electrode and collecting electrode. The maximum operation voltage U_{max} depends on the parameters which affect the formation of corona discharge in general. It increases with increasing temperature and moisture. The ESP is operated with 90–98% of the maximum operation voltage. The operation voltage U_{op} can thus be calculated with the following equation:

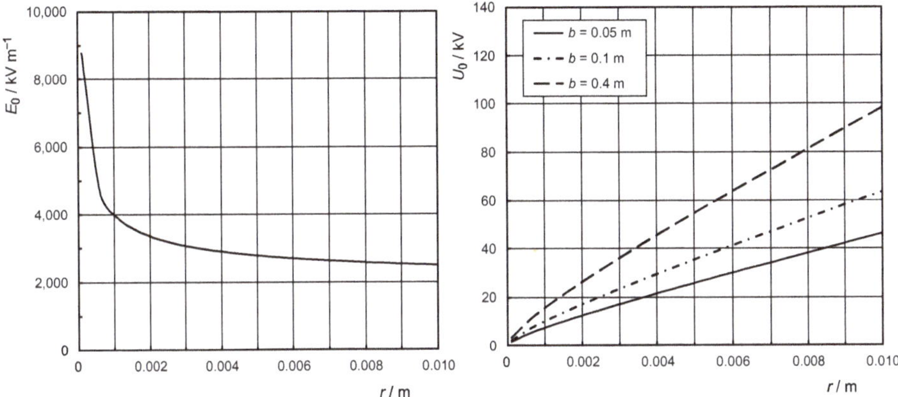

Fig. 6.17: (a) Onset field intensity E_0 as a function of the radius of the discharge electrode for gas density $\delta = 0.6$. (b) Onset voltage for $b = 0.05$, 0.1 and 0.4 m.

$$U_{op} = (0.9 - 0.98) \cdot U_{max} \tag{6.30}$$

Cooperman [102] developed eq. (6.31) for the calculation of the specific corona current I of plate-type ESPs. The specific corona current for tube-type ESPs can be calculated with eq. (6.32). Both equations take the ion mobility K and the operating voltage U_{op} in $kg^{0.5}\ m^{0.5}\ s^{-1}$ into account:

$$I\left[kg^{0.5}\ m^{0.5}\ s^{-2}\right] = \frac{U_{op} \cdot K \cdot \left(U_{op} - U_0\right)}{b^2 \cdot \ln\left(\frac{4 \cdot b}{r \cdot \pi}\right)} \tag{6.31}$$

$$I\left[kg^{0.5}\ m^{0.5}\ s^{-2}\right] = \frac{U_{op} \cdot 2 \cdot K \cdot \left(U_{op} - U_0\right)}{R^2 \cdot \ln\left(\frac{R}{r}\right)} \tag{6.32}$$

The ion mobility K for negative corona discharge is $K = 20\ m^{1.5}\ kg^{-0.5}$, and for positive corona discharge it is $K = 10\ m^{1.5}\ kg^{-0.5}$.

The relation of the specific corona current and the operation voltage U_{op}, usually denoted as the "I/U characteristics", is exemplarily shown in Fig. 6.18 for a plate-type ESP with the electrode distance $b = 0.10$ m (according to Fig. 6.15), the discharge electrode radius $r = 0.005$ m, the breakdown voltage $U_{max} = 80$ kV and the ion mobility of $K = 20\ m^{1.5}\ kg^{-0.5}$. The I/U curve starts at the corona onset voltage U_0. With further increase of the voltage, the specific corona current increases too. When the flashover limit is reached, the corona current increases rapidly while the voltage remains the same (or even breaks down). This point is denoted as the breakdown voltage U_{max} and represents the application limit of an ESP. The I/U characteristics are therefore highly relevant for the application of ESPs. The I/U characteristic is a function of the gas composition, gas temperature, gas pressure, particle concentration in gas, as well as the geometry of the electrode, voltage shape and polarity.

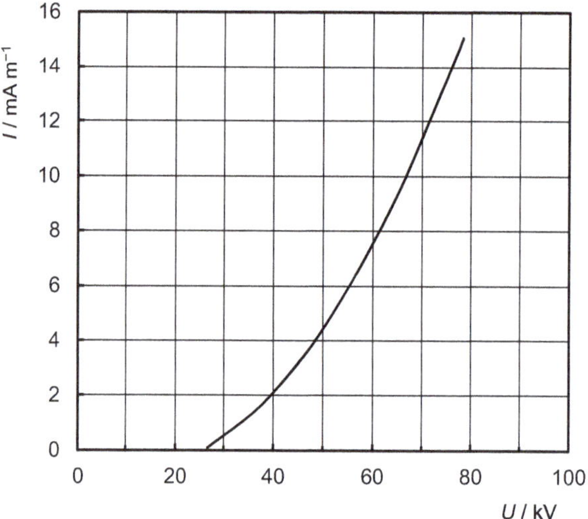

Fig. 6.18: Specific corona current I as a function of the operation voltage U_{op}, exemplarily shown for a plate-type ESP with an electrode distance b = 0.10 m, discharge electrode radius r = 0.005 m, breakdown voltage U_{max} = 80 kV and the ion mobility of K = 20 $m^{1.5}$ $kg^{-0.5}$.

Kalaschnikow [103] investigated the rate of particle migration $w(x)$ and derived a correlation for the precipitation field strength E_p. According to eq. (6.33), there is a proportionality between the precipitation field strength and the specific corona current. White [104] confirmed the applicability of this correlation:

$$E_p \left[\mathrm{kg}^{0.5}\ \mathrm{m}^{-0.5}\ \mathrm{s}^{-1}\right] = \left(2 \cdot \frac{I}{K}\right)^{\frac{1}{2}} \tag{6.33}$$

In addition to the precipitation field strength E_P, the discharge field strength E_a needs to be determined for the calculation of the rate of particle migration. Although not linear it is well estimated from the operation voltage U_{op} and the distance between the discharge electrode and collecting plate b, as shown in eq. (6.34). Finally, the particle migration rate $w(x)$, as shown in eq. (6.35), can be determined which is needed for latter calculations of the separation efficiency:

$$E_a \left[\mathrm{kg}^{0.5}\ \mathrm{m}^{-0.5}\ \mathrm{s}^{-1}\right] = \frac{U_{op}}{b} \tag{6.34}$$

$$w(x) \left[\mathrm{m\,s}^{-1}\right] = \frac{E_a \cdot E_p \cdot x}{4 \cdot \pi \cdot \eta} \tag{6.35}$$

For the design of plate-type ESPs, it is important to consider a minimum distance between the discharge electrodes. The minimum distance should be twice the distance of neighboring collector plates ($D_{min} = 2 \cdot b$) to provide maximum corona current, as illustrated in Fig. 6.15. If the distance between the discharge electrodes is too small, the corona current will be suppressed. For the minimum distance D_{min}, the corona current is already suppressed by 20% of the calculated value. Interaction will be negligible for conditions four times the distance of neighboring discharge electrodes.

6.3.3.4 Separation efficiency

The calculation of the separation efficiency is exemplarily shown for tube-type ESPs with the radius R. The number of particles migrating to the collector electrode within the differential length dl at a given rate of particle migration $w(x)$ equals the number of particles removed from the gas stream with the gas velocity v_g, as shown in the following equation:

$$2 \cdot R \cdot \pi \cdot w(x) \cdot c \cdot \mathrm{d}l = R^2 \cdot \pi \cdot v_g \cdot (c - (c + \mathrm{d}c)) \tag{6.36}$$

After separation of the variables, eq. (6.36) is rearranged to eq. (6.36a):

$$-\mathrm{d}c = \frac{2 \cdot w(x) \cdot c \cdot \mathrm{d}l}{R \cdot v_g} \tag{6.36a}$$

Integration from the inlet of the tube, indicated as indices E, to the length L, leads to the particle concentration c_L at the length L of

$$c_L = c_E \cdot e^{\frac{-2 \cdot L \cdot w(x)}{R \cdot v_g}} \tag{6.37}$$

The equation for the separation efficiency $T(x)$ according to Deutsch [105] can be calculated as follows:

$$T(x) = 1 - \frac{c(x)_L}{c(x)_E} = 1 - \exp(-a \cdot w(x)) \tag{6.38}$$

where a is the specific separation area and thus $a = A/F_{V,g}$ in s m^{-1}.

For plate-type ESPs, the separation efficiency $T(x)$ derived from the Deutsch equation is given as follows:

$$T(x) = 1 - \frac{c(x)_L}{c(x)_E} = 1 - e^{\frac{-L \cdot w(x)}{b \cdot v_g}} \tag{6.39}$$

and for tube-type precipitators the separation efficiency $T(x)$ is

$$T(x) = 1 - \frac{c(x)_L}{c(x)_E} = 1 - e^{\frac{-2 \cdot L \cdot w(x)}{R \cdot v_g}} \tag{6.39a}$$

Summation of the grade separation efficiency finally leads to the overall degree of separation E:

$$E = \int T(x) \cdot q(x) \cdot \mathrm{d}x$$

6.3.3.5 Theoretical energy consumption

According to the Stokes law, the friction force F for spherical particles is described by

$$F\left[\mathrm{kg\,m\,s^{-2}}\right] = 3 \cdot \pi \cdot \eta \cdot x \cdot w(x) \tag{6.40}$$

With the friction force, the energy consumption W can be expressed as

$$W = F \cdot s \tag{6.41}$$

with s being the distance of the particles to the collecting electrode. The distance b can be used instead of s. For given off-gas concentration $c_{\text{Off-gas}}$ in kg $\mathrm{m^{-3}}_{\text{STP,dry}}$, the number of particles N per cubic meter volume is then

$$N = \frac{6 \cdot c_{\text{Off-gas}}}{\pi \cdot \rho_p \cdot x^3} \tag{6.42}$$

resulting in the theoretical specific energy consumption $E_{\text{theo,spec}}$:

$$E_{\text{theo, spec}}\left[\mathrm{Ws\,m^{-3}}\right] = W \cdot N \tag{6.43}$$

Exercise 6.8: Design and performance of a plate-type ESP

For our off-gas specification summarized in Tab. 6.1, we have to design a plate-type ESP and determine the separation efficiency. If you hold on to the following suggested design guideline, the design of the ESP will be an ease. This guideline may be used for estimating ESP dimensions and separation performance, and for optimization and performance control.

Solver:

The ESP may be constructed as schematically shown in Fig. 6.19. (Comment: The number of sections, hoppers, etc. can of course vary.) There are four degrees of freedom that we need to define:

- the radius of the discharge electrode r,
- the distance between the discharge electrode and the collecting electrode b,
- the specific collection area (SCA) and the precipitation area A_{Prec}, respectively, and
- the operation mode of negative or positive corona.

However, let us go through the design guideline step by step.

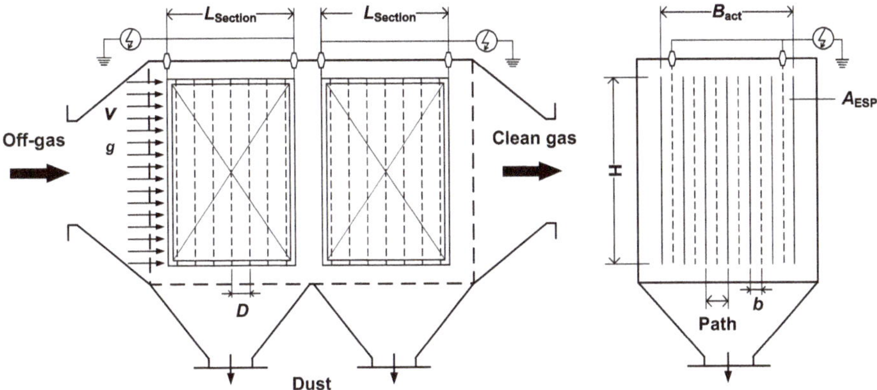

Fig. 6.19: Plate-type ESP as constructed for industrial purposes.

First step:
We start with the specification of the radius r of the discharge electrode. The recommended value can be up to 20 mm. From Fig. 6.17b, we see that the corona onset voltage U_0 is lower at a smaller radius of the discharge electrode. Let us fix the radius of the discharge electrode to be r = 0.005 m.

Second step:
Now we fix the distance between the discharge electrode and the collecting electrode, which we set to b = 0.05 m for our separation task. The distance b should not exceed 400 mm, since greater distance will require high onset voltage, as shown in Fig. 6.17b.

Third step:
With the defined radius of the discharge electrode r, the distance between the electrodes b and the relative gas density δ, we can calculate the corona onset field strength E_0 according to eq. (6.25):

$$\delta = \frac{293.15 \cdot 981}{1{,}013 \cdot 473.15} = 0.6$$

$$E_0 = 3{,}000 \cdot 0.6 + 90 \cdot \sqrt{\frac{0.6}{0.005}} = 2{,}786 \ \text{kV m}^{-1}$$

For the conversion into electromechanical SI units, we need to multiply the result with the conversion factor $1.05 \cdot 10^{-2}$:

$$E_0 = 2{,}786 \cdot 1.05 \cdot 10^{-2} = 29.25 \ \text{kg}^{0.5} \ \text{m}^{-0.5} \ \text{s}^{-1}$$

Fourth step:
With the corona onset field intensity E_0, we calculate the corona onset voltage U_0, according to eq. (6.28):

$$U_0 = 2{,}786 \cdot 0.005 \cdot \ln\left(\frac{2 \cdot 0.05}{0.005 \cdot 3.14}\right) = 25.79 \ \text{kV}$$

$$U_0 = 25.79 \cdot 1.05 \cdot 10^{-2} = 0.271 \ \text{kg}^{0.5} \ \text{m}^{0.5} \ \text{s}^{-1}$$

Fifth step:
The maximum operation voltage is about U_{max} = 6–8 kV cm^{-1} of the electrode distance between discharge electrode and collecting electrode. For the calculation of the maximum operation voltage, we decide for 8 kV cm^{-1}. U_{max} is then

$$U_{max} = 0.05 \cdot 800 = 40 \ \text{kV}$$

The operation voltage for 98% of U_{max} is

$$U_{op} = 0.98 \cdot 40 = 39.2 \ \text{kV}$$

For SI units, we get U_{op} = 0.412 kg$^{0.5}$ m$^{0.5}$ s^{-1}.

Sixth step:
The specific corona current I for negative corona (K = 20 m$^{1.5}$ kg$^{-0.5}$), according to eq. (6.31), is

$$I = \frac{0.412 \cdot 20 \cdot (0.412 - 0.271)}{0.05^2 \cdot \ln\left(\frac{4 \cdot 0.05}{0.005 \cdot \pi}\right)} = 182.17 \ \text{kg}^{0.5} \ \text{m}^{0.5} \ \text{s}^{-2}$$

and according to the Giorgi system, I is

$$\mathrm{I} = \frac{182.17}{9.49 \cdot 10^4} = 0.00192 \ \text{A m}^{-1}$$

For industrial particle precipitation, ESPs are usually operated with negative corona onset voltage, although the formation of ozone can occur in the case of oxygenic off-gas. Therefore, it is crucial that the ESP is designed to ensure that the ozone level does not exceed emission limits. Since the ozone generation rate is proportional to the operating voltage, and so is the separation efficiency, a trade-off needs to be considered. For our design exercise we will neglect the generation of ozone, but in practice (or for detailed design) keep this fact in mind! Research work of ozone generation in ESP can exemplarily be found in [106–110].

Seventh step:
The precipitation field strength E_p, again for negative corona, according to eq. (6.33) is

$$E_p = \left(2 \cdot \frac{182.17}{20}\right)^{\frac{1}{2}} = 4.27 \ \text{kg}^{0.5} \ \text{m}^{-0.5} \ \text{s}^{-1}$$

and the discharge field strength E_a according to eq. (6.34) is

$$E_a = \frac{0.412}{0.05} = 8.23 \ \text{kg}^{0.5} \ \text{m}^{-0.5} \ \text{s}^{-1}$$

Eighth step:
Now we can determine the particle migration velocity $w(x)$ with eq. (6.35) for every particle size x of our PSD. The results are listed in Tab. 6.9.

Ninth step:
For determination of the separation efficiency, we first need to specify the gas velocity of our off-gas and fix (assume) the precipitation area A_{Prec} for the derivation of the geometric dimension of the ESP elements, as done in the tenth step. The typical gas velocities through the body of the ESP are in the magnitude of 0.6–2.4 m s^{-1}. For our design purposes, let us assume a gas velocity of v_g = 1.3 m s^{-1}.

The typical range for the SCA in m^2 per 1,000 m^3 h^{-1} is between the magnitude of 11 and 45 (precipitation of dust with high resistivity needs higher SCA). Let us decide for an SCA of 40, resulting in the precipitation area A_{Prec} of 400 m^2.

If you rearrange the Deutsch equation and define the separation efficiency, you can also estimate the separation area. If we exemplarily aim for a particle concentration of our clean gas of $c_{Clean\ gas}$ = 10 mg m$^{-3}{}_{STP,\ dry}$ we need a separation efficiency of 99.7% (remember eq. (6.2)). With the rate of particle migration for particles of exemplarily 0.5 µm particle size of $w(x = 0.5\mu m)$ = 0.078 m s^{-1}, the corresponding precipitation area A_{Prec} is

$$A_{Prec} = -\frac{F_{V,g}}{w(x)} \cdot \ln(1 - T(x)) = -\frac{\frac{20{,}111}{3{,}600}}{0.078} \cdot \ln(1 - 0.997) = 408.5\ \text{m}^2$$

Tenth step: The geometric dimensions, the separation efficiency and the power demand
With the specified gas velocity and the volumetric flow rate, we can calculate the cross-sectional area perpendicular to the gas flow direction A_{ESP}, as shown in Fig. 6.19:

$$A_{ESP} = \frac{F_{V,g}}{v_g} = \frac{\frac{20{,}111}{3{,}600}}{1.5} = 4.3\ \text{m}^2$$

Next, we define the height of the electrodes. We assume, for example, H = 2.5 m, which results in the total width B of 1.72 m.

Comment: Depending on the operation conditions, the height of the electrodes can be significantly larger. Recommended values for the height of the electrodes are in a magnitude of 5–12 m.

We also need to specify the distance between the single discharge electrodes D. Remember, it must be at least two times the distance of the discharge electrode and the collecting electrode. In our case, we decide for D to be three times the distance of the discharge electrode and the collecting electrode:

$$D = 3 \cdot b = 3 \cdot 0.05 = 0.125\ \text{m}$$

The number of paths is

$$n_{Path} = \frac{B}{2 \cdot b} = \frac{1.72}{2 \cdot 0.05} = 17$$

With the number of paths, we get the actual sizing width of B_{act} = 1.7 m and the resulting gas velocity of v_g = 1.31 m s^{-1}, respectively. Now we can calculate the separation efficiency according to eq. (6.39). The results are listed in Tab. 6.9. The total separation efficiency is E = 99.75% and the particle concentration of the clean gas $c_{Clean\ gas}$ = 7.37 mg m^{-3}. Congratulations, we compare well with the emission limits (theoretically!).

Next, we need to determine the power requirement. Therefore, we need further geometric dimensions of the precipitation device. The effective length of the separation surface is derived from the precipitation area A_{Prec} and the number of paths:

$$L = \frac{A_{Prec}}{H \cdot n_{Path} \cdot 2} = \frac{400}{2.5 \cdot 17 \cdot 2} = 4.7\,\text{m}$$

The aspect ratio is the ratio of the effective length L to the height of the collector surface H (= height of electrodes). For high-efficiency ESPs, the aspect ratio is usually greater than 1. Aspect ratios in a magnitude of 1.3–1.5 are common and can sometimes be up to the magnitude of 2. The aspect ratio should be high enough to allow the collected particles to settle in the hopper before they are carried out by the gas flow.

The aspect ratio of our designed ESP is

$$\frac{L}{H} = \frac{4.7}{2.5} = 1.88$$

The number of discharge electrodes is

$$n_{Electrode} = \frac{L}{D} \cdot n_{Path} = \frac{4.7}{0.15} \cdot 17 = 533$$

It follows the total corona current I_{tot}

$$I_{tot} = n_{Electrode} \cdot H[\text{m}] \cdot I\left[\text{A}\,\text{m}^{-1}\right] = 533 \cdot 2.5 \cdot 0.00192 = 2.56\,\text{A}$$

and the total energy consumption E_{tot}

$$E_{tot} = I_{tot} \cdot U_{op} = 2.56 \cdot 39.2 = 100.5\,\text{kWh}$$

The specific energy consumption E_{sp} is then

$$E_{sp} = I_{tot} \cdot U_{op} \cdot \left[\frac{1{,}000}{10{,}000}\right] = 2.56 \cdot 39.2 \cdot \frac{1{,}000}{10{,}000} = 10.5\,\text{kWh}\ \ 1{,}000\ \text{m}^{-3}$$

The theoretical specific energy demand can be calculated according to eqs. (6.40)–(6.43) and is in a magnitude of $E_{theo,spec}$ = 2.1 Ws m^{-3} for our application. For reasons of comparison, we convert this value into $E_{theo,spec}$ = 0.58 Wh 1,000 m^{-3}. Confusing, isn't it? Actually not. The theoretical energy demand for particle precipitation in the ESP will always be just a fraction of the specific energy needed to operate the ESP. The number you are therefore interested in is

$$E_{sp} = 10.5\ \ \text{kWh}\ \ 1{,}000\ \text{m}^{-3}$$

That still does not explain the huge gap between theoretical and actual specific energy demand. Equation (6.35) just tells us that we need an appropriate discharge field intensity E_a and an appropriate precipitation field intensity E_p (deduced from the specific corona current) to suffice the minimum rate of migration, as needed in eq. (6.38) (Deutsch equation). Neither eq. (6.35) nor eq. (6.38) tell the particle that it must not run out of charge on his way along the ESP. The particle must frequently be updated with charge after the distance D, provided by the discharge electrodes arranged in series in the gas path.

Eleventh step: Construction of the equipment

Finally, the ESP needs to be constructed. Although our job is done, let us have a look on some recommendations for the geometric design of the ESP. As schematically shown in Fig. 6.19, the inlet and the outlet should be designed with expansion zones. This is because the gas flow across the precipitator needs to be homogeneously distributed to ensure good separation performance. The expansion inlet should contain perforated diffusor plates. The distance of $L_{Expansion}$ should be at least as long as the width of the inlet B_{Inlet} [111]. The inlet ducts should be straight or equipped with gas-rectifying vanes to ensure stratified gas flow. Pay attention to the aspect ratio, which should not be less than 1, or higher than 2. The precipitation area is commonly divided into sections. For our ESP we can exemplarily divide the precipitation length of 4.7 m into two sections, one section is $L_{Section}$ = 2.35 m, respectively.

Discussion: The calculation results are summarized in Tab. 6.9. The grade separation efficiency $T(x)$ and the PSD $Q_3(x)$ of the off-gas and the clean gas are pictured in Fig. 6.20a. Figure 6.20b depicts the probability distribution function Δq_3 of the off-gas and the clean gas, respectively.

The calculated overall degree of separation of our plate-type ESP is E = 99.75%. As we can see from Tab. 6.9, particles may be separated in a magnitude of 29.92 kg h^{-1}. The particle mass flow within the off-gas is reduced from 30 to 0.074 kg h^{-1}. The PM1 particle concentration (the sum of particle classes ≤1 µm) as well as the total particle concentration in the clean gas is 7.37 mg m$^{-3}_{STP}$, indicating that every particle with the particle size >1 µm may be separated. With the achieved ESP performance, we would meet the required emission limits.

Tab. 6.9: Calculation results of plate-type ESP design for the off-gas specification listed in Tab. 6.1.

PSD (µm)	x_{mean} (µm)	$w(x)$ (m s^{-1})	$T(x)$	$\dot{m}_{Prec}$ (kg h^{-1})	$\dot{m}_{Passed}$ (kg h^{-1})	Q_3 (clean gas)	$c_{Off\text{-}gas}$ (mg m^{-3})	c_{Prec} (mg m^{-3})	$c_{Clean\ gas}$ (mg m^{-3})
0.0	0.0	0.00	0.00	0.00	0.00	0.00	0.00	0.00	0.00
0.2	0.1	0.01	0.63	0.08	0.05	0.66	13.30	8.42	4.89
0.4	0.3	0.04	0.95	0.38	0.02	0.93	39.56	37.60	1.96
0.6	0.5	0.07	0.99	0.64	0.00	0.99	64.77	64.34	0.43
0.8	0.7	0.10	1.00	0.88	0.00	1.00	88.29	88.21	0.08
1.0	0.9	0.13	1.00	1.10	0.00	1.00	109.56	109.54	0.01
2.0	1.5	0.21	1.00	7.61	0.00	1.00	760.98	760.98	0.00
4.0	3.0	0.42	1.00	14.17	0.00	1.00	1,416.50	1,416.50	0.00
6.0	5.0	0.70	1.00	4.52	0.00	1.00	452.09	452.09	0.00
8.0	7.0	0.98	1.00	0.52	0.00	1.00	52.50	52.50	0.00
10.0	9.0	1.26	1.00	0.02	0.00	1.00	2.40	2.40	0.00
				Σ 29.93	Σ 0.07		Σ 2,999.96,0	Σ 2,992.58	Σ 7.37

The separation curve, as shown in Fig. 6.20a, indicates the good separation performance of the ESP. Even particles with 0.1 µm particle size may be separated with an efficiency of about 60%. The cutoff diameter is in the magnitude of x_T = 0.07 µm. The maximum values of the density distribution function, as pictured in Fig. 6.20b, shift from 2 µm from the off-gas to 0.1 µm of the clean gas. The respective q_3 value indicates that the relative amount of fines in our clean gas increased from 0.25 to 3.31.

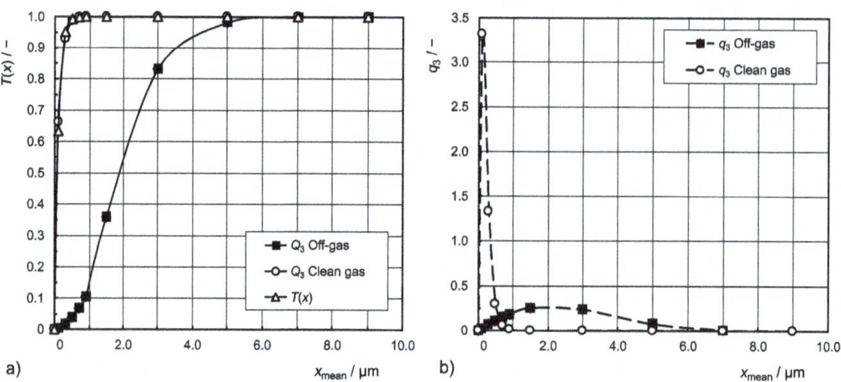

Fig. 6.20: (a) Separation efficiency $T(x)$ and the cumulative distribution function Q_3 of the off-gas and the clean gas. (b) Density distribution function q_3 of the off-gas and the clean gas. The results correspond to Exercise 6.8.

The total energy consumption needed for the separation is in a magnitude of E_{tot} = 120 kWh which is reasonable. Usually, ESPs require a relatively large amount of space. With the length of 4.7 m, the width of 1.7 m and the height of 2.5 m, the dimension of the ESP as designed based on our off-gas specification is pretty small. From an industrial point of view, we might call it a micro-ESP. But have in mind that we did this exercise with just 10,000 $m^3_{STP,dry}\ h^{-1}$. You will for sure get very different numbers for gas flow rates of several 100,000 $m^3_{STP,dry}\ h^{-1}$. (Comment: In industry, ESP dimensions can be up to 15 m height and 40 m length.)

Exercise 6.9: Design of a tube-type ESP

Now we are going to determine the size of a tube-type ESP to obtain the same separation performance (= 99.75%) as obtained for the plate-type ESP. We must consider that the tube-type ESP is operated at complete water saturation. Therefore, temperature and gas flow rate are different. For comparability, we decide for a tube radius of R = 0.05 m, and we adjust the discharge electrode radius to obtain nearly the same corona onset voltage. For the design of a tube-type ESP, the calculation steps are nearly the same as for the plate-type ESP.

Solver:

At complete water saturation, the off-gas temperature is Θ = 56 °C or T = 329.15 K. The relative gas density is then δ = 0.86. To obtain the same corona onset voltage as in Exercise 6.8 we decide for a discharge electrode radius of r = 0.0016 m, ending up in a corona onset field intensity of E_0 = 4,673 kV m^{-1} and a corona onset voltage of U_0 = 25.74 kV.

The corona onset voltage U_0 for tube-type ESP can be calculated according to eq. (6.29) and it is

$$U_0 = E_0 \cdot r \cdot \ln\left(\frac{R}{r}\right) = 4{,}673 \cdot 0.0016 \cdot \ln\left(\frac{0.05}{0.0016}\right) = 25.74\ \text{kV}$$

Converted into SI units we get

$$U_0 = 25.74 \cdot 1.05 \cdot 10^{-2} = 0.27\ \text{kg}^{0.5}\ \text{m}^{0.5}\ \text{s}^{-1}$$

The maximum operation voltage is about U_{max} = 6–8 kV cm^{-1} of the electrode distance between discharge electrode and collecting electrode. For the calculation, we choose the magnitude of 8 kV cm^{-1}. U_{max} is then

$$U_{max} = 0.05 \cdot 800 = 40\ \text{kV}$$

The operation voltage for 98% of U_{max} is

$$U_{op} = 0.98 \cdot 40 = 39.2\ \text{kV}$$

Converted into SI units, we get U_{op} = 0.412 $kg^{0.5}$ $m^{0.5}$ s^{-1}.

The specific corona current for the tube-type ESP, as calculated with eq. (6.32) (negative corona K = 20 $m^{1.5}$ $kg^{-0.5}$), is

$$I = \frac{U_{op} \cdot 2 \cdot K \cdot (U_{op} - U_0)}{R^2 \cdot \ln\left(\frac{R}{r}\right)} = \frac{0.412 \cdot 2 \cdot 20 \cdot (0.412 - 0.27)}{0.05^2 \cdot \ln\left(\frac{0.05}{0.0016}\right)} = 270.5\ \text{kg}^{0.5}\ \text{m}^{0.5}\ \text{s}^{-2}$$

$$I = \frac{213.9}{9.49 \cdot 10^4} = 0.0029\ \text{A m}^{-1}$$

The precipitation field intensity E_p is

$$E_p = \left(2 \cdot \frac{270.5}{20}\right)^{\frac{1}{2}} = 5.2\ \text{kg}^{0.5}\ \text{m}^{-0.5}\ \text{s}^{-1}$$

and the discharge field strength E_a according to eq. (6.34) is

$$E_a = \frac{0.412}{0.05} = 8.232\ \text{kg}^{0.5}\ \text{m}^{-0.5}\ \text{s}^{-1}$$

The calculation of the separation efficiency for tube-type ESP is given in eq. (6.39a). For comparability we decide for the same filter height of H = 2.5 m but may apply a higher gas velocity of, for example, v_g = 1.5 m s^{-1}.

The cross-sectional area of a single-tube CSA_{Tube} is given with

$$CSA_{Tube} = R^2 \cdot \pi = 0.05^2 \cdot 3.14 = 0.00785\ \text{m}^2$$

With the gas velocity v_g = 1.5 m s^{-1}, the number of tubes results from

$$n_{Tubes} = \frac{A_{ESP}}{CSA_{Tube}} = \frac{\frac{F_{V,g}}{v_g}}{CSA_{Tube}} = \frac{\frac{14{,}980}{1.5 \cdot 3{,}600}}{0.00785} = 353$$

The precipitation area is

$$A_{Prec} = 2 \cdot R \cdot \pi \cdot H \cdot n_{Tubes} = 2 \cdot 0.05 \cdot 3.14 \cdot 2.5 \cdot 353 = 277.3\ \text{m}^2$$

Now it is on you, whether you arrange the tubes as tube bundle or rectangular. This decision mainly depends on the local conditions. If you arrange them in dense pack design, you will need a rectangular cross-sectional area of 2.2 m · 1.6 m.

Independently from the design configuration, the total corona current is

$$I_{tot} = n_{Tubes} \cdot H[m] \cdot I\left[A\,m^{-1}\right] = 353 \cdot 2.5 \cdot 0.0029 = 2.56 \ \ A$$

resulting in the total energy consumption of

$$E_{tot} = I_{tot} \cdot U_{op} = 2.56 \cdot 39.2 = 100 \ \ kWh$$

and 9.9 kWh per 1,000 m^3, respectively.

! Discussion: For tube-type ESP, a smaller precipitation area A_{Prec} is needed to achieve the same precipitation performance as provided by the plate-type ESP. Space requirement is therefore less. The total energy consumption for the tube-type ESP is slightly lower compared to the plate-type ESP.

6.3.3.6 Summary

Although the ESP has been used for the precipitation of particles for almost a century, the basic design to meet a specific separation task still cannot be determined within the first run. Besides the acquired basics, a great deal of practical experience is needed in the ESP design. The intention behind our discussion was to provide you with the basic tools you need for discussing the performance of the installed equipment. Although the application of numerical simulations (CFD) may assist in understanding the phenomena and laws of precipitation physics, there is no computer program for ESP design available.

However, the ESP is a high-efficiency separation device reaching an overall precipitation efficiency of up to 99.9% over a wide range of particle sizes (~0.05–5 µm). As we were able to demonstrate with the exercises, the final dust concentration of our exhaust gas may theoretically be reduced to $c_{Clean\ gas}$ = 7.4 mg m^{-3} of PM1, in very compliance with the emission limits. But, as mentioned, theoretically operators fear the ESP because you must always keep an eye on it. When it works well, it will meet the target data. When you slip into an electrical breakdown, precipitation performance immediately breaks down too and you are left with the precipitation performance of a settling chamber.

Compared to the plate-type ESP, the size of a tube-type ESP turns out smaller for the same separation performance (compare the Deutsch equation for plate-type and for tube-type ESPs). The total energy consumption for the tube-type ESP is slightly less compared to the plate-type ESP. Opposite to the dry plate-type ESP, the tube-type ESP is operated in the wet mode. The specific electrical resistance of PM is, therefore, not an issue. It is also capable of separating liquid aerosols. While dry plate-type ESPs are substituted whenever possible, wet-type ESPs may contribute to off-gas purification in future.

6.3.4 Scrubbers

Scrubbers cover the precipitation of solid, liquid and gaseous pollutants. Pollutants in off-gas streams are removed with a scrubbing liquid, typically water, and are subsequently separated together with the liquid. In the nineteenth century, the first patent for a packed tower absorption scrubber occurred in industry. Almost nearly 100 years later, a separation efficiency of 98% for SO_2 in a flue gas scrubber was obtained. The first scrubber controlling gaseous and PM was patented in the year 1901 [112].

The operation principle of scrubbers is based on the interaction of the pollutant with the scrubbing liquid by inertial impaction, direct interception, and Brownian diffusion and/or absorption. Scrubbers are typically categorized into particulate collectors and gaseous emission control devices. A scrubber always consists of two main sections. In the first section, the pollutants are moved toward the scrubbing liquid to be finally captured. In the second section, the pollutants are separated with the liquid carrier from the gas stream. Accordingly, droplet separation in scrubbers has also to be designed accurately. For successful purification of off-gas streams, the complete process, including the consecutive wastewater treatment, needs to be considered. Due to the ability of precipitating particles and gaseous pollutants, the design of scrubbers must also consider the hydraulic boundaries for successful gas–liquid mass transfer.

The mechanism of separation depends on the type of species to be separated. While hydraulic mechanisms such as impaction, interception and diffusion are used for the separation of solid and liquid pollutants, chemical–physical mechanisms for absorption are responsible for the separation of gaseous pollutants. Although most scrubbers simultaneously operate as particulate collectors and absorbers, emission control of gaseous constituents by absorption will not be discussed in this chapter.

6.3.4.1 Mechanism of dust precipitation by scrubbing

To separate particles from a gas stream, the particles need to get trapped in the scrubbing liquid by wetting. A large liquid surface area is therefore needed, which can be provided by either dispersing the liquid phase in the gaseous phase (droplets), with a fine water film (created by a packing material) or by dispersing the gas into a continuous liquid phase (bubbles).

The precipitation basically consists of the following process steps [86]:

- Distribution/dispersion of the scrubbing liquid in the gaseous phase (or dispersion of the gaseous phase in the scrubbing liquid)
- Interaction of the particle and the scrubbing liquid
- Capture of the particle in the scrubbing liquid (or on the droplet surface)

Standard application of particle precipitation by scrubbing is carried out with droplets. In nature, it is demonstrated by the purifying properties of rain. The scrubbing liquid is dispersed in the gaseous phase by mechanical energy dissipation, e.g., via nozzles. The particles in the gas stream get captured by the droplets which are subsequently separated from the gas stream in a droplet separator. Thus, scrubber devices always consist of a dust separation and droplet separation section. The separation of the particles from the scrubbing liquid, e.g., by filtration or sedimentation, is commonly performed in a separate equipment. Performance and separation efficiencies are determined by the relative velocity of the scrubbing liquid and the gas phase. In general, the separation efficiency increases with increasing relative velocity, increasing particle size, increasing particle concentration and high-volume flow rate of the scrubbing liquid.

As already mentioned, the main mechanisms for particle precipitation in scrubbers are inertial impaction, direct interception and diffusion, as schematically shown in Fig. 6.21. (Comment: Electrical forces can also lead to precipitation but play a minor role. Electrical forces are rather important in the separation with fabric filters.) For convenience, the mechanism of dust precipitation will be explained for droplets although same mechanisms are applicable for fibers or bubbles.

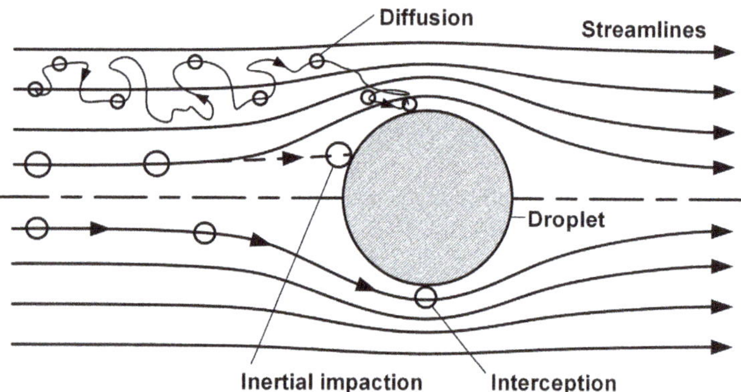

Fig. 6.21: Mechanism of dust precipitation: inertial impaction, interception and convective Brownian diffusion.

6.3.4.1.1 Inertial impaction

The most prevalent means of particulate removal is impaction. When off-gas moves toward a droplet, it flows along streamlines around the droplet, as schematically pictured in Fig. 6.21. In turn, the particles in the gas are accelerated, maintain their forward trajectory and impact the liquid droplet when the inertial force is sufficient. The surface of the droplet is penetrated and thus the particle gets enclosed and trapped by the liquid [112]. Particles larger than 5–10 μm in diameter are characterized with sufficient inertial force and are thus generally collected by impaction [113, 114]. In order to achieve high separation efficiency via impaction, high relative velocity between the particle and the liquid–gas interface is necessary [115]. High relative velocities in turn lead to high pressure drop. Inertial forces become significantly small when the particle size decreases, necessitating increase of the energy input with decreasing particle size to reach appropriate separation efficiency, which may in turn lead to uneconomic operation of the scrubber.

6.3.4.1.2 Direct interception

Particles with particle diameter of 0.1–1 µm are rarely captured by impaction due to insufficient inertia forces but can be captured by interception [116]. Due to their lower mass, the particles are carried by the gas streamline sufficiently close to the surface of the collecting droplet and will touch it when its center moves toward a distance of half of the droplet diameter $d_{dr}/2$ of the droplet [112]. The mechanism of interception is related to the droplet density. With increasing droplet density, the chance for interception increases. Creating high-density sprays of fine droplets thus increases precipitation by interception.

6.3.4.1.3 Diffusion

Diffusion is responsible for the separation of very small particles with less than 0.1 µm diameter. Diffusion is a result of Brownian motion. Random intersection of particles with fast-moving gas molecules causes the particles to deflect slightly from their original streamline [116]. Deflection may again bring the particles into contact with the liquid droplet. To make use of the diffusion mechanism, proper residence time between the gas and liquid droplets in the separation chamber is needed. High temperature difference between the gas stream and the scrubbing liquid increases the diffusion mechanism.

6.3.4.1.4 Other mechanisms: diffusiophoresis, thermophoresis and condensation

Intermolecular diffusion, also called diffusiophoresis, is a very weak force, just acting on extremely small particles with less than 0.05 µm in particle diameter. Diffusiophoresis occurs due to isolated and rapidly changing evaporation or condensation rate within the scrubbing liquid [112]. This force can be used to separate nanoparticles on applications wherein a hot dry gas stream is preconditioned to increase its humidity and forces the nanoparticles to accumulate at the droplets.

Thermophoresis describes a temperature gradient across a gas space from one side of a particle to the other side, forcing particles to move. Gas molecules on the side of higher temperature collide with the particle with higher energy than the gas molecules on the cooler side. Consequently, the particle moves slightly toward the cold side. The input of cold water can enhance thermophoresis; nevertheless, it is a relatively weak mechanism for particle separation.

The separation efficiency can be improved through measures that cause particle growth. The particle size can be increased by condensing vaporous scrubbing liquid on the particle surface. The technique is mainly limited by the residence time needed for condensation and is therefore rarely applied.

6.3.4.2 Separation efficiency of scrubbers

Three basic approaches according to Semrau, Calvert and Barth are known for the prediction of the separation efficiency of scrubbers. Semrau assumes in his model that the separation efficiency of a scrubber can be deduced from the amount of energy required [117–121]. The approach of Calvert [122–127] and Barth [122, 128] is based on the separation mechanism of particles on a single droplet. The latter approach by Barth was refined by Löffler [86] and Schuch [129] and is currently the most accepted, physics-based approach [68]. Calvert's model relies on experimental data, while a complete prediction without experimental data should be possible with the model of Barth.

6.3.4.2.1 Semrau model

The Semrau model correlates the energy input and the total degree of separation. The empirical model is mainly based on the characteristics of Venturi scrubbers. The overall degree of separation E is specified as follows [118, 120]:

$$E = 1 - \exp(-N_t) \tag{6.44}$$

where N_t is the number of transfer units. Semrau related the number of transfer units to the contacting power as follows:

$$N_t = \alpha \cdot P_t^{\beta} \tag{6.45}$$

where P_t is the total contacting power in kWh per 1,000 m^3 and α and β are fit parameters. The total contacting power is defined as the energy dissipated per unit volume of the treated gas. It depends on the mass flow rate of the off-gas $F_{m,g}$ and the mass flow rate of the scrubbing liquid $F_{m,l}$ as well as on the total pressure drop Δp and the nozzle pressure drop Δp_D. The total contacting power can be calculated as follows:

$$P_t = 96.17 \cdot \left(\Delta p + \Delta p_D \cdot \frac{F_{m,l}}{F_{m,g}} \right) \tag{6.46}$$

The fit parameters alpha and beta have to be determined experimentally. The prediction of the overall degree of separation is therefore not possible.

6.3.4.2.2 Barth model

The Barth model assumes that the scrubbing liquid is dispersed in the gas phase. The precipitation of particles is mainly controlled by inertial interception. The Barth model was corroborated by experimental investigations of Schuch [129]. Schuch was also responsible for the technical applicability of the model by delivering values for the extent of impaction. He suggested to divide the precipitation into three phases:

1. Precipitation of the particle on a single droplet
2. Ratio of solvent droplets to the purified gas volume
3. Decrease of the particle concentration

6.3.4.2.2.1 Precipitation of the particle on a single droplet

The degree of impaction η is defined by the ratio of the effective capture of cross-sectional area $(\pi\, e^2)/4$ to the spherical cross-sectional area of the droplet $(\pi\, d_{dr}{}^2)/4$, with e being the diameter of the interception area and d_{dr} the droplet diameter:

$$\eta = \left(\frac{e}{d_{dr}}\right)^2 \tag{6.47}$$

For scrubbers, it can be assumed that the adhesivity h, which is the rate of particles that adhere on a droplet to the particles that impact on a droplet, is 100% [68]. In this case, the separation efficiency of the single droplet $\varphi = \eta \cdot h$ equals the probability of interception h.

As already mentioned, the model is based on precipitation by inertia. The gravitational force is small compared to the inertial force and can thus be neglected. It follows that the separation efficiency is a function of the droplet diameter d_{dr}, the particle diameter x, the particle density ρ_P, the viscosity of the gas μ_g, the gas density ρ_g and the gas velocity. From these characteristics, the Stokes number , as shown in eq. (6.48), can be derived:

$$\psi = \frac{\rho_P \cdot v_g \cdot x^2}{18 \cdot \mu_g \cdot d_{dr}} \tag{6.48}$$

The Stokes number (= inertia parameter) specifies the particle mobility. The flow pattern is specified by the Reynolds number, according to eq. (6.49). The ratio of density is constant for one separation process:

$$Re = \frac{v_g \cdot d_{dr} \cdot \rho_g}{\mu_g} \tag{6.49}$$

The probability of interception correlates with the Stokes number, with the regression parameters a and b correcting the flow pattern. The probability of interception is given with the following equation:

$$\eta = \left(\frac{\psi}{\psi + a}\right)^b \tag{6.50}$$

Table 6.10 lists values for a and b for different flow conditions.

The separation efficiency of the single droplet increases with increasing Stokes number. High relative velocity, large particle diameter, small droplet diameter and low gas viscosity will thus increase the separation performance of the scrubber. Laminar flow conditions or low-density ratio decrease the separation efficiency.

Tab. 6.10: Parameters *a* and *b* for the prediction of the theoretical probability of interception for different Reynolds numbers (Re) [86].

Re	*a*	*b*
≫1	0.25	2
60–80	0.506	1.84
40	1.03	2.07
10–20	1.24	1.95
<1	0.65	3.7

6.3.4.2.2.2 Ratio of solvent droplets to the purified gas volume

The purified gas volume is calculated by multiplying the effective capture of cross-sectional area of a droplet with the distance passed. Relating the purified gas volume to the droplet volume, the specific purification volume m can be calculated, as follows (v_{rel} is the relative velocity):

$$m = \frac{3}{2 \cdot d_{dr}} \cdot \int_{t_1}^{t_2} \eta(t) \cdot v_{rel}(t) \cdot dt \tag{6.51}$$

A detailed explanation for the derivation and solution of eq. (6.51) is given in [86]. The integral value for the calculation of the specific purification volume approaches zero for very small droplets. This is because very small droplets will accelerate immediately to the velocity of the gas stream and thus the probability of impaction will decrease. Increasing diameter of the individual droplet will increase the probability of interception.

Each particle size correlates with an optimal droplet diameter at which the specific purification volume obtains a maximum [68]. Therefore, the droplet size distribution should be aligned to the particle size distribution.

6.3.4.2.2.3 Decrease of the particle concentration

The decrease of the particle concentration c within a differential section of the scrubber equals the specific purification volume m and the ratio L of the volumetric flow rate of the scrubbing liquid $F_{V,l}$ to the gas flow rate $F_{V,g}$, as follows:

$$L = \frac{F_{V,l}}{F_{V,g}} \tag{6.52}$$

$$-dc = c \cdot d(m \cdot L) \tag{6.53}$$

The integration of eq. (6.53) leads to the following solution for the particle concentration c with c_0 being the off-gas concentration:

$$c = c_0 \cdot \exp(-m \cdot L) \tag{6.54}$$

The product of m and L represents the ratio of the purified gas volume to the total gas volume.

The overall degree of separation E is given as follows:

$$E = 1 - \frac{c}{c_0} = 1 - \exp(-m \cdot L) \tag{6.55}$$

6.3.4.2.3 Calvert model

The model of Calvert [122, 123, 125–127, 130, 131] is similar to the Barth model. Missing the actual process conditions, Calvert described the precipitation by fitting experimentally obtained data. With his model, Calvert does not address the solution of the droplet's equation of motion, which however leads to a reduction in information [86]. The model mainly estimated the design data for Venturi scrubbers but can be used for rotary scrubbers if certain assumptions are made. Calvert primarily tailored the model to Venturi scrubbers because the determination of an average droplet size seems to be possible for Venturi scrubbers [132]. Calvert made use of the Sauter diameter $d_{1,2}$ derived from Nukiyama and Tanasawa [133], who modeled $d_{1,2}$ (ratio of droplet volume to droplet surface) as shown in the following equation:

$$d_{1,2} = 0.585 \cdot \sqrt{\frac{\sigma}{\rho_l \cdot v^2}} + 53.2 \cdot \left(\frac{F_{V,l}}{F_{V,g}}\right)^{1.5} \cdot \left(\frac{\mu_l}{\sqrt{\sigma \cdot \rho_l}}\right)^{0.45} \tag{6.56}$$

In eq. (6.56), σ is the surface tension in N m^{-1}, μ_l is the dynamic viscosity of the scrubbing liquid in kg m^{-1}s^{-1} and ρ_l is the density of the scrubbing liquid in kg m^{-3}.

The drag force during acceleration of the droplet is assumed to correspond with

$$c_W(Re) = \frac{55}{Re} \tag{6.57}$$

for $6 < Re < 400$.

The probability of interception is described via the approximation of numerical results as given in the following equation:

$$\eta = \left(\frac{\psi}{\psi + 0.35}\right)^2 \tag{6.58}$$

Since the scrubbing liquid is dispersed in the gas phase, droplets are formed after a certain length of path. Calvert considered this fact with the correction factor f. Hence, f is the ratio of the initial relative velocity to the gas velocity $f = v_{rel}/v_g$. The correction factor has been derived from experimental investigation to be between 0.25 and 0.5. Precipitation is already observed for low relative velocities. Under consideration of this empirical correction, the probability of interception (now equals with the collection efficiency) then corresponds with

$$\eta = \left(\frac{\psi \cdot f}{\psi \cdot f + 0.35} \right)^2 \tag{6.59}$$

The integral change of the particle concentration ($c_{\text{Clean gas}}/c_{\text{Off-gas}}$), derived from the specific purification volume, results in the following equation:

$$\ln\left(\frac{c_{\text{Clean gas}}}{c_{\text{Off-gas}}}\right) = \frac{2 \cdot d_{\text{dr}} \cdot \rho_{\text{l}} \cdot v_{\text{g}} \cdot F_{\text{V,l}} \cdot F(\psi,f)}{55 \cdot \mu_{\text{l}} \cdot F_{\text{V,g}}} \tag{6.60}$$

with $F(\psi,f)$ being

$$F(\psi,f) = \frac{-\psi \cdot f - 0.35 + 0.7 \cdot \ln\left(\frac{\psi \cdot f + 0.35}{0.35}\right) + \frac{0.1255}{\psi \cdot f + 0.35}}{\psi} \tag{6.61}$$

The separation efficiency $T(x)$ is given with

$$T(x) = 1 - \exp\left(\frac{-2 \cdot d_{\text{dr}} \cdot \rho_{\text{l}} \cdot v_{\text{g}} \cdot F_{\text{V,l}}}{55 \cdot \mu_{\text{l}} \cdot F_{\text{V,g}}} \cdot F(\psi,f')\right) \tag{6.62}$$

After drop formation, the value f is to be set equal to f'. A detailed overview and explanation of the derivation of these equations can be found in [86] and of course in Calvert's and Barth's studies. Anyway, Calvert managed to correlate the separation efficiency $T(x)$ with the pressure drop by using the simple approach as follows:

$$\Delta p = \left(0.85 \cdot \rho_{\text{l}} \cdot v_{\text{g}}^2 \cdot \frac{F_{\text{V,l}}}{F_{\text{V,g}}}\right) \tag{6.63}$$

And for the approximation of the function $F(\psi, f')$ in the region of $1 \le \psi \le 4$, Calvert suggested the following equation:

$$F(\psi,f') = 0.312 \cdot \psi \cdot f'^2 \tag{6.64}$$

The separation efficiency can then be evaluated with the following approach:

$$T(x) = 1 - \exp\left(-7.415 \cdot 10^{-4} \cdot \frac{\rho_{\text{p}} \cdot x^2}{\mu_{\text{g}}^2} \cdot f'^2 \cdot \Delta p\right) \tag{6.65}$$

The result of Calvert's model approach is remarkable since it does not include the droplet diameter anymore. The determination of the droplet diameter is thus no longer necessary. Evaluation of the separation efficiency as a function of the pressure drops makes Calvert's model similar to the model of Semrau.

The separation efficiency of rotary spray scrubbers, as given in eq. (6.66), may be derived from simplifying the flight path of the droplet to be equal to the distance R of the central origin of the scrubber and the scrubber wall (based on droplet dispersion by shot blast texturing):

$$T(x) = 1 - \exp\left(\frac{-3 \cdot F_{V,l} \cdot \eta \cdot R}{2 \cdot F_{V,g} \cdot d_{1,2}}\right) \tag{6.66}$$

6.3.4.2.4 Pressure drop

Modeling of the pressure drop is again based on the investigation of Venturi scrubbers. Improvement of the precipitation efficiency of Venturi scrubbers is either achieved by an increase of the gas velocity through the Venturi throat or by increasing the ratio of scrubbing liquid to the gas phase. In both cases, the pressure drop increases.

Empirically, the estimation of the pressure drop may be derived from the pressure drop in a tube flow according to

$$\Delta p = \frac{\xi \cdot \rho_g \cdot v_g{}^2}{2} \tag{6.67}$$

with ξ being the pressure drop coefficient and v_g the gas velocity in the Venturi throat, i.e., the ratio of volume flow rate to the cross-sectional area of the Venturi throat. The pressure drop coefficient ξ can be calculated as follows:

$$\xi = A + B \cdot \frac{F_{V,l}}{F_{V,g}} \cdot 1,000 \tag{6.68}$$

with the empirical factors A and B. The empirical factors A and B can either be constant within the boundaries of $0 \leq A \leq 0.5$ and $0 \leq B \leq 1.7$, or they can be used as functions. The first term in eq. (6.68) considers the gas flow, while the second term calculates the energy dissipation.

The actual pressure drop cannot be calculated with these equations, since the pressure drop depends on several factors, such as the wall roughness, scrubber geometry or two-phase flow. However, this estimation is sufficient for a quick design with paper and pen.

Several researches have been done to improve the estimation of the pressure drop. The mathematical model for the calculation of the pressure drop was, for example, improved by Boll [134]. He integrated equations for drop motion, momentum exchange and particle impaction on droplets with the assumption of uniform drop size and uniform liquid distribution. Azzopardi and coworkers have provided great insight into the mechanisms affecting the pressure drop [135–138]. They show that the pressure drop is composed of frictional pressure drop, accelerational pressure drop of the gas, accelerational pressure drop of the droplets, accelerational pressure drop of the film and gravitational (static) pressure drop. Viswanathan [139] experimentally investigated the liquid film characteristics in a Venturi scrubber. He developed an annular two-phase flow model for the prediction of the pressure drop that used a film-flow correlation to predict the liquid film thickness. A two-dimensional model for the prediction of the spatial droplet distribution and the collection efficiency in a Venturi scrubber was developed by Ananthanarayanan and Viswanathan [140]. The penetration length of the injected liquid was calculated with an empirical correlation. More recently, Rahimi et al. [141] developed a single-directional descriptive mathematical model to predict the pressure drop in Venturi scrubbers without neglecting heat and mass transfer phenomena.

However, the developed models range from simple correlations of experimental data through more valuable analytical models to multidimensional descriptions of the flow pattern, requiring complex numerical solutions. Nowadays fluid dynamic behavior of industrial equipment (e.g., the pressure drop) is predicted by CFD simulations. In particular, the mathematical modeling and numerical simulation of multiphase systems is challenging, and the subject of recent studies was done by Ananthanarayanan and Viswanathan [142], Pak and Chang [143], Ahmadvand and Talaie [144], Guerra et al. [145] and Manzano et al. [146].

6.3.4.3 Industrial scrubbers

Many different types of scrubbers can be found in industry. The current classification distinguishes between the basic design or between the liquid inlet design of scrubbers. However, a systematic classification of scrubbers is not possible, since the separation performance, energy input and pressure loss do not exactly correlate.

The following scrubbers are commonly used in the industry:
- Scrubbing towers (spray scrubbers, packed or tray scrubbers)
- Ejector spray scrubbers
- Self-induced spray scrubbers (or orifice scrubbers)
- Disintegrator scrubbers
- Rotary scrubbers
- Venturi scrubbers

The main operation parameters for scrubbers are:
- Particle size distribution
- Gas velocity or gas flow rate
- Liquid-to-gas ratio
- Drop size distribution
- Temperature and pressure drop

6.3.4.3.1 Scrubbing tower

The scrubbing tower is the simplest and oldest type of scrubbers. The gas is fed at low gas velocities (<2 m s^{-1}) usually in countercurrent direction to the scrubbing liquid. The scrubbing liquid is sprayed at several levels along the scrubber height whereby high residence time will ensure sufficient contact between the gas and the droplets. In addition to particle precipitation and of course absorptive effects, the gas often gets cooled in washing towers.

Several types of scrubbing towers have been in use. Figure 6.22 shows a schematic illustration of a spray tower, a packed tower and a tray tower.

The separation efficiency of spray towers is around 90% for particles larger than 5 µm and lies between 60% and 80% for particles with 3–5 µm in diameter. Below 3 µm, the separation efficiency decreases to less than 50%. The spray tower has low maintenance and investment cost.

The collection efficiency is increased by installing irregular packings. Packed towers can therefore be smaller in size than spray towers. Packed towers show higher pressure drop and are susceptible to plug with precipitated solid pollutants. Irregular packings, such as Raschig rings and Berl saddles, are seriously affected by plugging since particles accumulate in the dead zones. Regular packings (spheric packings) are more advantageous in particle precipitation. Plugging is avoided when mobile packings are used (e.g., mobile bed scrubbers). Low-density spheres build up a fluidized bed between two retaining grids. Increase of the gas velocity ensures the permanent movement of the spheric packings leading to a mutual purification. Scrubber packings are too large to serve as collecting bodies for any except very large particles. In the collection of fine particles, the packings primarily promote fluid turbulence and support the deposition of dust particles on droplets.

Tray towers have several perforated plates installed along the column height. Several geometrical shapes of trays, for example, sieve tray, bubble cap and valve trays can be used. The separation efficiency in tray towers for particles >5 µm is more than 97% but submicron particles cannot efficiently be removed. Investment and operating costs are moderately higher compared to spray towers.

The total amount of scrubbing liquid needed is derived from the hydraulic capacity of the tower (column), which is about 15–40 $m^3\ m^{-2}\ h^{-1}$ for scrubbers without installations and 25 $m^3\ m^{-2}\ h^{-1}$ for packed scrubbers and plate-type scrubbers.

6.3.4.3.2 Ejector spray scrubbers (jet scrubber)

In an ejector spray scrubber, a water jet (scrubbing liquid) accelerated with high pressure is injected via a nozzle in the throat of a Venturi tube. The high velocity of the water jet does induce droplet dispersion and provides the draught for moving the raw gas. The liquid velocity at the nozzle outlet lies between 25 and 35 $m\ s^{-1}$ resulting in a gas velocity of 10–20 $m\ s^{-1}$. The high velocities ensure proper mixing of the water and the off-gas. The transportation of the gas requires no fan; however, the specific water consumption related to the volumetric gas flow rate is quite high (5–50 $dm^3\ m^{-3}$). From an economic aspect, it is therefore suggested to reuse the water in a circular flow. Variation of the gas flow does not affect the operation behavior of the jet scrubbers negatively. Jet scrubbers are therefore often used when varying gas flow is to be expected

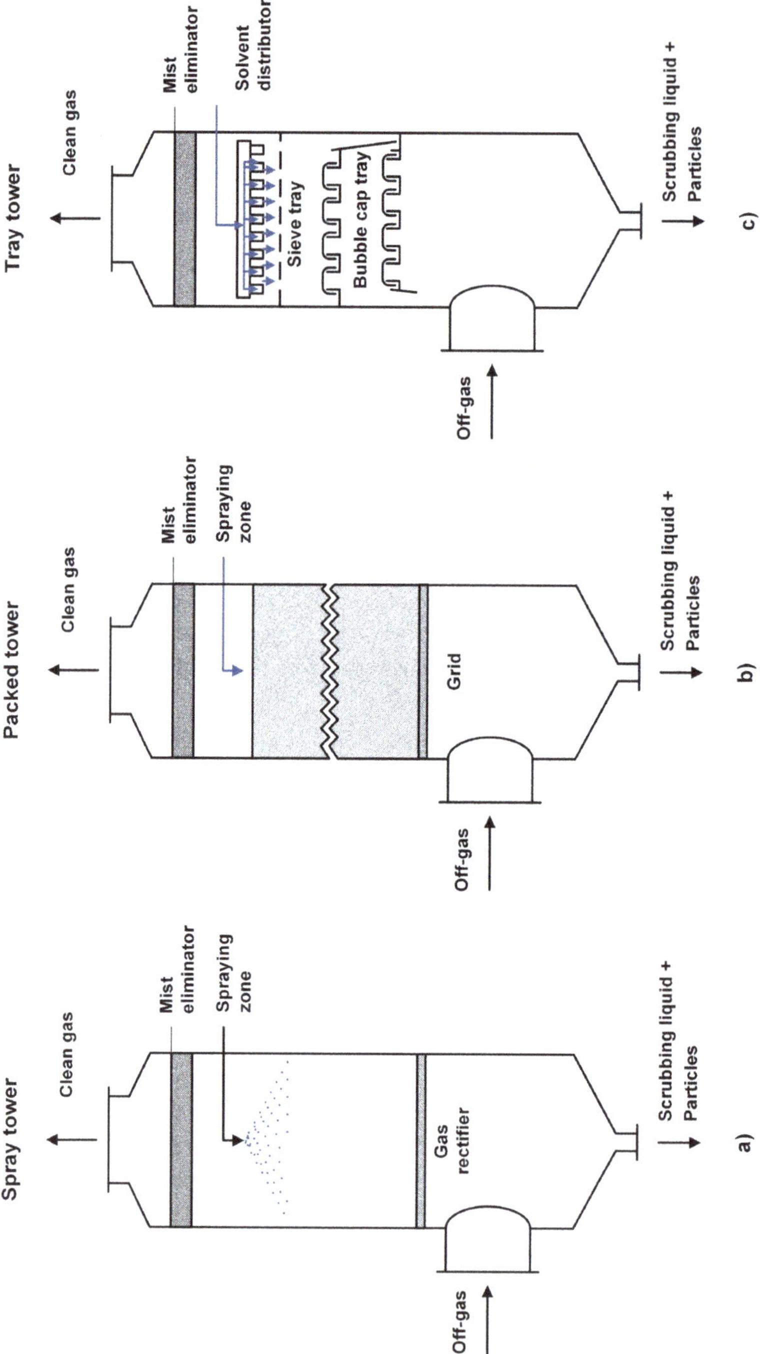

Fig. 6.22: Schematic illustration of scrubbing towers: (a) spray tower, (b) packed tower and (c) tray tower.

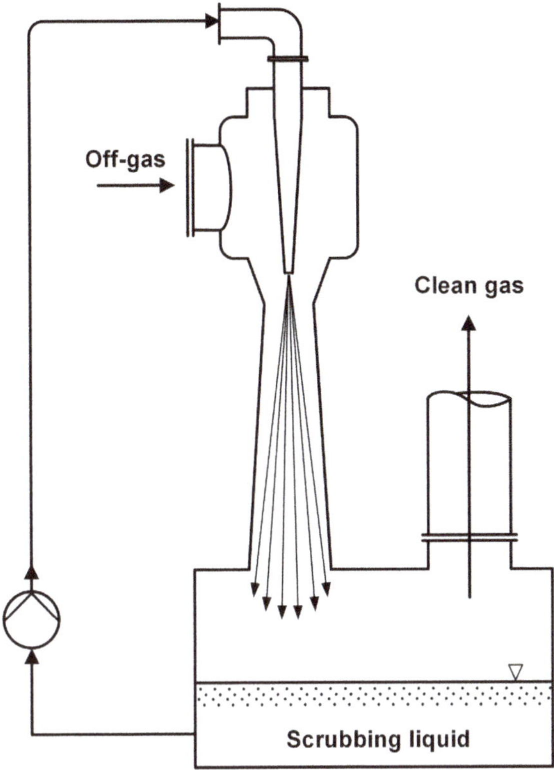

Fig. 6.23: Illustration of an ejector spray scrubber.

or when minimal pressure loss is needed [86]. Plugging of the nozzles can be a persistent maintenance problem. A schematic drawing of the ejector spray scrubber is shown in Fig. 6.23.

6.3.4.3.3 Self-induced spray scrubbers (orifice scrubbers)

In a self-induced spray scrubber, a tube or a duct of appropriate shape forms the gas–liquid mixing zone. The gas stream impinges on the surface of the scrubbing liquid at high velocity, whereby the gas stream provides the energy for droplet dissipation in the same manner as in a Venturi scrubber. The droplets and the particle-laden gas stream get mixed, and particles get captured by the droplets. Subsequently, the droplets hit a series of baffles and get separated from the gas stream again, as shown in Fig. 6.24.

The scrubbing liquid is recirculated from the entrainment section by gravity, whereby no pump is needed for this process. The effort for maintenance is thus low. Self-induced spray scrubbers can be built as high-energy units and low-energy units. Self-induced spray scrubbers are less flexible to fluctuations in the gas flow rate.

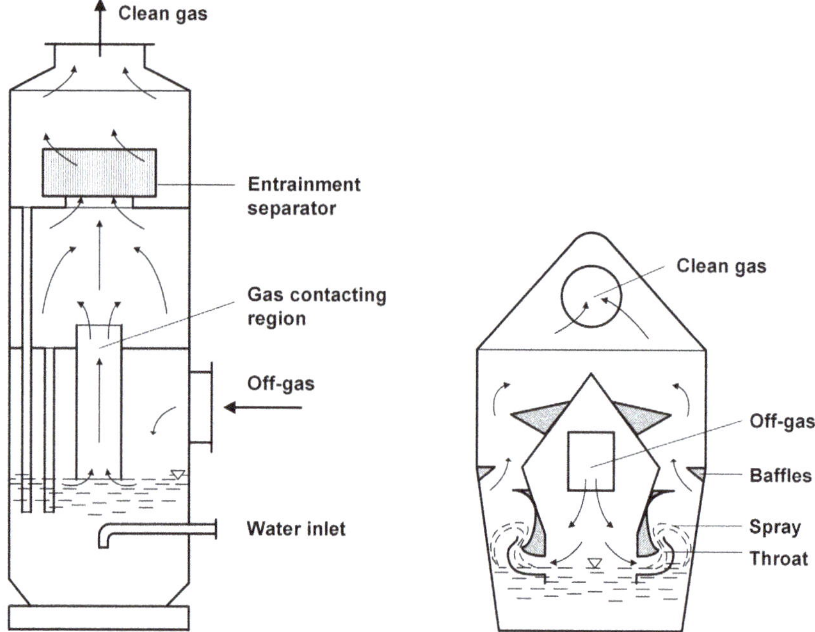

Fig. 6.24: Self-induced spray scrubber.

Frothing materials cannot be separated with this device. The gas velocity is usually higher than 10 m s^{-1}, leading to pressure drops of more than 15 mbar [86]. Pressure drop increases linearly with increasing gas flow rate. It also depends on the water level in the dispersing zone of the scrubber. Approximately 1–3 dm^3 of the scrubbing liquid is needed for purification of 1 m^3 raw gas. This type of scrubber can effectively remove particles ≥2 μm in diameter with a separation efficiency of 80–99%.

6.3.4.3.4 Rotary scrubber

In a rotary scrubber, the scrubbing liquid is dispersed by centrifugal force of a rotating disk, as pictured in Fig. 6.25. The particle-laden gas moves tangentially through the scrubber, whereby larger particles get additionally separated by the centrifugal force of the gas flow. The solvent-to-gas ratio is approximately 1–5 dm^3 m^{-3}. Rotary scrubbers are very flexible regarding the volume flow and particle concentration of the raw gas since droplet size distribution can be adjusted by the volumetric flow rate of the scrubbing liquid and independently by the rotational speed of the disk.

6.3.4.3.5 Venturi scrubbers

The Venturi scrubber is widely used in industry because of its high separation efficiency while investment and maintenance costs are low. A classic Venturi scrubber

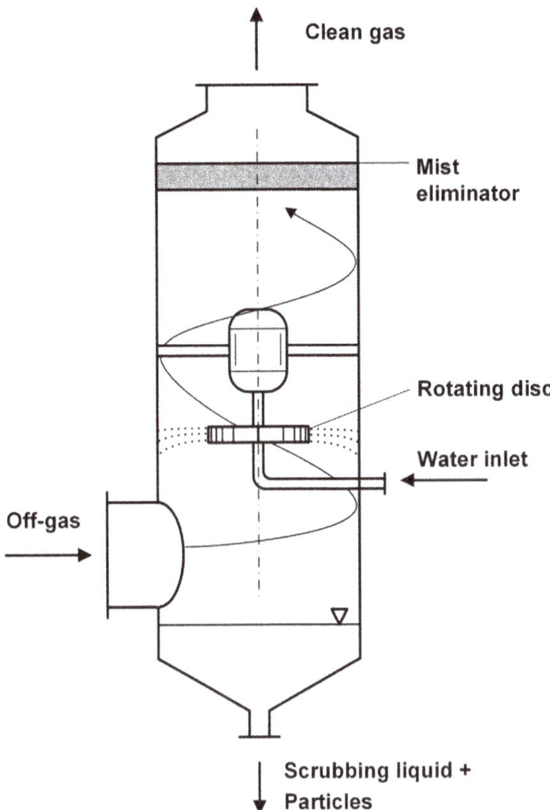

Fig. 6.25: Rotary scrubber.

can be divided into three sections: the convergent section (confusor), the throat and the diffuser, as shown in Fig. 6.26. In the convergent section, gas acceleration takes place. The scrubbing liquid is injected in the throat where it is dispersed due to the high gas velocity and turbulences. In some Venturi designs, the gas velocity can reach up to 150 m s^{-1} and ensures particle collection down to 0.1 µm at high efficiency. The high gas velocity is accompanied with pressure drops up to 200 mbar. The particles in the gas stream collide with the liquid droplets in the throat and diffusor section mainly due to inertial impaction. Interception and Brownian diffusion are very weak in comparison with the inertial impaction. Apart from the interaction of droplets and particles, the gas stream is decelerated in the diffusor section causing agglomeration of the droplets. The Venturi scrubber is inevitably followed by an entrainment separator, for instance, a cyclone that separates the dust-laden drops from the gas stream. The geometrical cross section of a Venturi scrubber can either be rectangular or circular. The scrubbing liquid can be injected as film, jet or spray via pumps (forced feed method) or via pressure difference composed of hydrostatic pressure of the liquid and static pressure of the gas (self-priming method) [147]. The throat is sometimes fitted with a refractory lining to resist abrasion by particles. Adjustable throat contactors under manual or automatic control permit maintaining a constant pressure drop and constant efficiency under conditions of varying gas flow.

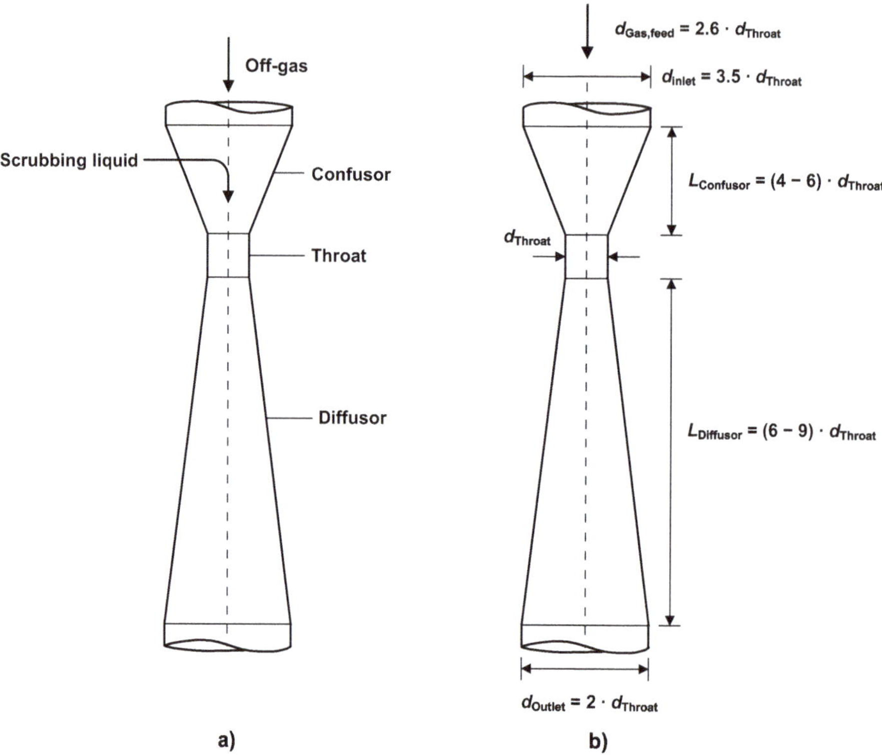

Fig. 6.26: (a) Venturi scrubber and (b) Venturi scrubber with recommended length ratios.

6.3.4.4 Scrubber design

Table 6.11 gives an overview of the main characteristics of different types of scrubbers. The table may be useful to find the best suitable scrubber type for the required separation efficiency of a specific off-gas purification task.

Tab. 6.11: Characteristics of scrubbers [148].

	Spray tower	Water jet scrubber	Orifice scrubber	Rotary scrubber	Venturi scrubber
x_T (µm) at $\rho_P = 2.42$ g cm^{-3}	0.7–1.5	0.8–0.9	0.6–0.9	0.1–0.5	0.05–0.2
Separation efficiency (%)	90	95	90	–	95
Relative velocity (m s^{-1})	1	10–25	8–20	25–70	40–150
Pressure drop (mbar)	2–25	–	15–28	4–10	30–200
Liquid to gas ratio (dm^3 m^{-3})	0.05–5	5–20	Undef.	1–3	0.5–5
Power input (kWh 1,000 m^{-3})	0.2–1.5	1.2–3	1–2	2–6	1.5–6

Exercise 6.10: Design of a Venturi scrubber

For the purification of our off-gas, we first need to find the best suitable scrubber technology. The data listed in Tab. 6.11 can be used as a rough selection guide. Due to the huge amount of fines in our off-gas, it may be the best choice to use a Venturi scrubber. Keep in mind that the choice of scrubber type does not solely depend on the minimum particle diameter needed to be precipitated but also on the local conditions. How much space is available? What about the peripheral equipment (such as fans, pumps and piping) and the power supply? For our exercise let us neglect the local conditions, we just focus on the separation task.

Since the temperature of the gas inflow and the washing liquid are different, you need to consider the additional amount of water in the gas stream, evaporated from the washing liquid. Additionally, you need to consider that there will be a lower pressure on the gas outlet (after the Venturi scrubber), which is responsible for further water evaporation and increased amount of water in the off-gas. The "lazy student" will ignore this fact, but we will take care of it, as more "lazy student" approaches are to come. Let us assume that the evaporated amount of water is $F_{m,evap} = 853.8$ kg h^{-1}. (Comment: This is again a prepared "lucky punch"; by guessing you will need several trials.) Now we can calculate the actual volumetric flow rate of the off-gas stream. You already did similar calculations in Chapter 2; therefore, we refrain from showing the calculation steps again. You can find the results in Tab. 6.12.

Tab. 6.12: Off-gas specification due to evaporation in the Venturi scrubber and operation conditions.

Off-gas specification			Dust	3,000	mg m$^{-3}_{STP,dry}$
p_{stat}	980	hPa	Dust, PSD		
p_{dyn}	981.095	hPa	x_{ol} (µm)	Q_3	c (mg m$^{-3}_{STP,dry}$)
T	200	°C	0.2	0.004	13.30
			0.4	0.018	39.56
N_2	78	%	0.6	0.039	64.77
O_2	12	%	0.8	0.069	88.29
CO_2	10	%	1.0	0.105	109.56
H_2O	100	g Nm^{-3}	2.0	0.359	760.98
O_2-ref.	3	%	4.0	0.831	1,416.50
NO_x	300	ppm	6.0	0.982	452.09
SO_x	300	ppm	8.0	0.999	52.50
			10.0	1.000	2.40

Tab. 6.12 (continued)

Off-gas specification			Dust	3,000	mg $m^{-3}_{STP,dry}$
$F_{V,g}$	10,000.0	$m^3_{STP,dry}$ h^{-1}			
	12,307.0	$m^3_{STP,humid}$ h^{-1}	$F_{m,Evap.}$	853.8	kg h^{-1}
	17,871.4	am^3 h^{-1}	T_{dp}	54.98	°C
$\rho_{g,dry}$	1.343	kg $m^{-3}_{STP,dry}$	X_{H2O},out	0.185	kg m^{-3}_{STP}
$\rho_{g,humid}$	1.242	kg m^{-3} $_{STP,humd}$	Antoine constants (P in hPa):		
$\rho_{g,actual}$	0.855	kg am^{-3}	H_2O-A	8.196	
μ_g	$2{\cdot}10^{-5}$	Pa s	H_2O-B	1,730.462	
ρ_{Liquid}	1,000.0	kg m^{-3}	H_2O-C	233.426	
μ_l	0.0015	Pa s	$p_{H_2O}(T_{dp})$	157	hPa
σ	0.0670	N m^{-1}			
ρ_P	2,000	kg m^{-3}	Δp	143	hPa

Solver 1: Calvert's simplification (or the "lazy student" approach)
With the off-gas specification listed in Tab. 6.12, we can determine the separation efficiency of the Venturi scrubber using eq. (6.65), or the "lazy student" approach. Therefore, we just need to determine the pressure loss according to eq. (6.63), if we assume that the gas velocity in the Venturi throat is v_g = 150 m s^{-1} and the ratio L/G = 1:

$$F_{V,l} = F_{V,g} \cdot \frac{\rho_{g,\,STP,dry}}{\rho_l} \cdot 1 = 10{,}000\left[m^3{}_{STP,dry}\ h^{-1}\right] \cdot \frac{1.343\ \left[kg\ m^{-3}{}_{STP,dry}\right]}{1{,}000\left[kg\ m^{-3}\right]} \cdot 1 = 13.43\ m^3\ h^{-1}$$

$$\Delta p = 0.85 \cdot \rho_l \cdot {v_g}^2 \cdot \frac{F_{V,l}}{F_{V,g}} = 0.85 \cdot 1{,}000 \cdot 150^2 \cdot \frac{13.43}{17{,}871.4} = 14.371\ \text{Pa}$$

Now we can calculate the separation efficiency $T(x)$ with the correction factor f' = 0.35:

$$T(x) = 1 - \exp\left(-7.415 \cdot 10^{-4} \cdot \frac{\rho_P \cdot x^2}{\mu_g^2} \cdot f'^2 \cdot \Delta p\right)$$

The results are listed in Tab. 6.13. The curve of the separation efficiency is shown in Fig. 6.27.

Tab. 6.13: Separation efficiency and particle concentration as predicted with Calvert's simplified approach according to eqs. (6.63) and (6.65).

$T(x)$	$\dot{m}_{Prec}$ (kg h^{-1})	$\dot{m}_{Passed}$ (kg h^{-1})	Q_3 (clean gas)	$c_{Off\text{-}gas}$ (mg m^{-3})	c_{Prec} (mg m^{-3})	$c_{Clean\ gas}$ (mg m^{-3})
0.00	0.00	0.00	0.00	0.00	0.00	0.00
0.06	0.01	0.12	0.24	13.30	0.84	12.46
0.44	0.18	0.22	0.67	39.56	17.57	21.99
0.80	0.52	0.13	0.92	64.77	52.10	12.67
0.96	0.85	0.04	0.99	88.29	84.69	3.61
0.99	1.09	0.01	1.00	109.56	109.00	0.55
1.00	7.61	0.00	1.00	760.98	760.98	0.00

Tab. 6.13 (continued)

$T(x)$	$\dot{m}_{Prec}$ (kg h^{-1})	$\dot{m}_{Passed}$ (kg h^{-1})	Q_3 (clean gas)	$c_{Off\text{-}gas}$ (mg m^{-3})	c_{Prec} (mg m^{-3})	$c_{Clean\ gas}$ (mg m^{-3})
1.00	14.17	0.00	1.00	1,416.50	1,416.50	0.00
1.00	4.52	0.00	1.00	452.09	452.09	0.00
1.00	0.52	0.00	1.00	52.50	52.50	0.00
1.00	0.02	0.00	1.00	2.40	2.40	0.00
				Σ 2,999.96	Σ 2,948.68	Σ 51.28

Discussion: From the calculation results listed in Tab. 6.13, we can evaluate the overall degree of separation which is in a magnitude of E = 0.983 and the penetration fraction P = 0.017, respectively. We end up in a clean gas concentration of $c_{Clean\ gas}$ = 51.3 mg m^{-3}. The cutoff particle diameter is x_T = 0.326 µm.

Solver 2: The separation efficiency based on the droplet diameter

For the evaluation of the separation efficiency of the Venturi scrubber, we first need to calculate the Sauter mean diameter according to eq. (6.56), which is

$$d_{1,2} = 0.585 \cdot \sqrt{\frac{\sigma}{\rho_l \cdot v_g^2}} + 53.2 \cdot \left(\frac{\dot{V}_{V,l}}{\dot{V}_{V,g}}\right)^{1.5} \cdot \left(\frac{\mu_l}{\sqrt{\sigma \cdot \rho_l}}\right)^{0.45} = 5.46 \cdot 10^{-5}\ \text{m}$$

With $d_{1,2}$ we subsequently calculate ψ with the aid of eq. (6.48) and with ψ we calculate $F(\psi,f)$ based on eq. (6.61). For f we again use the magnitude of 0.35. The separation efficiency can thus be calculated with eq. (6.62). The results are summarized in Tab. 6.14, and the separation efficiency curve, as well as the cumulative distribution Q_3 of the off-gas and the clean gas are shown in Fig. 6.27a. Additionally, the density distribution function q_3 is shown in Fig. 6.27b.

Tab. 6.14: Separation efficiency and particle concentration as predicted with eq. (6.62).

ψ	$F(\psi,f)$	T(x)	$\dot{m}_{Prec}$ (kg h^{-1})	$\dot{m}_{Passed}$ (kg h^{-1})	Q_3 (clean gas)	$c_{Off\text{-}gas}$ (mg m^{-3})	c_{Prec} (mg m^{-3})	$c_{Clean\ gas}$ (mg m^{-3})
0.00	0.00	0.00	0.00	0.00	0.00	0.00	0.00	0.00
0.15	−0.05	0.06	0.01	0.12	0.09	13.30	0.84	12.46
1.37	0.05	0.46	0.18	0.22	0.23	39.56	18.01	21.55
3.81	0.13	0.78	0.50	0.14	0.33	64.77	50.30	14.47
7.46	0.19	0.88	0.78	0.10	0.41	88.29	77.91	10.38
12.33	0.23	0.92	1.01	0.08	0.46	109.56	101.18	8.38
34.26	0.29	0.96	7.31	0.30	0.67	760.98	730.58	30.39
137.03	0.33	0.97	13.80	0.36	0.92	1,416.50	1,380.47	36.03
380.65	0.34	0.98	4.42	0.10	0.99	452.09	442.11	9.98
746.07	0.34	0.98	0.51	0.01	1.00	52.50	51.39	1.10
1,233.30	0.35	0.98	0.02	0.00	1.00	2.40	2.35	0.05
						Σ 2,999.96	Σ 2,855.16	Σ 144.80

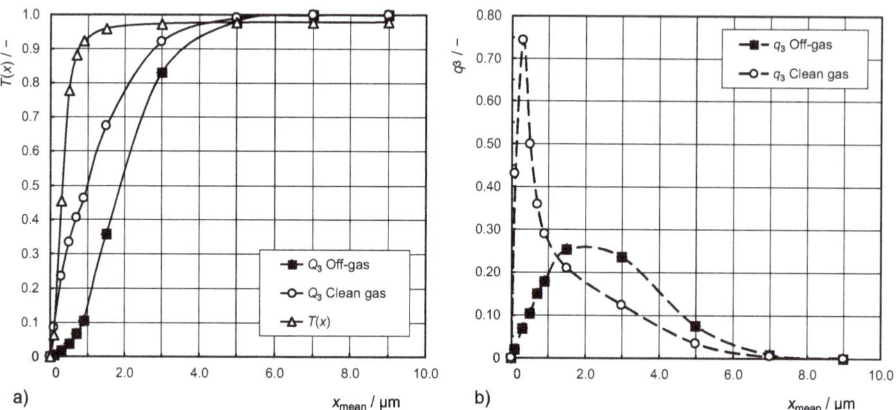

Fig. 6.27: (a) Separation efficiency $T(x)$ and the cumulative distribution function Q_3 of the off-gas and the clean gas. (b) Density distribution function q_3 of the off-gas and the clean gas. The results correspond to Exercise 6.10, Solver 2.

Discussion: The overall degree of separation as predicted based on the Sauter mean diameter is in a magnitude of $E = 0.952$. The penetration factor is $P = 0.048$, respectively. We end up in a clean gas concentration of $c_{\text{Clean gas}} = 144.8\ \text{mg m}^{-3}$. The cutoff diameter is $x_T = 0.318\ \mu\text{m}$.

Solver 3: "Super lazy student" approach:
For the "super lazy student" approach, we need to ask the question whether the theoretical probability of interception equals the separation efficiency. To answer this question, we need to predict the probability of interception η. For the parameters a and b, which we need for the prediction of the theoretical probability of interception, we use the values $a = 0.25$ and $b = 2$, as recommended in Tab. 6.10 for Re $\gg$ 1 (for our off-gas specification Re = 349 calculated with $d_{1,2}$). The predicted separation efficiency curve is shown in Fig. 6.28.

Discussion: Yes it does! The curve of $\eta(x)$ is similar to the separation curve as calculated based on $d_{1,2}$ and as calculated via Calvert's simplified approach, as shown in Fig. 6.28. If you consider that under real conditions we will not reach the separation efficiency of 100% and if it is needed as rough and quick estimation, do not be shy to use this approach. Actually, it delivers more plausible results for the ratio of $L/G = 1$ than Calvert's simplified approach (or the "lazy student" approach) does. (Comment: For the case $L/G = 1.5$, it isn't the case anymore as you can see in Exercise 6.11.)

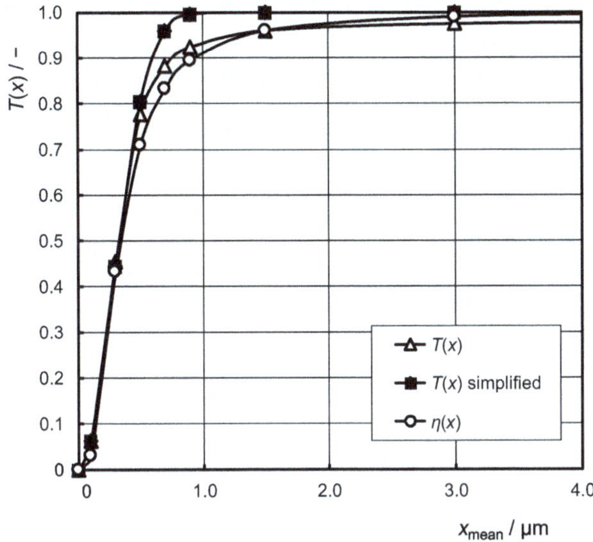

Fig. 6.28: Comparison of separation efficiency curves predicted with three different approaches.

6.3.4.5 Summary

We did apply different approaches for the estimation of the separation efficiency of Venturi scrubbers. The separation efficiency curves for all methods are compared in Fig. 6.28. All approaches show similar results in the particle range of ≤0.5 µm. In the particle range of 0.7–1.5 µm, the separation curves show higher deviation. The best overall degree of separation in a magnitude of E = 98.3% is given with Calvert's simplified approach, which does not mean that it is the most accurate. For a quick and rough estimation of the separation efficiency, all approaches are suitable, the simplest is the estimation of the theoretical probability of interception η though.

However, we suggest to apply the approach for the prediction of the separation efficiency based on the droplet diameter $d_{1,2}$. With an overall degree of separation in a magnitude of E = 95.2% we result in a concentration of the purified gas of $c_{\text{Clean gas}}$ = 144.8 mg m^{-3}. We cannot comply with the emission limits. Can we do a better job?

Since we are already at the upper recommended limit of the gas velocity (v_g = 150 m s^{-1}), we cannot increase the separation efficiency with the aid of the gas velocity (lower velocity will decrease the separation efficiency). Another possibility to increase the separation efficiency is to increase the mass flow rate of the scrubbing liquid (= water). Let us take a closer look at this fact in Exercise 6.11.

Exercise 6.11: The effect of the *L/G* ratio

If we increase the *L/G* ratio to a magnitude of 1.5, pressure drop and thus the evaporated amount of water increase result in a higher actual volumetric flow rate. An increase of the *L/G* ratio to 1.5 results in a pressure drop of Δp = 201 hPa and an evaporated amount of water in a magnitude of $F_{m,evap} = 863\ \text{kg h}^{-1}$. The actual volumetric flow rate is thus 19,135.5 am^3 h^{-1}. Following the approach for the prediction of the separation performance based on the Sauter mean diameter, we get

$$F_{V,l} = F_{V,g} \cdot \frac{\rho_{g,dry}}{\rho_l} \cdot 1.5 = 10{,}000 \left[\text{m}^3\ \text{h}^{-1}\right]_{STP,dry} \cdot \frac{1.343\ \left[\text{kg m}^{-3}\right]_{STP,dry}}{1{,}000 [\text{kg m}^{-3}]} \cdot 1.5 = 20.14\ \text{m}^3\ \text{h}^{-1}$$

$$\Delta p = 0.85 \cdot \rho_l \cdot v_g^2 \cdot \frac{F_{V,l}}{F_{V,g}} = 0.85 \cdot 1{,}000 \cdot 150^2 \cdot \frac{20.14}{19{,}135.5} = 20{,}132\ \text{Pa}$$

The Sauter mean diameter $d_{1,2}$ is

$$d_{1,2} = 0.585 \cdot \sqrt{\frac{\sigma}{\rho_l \cdot v_g^2}} + 53.2 \cdot \left(\frac{F_{V,l}}{F_{V,g}}\right)^{1.5} \cdot \left(\frac{\mu_l}{\sqrt{\sigma \cdot \rho_l}}\right)^{0.45} = 6.97 \cdot 10^{-5}\ \text{m}$$

With a ratio of *L/G* = 1.5, the overall degree of separation is E = 0.984 and the concentration in the clean gas $c_{Clean\ gas} = 47.1\ \text{mg m}^{-3}$. The separation efficiency curves for *L/G* = 1 and *L/G* = 1.5 are compared in Fig. 6.29.

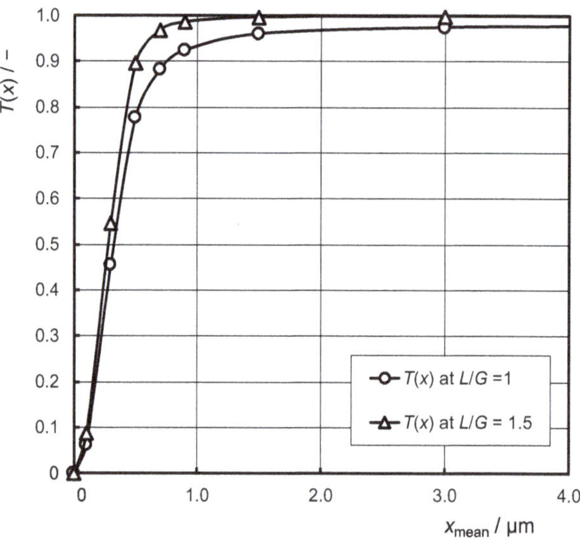

Fig. 6.29: Effect of the ratio *L/G* on the separation efficiency shown for *L/G* = 1 and 1.5.

Discussion: By increasing the amount of the washing liquid, the final concentration in the clean gas **could** be reduced by a factor of 3 from $c_{\text{Clean gas}}$ = 144.8 mg m^{-3} with the ratio of L/G = 1 to $c_{\text{Clean gas}}$ = 47.1 mg m^{-3} with L/G being 1.5. We still cannot meet the emission limits, but we are pretty close. We should keep in mind that we have to get a really bad off-gas under control.

Exercise 6.12: Geometric design of the scrubber

Once we decide for a Venturi scrubber, we can start with the geometric design of it. If you stick to the recommended length ratios shown in Fig. 6.26b, the geometric design is an ease since it is only based on the gas velocity in the Venturi throat and the volumetric flow rate of our off-gas.

Solver:

The throat diameter is

$$d_{\text{Throat}} = \left(\frac{F_{V,g} \cdot 4}{v_g \cdot \pi}\right)^{0.5} = \left(\frac{\left(\frac{17{,}871.4}{3{,}600}\right) \cdot 4}{3.14 \cdot 150}\right)^{0.5} = 0.205 \text{ m}$$

For the calculation of the confusor length we chose factor 4:

$$L_{\text{Confusor}} = 4 \cdot d_{\text{Throat}} = 4 \cdot 0.205 = 0.821 \text{ m}$$

and for the diffusor length we chose factor 6

$$L_{\text{Diffusor}} = 6 \cdot d_{\text{Throat}} = 6 \cdot 0.205 = 1.232 \text{ m}$$

The total length of the Venturi scrubber is thus

$$L_{\text{Tot}} = L_{\text{Confusor}} + L_{\text{Diffusor}} = 0.821 + 1.232 = 2.053 \text{ m}$$

The inlet diameter is

$$d_{\text{Inlet}} = 3.5 \cdot d_{\text{Throat}} = 3.5 \cdot 0.205 = 0.719 \text{ m}$$

and the diameter for the gas feed is

$$d_{\text{Gas, feed}} = 2.6 \cdot d_{\text{Throat}} = 2.6 \cdot 0.205 = 0.534 \text{ m}$$

Finally, the outlet diameter is

$$d_{\text{Out}} = 2 \cdot d_{\text{Throat}} = 2 \cdot 0.205 = 0.411 \text{ m}$$

Discussion: With a total length of 2 m and a maximum diameter size of 0.72 m, the Venturi scrubber will need very little space.

6.3.5 Dust precipitation by filtration

Filtration of PM is realized by a solid, porous barrier, which is permeable for the carrier gas but retains particles. In practice, the particle-laden off-gas flows through a filter system, with the filter material forming the mechanical barrier for the particles. The filter material is decisive to separate certain particle fractions from the off-gas. The filtration process can be described by different separation principles as the principle of size exclusion or the principle of precipitation by inertia, electrical (electrostatic) forces and molecular diffusion. The main mechanisms are the same as during scrubbing, except that fibers or granular materials (loose or sintered) are responsible for particle precipitation, instead of droplets.

- Principle of size exclusion:
  When the pore or mesh size of the filter material is smaller than the particle size, particles precipitate on the filter surface. A so-called filter cake is formed over time. This precipitation mechanism is also called surface filtration and is used especially for the precipitation of coarse particles >5 µm.
- Principle of diffusion and interception by impaction with fibers:
  For decreasing particle size (especially <1 µm), it becomes more challenging to remove fines on the filter surface. Usually, fines are removed by interception within the filter material and adhesion or Van der Waals interaction on thin filter fibers.

In general, filtration does not perform very well with off-gas containing particles larger than 10 µm especially at high dust load conditions. Filters are commonly used to control fine PM emissions at lower dust loads where high separation performance is needed. Clean gas concentrations below 0.5 mg m^{-3} can be achieved by filtration even for submicron particles. Based on the type of construction and the mode of action it is distinguished between deep bed filters, cleanable filters, and granular bed filters.

For the design of a filter system, the engineer has to make a reasonable choice on both, the design of the filter system and the filter material. Latter highly depends on the scope of application as well as the particle size to be precipitated. In addition to the separation efficiency, as calculated with eq. (6.2) or (6.5), the permeability of the filter material plays a major role for the design of filters. The permeability equals the penetration factor as given in eq. (6.7) and should be considered in terms of the pressure drop since the resulting pressure drop of a certain filter material is prevailing. In general, the thickness of a filter material increases the separation efficiency but decreases its permeability and consequently increases the energy demand.

6.3.5.1 Deep bed filters

Deep bed filters consists of fibers with a large void fraction ε (>90%) in which the particles can accumulate. Deep bed filters are mainly used for precipitating particles from low laden gas and they are operated at high gas flow rates (0.01–3 m s^{-1}, up to 10 m s^{-1}).

High-efficiency air purification is the main field of application for deep bed filters. Clean gas concentrations of ≪1 mg m^{-3} can be achieved. Particles adhere inside the filter by interception with the filter material. Particle transport is mainly controlled by diffusion and inertia. Cleaning of laden filters is commonly not applied; therefore, depleted deep bed filters are usually disposed of after reaching a maximum pressure drop.

The thickness of the fibrous layer, the type of fiber and the void fraction of the fabric control the precipitation process. For the design of fiber filters, the precipitation efficiency of the single fiber $\varphi(x)$ is needed, which is the product of the probability of interception η and the adhesivity h, as follows:

$$\varphi(x) = \eta \cdot h \tag{6.69}$$

The relationship between the separation efficiency of a fiber layer and the separation efficiency of the single fiber results from the mass balance of a fibrous layer with the thickness z with D_F being the diameter of the fiber. (Comment: D_F commonly lies between 5 and 50 µm):

$$-\frac{\mathrm{d}c}{c} = \frac{4}{\pi} \cdot \frac{\varepsilon - 1}{\varepsilon} \cdot \frac{\varphi(x)}{D_F} \cdot \mathrm{d}z \tag{6.70}$$

The velocity between the fiber layers for the model filter is v_g while the velocity for the irregular fiber structure is assumed to be $v = v_g/\varepsilon$. The velocity in real filters may be lower than v_g/ε but for filters with a high density of fibers this assumption is applicable. However, by integrating eq. (6.70), the grade separation efficiency $T(x)$ can be derived to be

$$T(x) = 1 - \exp(-f' \cdot \varphi(x)) \tag{6.71}$$

with the form factor f' being

$$f' = \frac{4 \cdot (1 - \varepsilon) \cdot z}{\pi \cdot \varepsilon \cdot D_F} \tag{6.72a}$$

With the specific weight of the filter material m_F in g m^{-2} and the density of the fiber ρ_F, the form factor f' can also be calculated as follows:

$$m_F = \rho_F \cdot (1 - \varepsilon) \cdot z \tag{6.73}$$

$$f' = \frac{4 \cdot m_F}{\pi \cdot \varepsilon \cdot D_F \cdot \rho_F} \tag{6.72b}$$

As already mentioned, particle transport is mainly controlled by diffusion and inertia. With increasing gas velocity and increasing particle size, flow-line interception becomes predominant over diffusion. Diffusion effects apply to gas velocities of $v_g < 0.1$ m s^{-1} and the particle size of $x < 0.5$ µm. In order to predict the filter efficiency, there are several theoretical and empirical approaches available to describe the probability of interception depending on the predominating filtration mechanism.

Stechkina and Fuchs [149] give the following approximation for the probability of interception η as controlled by flow-line interception:

$$\eta = \frac{(1-R)^{-1} - (1+R) + 2\cdot(1+R)\cdot\ln(1+R)}{2\cdot H} \tag{6.74}$$

with H being the hydrodynamic factor and

$$R = \frac{x}{D_\mathrm{F}} \tag{6.75}$$

Lee and Liu [150] approximate the probability of interception η, as controlled by diffusion and flow-line interception, according to

$$\eta = 1.6\cdot\sqrt[3]{\frac{1-\alpha}{H_\mathrm{Ku}}}\cdot\frac{1}{\sqrt[3]{Pe^2}} + 0.6\cdot\left(\frac{x}{D_\mathrm{F}}\right)^2\cdot\frac{1-\alpha}{H_\mathrm{Ku}\cdot\left(1+\frac{x}{D_\mathrm{F}}\right)} \tag{6.76}$$

For the hydrodynamic factor H_Ku according to Kuwabara [151], Lee and Liu suggested the approach as shown in eq. (6.77), with α being the packing density of the fiber layer. (Comment: The theory of Kuwabara [151] as well as the theory of Happel [152] are the two most commonly used approximations for the description of flow around fibers):

$$H_\mathrm{Ku} = -0.5\cdot\ln\alpha - 0.75 + \alpha - 0.25\cdot\alpha^2 \tag{6.77}$$

The Peclet number Pe is

$$\mathrm{Pe} = v_\mathrm{g}\cdot\frac{D_\mathrm{F}}{D_\mathrm{P}} \tag{6.78}$$

with the particle diffusion coefficient D_P being

$$D_\mathrm{P} = \frac{k\cdot T}{3\cdot\pi\cdot x\cdot\mu}\cdot\mathrm{Cu} \tag{6.79}$$

The Boltzmann-constant k is in the magnitude of $1.381\cdot10^{-23}$ in J K^{-1}. Cu is the Cunningham correction factor and can be calculated as follows [100]:

$$\mathrm{Cu} = 1 + \frac{1.7\cdot\lambda}{x} \tag{6.80}$$

The mean free path of gas molecules λ (about 0.1 µm at STP) is

$$\lambda = \frac{k\cdot T}{4\cdot\pi\cdot\left(\frac{x}{2}\right)^2\cdot p\cdot\sqrt{2}} \tag{6.81}$$

The approximation of Lee and Liu shows good agreement with experimental data for velocities between 0.01 and 0.3 m s^{-1} and particle sizes of 0.05–1.3 µm.

Hiller [153] proposed an approximation to enable a quick estimation of the probability of interception and the amount of adhesion by diffusion, inertia and electrostatic effects. The approximation does not consider the influence of gravity. With eq. (6.82) the probability of interception can be approximated with an error of <5%:

$$\eta = 1.03 + (0.5 \cdot \mathrm{Re} - 1.5) \cdot 0.85^{\psi + 0.5} \tag{6.82}$$

The Reynolds-number (Re) for fibers ($\varepsilon = 1$) is

$$\mathrm{Re} = \frac{v_g \cdot D_F \cdot \rho_g}{\mu_g} \tag{6.83}$$

The inertia parameter ψ can be calculated in the same way as for droplets, but instead of the droplet diameter the fiber diameter D_F must be considered, as follows:

$$\psi = \frac{\rho_p \cdot v_g \cdot x^2}{18 \cdot \mu_g \cdot D_F} \tag{6.84}$$

The range of validity for the approximation is

- $1 < \psi < 10$
- $< \mathrm{Re} < 1$
- $\eta > 0.1$.

For the adhesivity h, Hiller [153] proposed the following approximation:

$$h = 1.368 \cdot \psi^{-1.09} \cdot \mathrm{Re}^{-0.37} \tag{6.85}$$

with the range of validity of

- $1 < \psi < 20$
- $< \mathrm{Re} < 1$
- $h > 0.1$.

The theoretical approaches may greatly overestimate experimentally obtained data. This is mainly due to the fact that the models do not consider the real filter structure but assume an even structured cylinder system. In real filter systems, however, the cavities are statistically distributed, which results in different gas flow rates through the filter system. Consequently, the total separation efficiency in the real filter system will be lower than estimated with a uniform fiber structure. Benarie [154] assumes that the distribution of the characteristic parameter describing the real filter structure can be approximated by a logarithmic normal distribution function. This results in a reduction in the estimated separation efficiency due to the geometric standard deviation σ as follows:

$$lg\left(\frac{\varphi_0}{\varphi_r}\right) = 10.36 \cdot \log^2 \cdot \sigma \tag{6.86}$$

with φ_0 being the precipitation efficiency of the single fiber for the even structured model filter system and φ_r being the corrected precipitation efficiency of the single fiber for the real filter structure. The standard deviation was empirically determined to be $\lg \sigma = 0.23$ [155].

With the help of the mentioned model, the initial state of dedusting can be approximated. Due to increasing flow-line interception the collection efficiency increases with increasing dust load. Probability of interception is increased by the individual fiber plus the adhering particles. Theoretical approximation of increasing collection efficiency has not been developed yet.

The specification of the fiber depends on the field of application. Table 6.15 exemplarily shows fiber specifications commonly used in filtration technology.

Tab. 6.15: Fiber specification depending on the field of application.

Fiber specification	Deep bed filters	Cleanable filters	
	Coarse dust precipitation	Suspended solids precipitation	
D_F (µm)	50–100	1–5	10–30
z (mm)	10–30	1–3	1.5–3
E	>99	>99	>90

6.3.5.1.1 Pressure drop of fiber layer

Operation costs of filter systems are mainly depending on the pressure drop. Pressure drop increases with increasing particle precipitation. For the determination of the pressure drop the pressure drop of the "unloaded" fiber layer is first calculated and then the influence of the precipitated particles is considered.

For the description of the pressure drop of the unloaded fiber layer, there is the theory of the flow through pores (Canal model) and the theory of the drag force yield from the flow acting on a cylinder (flow resistance model). The flow resistance model seems to be more suitable for describing the pressure loss for the deep bed filter. However, according to the flow resistance model, the resistance force W of a flowed body is given with

$$W = \frac{\rho_g \cdot v_g^2 \cdot c_W \cdot A_F}{2} \tag{6.87}$$

with the drag coefficient c_W. A_F is the projection surface of the fiber perpendicular to the flow direction and can be calculated as follows:

$$A_F = L_V \cdot D_F \cdot A \cdot z \tag{6.88}$$

The length of a fiber per volume fraction L_V can be calculated with

$$L_V = \frac{(1-\varepsilon) \cdot 4}{\pi \cdot D_F^2} \tag{6.89}$$

With eqs. (6.88) and (6.89), the resistance force is then given as follows:

$$\frac{W}{A \cdot z} = \frac{\Delta p}{z} = \frac{2 \cdot \rho_g \cdot v_g^2 \cdot (1-\varepsilon)}{\varepsilon^2 \cdot D_F \cdot \pi} \tag{6.90}$$

According to Prandtl [156] c_W for Re < 1 is

$$c_w = \frac{8 \cdot \pi}{\mathrm{Re} \cdot (2 - \ln(\mathrm{Re}))} \tag{6.91}$$

With the fiber Reynolds number considering the void fraction ε

$$\mathrm{Re} = \frac{v_g \cdot D_F \cdot \rho_g}{\mu_g \cdot \varepsilon} \tag{6.92}$$

a linear correlation of the pressure drop and the gas velocity results as follows:

$$\frac{\Delta p}{z} = \frac{16 \cdot v_g \cdot \mu_g \cdot (1-\varepsilon)}{(2 - \ln(\mathrm{Re})) \cdot D_F \cdot \varepsilon} \tag{6.93}$$

For the technically relevant area of 1 < Re < 50 and $\varepsilon \approx 1$, the drag coefficient can be approximated according to eq. (6.94) [157], whereas the error is less or equal to 10%:

$$c_W = \frac{10}{\mathrm{Re}} + 1.5 \tag{6.94}$$

The pressure drop increases with time as a result of dust precipitation. An estimation of the increase of the pressured drop is hardly possible and needs to be determined experimentally. In addition to the structure of the fiber layer, the dust properties in particular affect the increase of pressure drop.

6.3.5.2 Cleanable filters (fabric filters)

Cleanable filters are commonly constructed as bag filters, arranged in filter bag houses as schematically shown in Fig. 6.30. The particles are removed from the off-gas by passing the gas through a fabric, which can, for instance, be woven cloth, felt, fleece, stapled fibers or a mixture of several fabrics. Cleanable filters are applied in purifying off-gas with higher dust concentrations suitable for particle size of fines. The precipitated particles are initially not removed from the surface of the fabric whereby a filter cake is built over time and the precipitated particles become the effective filter medium itself, enabling collection efficiencies of $E > 99.9\%$. Surface filtration is mainly based on the interception of particles and the filter cake. The filter cake leads to an increase of pressure drop and needs to be removed periodically. Clean gas concentrations are highest immediately after regeneration and decrease usually upon the build-up of the filter cake. Standard filter bag houses are either equipped with cylindric tubes or oval bags.

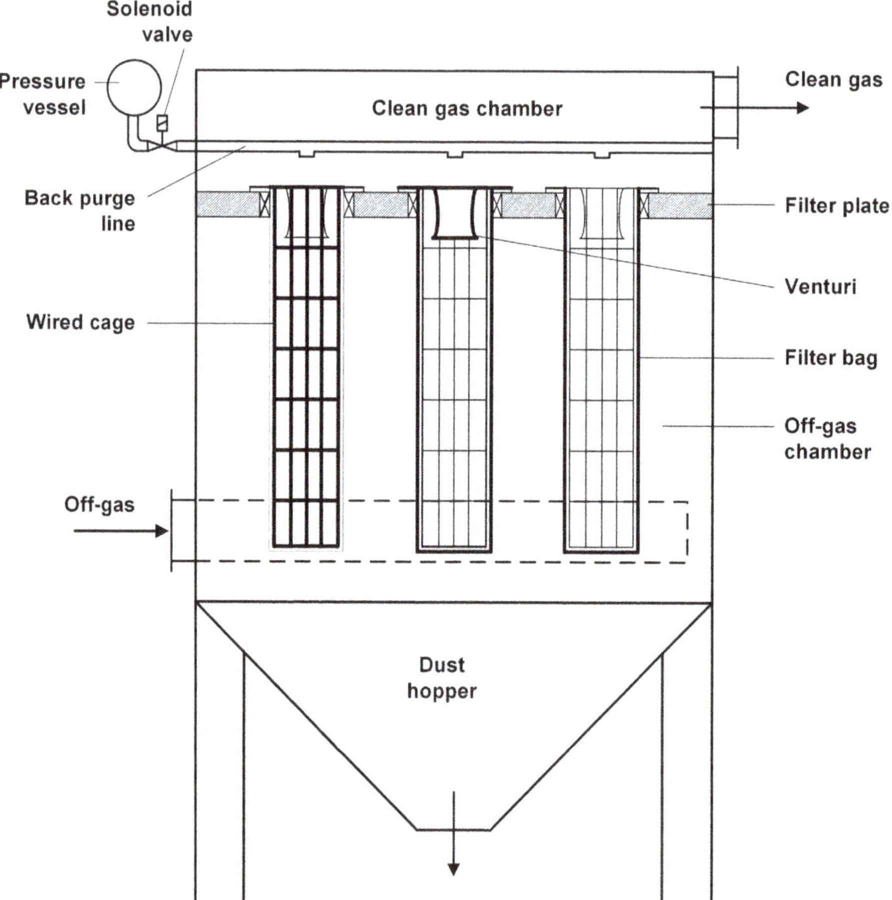

Fig. 6.30: Schematic drawing of a filter bag house.

Precipitation with cleanable filters is divided into two phases. The fabric has to be conditioned during the startup period. Within this period surface filtration is comparable with deep bed filtration. Precipitation efficiency is beneath the value achieved in the second phase of surface filtration when collection is controlled by the sieving properties of the filter cake itself.

6.3.5.2.1 Precipitation characteristics and separation efficiency

Precipitation is controlled by the specific gas flow rate, the particle properties, the particle size distribution, the dust concentration in the off-gas and the interaction between particles and the fabric.

The specific gas flow rate is determined by the fabric properties and the dust properties. In general, the clean gas concentration increases with increasing specific gas flow rate. Increasing the gas flow rate will finally lead to the formation of pin holes for values of the gas flow rate above 110 m h^{-1} (0.027 m s^{-1}) [158]. Pin holes cause straight penetration of particles decreasing the separation efficiency. Specific gas flow rates above 170 m h^{-1} (0.047 m s^{-1}) lead to a distinct formation of pin holes [158]. The seepage of precipitated particles mainly occurs during the cleaning step of the fabrics via pressure shock [159, 160].

Increasing the temperature results in an increased viscosity of the gas. Low filter load decreases the precipitation efficiency then. At high dust load, the viscosity effect is negligible.

By increasing the moisture of the off-gas, the precipitation efficiency can be improved due to the agglomeration of particles.

The particle concentration of the off-gas affects the separation performance at direct particle penetration (at start-up), while electrostatic effects can significantly improve the operating behavior by improving the agglomeration behavior and by reducing the formation of pinholes.

Particle characteristics are of decisive importance in fabric filter design. In addition to the particle size, as the primary particle characteristic, agglomeration, electrical charging and polarity, chemical reactivity and the interaction between particles and filter medium influence the separation performance [161]. With decreasing particle size, the precipitation efficiency decreases since particle penetration of smaller particles increases.

The separation efficiency for cleanable filters can be estimated according to the following equation:

$$E = 1 - \exp\left(-\frac{4 \cdot z \cdot (1-\varepsilon)}{\varepsilon \cdot \pi \cdot D_F}\right) \tag{6.95}$$

The collection efficiency increases with increasing layer thickness z, decreasing diameter of the fiber D_F and decreasing void fraction of the filter layer ε. High precipitation efficiency therefore demands low air permeability and high area density.

6.3.5.2.2 Pressure drop

Analogous to deep bed filters the pressure drop is estimated empirically. Optimum operating conditions are based on the experience with the dust to be handled, the fabric and the cleaning system. It is assumed that the total pressure drop is made up of the pressure drop of the filter medium and the pressure drop of the filter cake. Since filter perfusion is realized at Re < 1, the pressure drop can be calculated according to an empiric correlation based on Darcy's law, as follows:

$$\Delta p[\mathrm{Pa}] = K_1 \cdot \mu \cdot v_F + K_2 \cdot \mu \cdot W \cdot v_F \tag{6.96}$$

K_1 in m^{-1} is the residual resistance of the filter medium with the precipitated particles, K_2 in m kg^{-1} is the specific resistance of the filter cake, W is the areal dust density in g m^{-2} and v_F is the specific rate of filtration, or the actual gas permeability in m h^{-1}, respectively. K is often summarized to $K'_1 = K_1 \cdot \mu$ and $K'_2 = K_2 \cdot \mu$. There are several approaches for the calculation of the resistance values K'_1 and K'_2 for different boundary conditions available. Dennis and Klemm [162] exemplarily suggest eq. (6.97) for the calculation of K_1, with the pressure in the pressure vessel $p_{Res.}$ in kPa:

$$K'_2\left[\mathrm{Pa\ h\,m^{-1}}\right] = 1,850 \cdot p_{Res}{}^{-0.66} \tag{6.97}$$

Strangert [163] proposed eq. (6.98) for the calculation of K_2 for the velocities of v_g = 60–162 m h^{-1} and the mean particles sizes x_{50} of 0.3–50 μm:

$$K'_2\left[\mathrm{Pa\ m\ h\ kg^{-1}}\right] = \frac{1.25}{x_{50}} \tag{6.98}$$

For convenience, the pressure drop can be approximated as a function of the gas velocity according to

$$\Delta p \sim v_g{}^{1.5-2} \tag{6.99}$$

In practice, the pressure drop plays a major role and may be as important as the separation efficiency. If the filter medium is not sufficiently cleaned, precipitated particles are responsible for a constant increase of the pressure drop, which in turn may lead to operation disruption. The pressure drop over the filtration time is schematically shown in Fig. 6.31. Due to the precipitation of particles and the formation of the filter cake, pressure drop increases over time. After reaching a maximum pressure drop Δp_{max}, the filter medium is cleaned, and the pressure drop falls to the residual pressure Δp_R. Due to residual particles, the initial pressure drop Δp_0 cannot be reached anymore; however, this fact is favorable for the further dust precipitation periods. The residual pressure drop Δp_R over time indicates whether the stable operating conditions are achieved or whether the filter medium is increasingly blocked.

6.3.5.2.3 Cleaning mechanism and filter designs

The precipitated particles need to be removed from the fabrics frequently. The filter chamber to be cleaned is either shut off from the off-gas supply (offline operation) and is cleaned mechanically or pneumatically or the filter bags are cleaned during operation (online operation) by reverse flow or reverse pulse.

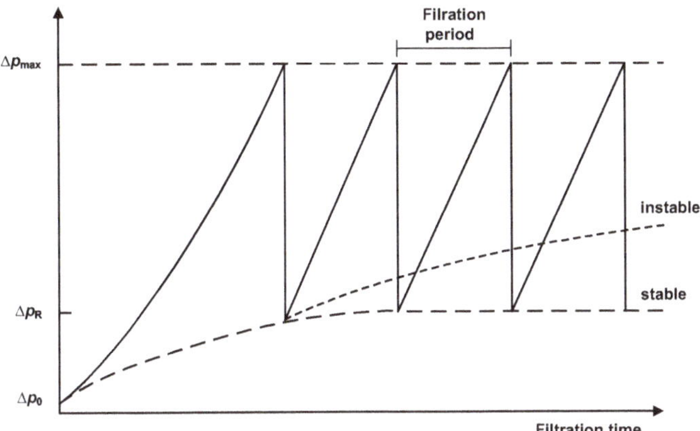

Fig. 6.31: Pressure drop versus filtration time of a cleanable filter.

Current fabric filter designs can be divided into three types, depending on the method of cleaning preferred:

- mechanically cleaned filters (shaking, vibrations),
- reverse-flow cleaned filters and
- reverse-pulse cleaned filters.

The mechanically cleaned filter (shaker cleaned filter) is the earliest form of filter bag houses. The tube bag or oval bag is installed upside down and is fixed in the filter plate, which separates the lower off-gas feed chamber from the upper clean gas chamber. The bag supports connect the filter bag with a shaking mechanism at the upper end and suspend it. The off-gas flows upward into the filter bag and particles precipitate on the inside surface of the bags. During operation, the pressure drop rises to an upper limit. Then the gas flow is stopped, and the shaker is operated, giving a whipping motion to the bags. The removed particles fall into dust hoppers located below the filter plate. For continuous operation, the filters must be constructed with multiple compartments. Individual compartments are sequentially taken off-line for cleaning. The bags must be made of woven fabrics (or needled felts with a woven background material) to withstand the stretching involved in shaking.

Reverse-flow cleaned filters are similarly constructed as mechanically cleaned filters except for the shaker. For cleaning purpose, a fan is used to force clean gas through the bags against the actual flow direction whereby the bags partially collapse, and the precipitated particles are removed. A complete collapse of the bags is prevented by ring or cage stabilizers mounted along the bags. Excessive stress has to be avoided by gentle collapsing and reinflation of the bags. Reverse gas flow is of similar velocity as the filtration velocity.

In reverse-pulse cleaned filters the filter bag is supported with a wire cage. A Venturi nozzle is located in the clean gas outlet, as schematically shown in Fig. 6.32. For cleaning, a high-velocity air jet of low pressure (0.5–7 bar) is directed through the Venturi nozzle into the filter bags. As a result, a secondary flow of clean gas is induced that passes the fabric to the off-gas side. The high-velocity air jet is injected as a sudden, short pulse (~0.3 s) from a compressed air duct through a solenoid valve. The filter bag is expanded, and the dust layer is removed. During this process, precipitated particles can be transported into the cleaned gas and may increase its particle concentration. Filtration rates and storage pressure therefore determine the clean gas concentration to a very large extent. The cleaning action of reverse-pulse jet filters is very effective and may remove the dust from the surface completely. Consequently, the fabric itself should serve as the principal filter medium. Woven fabrics are unsuitable for that necessity. Felts are preferred, ensuring an adequate separation efficiency until the dust layer has formed again.

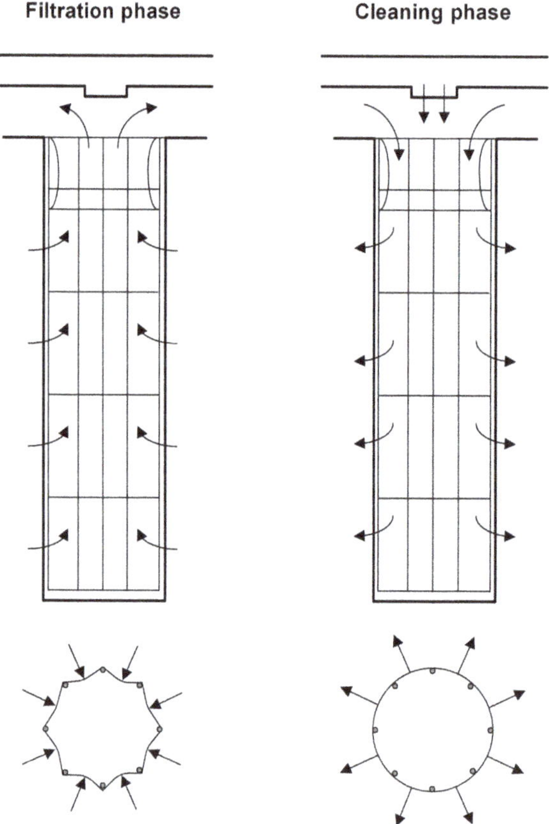

Fig. 6.32: Filtration phase and cleaning phase of a bag filter with pulse jet cleaning.

6.3.5.2.4 Design of cleanable filters

The prevailing parameter for the design of cleanable filter is the specific rate of filtration v_F. It is the ratio of the gas flow rate and the total filtration area needed, whereby the aim of the filter design is to evaluate the total filtration area. It is important to distinguish between online and off-line operation mode. The online operation does not limit the total filtration area. Off-line operation needs a correction of the specific rate of filtration since single compartments are frequently separated from the raw gas resulting in a reduction of the total filter area.

The specific rate of filtration is determined by correcting (multiplying) the basic rate of filtration $v_{F,0}$ with empirical correction parameter. The basic rate of filtration depends on particle properties and the type of cleaning. Most manufacturers have their own comprehensive database collecting values of the basic rate of filtration and the correction parameter.

The design algorithm of Flatt [161] gives access to the rate of filtration v_F. The algorithm corrects the basic rate of filtration $v_{F,0}$ with eight parameters A–H, as follows:

$$v_F = v_{F,0} \cdot A_n \cdot B \cdot C \cdot D \cdot E \cdot F \cdot G \cdot H \tag{6.100}$$

The correction parameters take the following influencing variables into account:

A_n Type of filter
B Field of application
C Particle size distribution
D Off-gas dust content
E Off-gas temperature
F Bulk density
G Filtration conditions
H Climatic parameter

In Tab. 6.16, an excerpt of Flatt's list for the basic rate of filtration $v_{F,0}$ of several dust types is given.

Tables 6.17–6.24 list values for the correction parameters A–H as proposed by Flatt.

Tab. 6.16: Basic rate of filtration for several dust types according to Flatt [161].

Dust type	Basic rate of filtration $v_{F,0}$ ($m^3\,m^{-2}\,min^{-1}$)
Activated carbon	2.5–2.8
Fly ash	2.7–3.0
Iron oxide dust	1.5–2.0
Blast furnace slag	1.8–2.3
Wooden particles	3.8–4.1
Dust from nonferrous smelter	1.8–2.4
Quartzite	2.1–2.8
Soot	1.1–2.1
Soda	3.6–4.0
Zinc coating dust	1.2–2.0
Cement dust	2.4–3.0
Zinc oxide	1.8–2.9

Tab. 6.17: Correction parameter A_n (type of filter).

Type of cleaning	Design	Value
Jet filter	Tube	1
Jet filter	Oval	0.5
Vibrating and counter-flow filter	Tube	0.65
Vibrating and counter-flow filter	Oval	0.45
Direct impulse filter	Single tube	1.5
Direct impulse filter	Series of tubes	1.3

Tab. 6.18: Correction parameter B (field of application).

Field of application	Value
Dedusting of conveyors, filling devices	1
Dedusting of crushing, disintegration, pneumatic transport processes	0.9
Dedusting of kilns, dryers, reactors	0.8

Tab. 6.19: Correction parameter C (particle size distribution).

Particle size (μm)	Value
$x > 100$	1.2
$50 < x < 100$	1.1
$10 < x < 50$	1.0
$5 < x < 10$	0.9
$2 < x < 5$	0.8
$x < 2$	0.7

Tab. 6.20: Formula for the correction parameter D (off-gas dust content).

Formula	Limit
$D = \frac{1.4 + 0.013 \cdot c_{\text{Off-gas}}}{1 + 0.02 \cdot c_{\text{Off-gas}}}$	Upper limit: 200 g m^{-3}

Tab. 6.21: Formula for the correction parameter E (off-gas temperature).

Formula	Limit
$E = 0.548 + \frac{23.14}{T}$	20–250 °C

Tab. 6.22: Correction parameter *F* (bulk density).

Bulk density (kg m^{-3})	Value
>600	1
400–600	0.95
200–400	0.85
<200	0.65

Tab. 6.23: Correction parameter *G* (filtration conditions).

Basic filtration velocity (m^3 m^{-2} min^{-1})	Value
<1.4	1.0
1.8	0.95
2.2	0.90
2.6	0.85
3.0	0.80
3.4	0.75
3.8	0.70
4.2	0.65

Tab. 6.24: Correction parameter *H* (climatic conditions: important for hygroscopic dust and dust from food industry).

Property	Value
Extremely humid climate	0.8

6.3.5.2.5 Filter material and standards

There is a huge variety of different filter materials available on the market to meet specific requirements. Depending on the function (separation of gas, liquid and solid materials), there is a standardized test procedure to measure and classify the performance of available filter material. Especially for PM a lot of effort was done to classify the filter material in order to meet the growing requirements of emission control. ISO 16890 is the international test standard implemented to replace EN779. As the previous standard EN779 was based on average-sized particles of 0.4 µm rating the efficiency in five fine filter classes (M5–F9) and coarse particle filter classes G1–G4, current ISO 16890 distinguishes between the four particle size fractions PM1 (0.3–1 µm), PM2.5 (0.3–2.5 µm), PM10 (0.3–10 µm) and coarse dust (>10 µm). An overview of major changes of the EN 779 is given in Tab. 6.25.

For fine particles smaller than 1 µm, high-performance filter systems are used. In this case, EN 1822 distinguishes between three groups of filter classes:

Tab. 6.25: Overview of major changes of the EN 779:2012 and ISO 16890 test and classification procedure.

	EN 779:2012	ISO 16890:2017
Particle size for classification	0.4 μm	0.3–1 μm (PM1)
		0.3–2.5 μm (PM2.5)
		0.3–10 μm (PM10)
Test aerosol	Di-ethylhexyl sebacate (DEHS)	DEHS for 0.3–1 μm
		Potassium chloride (KCL) for 2.5 and 10 μm
Efficiency of discharge	Comparison of sample and filter	Average efficiency of treated and untreated (conditioned) filter
Dust feed for classification	Incremental dust feed	Classification without dust feed
Test dust for ISO coarse and energy efficiency	AC fine (ASHRAE)	ISO fine
Dust feed	70 mg m^{-3}	140 mg m^{-3}
Test final differential pressure	G1, G2, G3, G4 = 250 Pa	PM10 < 50% = 200 Pa
	M5, M6, F7, F8, F9 = 450 Pa	PM10 ≥ 50% = 300 Pa
Classification	G1 to G4	ISO coarse
	M5 to M6	PM10
	F7 to F9	PM2.5
		PM1

- EPA filter (efficient particulate air filter)
- HEPA filter (high-efficient particulate air filter)
- ULPA filter (ultra-low penetration air)

Operation characteristics and efficiency of filter bag houses widely depend on the quality of the filter fabric. Cost of the filter bags represents a significant part of the total cost of investment, typically 10–15%. Filter fabrics are mainly made of needled felts or fleece with woven fibers providing the background material. The quality of woven fabrics is determined by the number of mesh/length. Air permeability is exemplarily given with the standard DIN EN ISO 9237, recommending woven fabrics with a specific mass of 200–450 g m^{-2} and an air permeability of 600–1,800 $m^3\ m^{-2}\ h^{-1}$ at a pressure drop of 196 Pa. High values of mesh/length enable efficient collection but low filtration velocity of 30–60 $m^3\ m^{-2}\ h^{-1}$. Felts and needled felts have become the main fabrics in off-gas purification. Opposite to woven fabrics felts, needled felts are specified by a high collection efficiency even at start-up conditions, although the void fraction is higher. High air permeability enables high filtration velocities of 100–200 $m^3\ m^{-2}\ h^{-1}$. Specification is standardized in national and international standards (DIN EN 12127 and 29073, DIN EN ISO 5084 and DIN 53885, DIN EN ISO 9237, DIN EN ISO 13937-1 to -4, DIN EN ISO 13938-1 to -2, DIN EN ISO 9073-4, DIN EN ISO 12947-1 to 4, etc.)

The specification of the fiber material must correspond with

- the chemical and mechanical stability and temperature stability,
- high tensile strength and
- high air permeability and high collection efficiency.

Polyester is the preferred fiber material for temperatures up to 150 °C. Polyamides, Teflon and glass fibers cover a temperature region of up to 300 °C. Dedusting above 300 °C needs metal wire mesh and mineral fibers.

6.3.5.2.6 Fiber materials

– Cotton

Cotton is a low-cost fiber material. Low temperature resistance limits application to less than 90 °C. The resistance to acidic conditions is worse than the resistance to basicity. The application of cotton is very limited.

– Wool

The fiber roughness of wool enables high collection efficiency. Manufacturing of woven fabrics and needle felts is common. The upper temperature limit is 100 °C. The basicity resistance is limited.

– PVC

PVC fibers are specified by a high chemical stability. Both woven and needle felt fabrics are manufactured. Temperature stability is limited; therefore, operation temperature shall not exceed 65 °C.

– Polypropylene

The fiber material is characterized by high chemical and mechanical stability. Woven fabrics and felts are manufactured. The upper temperature limit is 100 °C.

– Polyamide (PA)

PA fibers have high tensile strength and chafing stability as well as high ductility. PA fibers are hydrophobic. Woven fabrics and felt fabrics are manufactured. The properties of PA fibers are comparable with cotton.

– Polyamide HT
The fiber is specified by high chemical stability. It is a flame retardant. Woven fabrics and felt fabrics are manufactured. Acidic corrosion below dew point temperature causes a loss of consistence. Dry gas dedusting is possible up to 200 °C.

– Polyacrylonitrile
The fiber allows application temperatures of up to 135 °C. Woven fabrics and felt fabrics are manufactured. It has sufficient resistance to acidic and alkaline conditions. The mechanical properties are comparable to cotton.

– Polyester
Polyester fibers are characterized by high tensile strength and temperature resistivity. Woven fabrics and needle felt fabrics are manufactured. The upper temperature limit is 140 °C.

– Polytetrafluoroethylene (Teflon)
Teflon fibers are characterized by excellent chemical stability exceeding the temperature limit of 220 °C. Woven fabrics and felt fabrics are manufactured. The tensile strength is limited which is the major disadvantage of Teflon fibers.

6.3.5.3 Granular bed filters

In a granular bed filter, the off-gas is cleaned by flowing through a granular filter layer. The granular filter layer can consist of different granular materials of various types and sizes. Typically, materials as gravel, sand, ceramics, activated carbon or different types of packing are used in industry. Granular bed filters are preferably used for purifying hot gas, abrasive gas, and reactive dust. They can be designed as fixed beds, fluidized beds and moving granular beds. The separation efficiency of a fixed bed granular filter is higher than 99% with the disadvantage of the necessity of periodical cleaning. A fluidized bed filter provides continuous operation, but the separation efficiency of smaller particles is less compared to a fixed bed granular filter. Moving granular beds are typically operated continuously with high filtration efficiencies (over 97%) and small pressure losses.

Precipitation mechanism of granular bed filters are rather complex and depend on many influencing variables. Therefore, and due to the fact that this type of filter is rarely used, the design of granular bed filters is not discussed. A comprehensive overview of the design of granular bed filters can be found in [86].

Exercise 6.13: Design of a filter bag house
We already learned that our off-gas seemingly originates from a combustion process. Fly ash is therefore the main PM we need to take care of. For the design of the filter bag house and estimation of the precipitation efficiency we make use of the simple design algorithm and values as proposed from Flatt. The cleaning of the single tube filters should be performed via pressure jet.

Comment: Detailed design of filters is based on empirical values. Correction factors are therefore up to the experience.

Solver:
For fly ash the basic rate of filtration is proposed to be $v_{F,0}$ = 2.8 m^3 m^{-2} min^{-1}. The correction factors used for the calculation of the specific rate of filtration are listed in Tab. 6.26. The correction factors result from the intended design of the filter bag house and our off-gas specification.

Tab. 6.26: Correction parameters for the design of the filter bag house resulting from the off-gas specification as given in Tab. 6.1.

Correction parameter	Value
A_n	1.0
B	0.8
C	0.8
D	1.36
E	0.66
F	0.95
G	0.825
H	0.80

The actual rate of filtration v_F, as calculated with eq. (6.100), is

$$v_F = 2.8 \cdot 1.0 \cdot 0.8 \cdot 0.8 \cdot 1.36 \cdot 0.66 \cdot 0.95 \cdot 0.825 \cdot 0.80 = 1.01 \ \text{m}^3\ \text{m}^{-2}\ \text{min}^{-1}$$

The total filter area needed for our separation task is calculated with the volumetric flow rate $F_{V,g}$ and the actual rate of filtration v_F

$$A = \frac{F_{V,g}}{v_F} = \frac{20{,}111\left[\text{m}^3\ \text{h}^{-1}\right]}{1.01\left[\text{m}^3\ \text{m}^{-2}\ \text{min}^{-1}\right] \cdot 60} = 331.9\ \text{m}^2$$

The length of a single filter tube L_{Tube} is chosen to be 2 m with the tube diameter d_{Tube} of 0.16 m. (To make a comment on the length of the filter tubes: It mainly depends on the available space in a plant. Filter tubes with a length up to 9 m haven been successfully installed and operated). The number of filter bags results from the filter area of a single filter bag:

$$A_{Tube} = d_{Tube} \cdot \pi \cdot L_{Tube} = 0.16 \cdot 3.14 \cdot 2 = 1\ \text{m}^2$$

$$n_{Tubes} = \frac{A}{A_{Tube}} = \frac{331.9}{1} = 332$$

If we consider a distance of 4 cm between the tubes and arrange 20 tubes along the width, we end up in a total width of 4 m and a total length of 3.4 m for our filter bag house.

Based on Tab. 6.15 we choose the following values for the specification of the fiber (layer):

- Z = 1 mm
- E = 0.94
- D_F = 5 μm

According to eq. (6.95), the separation efficiency E is

$$E = 1 - \exp\left(-\frac{4 \cdot z \cdot (1-\varepsilon)}{\varepsilon \cdot \pi \cdot D_F}\right) = 1 - \exp\left(-\frac{4 \cdot 0.001 \cdot (1-0.94)}{0.94 \cdot 3.14 \cdot 5 \cdot 10^{-6}}\right) = 0.999999$$

For the evaluation of the grade separation efficiency, we first need to calculate the form factor. Therefore, we use the specific weight of m = 400 g m^{-2} and the fiber density of 950 kg m^{-2}. According to Steckina and Fuchs (eqs. (6.74) and (6.75)), the curve of the grade separation efficiency $T(x)$ considering single flow-line interception effects is depicted in Fig. 6.32a. (Comment: For the calculation of the hydrodynamic factor H we use eq. (6.77) and set α to be 0.05.)

Discussion: The calculation results are summarized in Tab. 6.27. The grade separation efficiency $T(x)$ and the PSD $Q_3(x)$ of the off-gas and the clean gas are pictured in Fig. 6.33a. Figure 6.33b depicts the probability distribution function Δq_3 of the off-gas and the clean gas, respectively.

Tab. 6.27: Calculation results of cleanable filter for the off-gas specification listed in Tab. 6.1.

PSD (μm)	x_{mean} (μm)	$T(x)$	$\dot{m}_{Prec}$ (kg h^{-1})	$\dot{m}_{Passed}$ (kg h^{-1})	Q_3 (clean gas)	$c_{Off\text{-}gas}$ (mg m^{-3})	c_{Prec} (mg m^{-3})	$c_{Clean\ gas}$ (mg m^{-3})
0.00	0.00	0.00	0.00	0.00	0.00	0.00	0.00	0.00
0.20	0.10	0.95	0.13	0.01	0.99	13.30	12.59	0.72
0.40	0.30	1.00	0.40	0.00	1.00	39.56	39.55	0.00
0.60	0.50	1.00	0.65	0.00	1.00	64.77	64.77	0.00
0.80	0.70	1.00	0.88	0.00	1.00	88.29	88.29	0.00
1.00	0.90	1.00	1.10	0.00	1.00	109.56	109.56	0.00
2.00	1.50	1.00	7.61	0.00	1.00	760.98	760.98	0.00
4.00	3.00	1.00	14.17	0.00	1.00	1,416.50	1,416.50	0.00
6.00	5.00	1.00	4.52	0.00	1.00	452.09	452.09	0.00
8.00	7.00	1.00	0.52	0.00	1.00	52.50	52.50	0.00
10.00	9.00	1.00	0.02	0.00	1.00	2.40	2.40	0.00
			Σ 29.99	Σ 0.01		Σ 2,999.96	Σ 2,999.23	Σ 0.72

The separation efficiency of our cleanable filter according to eq. (6.2) is E = 99.97% and the penetration factor is P = 0.03%, respectively. As we can see from Tab. 6.27, particles may be separated in a magnitude of 29.99 kg h^{-1}. The particle mass flow within the off gas is reduced from 30 to 0.01 kg h^{-1}. The particle concentration in the clean gas is calculated to be $c_{Clean\ gas}$ = 0.72 mg m^{-3}_{STP}, indicating that we would meet the required emissions limits by using cleanable filter system.

The separation curve, as shown in Fig. 6.33a, indicates the excellent separation performance of filters. The cutoff diameter is in a magnitude of x_T = 0.024 μm. The maximum values of the density distribution function, as pictured in Fig. 6.33b, shift from 2 μm from the off-gas to 0.1 μm of the clean gas. The respective q_3 value indicates that the relative amount of fines in our clean gas increased from 0.25 to 5.

For the estimation of the separation efficiency, the fiber diameter, the layer thickness, and the void fraction are crucial. By evaluating the actual rate of filtration v_F, the needed filter area can easily be calculated. For safety reasons we may add 20% to the calculated filter area without getting into financial

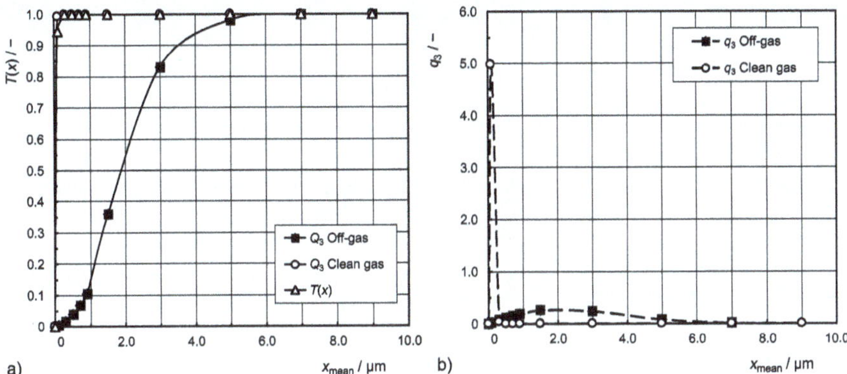

Fig. 6.33: (a) Separation efficiency $T(x)$ and the cumulative distribution function Q_3 of the off-gas and the clean gas. (b) Density distribution function q_3 of the off-gas and the clean gas. The results correspond to Exercise 6.13.

embarrassment. For the sake of convenience, we calculate the grade separation efficiency with the approximation of Stechkina and Fuchs, which considers only flow-line interception effects. The approximation according to Lee and Liu would consider both diffusion and flow-line interception. Considering both diffusion and interception, the grade separation curve would show a minimum between 0.1 and 1 µm, as exemplarily shown in Fig. 6.34. The drop of the separation efficiency within this region is typical for high-efficiency separation devices since diffusion effects are still too low and separation effects of diffusion will act for smaller particles size, while separation by interception is too low for small particles, which leads to a drop of the separation performance.

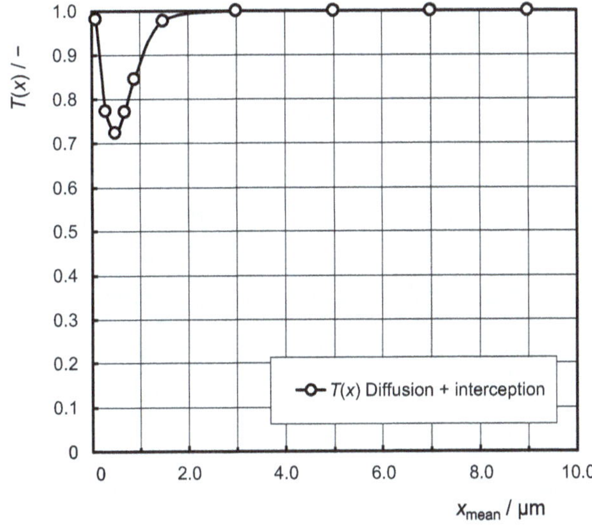

Fig. 6.34: Drop of the grade separation efficiency $T(x)$ within the area of 0.1–1 µm. $T(x)$ is calculated with the approximation of Lee and Liu.

In case of separation of reasonable amount of fine particles (<1 μm) diffusion is the separation mechanism to choose and to be considered in the separation model. Practically, this effect can be enhanced by choosing the filter material with higher content of fine fibers.

Exercise 6.14: Effect of the fabric specification

If we increase the fiber diameter of our fabric from 10 to 30 μm (fabric thickness and void fraction remain the same as for Exercise 6.13), the separation efficiency will decrease to a magnitude of $E = 93.3\%$. On the other hand, if we increase the fabric thickness to 1.5 mm, the calculated separation efficiency is $E = 99.99\%$ for the case that the fiber diameter is 10 μm and the void fraction is 0.94. The effect of the specific weight of the filter material m_F on the separation performance is shown in Fig. 6.35. For coarse particles, a specific filter weight of about $m_F = 100$ g m^{-2} is sufficient, while fines demand higher values. To conclude, for effective precipitation of fines you need smaller fiber diameter, a thicker fabric layer with higher specific weight and high void fractions. Keep the pressure drop in mind!

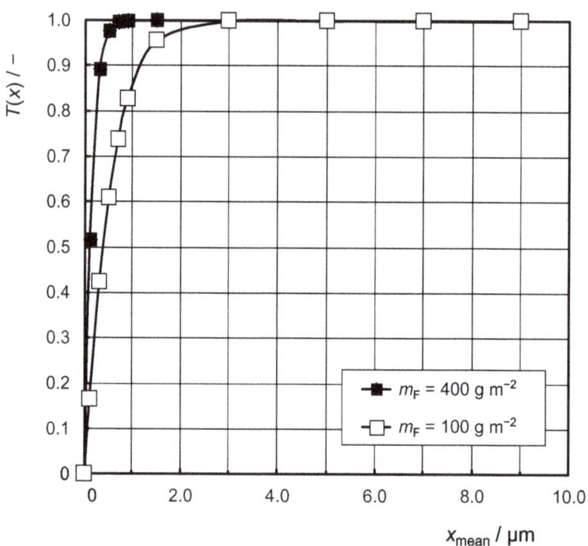

Fig. 6.35: Effect of the specific weight of the filter material m_F on the grade separation efficiency $T(x)$.

Exercise 6.15: Pressure drop

For the determination of the pressure drop, we use eqs. (6.96)–(6.98). We set the reservoir pressure of the compressed air to be 400 kPa, whereas K_1 is

$$K'_2 = 1,850.400^{-0.66} = 35.47 \text{ Pa h m}^{-1}$$

The specific filter resistance K_2 is (x_{50} in μm)

$$K'_2 = \frac{1.25}{2.5} = 0.5 \text{ Pa m h kg}^{-1}$$

We assume that the specific areal dust density is $W = 1{,}500$ g m^{-2} (i.e., a representative value) and thus the pressure drop is

$$\Delta p = K'_1 \cdot v_F + K'_2 \cdot W \cdot v_F = 35.47 \cdot 1.01 \cdot 60 + 0.5 \cdot \frac{1{,}500}{1{,}000} \cdot 1.01 \cdot 60 = 2{,}195\,\text{Pa}$$

Discussion: K_1 and K_2 depend on many parameters and are usually determined experimentally. Therefore, the correlations used to determine K_1 and K_2 should be used carefully.

!

6.3.5.4 Summary

The final dust concentration of our exhaust gas may theoretically be reduced to $c_{\text{Clean gas}} = 0.7$ mg m^{-3}, in very compliance with the emission limits. With the exercise we were able to demonstrate the excellent separation performance of filters reaching an overall separation efficiency of up to 99.99% over a wide range of particle sizes. Separation performance of filters can be affected by the filter thickness, the fiber diameter and the void fraction.

However, the filter design is based on empirical data. Without these data, we can only estimate the required total filter area needed for our separation task and the separation performance. Detailed design of filters will need manufacturers know-how (and database). So do not hesitate to contact the manufacturer after having done calculations yourself. When they register your basic expertise, they will act with more care.

6.4 Conclusion

As a matter of fact, particle precipitation is still a major challenge in modern off-gas purification. Many processes cause the formation of an increased amount of submicronic particles. Airborne particles with low sedimentation properties are a significant health risk because of their respiratory properties, high catalytic activity due to several photocatalytic gas-phase reactions, and adsorptive properties for gaseous pollutants.

Modern off-gas purification therefore has to pay increased attention to efficient precipitation of submicronic particles.

ESP and filtration belong to high-efficiency filtration equipment, whereas cyclones and scrubbers, except the Venturi scrubber, are rather used as preseparation devices for coarse particles. Filtration techniques can cover a wide range of efficient particle precipitation in industrial application. Dust collection by filtration is increasingly preferred over electrostatic precipitation. Even in the field of waste incineration, filtration is the state of the art meanwhile. Is therefore any field of application left for ESP, cyclones or scrubbers in modern off-gas purification? As demonstrated by several examples, it is up to you to find the best suitable separation device for a specific separation task, and this chapter will hopefully support your choice.

7 Thermal and catalytic off-gas purification

Thermal and catalytic off-gas purification technologies are based on chemical reactions in the gas phase to convert hazardous gaseous air pollutants into nontoxic species. The term “nontoxic” mainly refers to species that naturally occur in the air such as carbon dioxide (CO_2, in minor amounts), water (H_2O) and nitrogen (N_2). In classical thermal and catalytic off-gas purification, the chemical reactions may be classified into three types: redox reactions such as (i) oxidation, (ii) reduction, and (iii) decomposition (detoxification) reactions. All three categories are important in off-gas purification practice and are used according to the respective application case.

7.1 Introduction to thermal and catalytic off-gas purification

The main target of classical off-gas purification via chemical conversion in the gas phase is to reduce pollutant concentrations and therefore pollutant activity that has a harmful effect on the environment; not to make the pollutants accessible for further utilization. In most cases, pollutant concentrations are too low to make separation and further utilization economically viable anyway. In other words, the pollutants are rendered harmless by destroying them.

An exception in this regard is CO_2, which takes on a special position. Hazardous gaseous pollutants present in low concentrations in the off-gas (e.g., hydrocarbons) can be converted to CO_2 by oxidation. Due to the low concentration, CO_2 is considered a harmless component in the purified gas. In addition to that, CO_2 may occur in high concentrations in off-gas streams from CO_2-rich industries such as the cement or iron and steel industries [164, 165] or from combustion processes (e.g., for power generation [166] or waste incineration [167]). In terms of the high quantity of CO_2 arising from these processes, it can play a significant role in the climate balance and become economically viable as an important building block for the synthesis of chemicals, fuels, building materials and a large number of carbon-containing products [168, 169]. However, the focus of this book is on the removal of harmful gaseous pollutants by redox reactions and detoxification. The currently increasingly important topic of carbon capture and utilization that addresses the use of CO_2 in concentrated form from CO_2-rich process gas or off-gas for the production of carbonaceous products in chemical and technical biological processes is mentioned here, but not addressed in detail. Reference should be made to special literature [170–173].

In general, gaseous pollutants may have different properties. Apart from being harmful to the environment and health, they may act

- acidic (e.g., HCl, HF and SO_2)
- caustic (basic) (e.g., NH_3)
- oxidizing (e.g., NO_2, NO and SO_2) or
- reducing (e.g., SO_2, NO, NH_3 and $C_nH_mO_rS_sN_t$)

https://doi.org/10.1515/9783110763928-007

Thermal and catalytic off-gas purification technologies mainly tackle the oxidizing and reducing properties of pollutants through chemical conversion in redox reactions. Acidic and caustic properties can be exploited in absorption processes (scrubbing), where, for instance, neutralization reactions are used to increase the efficiency of the overall separation task.

Several manufacturing processes and facilities are accompanied by the formation of gaseous pollutants with oxidizing or reducing properties. During combustion processes, for instance, nitrogen and oxygen from air form dinitrogen oxide (N_2O) and nitrogen oxide (NO), the so-called thermally formed nitrogen oxides (NO_x) [174]. Nitrogen oxides also originate from nitrogen-containing fuels or from decomposition products of nitric acid (e.g., pickling [175]). Nitrogen oxides have both oxidizing and reducing properties. Under ambient conditions, NO_x is oxidized with oxygen from air. The acid rain potential of nitrogen oxides is significant. Due to nitrogen eutrophication, the ecological relevance of nitrogen oxides underlines the necessity of removal from off-gas. Nitrogen oxides are preferably removed from off-gas by reduction processes that convert them into nitrogen and water. This is achieved by what is probably the best-known reduction process in environmental technology, the selective catalytic reduction (SCR) of NO_x with ammonia or its derivative urea.

Fuels that contain sulfur form sulfur dioxide (SO_2) during combustion [174]. SO_2 also has oxidizing and reducing properties. In a humid atmosphere, it is oxidized with oxygen from air to sulfuric acid. It is thus, another significant constituent responsible for acid rain. Reduction of SO_2 with dihydrogen sulfide (H_2S), well-known from the Claus process [176], may be applied in off-gas purification too. The conventional Claus process is not regarded as an off-gas purification process in the original sense of the term. Its main objective is the recovery of sulfur from acidic gas streams that contain high concentrations of hydrogen sulfide.

Manufacturing operations in which solvents are used may also cause partial emission of these. These gaseous solvent-based pollutants are mainly hydrocarbons [177]. The atmospheric interaction of solvents needs attention, even though emission of solvents is limited. The interaction of solvents with the atmosphere can be manifold and is usually detrimental (e.g., ground-level ozone and destruction of the ozone layer). Emission control is possible by several techniques including value-added recycling processes and destruction processes. Latter techniques are predominately carried out by oxidation, namely combustion.

Due to human-induced formation, the concentration of ground-level ozone (O_3) has increased in recent decades to such an extent that it has become a significant environmental issue [178]. To decrease the level of ozone, it may be detoxified via degradation to oxygen.

Thus, prominent examples for off-gas purification techniques by chemical reactions in the gas phase are (i) the oxidation of hydrocarbons C_nH_m to carbon dioxide and water vapor, (ii) the reduction of nitrogen oxides, NO and NO_2, specified as NO_x, to nitrogen and oxygen, and (iii) the detoxification of ozone (O_3) to oxygen (O_2). The reactions can either be carried out at elevated temperatures without a catalyst (*thermal*) or in the presence of a catalyst (*catalytic*). Thus, oxidation and reduction processes in off-gas purification and emission control cover the fields of:

- thermal oxidation (combustion) and catalytic oxidation (catalytic combustion),
- thermal reduction and catalytic reduction and
- detoxification

of pollutants according to their individual properties (Tab. 7.1). The reactions can occur in flames, in combustion chambers or flameless over catalysts. They can be carried out without further auxiliaries or with the aid of auxiliary air or auxiliary fuel and additional reactants – oxidizing or reducing agents.

Tab. 7.1: Representative chemical reactions in off-gas purification.

Oxidation reactions	
Hydrocarbon combustion	$C_mH_nO_p + \left(m + \frac{n}{4} - \frac{p}{2}\right) O_2 \rightleftharpoons m\, CO_2 + \frac{n}{2} H_2O$
Oxidation of H_2S	$H_2S + 1.5\, O_2 \rightarrow SO_2 + H_2O$
Oxidation of sulfur	$S + O_2 \rightarrow SO_2$
Reduction reactions	
Selective catalytic nitrogen oxide reduction (SCR)	$4\, NO + 4\, NH_3 + O_2 \rightarrow 4\, N_2 + 6\, H_2O$ $NO + NO_2 + 2\, NH_3 \rightarrow 2\, N_2 + 3\, H_2O$
Nonselective catalytic nitrogen oxide reduction (NSCR)	$NO + CO \rightarrow \frac{1}{2} N_2 + CO_2$ $2\left(m + \frac{n}{4}\right) NO + C_mH_n \rightleftharpoons \left(m + \frac{n}{4}\right) N_2 + \frac{n}{2} H_2O + m\, CO_2$
Detoxification reactions	
Ozone decomposition	$2\, O_3 \rightarrow 3\, O_2$

A redox (reduction–oxidation) reaction is a chemical reaction, which involves transfer of electrons between two species and thus results in a change of the oxidation number of atoms or ions by gaining or losing an electron. Oxidation defines a reaction that takes place with the release of electrons. Conversely, reductions are reactions in which electrons are collected by a constituent. Thus, oxidations and reductions are coupled, and the overall reaction is called a redox reaction. Oxidation is always associated with an increase in the oxidation number of an atom (e.g., oxidation of CH_4 (oxidation number −4) to CO_2 (oxidation number +4)), whereas reduction leads to a decrease in the oxidation number of an atom (e.g., reduction of NO (oxidation number −2) to N_2 (oxidation number 0)).

Exercise 7.1: Oxidizing and reducing properties of SO_2

SO_2 has both oxidizing and reducing properties. With H_2S it is reduced to sulfur. With oxygen it is oxidized to SO_3. State the reaction equations for these two reactions and determine the change in oxidation number.

Solver:

$$\text{Reduction of } SO_2\text{: } SO_2 + 2\,H_2S \rightleftharpoons 3\,S + 2\,H_2O$$

Change in oxidation number from +4 (SO_2) to 0 (S)
(Note: this reaction is thermodynamically not feasible at temperatures ≤2,400 °C)

$$\text{Oxidation of } SO_2\text{: } 2\,SO_2 + O_2 \rightleftharpoons 2\,SO_3$$

Change in oxidation number from +4 (SO_2) to +6 (SO_3)

Explanation: The oxidation number of oxygen is −2, the oxidation number of species in their elemental form is always 0.

At a specified temperature, the rate of reactant conversion may be raised by the help of catalysts. Due to increased reaction rates with catalysts, pollutant conversion by chemical reactions can be accomplished at lower temperatures and at shorter residence times, thus in smaller reactors; compared to noncatalytic thermal processes. Off-gas purification catalysts are used in the same way in automobiles as in power plants or combustion processes. A distinction is made between nonselective catalysts for the simultaneous conversion of as many components as possible and selective catalysts for the conversion of one or as few components as possible.

Sometimes several off-gas purification techniques can be combined in one operational step. In general, pollutants of either oxidizing or reducing properties are emission-determining. Consequently, the technology and the process design need to be adjusted to the requirements of the respective emission conditions. Apart from the characteristics of the off-gas in general and the pollutants in specific, the desired degree of control and the required purity are decisive for the choice of off-gas purification technologies. In general, changes to the emission-causing processes to avoid pollutant emission (primary measures) should be aimed for and always be preferred. In this regard, nitrogen oxides provide a good example of pollutants whose emissions can be controlled via precise process adjustment in a first step and further be handled by reactive after-treatment in a consecutive step.

In the following chapter, the fundamentals of thermal and catalytic off-gas purification processes in the gas phase are explained. The most important applications are given as examples, with the focus on the elaboration of calculation algorithms and providing numerous exercise examples for self-study. In the calculated exercises, the basic calculation algorithm is presented on the basis of simple relations. Simplifications are shown which can be transferred and extended to more complex tasks at any time (e.g., first-order reactions versus more complex reaction kinetics and constant-volume reactions versus varying-volume reactions). Reference is made to further basic literature, especially textbooks and handbooks, for more detailed information as well as to current research work, whereby these are to be regarded as exemplary and no claim is made to completeness.

For all thermal and catalytic off-gas purification technologies that we discuss in the following sections, simplified design equations will be developed with the primary purpose of illustrating the most important factors that influence their behavior. Rules of thumb are given to facilitate initial estimates. For further estimations, refer to Woods [179] who provides a compilation of rules of thumb in engineering practice in his textbook. All off-gas purification processes based on chemical reactions have the fundamentals of chemical reaction engineering (CRE) in common. Classic textbooks on fundamentals of CRE are given as follows [180–182]. The following is a brief summary of the most important principles.

7.2 Chemical reaction engineering fundamentals

Some basic CRE considerations are fundamental for the discussion of off-gas purification technologies based on chemical reactions in the gas phase. These include the basics of the chemical thermodynamics, the reaction kinetics of the respective chemical reaction and the classification of chemical reactions in general as well as catalysts, reactors and process design.

7.2.1 Chemical thermodynamics

The enthalpy of reaction $\Delta_R H$ and the free enthalpy of reaction $\Delta_R G$ are important quantities to set up energy balances, to assess the extent of chemical reactions and to evaluate the feasibility of reactions. For an open system (the system can interact with its surroundings), it is possible to calculate the enthalpy of reaction based on the enthalpy of formation of the constituents of a chemical reaction (important: the reaction has to be carried out at constant pressure).

7.2.1.1 Enthalpy of reaction

In general, chemical reactions can take place with the release or the absorption of heat. If heat is released during a chemical reaction, this is referred to as an exothermic reaction; if, on the other hand, heat is absorbed, the reaction is endothermic. The amount of heat released or absorbed during chemical conversion is referred to as the enthalpy of reaction or reaction energy $\Delta_R H$ (also heat of reaction) and is generally related to the amount of reactant converted. Utilization of the reaction energy of exothermic reactions enables thermal recycling to be achieved and not the recycling of material. The latter can only be accomplished by mass transfer unit operations such as absorption, adsorption or membrane processes.

The standard enthalpy of reaction $\Delta_R H^0$ in kJ mol^{-1} can be calculated at standard conditions via the standard enthalpy of formation $\Delta_f H^0$ of the reactants (educts, starting materials) and products of a reaction. The standard enthalpy of formation of a compound refers to the formation of 1 mol of the respective compound from its elements at standard conditions. To make it comparable, it is mainly indicated at $\theta = 25$ °C and $P = 100$ (or 101.3) kPa for the most stable modification of a compound. It is described with the symbol $\Delta_f H^0$ in kJ mol^{-1}. The standard enthalpy of formation can be found in tables and databases for a multitude of compounds. It is possible to use the standard enthalpy of formation to calculate the enthalpy of reaction, also if the reaction is not carried out at standard conditions since the temperature deviation of the (specific heat of) compounds is similar. Therefore, the 1 Ulich's approximation is applicable [183, 184]. Equation (7.1) applies for the standard enthalpy of reaction $\Delta_R H^0$. Alternatively, reaction enthalpies can be determined experimentally by direct calorimetric measurements.

$$\Delta_R H^0 = \sum \nu_{\text{products}} \cdot \Delta_f H^0_{\text{products}} - \sum \nu_{\text{reactants}} \cdot \Delta_f H^0_{\text{reactants}} \tag{7.1}$$

The enthalpy of reaction of exothermic reactions (the reaction releases heat) is negative ($\Delta_R H^0 < 0$), and for endothermic reactions (the reaction needs heat to proceed), it is positive ($\Delta_R H^0 > 0$). If the possibility of a multistep reaction with identical initial reactants and final products exists, the total of all enthalpies of reaction of each stage equals the enthalpy of the overall reaction in case of a direct reaction (Hess's law). The reaction system changes from the same initial state to the same final state; regardless of the way in which this final state was achieved.

Standard state, standard conditions

Within chemical thermodynamics, reference is often made to the so-called "standard state" with many different linguistic customs and therefore often confusion. Standard state in the narrowest sense is a state of fixed pressure (standard pressure P^0) and pure components. In this narrowest formulation, temperature is not part of the standard state – although some sources do include temperature in the standard state (e.g., $\theta_0 = 0$ °C). The standard pressure originally suggested by the International Union of Pure and Applied Chemistry (IUPAC) is $P_0 = 101.3$ kPa (= 1 atm). Since 1982 the value recommended by IUPAC for tabulating thermodynamic data is $P_0 = 100$ kPa [185]; even this is only a suggestion and can be chosen differently. The frequently chosen temperature of $\theta = 25$ °C is often referred to as the normal temperature. In any case, the values of P_0 and T_0 need to be specified. Omission of specifying the standard state inevitably leads to ambiguity in the values of quantities and thus, errors in calculation. For our purposes the standard state (standard temperature and pressure (STP)) is specified in terms of standard temperature $\theta_0 = 0$ °C = 273.15 K and standard pressure $P_0 = 101.3$ kPa, unless otherwise stated. Note that flow meters that are calibrated in gas volumes per unit time often refer to "standard" volumes at 25 °C, not 0 °C.

Exercise 7.2: Standard enthalpy of reaction – combustion of methane
Let us determine the standard enthalpy of reaction $\Delta_R H^0$ of the oxidation of methane (CH_4) with oxygen (= combustion of methane):

$$CH_4 + 2\,O_2 \rightleftharpoons CO_2 + 2\,H_2O$$

The exercise shall be calculated for the reference state $P = 100$ kPa and $\theta = 25$ °C. The enthalpies of formation $\Delta_f H^0$ of the respective species at the conditions stated are available in the literature [186]; Tab. 7.2.

Tab. 7.2: Substance data for Exercise 7.2.

Compound	$\Delta_f H^0$ (kJ mol^{-1})
$CH_{4(g)}$	−74.6
$O_{2(g)}$	0
$CO_{2(g)}$	−393.5
$H_2O_{(g)}$	−241.8
$H_2O_{(l)}$	−285.8
$\Delta_{vap} H^0_{H_2O}$	43.99

Solver:

$$CH_{4(g)} + 2\,O_{2(g)} \rightleftharpoons CO_{2(g)} + 2\,H_2O_{(g)}$$

$$\Delta_R H^0 = \Delta_R H^0_{(g)} = \left(\nu_{CO_2} \cdot \Delta_f H^0_{CO_{2(g)}} + \nu_{H_2O} \cdot \Delta_f H^0_{H_2O_{(g)}}\right) - \left(\nu_{CH_4} \cdot \Delta_f H^0_{CH_{4(g)}} + \nu_{O_2} \cdot \Delta_f H^0_{O_{2(g)}}\right)$$

$$= (1 \cdot (-393.5) + 2 \cdot (-241.8)) - (1 \cdot (-74.6) + 2 \cdot 0) = -802.5 \text{ kJ mol}^{-1}$$

Thus, the combustion of methane is highly exothermic.

As a supplement (preview to Chapter 7.3): Investigate the link between enthalpy of reaction $\Delta_R H^0$, calorific values (lower calorific value LCV and higher calorific value HCV) and the enthalpy of vaporization of water $\Delta_{vap} H^0_{H_2O}$.

The result of the standard enthalpy of reaction $\Delta_R H^0$ has to be identical to the LCV in kJ m^{-3} of the combustion of methane, since water is formed in the gaseous state ($\Delta_f H^0_{H_2O_{(g)}}$). Examine this statement.

Check:
We know that the molar volume of an ideal gas at $P = 101.3$ kPa and $\theta_0 = 0$ °C is MV = 22.41 dm^3 mol^{-1}. At P = 100 kPa and $\theta_0 = 0$ °C, MV = 22.71 dm^3 mol^{-1} [185]. We calculated the enthalpy of reaction $\Delta_R H^0$ at P = 100 kPa and $\theta = 25$ °C. With the assumption of ideal behavior in the gas phase, we can apply the perfect gas law (eq. (2.2), also ideal gas law) to determine the molar volume at these conditions.

At $P = 101.3$ kPa and $\theta_0 = 0$ °C: 1 mol CH_4 = 22.41 dm^3.

With $\frac{p_1 \cdot MV_1}{T_1} = \frac{p_2 \cdot MV_2}{T_2}$, the molar volume at $P = 101.3$ kPa and $\theta = 25$ °C can be determined:

$$MV_{100\text{ kPa},\,25\text{ °C}} = MV_2 = \frac{p_1 \cdot MV_1 \cdot T_2}{T_1 \cdot p_2} = \frac{101.3 \cdot 22.41 \cdot 298.15}{273.15 \cdot 100} = 24.78 \text{ dm}^3$$

Thus, at $P = 100$ kPa and $\theta = 25$ °C: 1 mol CH_4 = 24.78 dm^3 and 1 m^3 CH_4 = 40.36 mol.

Now, we can determine the standard enthalpy of reaction $\Delta_R H^0$ at STP (100 kPa and 25 °C) per kg CH_4 and per m^3 CH_4:

$$\Delta_R H^0 \left[\text{kJ kg}^{-1}\right] = \frac{\Delta_R H^0}{MM_{CH_4}} = \frac{-802.5 \cdot 1{,}000}{16.04} = -50{,}031 \text{ kJ kg}^{-1}$$

and

$$\Delta_R H^0 \left[\text{kJ m}^{-3}\right] = n_{CH_4} \cdot \Delta_R H^0 = 40.36 \cdot (-802.5) = -32{,}386 \text{ kJ m}^{-3}$$

This corresponds to the LCV of methane.

For comparison: LCV (25 °C, literature [187, 188]) = 50,010 kJ kg^{-1}.

If we use the standard enthalpy of formation of water in liquid state $\Delta_f H^0_{H_2O_{(l)}}$, we get the HCV of the combustion of methane:

$$CH_{4(g)} + 2\,O_{2(g)} \rightleftharpoons CO_{2(g)} + 2\,H_2O_{(l)}$$

$$\Delta_R H^0 = \Delta_R H^0_{(l)} = \left(\nu_{CO_2} \cdot \Delta_f H^0_{CO_{2(g)}} + \nu_{H_2O} \cdot \Delta_f H^0_{H_2O_{(l)}}\right) - \left(\nu_{CH_4} \cdot \Delta_f H^0_{CH_{4(g)}} + \nu_{O_2} \cdot \Delta_f H^0_{O_{2(g)}}\right)$$

$$= (1 \cdot (-393.5) + 2 \cdot (-285.8)) - (1 \cdot (-74.6) + 2 \cdot 0) = -890.5 \text{ kJ mol}^{-1}$$

The difference between the standard enthalpies of formation of gaseous (vaporous) and liquid water has to equal the enthalpy of vaporization of water $\Delta_{vap} H^0_{H_2O}$. Note that we have to take into consideration that 2 mol of water have been vaporized in the reaction:

$$\Delta_{vap} H^0_{H_2O} = \frac{\left(\Delta_R H^0_{(g)} - \Delta_R H^0_{(l)}\right)}{2} = \frac{(-8025) - (-8905))}{2} = 44 \text{ kJ mol}^{-1}$$

Attention to units

The enthalpy of formation is given for 1 mol of the respective component. Therefore, the unit is kJ mol^{-1}. The enthalpy of reaction is also given in kJ mol^{-1}. However, you must pay attention to what it is referred to. In the case of oxidation of methane, as in Exercise 7.2, it is simple: the numerical value refers to one mole of methane ($\Delta_R H^0 = -890.5$ kJ mol^{-1} CH_4). However, reaction equations are often represented in different ways to obtain integer stoichiometric coefficients. The enthalpy of reaction must therefore be converted to relate it to 1 mol of the starting material. In any case, it is convenient to express the enthalpy of reaction (and the Gibbs free enthalpy of reaction) in kJ mol^{-1} formula conversion.

Exercise 7.3: Standard enthalpy of reaction – oxidation of ammonia

The oxidation of ammonia (NH_3) to nitrogen and water can be expressed with the following reaction equation:

$$NH_3 + 0.75\,O_2 \rightarrow 0.5\,N_2 + 1.5\,H_2O$$

At 100 °C, the standard enthalpy of reaction is $\Delta_R H^0 = -316.3$ kJ mol^{-1} NH_3

To obtain integer stoichiometric coefficients, the reaction equation can also be written as

$$4\,NH_3 + 3\,O_2 \rightarrow 2\,N_2 + 6\,H_2O$$

For the reaction enthalpy according to this reaction equation, the following standard enthalpy of reaction is stated [189]: $\Delta_R H^0 = -1{,}265.4$ kJ mol^{-1}.

Discuss the results.

Solver:
$\Delta_R H^0 = -1{,}265.4\ \text{kJ mol}^{-1}$ gives the enthalpy of reaction per four moles of NH_3 (the fourfold amount as before) or more precisely per mole formula conversion:

$$\Delta_R H^0 = -1{,}265.4\ \text{kJ mol}^{-1}\ \text{formula conversion} = -1{,}265.4\ \text{kJ per 4 mol}\ NH_3 = \frac{-1{,}265.4\ \text{kJ mol}^{-1}}{4}$$

$$= -316.3\ \text{kJ mol}^{-1}\ NH_3$$

7.2.1.2 Gibbs free enthalpy of reaction

The maximum work, which can be delivered by a system, is called capacity to do work. In chemistry, the capacity to do work by conversion of substances is described with the Gibbs free enthalpy. The change in Gibbs free enthalpy ΔG equals the difference between the free enthalpy (work capacity) of the final and the initial state. For the case that the change in Gibbs free enthalpy is negative, the reaction may occur spontaneously (if the kinetics agree; a restriction is inhibition, which has to be overcome by supply of activation energy). If the Gibbs free enthalpy of reaction is positive, the system has to be supplied with work to carry out a reaction.

Chemical reactions strive towards a dynamic equilibrium, the chemical equilibrium, in which reactants and products are present and in which the composition of the reaction mixture no longer changes when viewed from the outside. In some cases, the concentration of the products at equilibrium is so high that one can speak of complete chemical conversion; an irreversible reaction. In practice, however, many reactions do not run to completeness so that unreacted reactants and products are present at equilibrium. We call this a reversible reaction. With the law of mass action, the ratio of the amounts of products and unreacted reactants at equilibrium is described by the equilibrium constant K. An equilibrium constant, which is calculated from activities or fugacities, is called thermodynamic equilibrium constant (eq. (7.2)). Under the condition of ideal behavior, partial pressures p_i (for gas-phase reactions) or molar concentrations c_i (for liquid-phase reactions) can be used for approximate determination of the equilibrium constant of a reaction (K_p and K_c, respectively):

$$K = \frac{\prod a^{\nu}_{\text{products}}}{\prod a^{\nu}_{\text{reactants}}} \tag{7.2}$$

Information about the status of the reaction equilibrium is not only given by the equilibrium constant derived from experimental data but also by the free standard enthalpy of reaction, also called Gibbs free standard energy of reaction, $\Delta_R G^0$. (Strictly speaking we talk about the free standard enthalpy of reaction, $\Delta_R G^0$, while the free standard energy is actually rather called the "Helmholtz energy". The thermodynamics people would punish us for mixing up free energy and enthalpy of reactions.) It can be calculated from the Gibbs free standard enthalpies of formation $\Delta_f G^0$ (but different to $\Delta_R H^0$ for activity $a = 1$, eq. (7.3)), which are tabulated for 25 °C in the literature [186]:

$$\Delta_R G^0 = \sum \nu_{\text{products}} \cdot \Delta_f G^0_{\text{products}} - \sum \nu_{\text{reactants}} \cdot \Delta_f G^0_{\text{reactants}} \tag{7.3}$$

Equation (7.4) states one of the most important thermodynamic relations. With its help, the equilibrium constant of any chemical reaction can be calculated from tabulated thermodynamic data and consequently the composition of the reaction mixture at equilibrium can be predicted:

$$\Delta_R G^0 = -R \cdot T \cdot \ln K \tag{7.4}$$

In reactions with a strongly negative free standard enthalpy of reaction ($\Delta_R G^0 < -60$ kJ mol^{-1}), the reactants can be completely converted to the products (= complete reaction). Chemical reactions with a strongly positive free standard enthalpy of reaction ($\Delta_R G^0 > 60$ kJ mol^{-1}) cannot take place at all. Chemical reactions with free standard enthalpies of reaction in the range between these two limit values are equilibrium reactions in which either the products (-60 kJ mol$^{-1} < \Delta_R G^0 < 0$) or the starting materials ($0 < \Delta_R G^0 < 60$ kJ mol^{-1}) predominate at equilibrium. Thus, determining the free standard enthalpy of reaction allows to assess the maximum achievable extent of a reaction process (but does not say anything about whether you will observe the reaction to occur or not; that is why we also need the kinetics).

Exercise 7.4: Gibbs free standard enthalpy of reaction – combustion of methane
Determine the free standard enthalpy of reaction $\Delta_R G^0$ of the oxidation (combustion) of methane:

$$CH_4 + 2\,O_2 \rightleftharpoons CO_2 + 2\,H_2O$$

The exercise is calculated for the reference state $P = 100$ kPa and $\theta = 25$ °C. The free standard enthalpies of formation of the respective species are taken from [186] (Tab. 7.3).

Tab. 7.3: Substance data for Exercise 7.4.

Compound	$\Delta_f G^0$ (kJ mol^{-1})
CH_4	-50.5
O_2	0
CO_2	−394.4
H_2O	−228.6

Solver:

$$\begin{aligned}\Delta_R G^0 &= \left(\nu_{CO_2} \cdot \Delta_f G^0_{CO_2} + \nu_{H_2O} \cdot \Delta_f G^0_{H_2O}\right) - \left(\nu_{CH_4} \cdot \Delta_f G^0_{CH_4} - \nu_{O_2} \cdot \Delta_f G^0_{O_2}\right) \\ &= (1 \cdot (-394.4) + 2 \cdot (-228.6)) - (1 \cdot (-50.5) + 2 \cdot 0) = -801.1 \ \text{kJ mol}^{-1}\end{aligned}$$

From the result, we can deduce that methane is completely oxidized (combusted) with oxygen at 25 °C. In this case, the reaction equation can be written as

$$CH_4 + 2\,O_2 \rightarrow CO_2 + 2\,H_2O$$

Note: In the course of this book, reactions are written as equilibrium reactions indicated by an equilibrium reaction arrow (⇌), unless the respective free standard enthalpy of reaction is $\Delta_R G^0 < -60$ kJ mol^{-1} and the reaction clearly irreversible. This is indicated by a single reaction arrow (→).

An essential difference of the free standard enthalpy of reaction to the standard enthalpy of reaction is its, for many reactions distinct, temperature dependence (underlining, why we prefer an optimum temperature in the office to keep labor performance at maximum level). The temperature dependence of the equilibrium constant is given by the law of van't Hoff (eq. (7.5)). If the equilibrium constant K_1 is known for a temperature T_1, the equilibrium constant K_2 can be determined for any desired temperature T_2 (eq. (7.6), provided that $\Delta_R H^0$ and the standard entropy of reaction $\Delta_R S^0$ are constant for the observed temperature range; an acceptable simplification, as temperature dependence of the specific heat *Cp* of substances is nearly the same for reactants and products. Through this relationship it is possible to calculate the equilibrium constant at any temperature, if it is known for the tabulated normal temperature of 25 °C ($T = 298.15$ K):

$$\frac{d(\ln K)}{dT} = \frac{\Delta H_R}{R \cdot T^2} \tag{7.5}$$

$$\ln K_2 = \frac{\Delta_R H^0}{R} \cdot \left(\frac{1}{T_1} - \frac{1}{T_2}\right) + \ln K_1 \tag{7.6}$$

7.2.1.3 Principle of Le Chatelier

Le Chatelier's principle, also called the principle of least constraint, qualitatively describes the dependence of the chemical equilibrium on external conditions: temperature, pressure and changes in concentration. Quantitatively every system strives for a state of minimal Gibbs free enthalpy (eq. (7.7), at equilibrium $\Delta_R G = 0$):

$$\Delta_R G = \Delta_R G^0 + R \cdot T \cdot \ln K \tag{7.7}$$

If a disturbance is exerted on a chemical reaction in equilibrium, then it reacts in such a way that the effect of this disturbance will be compensated. This means that, in case of exothermic reactions, additional heat supply shifts the chemical equilibrium in the direction of the reactants, and in case of endothermic reactions, in the direction of the products (but be careful with that; ΔG is made up of ΔH and ΔS). In case of a reaction associated with a change in volume (gas-phase reaction), an increase in pressure leads to a reduction in the relative reactant conversion. Conversely, in the case of a reaction associated with a decrease in volume, an increase in pressure increases the relative conversion. When the amount of a reactant is increased in relation to its reaction partner, or when a product is removed from the reaction mixture, the state of the chemical equilibrium shifts to the side of the products. In summary, we can say that the chemical equilibrium can be shifted to the side of the products, if the influencing factors temperature, pressure and reactant concentration are chosen with care.

Applied to combustion reactions, the temperature dependence of the chemical equilibrium means that, since combustion reactions are exothermic, the lower the temperature, the more the chemical equilibrium shifts to the side of the products. This has to be taken into account in thermal and catalytic off-gas purification where complete conversion of the combustible is aimed.

Exercise 7.5: Reaction equilibrium – temperature dependence

Hydrogen and oxygen react in a spontaneous combustion reaction, the so-called oxyhydrogen or bang-gas reaction:

$$2\,H_2 + O_2 \rightleftharpoons 2\,H_2O$$

Identify why combustion of hydrogen is an exothermic reaction process by monitoring the equilibrium water fraction at various reaction temperatures (Tab. 7.4). Note: the measurements were carried out with the same initial reactant concentrations (hydrogen and oxygen in stoichiometric ratio).

Tab. 7.4: Equilibrium water fractions (mol%) of the oxyhydrogen reaction at various reaction temperatures (Exercise 7.5).

θ (°C)	**H_2O (%)**
2,200	99.0
2,400	97.5
2,600	94.7
2,800	90.0

Solver:

The temperature increase shifts the equilibrium in favor of the reactants H_2 and O_2 resulting in a lower water content at equilibrium. From that, we know that during the combustion of H_2 heat is released (= exothermic reaction).

Literature value [186]: $\Delta_R H^0_{2,500\,°C} = -503.2\ \text{kJ mol}^{-1}$ formula conversion

Temperature dependence:

$\Delta_R H^0_{1,800\,°C} = -502.6\ \text{kJ mol}^{-1}$ formula conversion and

$\Delta_R H^0_{2,800\,°C} = -503.2\ \text{kJ mol}^{-1}$ formula conversion

---

The standard enthalpy of reaction also depends on the temperature, but to a much lesser extent than the free standard enthalpy of reaction (Gibbs free standard enthalpy). It can therefore be regarded as constant within a limited temperature range.

7.2.2 Reaction kinetics

In order to design off-gas purification processes based on chemical reactions, knowledge of the rate at which a chemical reaction takes place is essential. This so-called reaction rate $-r_A$ is defined as the change of number of moles N_A of a reacting species A per time t and reference basis (e.g., unit reaction volume V). The reaction rate depends on temperature and on pressure and concentration of the reactants and products of the reaction (eqs. (7.8) and (7.9)):

$$-r_A = -\frac{dN_A}{dt \cdot V} = f(\text{temperature-dependent term, concentration-dependent term}) \quad (7.8)$$

Note: In CRE textbooks the number of moles of a species A is preferably given with N_A. However, it can also be written with n_A.

For constant-volume reactions (V = const.):

$$-r_A = -\frac{dc_A}{dt} \quad (7.9)$$

In catalytic systems, the reaction rate can be expressed in various ways differing in the basis, e.g., unit mass of catalyst M_{cat} (eq. (7.10)), unit catalyst surface S_{cat} (eq. (7.11)), unit volume of the catalyst V_{cat} (eq. (7.12)) and unit reactor volume V_R (eq. (7.13)), which can be converted into each other (eq. (7.14)):

$$-r'_A = -\frac{dN_A}{dt \cdot M_{cat}} \quad (7.10)$$

$$-r''_A = -\frac{dN_A}{dt \cdot S_{cat}} \quad (7.11)$$

$$-r'''_A = -\frac{dN_A}{dt \cdot V_{cat}} \quad (7.12)$$

$$-r''''_A = -\frac{dN_A}{dt \cdot V_R} \quad (7.13)$$

and

$$r_A \cdot V = r'_A \cdot M_{cat} = r''_A \cdot S_{cat} = r'''_A \cdot V_{cat} = r''''_A \cdot V_R \quad (7.14)$$

7.2.2.1 Rate laws

Rate laws enable the progress of a reaction to be quantitatively described in dependence of reactant and product concentrations, temperature and – in the case of heterogeneous and heterogeneous catalytic reactions – adsorption and desorption characteristics of the chemical species involved. Rate laws provide the basic prerequisite to determine the size of a reactor in order to achieve a specified relative reactant conversion X_A (eqs. (7.15) and (7.16)). ε_A denominates the volumetric factor and accounts for the change of reaction volume due to the reaction stoichiometry (see information: eqs. (7.17) and (7.18)):

$$X_A = \frac{c_{A,0} - c_A \cdot \left(\frac{T \cdot P_0}{T_0 \cdot P}\right)}{c_{A,0} + \varepsilon_A \cdot c_A \cdot \left(\frac{T \cdot P_0}{T_0 \cdot P}\right)} \tag{7.15}$$

In case of constant-volume reactions ($\varepsilon_A = 0$), eq. (7.15) simplifies to the following equation:

$$X_A = \frac{c_{A,0} - c_A}{c_{A,0}} \tag{7.16}$$

The relative conversion can also be determined from molar flow rates, mass flow rates and mass fractions of a reactant A at the start of reaction ($t = 0$ or $\tau = 0$) and after a certain reaction time (t or τ).

Varying-volume reactions

Whereas in liquid-phase reactions the volume change of the reaction mixture can be neglected and the pressure does not have to be taken into account, in gas-phase reactions, as in the case of off-gas purification in the gas phase, both parameters have to be considered in reactor design. Provided that the reaction volume changes linearly with the relative conversion X_A of reactions having a single stoichiometry, a volumetric factor ε_A (= expansion factor) can be introduced at constant pressure operation (eq. (7.17)). ε_A considers the fractional change in reaction volume V when the reaction is complete (eq. (7.18), with the initial reaction volume V_0):

$$V = V_0 \cdot (1 + \varepsilon_A \cdot X_A) \tag{7.17}$$

$$\varepsilon_A = \frac{V_{X_A=1} - V_{X_A=0}}{V_{X_A=0}} \tag{7.18}$$

With the help of the volumetric factor, nonstoichiometric ratios of reactants and inert substances can also be dealt with in chemical processes.

Exercise 7.6: Varying-volume reactions – combustion of methane versus combustion of propane

Discuss the change of reaction volume for (a) the combustion of methane and (b) the combustion of propane (C_3H_8).

Solver:

(a) Combustion of methane: $CH_4 + 2\,O_2 \rightarrow CO_2 + 2\,H_2O$

For constant temperature and constant pressure operation, the volumes of the reactants and products are proportional to their molar amounts (assumption: the ideal gas law is valid):

$$p \cdot V = n \cdot R \cdot T => V = \frac{n \cdot R \cdot T}{p} => V \sim n$$

Thus, the volume of an undiluted stoichiometric feed of methane and oxygen ($V_{X_{CH_4}=0}$) can be approximated with:

1 volume unit CH_4 + 2 volume units O_2 = 3 volume units

The volume of the reaction products CO_2 and water at complete methane conversion ($V_{X_{CH_4}=1}$) is:

1 volume unit CO_2 + 2 volume units H_2O = 3 volume units

With this reaction equation-related basics, we can calculate the volumetric factor:

$$\varepsilon_{CH_4} = \frac{V_{X_{CH_4}=1} - V_{X_{CH_4}=0}}{V_{X_{CH_4}=0}} = \frac{3-3}{3} = 0$$

This means that the reaction volume does not change when methane is combusted.

(b) Combustion of propane: $C_3H_8 + 5\,O_2 \longrightarrow 3\,CO_2 + 4\,H_2O$

The volume of an undiluted stoichiometric feed of propane and oxygen ($V_{X_{C_3H_8}=0}$) can be approximated with:

1 volume unit C_3H_8 + 5 volume units O_2 = 6 volume units

The volume of the reaction products CO_2 and water at complete propane conversion ($V_{X_{C_3H_8}=1}$) is:

3 volume units CO_2 + 4 volume units H_2O = 7 volume units

For propane combustion, we get a volumetric factor of

$$\varepsilon_{C_3H_8} = \frac{V_{X_{C_3H_8}=1} - V_{X_{C_3H_8}=0}}{V_{X_{C_3H_8}=0}} = \frac{7-6}{6} = \frac{1}{6}$$

This means that the reaction volume increases when propane is combusted. After complete propane conversion, the reaction volume is increased by one-sixth.

Supplement 1: Do not panic when, based on your reaction equation, ε_A has a negative signum. It just confirms that the number of reaction products is less than the number of reactants. As a consequence, the volume will drop during conversion (and you will have fun with increasing the pressure, as stated before).

Supplement 2:
The volumetric factor can also be calculated in a second way:
For a general reaction a A + b B → c C + d D, first, a key component is selected to which the stoichiometric coefficients are normalized. In our example procedure this is reactant A:

$$\frac{a}{a}\mathrm{A} + \frac{b}{a}\mathrm{B} \longrightarrow \frac{c}{a}\mathrm{C} + \frac{d}{a}\mathrm{D}$$

Next, the change in the total number of moles in relation to the number of moles of the key component (in our case component A) is calculated with eq. (7.19):

$$\sigma_A = \frac{d}{a} + \frac{c}{a} - \frac{b}{a} - 1 \tag{7.19}$$

With the molar feed fraction of the key component $y_{A,0}$, the volumetric factor can be determined (eq. (7.20)):

$$\varepsilon_A = y_{A,0} \cdot \sigma_A \tag{7.20}$$

Applied to the combustion of propane:

$$\sigma_{C_3H_8} = \frac{3}{1} + \frac{4}{1} - \frac{5}{1} - 1 = 1$$

$$\varepsilon_{C_3H_8} = y_{C_3H_8,0} \cdot \sigma_{C_3H_8}\,\frac{1}{(1+5)} \cdot 1 = \frac{1}{6} \cdot 1 = \frac{1}{6}$$

Exercise 7.7: Effect of inert gas on varying-volume reactions – combustion of propane
Compare the change of reaction volume for the combustion of propane: (a) with pure oxygen (= no inert gas) and (b) with air (= with inert gas nitrogen). In both cases, oxygen is used in stoichiometric amount to propane.

Combustion of propane: $C_3H_8 + 5\,O_2 \longrightarrow 3\,CO_2 + 4\,H_2O$

(a) From Exercise 7.6 we know that, when propane is completely combusted in pure oxygen, the reaction volume increases by one-sixth ($\varepsilon_{C_3H_8} = \frac{1}{6} = 0.17$).
(b) The composition of dry air can be approximated with 21 vol% O_2 (= 21 mol%) and 79 vol% N_2 (= 79 mol%). Consequently, we feed 79% of inert gas to the combustion process. The molar fraction of propane in the feed is

$$y_{C_3H_8,0} = \frac{1(C_3H_8)}{1(C_3H_8) + 5(O_2) + \frac{5\cdot 0.79}{0.21}(N_2)} = 0.04$$

$$\varepsilon_{C_3H_8} = y_{C_3H_8,0} \cdot \sigma_{C_3H_8} = 0.04 \cdot 1 = 0.04$$

We can see that the presence of inert gas considerably reduces the volume increase during the reaction. The reaction volume increases by only four-tenths relative to the initial volume.

Reactions can be accompanied by an increase ($\varepsilon_A > 0$) or a decrease ($\varepsilon_A < 0$) in reaction volume. In both cases, the extent is mitigated by the presence of an inert component (e.g., nitrogen in air).

For a single-phase reaction of the type $a\ A + b\ B \rightleftharpoons c\ C + d\ D$, the resulting rate law (also kinetic law) may be expressed via a simple power-law approach with eq. (7.21):

$$-r_A = k_1 \cdot c_A^{\alpha} \cdot c_B^{\beta} - k_2 \cdot c_C^{\gamma} \cdot c_D^{\delta} \tag{7.21}$$

In the power rate law in eq. (7.21), k_1 and k_2 denote the reaction rate constants of the forward and reverse (backward) reaction (very important for representing the temperature-dependent terms of the rate of forward reaction and the rate of backward reaction). In the concentration-dependent terms, α, β, γ and δ are the orders of the reaction with respect to the corresponding reactant or product. For an elementary reaction, the reaction orders in the rate law can be derived from the stoichiometric coefficients of the respective species in the reaction equation directly. This is valid for elementary reactions only. For all other reactions the rate law has to be experimentally determined. An elementary reaction is a special type of chemical reaction, in which the products are formed from the reactants in one direct step according to the reaction equation.

A power rate law expression is simple, but in principle purely empirical and valid only in the range of conditions for which it was established (e.g., temperature, pressure, reactant concentrations and catalyst). However, the basic goal is to find a rate law that is as simple as possible, but adequately describes the experimental kinetic data. Simple rate laws reduce the computational effort in the subsequent design steps for reactor sizing considerably. The simplest case of a power rate law is given in eq. (7.22) for an irreversible first-order reaction:

$$-r_A = k \cdot c_A \tag{7.22}$$

In some cases, heterogeneous and heterogeneous catalytic reactions may be simply treated as homogeneous reactions. In such cases, pseudo-homogeneous rate laws can be postulated. The major drawback of this approach is that all factors associated with adsorption and reaction mechanisms on the surface of the solid catalyst are ignored. Thus, power law kinetics are often seen as a means of obtaining preliminary values of the kinetic parameters for heterogeneous catalytic reactions. Rate laws for heterogeneous reactions also account for adsorption and desorption of reactants and products and may adopt various forms containing adsorption (e.g., K_A and K_B) and desorption constants in addition to reaction rate constants (e.g., eq. (7.23)):

$$-r_A = \frac{k \cdot K_A \cdot K_B \cdot p_A \cdot p_B}{1 + K_A \cdot p_A + K_B \cdot p_B} \tag{7.23}$$

One of the most commonly used methods to describe heterogeneous catalytic reactions is the Langmuir–Hinshelwood–Hougen-Watson (LHHW) approach. The rate law is first derived in terms of surface concentrations of adsorbed species and vacant catalyst sites. In general, surface concentrations are not directly measurable. Therefore, surface concentrations are related to bulk (fluid) reactant concentrations that are measurable. The LHHW approach is a bimolecular approach that is based on the principle that two reactant molecules (A and B) are adsorbed on catalytically active sites before they react with each other in the adsorbed state. The Eley–Rideal (ER) approach is another model for heterogeneous catalytic reactions. It describes a reaction mechanism in which one reactant (A) is adsorbed while the second reactant (B) remains in the gas phase. Consequently, the reaction occurs when adsorbed species A react with passing gas molecules of B in the fluid phase (e.g., the off-gas).

When gas-phase reactions are considered, concentrations are preferably expressed in terms of partial pressures (p_A, p_B) in the rate law (e.g., eq. (7.23)). For ideal systems, the ideal gas law (also perfect gas law, eq. (2.2)) can be applied. Consequently, along with Dalton's law (additive pressures) and Amagat's law (additive volumes), the molar fraction y_A or molar concentration c_A of a species A in a gas mixture can be obtained from its partial pressure (eq. (7.24)):

$$c_A = \frac{n_A}{V} = \frac{P \cdot y_A}{R \cdot T} \tag{7.24}$$

Such rate laws derived from mechanistic models that simulate the actual surface phenomena during heterogeneous catalytic processes are to be preferred for reactions involving solid catalysts, when a precise description of the actual reaction mechanism is important. However, for estimation and preliminary design of reactors and process concepts, power rate law models are a good approximation.

i

Discussion of rate laws – SCR of NO

SCR of NO with NH_3 (so-called NH_3-SCR) is one of the most prominent reactions in catalytic automobile off-gas purification and industrial off-gas denoxing and has therefore been extensively studied:

$$4\,NO + 4\,NH_3 + O_2 \rightarrow 4\,N_2 + 6\,H_2O$$

In general, the rate of reaction depends on NO_x, NH_3, O_2 and H_2O concentrations. The reaction follows very complex kinetics and the reaction rate can only be described by a mix of different types of reaction kinetics as reported by Roduit et al. [190]. It was found that at low temperatures, the reaction can be described by Langmuir–Hinshelwood kinetics, while at medium temperatures the reaction follows an Eley–Rideal-type kinetics. Consequently, it can be assumed that in the general operation range of 300–400 °C, the reaction kinetics of Eley–Rideal type meets reality best. For a pure Eley–Rideal mechanism, the following rate law was postulated [191]:

$$-r_{NO} = k \cdot p_{NO} \cdot \frac{K_{NH_3} \cdot p_{NH_3}}{1 + K_{NH_3} \cdot p_{NH_3}} \cdot \frac{K_{O_2} \cdot p_{O_2}}{1 + K_{O_2} \cdot p_{O_2}}$$

However, a large number of heterogeneous catalysts and reaction conditions are described in the literature, and various rate laws have been postulated [192, 193]. The simplest approach is to express the rate of NO conversion by a power law (with either gas-phase concentrations or partial pressures):

$$-r_{NO} = k \cdot c_{NO}^{\alpha} \cdot c_{NH_3}^{\beta} \cdot c_{O_2}^{\gamma}$$

Depending on the process conditions and the nature of the catalyst, different reaction orders are reported in the literature. When the rate of reaction is independent of NH_3 concentration (because NH_3 acts much faster than NO) and the impact of O_2 is negligible, the rate law changes to a first-order rate law; as described by Li et al. [194] for Fe–Mo/ZSM-5 catalysts:

$$-r_{NO} = k \cdot c_{NO}$$

For simplicity, the SCR reaction is often regarded as an irreversible first-order reaction with respect to the limiting component (the component that is not used in excess) [191]. The first-order reaction approach is also used as the basis for the SCR calculations in Section 7.3.

7.2.2.2 Temperature dependence

Since chemical reactions are related to the frequency of collision and the relative velocity of the molecules in the reaction mixture, the reaction rate is, as previously discussed, proportional to the concentration of the reactants, and it depends on the state of aggregation (possibility of diffusion), pressure and catalyst. In addition to that, it especially depends on the temperature. In a specified temperature range, the Arrhenius' equation (eq. (7.25)) allows to quantitatively account for the effect of temperature on the rate of reaction by providing a quantitative relationship between the reaction rate constant k, the activation energy E_A, the pre-exponential factor (also called frequency factor) A and the temperature T:

$$k = A \cdot e^{-\frac{E_A}{RT}} \tag{7.25}$$

When the reaction rate constant is plotted logarithmically ($\ln k$) against the reciprocal of the temperature ($1/T$) in K^{-1}, the exponentially temperature-dependent rate constant is mapped onto a straight line (linearization of the function, Arrhenius plot). According to the usually quite different values of E_A the exponential function term is considerably more temperature-sensitive than the pre-exponential factor. Significant deviation from a straight line of a ln k–T^{-1} diagram, due to the temperature dependency of the pre-exponential factor A or the activation energy E_A, can only be expected if a large temperature interval is investigated ($\Delta T > 100$ K); or for reactions following a complex temperature-dependent reaction mechanism.

Nonlinear curves of ln k–T^{-1} diagrams are an indication for the existence of complex reaction mechanisms or a change of the dominant mechanism within the investigated temperature range. Examples of such behavior are shown in Fig. 7.1. If a complex reaction consists of two parallel reactions with different activation energies, it results in a curve like curve A (e.g., one reaction dominates at high temperatures and another reaction dominates at low temperatures). The sudden change in curve B is based on autoignition at a certain autoignition temperature. This can be, for example, a branched chain reaction, which proceeds explosively above this autoignition temperature (if you register this shape you must either try to keep temperature below this critical temperature, or you are recommended to jump through the emergency exit to avoid a fatal physical impact). Below this temperature a slower, molecular reaction dominates. The curve C could be the result of the decreasing efficiency of a heterogeneous catalyst or the catalyst's destruction with increasing temperature. It may also indicate that your reaction shifts into unwanted temperature-dependent decomposition of any constituent. According to Schmid and Sapunov

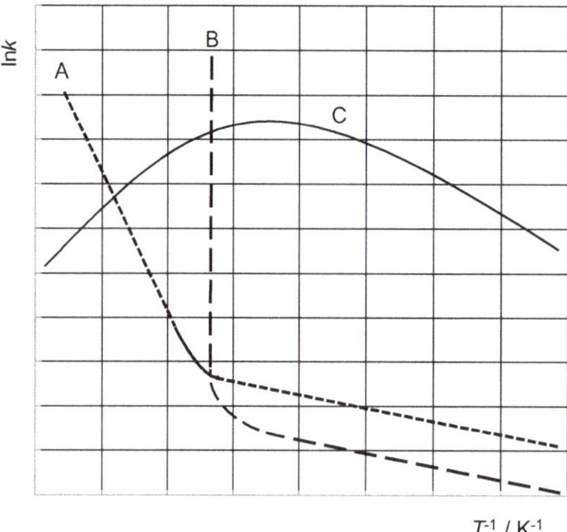

Fig. 7.1: Arrhenius plots for complex reaction mechanisms: (A) two parallel reactions with different activation energies, (B) branched chain reaction, which proceeds explosively above an autoignition temperature and (C) heterogeneous catalytic reaction with decreasing efficiency of the catalyst at elevated temperatures.

[195] a downward concave curve indicates a parallel reaction and an upward concave curve a change of the rate-determining step of a consecutive reaction.

7.2.2.3 Summary

The selection of the temperature at which a chemical reaction is to take place has to be considered in two aspects. On the one hand, a higher reaction temperature allows a higher reaction rate and thus a shorter residence time and a smaller reactor, which is beneficial from an economic and space-saving point of view. On the other hand, the temperature dependence of the chemical equilibrium must be taken into account. For example, (excessively) high reaction temperatures are disadvantageous in exothermic reactions since they shift the equilibrium to the side of the starting materials. In heterogeneous catalytic reactions, the temperature stability of the catalyst determines the maximum operation temperature and needs to be considered too.

Exercise 7.8: Selection of reaction temperature – chemical equilibrium versus reaction rate

The oxidation of NO to NO_2 is considered a key step in NO_x elimination as nitrogen oxides, mainly NO, are among the major air pollutants. The oxidation of NO with oxygen to NO_2 is an exothermic reaction. The chemical equilibrium shows a pronounced temperature dependency. The thermodynamic data are given in the table below [189]. The reaction kinetics are well investigated with a series of different catalysts [196] and without catalysts [197]. Mulla et al. [198], for instance, investigated the reaction kinetics on Pt catalysts including catalyst deactivation. The rate of NO conversion increases with increasing temperature. Let us discuss the selection of the reaction temperature:

$$2\,NO + O_2 \rightleftharpoons 2\,NO_2$$

Tab. 7.5: Substance data of NO oxidation to $NO_2 (2\,NO + O_2 \rightleftharpoons 2\,NO_2)$ for Exercise 7.8 [189].

θ (°C)	$\Delta_R H^0$ (kJ mol^{-1})	$\Delta_R G^0$ (kJ mol^{-1})
100	−115.4	−59.8
200	−116.2	−44.8
300	−116.6	−29.6
400	−116.8	−14.4
500	−116.7	0.08
600	−116.5	16.0
1,000	−115.1	76.4

Discussion:

Basically, the reaction rate increases by raising the temperature. This shortens the time until the reaction equilibrium is reached. However, the reaction temperature has a decisive influence on the position of the equilibrium. At 100 °C, the reaction of NO to NO_2 is complete, whereas at 100 °C, NO is no longer oxidized to NO_2. Up to approximately 500 °C, the reaction of NO to NO_2 is thermodynamically favored.

7.2.3 Types of reaction

Chemical reactions may be classified according to the number of phases involved: homogeneous and heterogeneous reactions. Homogeneous reactions are reactions that take place in a uniform phase, in the case of off-gas purification in a uniform gas phase (homogeneous gas-phase reactions). Heterogeneous reactions are reactions that take place in several phases. In the case of off-gas purification processes, this is mainly the case with catalytic processes, in which the gas-phase reaction takes place with the aid of a solid catalyst (heterogeneous catalytic gas-phase reactions).

7.2.3.1 Homogeneous gas-phase reactions

Homogeneous gas-phase reactions are based on three process steps. These are (i) the preparation of the feed material – the off-gas and auxiliary reactants, (ii) mixing of the feed material by either diffusion or convection, and (iii) subsequently the reaction of the reactants. The design parameters temperature, residence time and turbulence (the "three T") determine the overall rate of the reaction process. For exothermic reactions, the reaction temperature is generally so high that the chemical reaction is not the rate-determining step. Thus, the main task is to optimally design mixing of the reactants [199].

7.2.3.2 Heterogeneous catalytic gas-phase reactions

In heterogeneous catalytic gas-phase reactions, the gaseous reactants are present in a different phase from the solid catalyst. Therefore, physical transport processes to and from the catalyst are important in addition to the actual chemical reaction step. The chemical reaction either proceeds directly on the catalyst surface or within the immediate vicinity of the catalyst surface, suggesting a large effect of the surface area. A large surface area can be technically realized in different ways, for example, by high porosity of the catalyst. Usually heterogeneous catalytic reactions proceed sequentially in multiple steps (Fig. 7.2). The slowest step in the following sequence is the rate-determining step:

- convective transport of the gaseous reactants from the bulk phase to the boundary layer of the solid catalyst (1);
- diffusion of the reactants through the boundary layer to the geometric surface of the catalyst (2);
- diffusion of the reactants through the pores to the active inner surface of the catalyst (3);
- adsorption on the catalytically active site (4);
- chemical reaction (5);
- desorption of the products from the surface of the catalyst (6);
- diffusion of the products back through the pores to the external geometric surface (7);
- diffusion of the products through the boundary layer (8); and
- convective transport of the products from the boundary layer in the bulk phase (9).

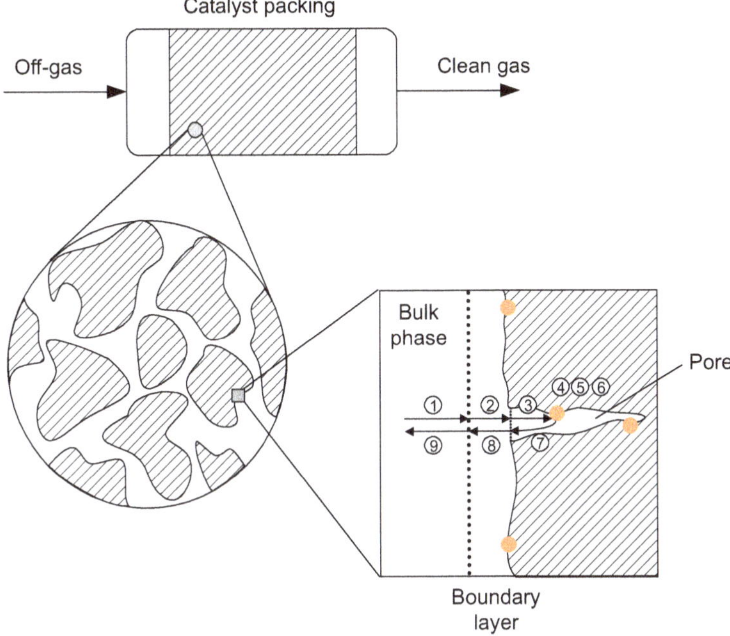

Fig. 7.2: Packed-bed reactor with a porous catalyst with illustration of the individual steps of heterogeneous catalytic gas-phase reactions: (1) convective transport of the gaseous reactants from the bulk phase to the boundary layer of the solid catalyst; (2) diffusion of the reactants through the boundary layer to the geometric surface of the catalyst; (3) diffusion of the reactants through the pores to the active inner surface of the catalyst; (4) adsorption on the catalytically active site; (5) chemical reaction; (6) desorption of the products from the surface of the catalyst; (7) diffusion of the products back through the pores to the external geometric surface; (8) diffusion of the products through the boundary layer; (9) convective transport of the products in the bulk phase.

Generally, in heterogeneous catalysis we have to distinguish between

- parallel processing of mass transfer and reaction $r_{\text{total}} = \Sigma\, r_i$ and
- sequential processing of mass transfer and reaction $r_{\text{total}} = r_1 = r_2 = \ldots$

For parallel processing, the overall conversion is higher than the conversion of the single step. For sequential processing, the slowest step in the sequence determines the overall conversion for steady-state operation. When combining mass transfer (eq. (7.26)) and first-order reaction (eq. (7.27)) the same basis has to be applied. Favorably, the reaction rate should be referred to the conversion by unit surface area of the catalyst (S_{cat}) or to the conversion by unit catalyst mass instead of using the reactor volume as reference basis:

$$r_{\text{mass transfer}}\colon -r_g = -\frac{1}{S_{\text{cat}}} \cdot \frac{dN_A}{dt} \tag{7.26}$$

$$r_{\text{reaction}}:\ -r_S = -r'' = -\frac{1}{S_{\text{cat}}} \cdot \frac{dN_A}{dt} \tag{7.27}$$

A mass balance over a section of the catalyst channel results for first-order reactions in the following relationship for the reciprocal of the overall reaction rate constant k_{total} [eq. (7.28), also see Exercise 7.9]. β_g is the mass transfer coefficient in the gas film and $(\varepsilon \cdot k_S)$ is an apparent chemical reaction rate constant normalized to the catalyst surface area in which ε is the effectiveness factor (eq. (7.29)). The concept of effectiveness factor is utilized when the effect of pore diffusion on the overall rate of a reaction needs to be taken into account:

$$\frac{1}{k_{\text{total}}} = \frac{1}{\beta_g} + \frac{1}{\varepsilon \cdot k_S} \tag{7.28}$$

$$\varepsilon = \frac{\text{actual rate of reaction}}{\text{rate of reaction if pore diffusion was infinitively fast}} \tag{7.29}$$

Exercise 7.9: Interaction of mass transfer and heterogeneous catalytic surface reaction

Discuss the interaction of mass transfer and surface reaction for irreversible first-order reactions. Develop the overall rate for this linear process. According to the schematic sketch (Fig. 7.3), gaseous reactant A diffuses from a carrier gas (bulk phase) to the surface of the solid matter B and reacts under formation of the gaseous reaction product R, which diffuses then back into the carrier gas. Thus, two steps in series are involved – mass transfer of gaseous reactant A to the surface of the solid matter (catalyst), followed by reaction of A at the surface of the catalyst to product R.

Note: the effect of pore diffusion on the overall rate of a reaction does not have to be taken into account:

$$A_{(g)} \overset{B_{\text{solid}}}{\longrightarrow} R_{(g)}$$

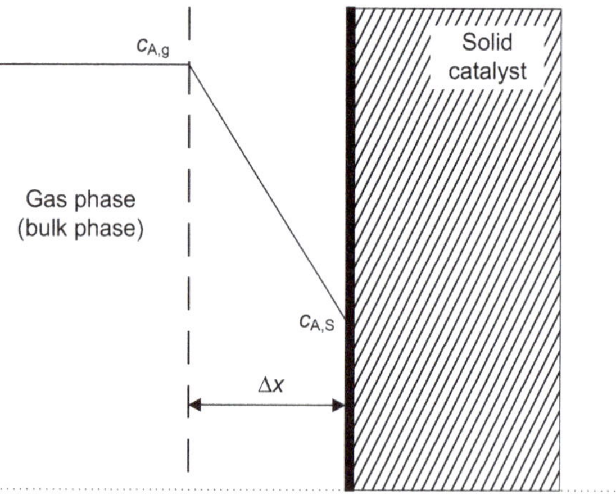

Fig. 7.3: Schematic sketch of mass transfer and heterogeneous catalytic surface reaction $A_{(g)} \overset{B_{\text{solid}}}{\longrightarrow} R_{(\text{gas})}$.

Solver:
To get an overall rate expression, we write the individual rate steps on the same basis; in our case per unit catalyst surface. For mass transfer, we assume that the film theory (see Chapter 8.5.1) applies:

Mass transfer term: For the flux of reactant A in the gas phase to the surface of the solid catalyst by convective diffusion we get

$$-r_g = -\frac{1}{S_{cat}} \cdot \frac{dN_A}{dt} = D \cdot \frac{\Delta c}{\Delta x} = \frac{D}{\Delta x} \cdot (c_{A,g} - c_{A,S}) = -\beta_g \cdot (c_{A,g} - c_{A,S})$$

Reaction term: The reaction is first order with respect to A. Thus, for the rate law (normalized equivalent reaction rate, stoichiometric coefficient of A $\nu_A = -1$) based on unit catalyst surface we get

$$r_S = -\frac{r''_A}{\nu_A} = -\frac{1}{\nu_A \cdot S_{cat}} \cdot \frac{dN_A}{dt} = k_S \cdot c_{A,S}$$

At steady-state conditions, the flow rate to the surface r_g is equal to the reaction rate at the surface of the solid catalyst r_S (= steps in series):

$$r_g = r_S \text{ and } \beta_g \cdot (c_{A,g} - c_{A,S}) = k_S \cdot c_{A,S}$$

Due to the steady-state condition the nonmeasurable concentration of gaseous reactant A at the interface $c_{A,S}$ can be eliminated:

$$c_{A,S} = \frac{\beta_g \cdot c_g}{k_S + \beta_g}$$

$$-r_g = -r_S = -\frac{1}{S} \cdot \frac{dN_A}{dt} = \frac{\beta_g \cdot k_S}{\beta_g + k_S} \cdot c_{A,g} = \frac{1}{\frac{1}{\beta_g} + \frac{1}{k_S}} \cdot c_{A,g} = k_{total} \cdot c_{A,g}$$

!

The resistance of mass transfer $1/\beta_g$ and the resistance of reaction $1/k_S$ can be added, although only for the condition that mass transfer and reaction rate occur in series and that the reaction is of first order and the mass transfer rate is a linear function of the driving force.

It is an ease to check for mass transfer or reaction control. If conversion increases when you increase the velocity of reactants, you are mass transfer-controlled. When conversion increases by increasing the temperature, you are kinetically controlled. You can easily identify the type of rate control, when performing and interpreting an Arrhenius plot. $E_A < 40$ kJ mol^{-1} strongly indicates mass transfer control. $E_A \gg 40$ kJ mol^{-1} strongly indicates reaction control.

The simplest method to represent data for gas-film mass transfer coefficients is achieved by relating the Sherwood number (Sh) to the Reynolds number (Re) and the Schmidt number (Sc). The Sherwood number (eq. (7.30)) describes the ratio of the convective mass transfer to the rate of diffusive mass transport. It, thus, forms as a dimensionless mass transfer coefficient:

$$\mathrm{Sh} = f(\mathrm{Re}, \mathrm{Sc}) = \frac{\beta_g \cdot d_H}{D} \tag{7.30}$$

There are numerous books available that list dimensionless groups for chemical engineering processes; [200–202] are some of them, for example. The following dimensionless groups apply to mass transfer (gas-solid) in laminar and turbulent flow.

For turbulent flow mass transfer in circular pipes, the following Sherwood correlation applies for gases and liquids (eq. (7.31)). Furthermore, it is mentioned that the data for gases only may be better correlated with 0.44 replacing 0.33 as the exponent on Sc [201, 202]:

$$\mathrm{Sh} = 0.023 \cdot \mathrm{Re}^{0.8} \cdot \mathrm{Sc}^{0.33} \tag{7.31}$$

Sherwood correlations for laminar flow are given in eq. (7.32) (fully developed parabolic velocity profile) and eq. (7.33) (approximate solution) with Graetz numbers (Gz) as stated in eq. (7.34) [201]:

for

$$\mathrm{GZ} < 16\colon\ \mathrm{Sh} = 3.66 + \frac{0.085 \cdot \mathrm{Gz}}{1 + 0.047 \cdot \mathrm{Gz}^{0.667}} \tag{7.32}$$

for

$$\mathrm{GZ} > 16\colon\ \mathrm{Sh} = 1.86 \cdot \sqrt[3]{\mathrm{Gz}} \tag{7.33}$$

$$\mathrm{Gz} = \mathrm{Re} \cdot \mathrm{Sc} \cdot \frac{d_\mathrm{H}}{L_\mathrm{cat}} \tag{7.34}$$

For noncircular ducts, the hydraulic diameter d_H can be calculated with the following equation [201]:

$$d_\mathrm{H} = \frac{4 \cdot \text{cross-sectional area}}{\text{perimeter}} \tag{7.35}$$

For laminar flow in tubular coils, the following Sherwood correlations for Dean numbers (De, eq. (7.38)) less than 10 (eq. (7.36)) and larger than 10 (eq. (7.37)) apply:

for

$$\mathrm{De} < 10\colon\ \mathrm{Sh} = \mathrm{Sh}_\text{straight tube} + 10 \cdot \mathrm{De} \tag{7.36}$$

for

$$\mathrm{De} > 10\colon\ \mathrm{Sh} = \mathrm{Sh}_\text{straight tube} \cdot (1 + 1.4 \cdot \mathrm{De}) \tag{7.37}$$

with

$$\mathrm{De} = \mathrm{Re} \cdot \frac{d_\mathrm{H}}{D_\text{module}} \tag{7.38}$$

7.2.4 Heterogeneous catalysis

In 1894, Ostwald found out that small amounts of substances can accelerate chemical reactions without being consumed. Since then, catalysis and catalysis research have developed into an independent discipline within the field of CRE in general and nowadays in off-gas purification in specific. A comprehensive presentation of the field of heterogeneous catalysis is given in the textbooks [191, 203, 204].

The reaction rate of noncatalytic reactions is distinctively temperature-dependent. However, different mechanisms due to catalysis can enable reactions at operation conditions – lower temperatures – at which noncatalytic reactions would not be observed. Qualitatively, this difference can be explained by comparing the activation energies (Fig. 7.4).

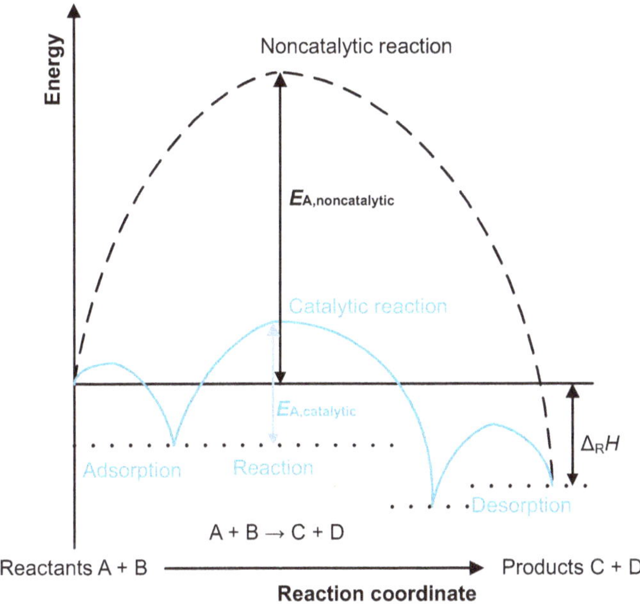

Fig. 7.4: Qualitative comparison of the energy barrier E_A (activation energy) of a catalytic and a noncatalytic chemical reaction.

As Fig. 7.4 shows, for noncatalytic reactions a considerably higher energy barrier has to be overcome to get the reaction started. Several steps contribute to the heterogeneous catalytic reaction. Nevertheless, in both cases the equilibrium between reactants and products is reached according to the standard free enthalpy of reaction $\Delta_R G^0$. Catalysts influence the reaction rate (in general they increase it), but not the position of the chemical equilibrium.

! Note that reactions that cannot take place due to limitations in chemical thermodynamics cannot be initiated by a catalyst either.

7.2.4.1 Selectivity and yield of a chemical reaction

Catalysis has proven to be an important tool to improve yield Y, selectivity S and rate of chemical reactions. As only those reaction(s) should occur on the catalyst that leads to the intended reaction products, a high selectivity is crucial. For reactant A that is converted via two parallel reactions (eqs. (7.39) and (7.40)) with product R being the desired reaction product, the definitions of yield (eq. (7.41)) and selectivity (eq. (7.42)) are shown:

Reactions

$$\nu_A A \rightleftharpoons \nu_R R + \nu_X X \tag{7.39}$$

and

$$\nu_A A \rightleftharpoons \nu_P P + \nu_Y Y \tag{7.40}$$

Yield Y of product R from reactant A:

$$Y_{RA} = \frac{n_R - n_{R,0}}{n_{A,0}} \cdot \frac{|\nu_A|}{|\nu_R|} \tag{7.41}$$

Selectivity S toward product R from reactant A:

$$S_{RA} = \frac{n_R - n_{R,0}}{n_{A,0} - n_A} \cdot \frac{|\nu_A|}{|\nu_R|} = \frac{Y_{RA}}{X_A} \tag{7.42}$$

The activity of a catalyst is expressed by its turnover frequency (TOF), which is defined by the number of molecules that react per active site of the catalyst in unit time (eq. (7.43), IUPAC definition [205]). The term is not always used uniformly and can therefore lead to misunderstandings [206]. According to IUPAC, the terms TOF and turnover number (TON) seem to have one and the same meaning. However, the term TON is sometimes also used for the number of molecules reacting per number of catalytically active sites. Moreover, the number of molecules of product may be used instead of the reacting molecules. For this reason, it is always important to indicate how the respective parameter was calculated:

$$\text{TOF} = \frac{\text{number of moles reacting}}{\text{number of active sites} \cdot \text{time}} \tag{7.43}$$

The operation capacity of heterogeneous catalysts is expressed by the ratio of the volumetric flow rate (often at STP) of the off-gas and the catalyst volume, yielding the catalyst volume-based space velocity (SV) in $m^3_{STP}\ h^{-1}\ m^{-3}$ (eq. (7.44)). In practice, the usual space velocities lie between 10,000 and 50,000 m^3_{STP} h^{-1} m^{-3}, depending on the desired purity of the purified off-gas, the temperature and the pressure during operation [207]:

$$SV = \frac{F_{V,0}}{V_{cat}} \tag{7.44}$$

7.2.4.2 Heterogeneous catalysts in off-gas purification

Heterogeneous catalysts are generally used as fine particles, powders and granules. In these heterogeneous catalysts, the catalytically active component may be

- deposited on a solid support (= supported catalysts) or
- used in bulk form (= unsupported catalysts).

In off-gas purification, mainly supported catalysts are used. In addition to the actual catalytically active component, these consist of a porous support material on which the catalytically active component is applied and usually additional promoters. As in heterogeneous catalysis the reaction either proceeds directly on the catalyst surface or within the immediate vicinity of the catalyst surface, a large surface area and thus high porosity of the catalyst is beneficial. High-porosity support materials with a specific surface area of some 100 m^2 g^{-1} catalyst mass are preferably used. Support materials are made of ceramics and metals, onto which the catalytically active substances, usually noble metals, are applied in fine distribution in concentrations of 0.1–0.5 wt%. In addition to ensuring a high specific surface area, support materials bring another advantage with them. As the catalytic material is usually expensive, fine distribution of the active component on the support material has a clear economic advantage.

Promoters are substances that are added in small amounts to the solid catalyst to improve its performance – activity, selectivity and catalyst lifetime – for the intended reaction process. A promoter either augments a desired reaction or suppresses an undesired reaction path. It has little or no catalytic effect by itself. Some promoters may also interact with active components of catalysts causing changes in the electronic or crystal structures of the catalytically active component. Commonly used promoters include metallic ions that are incorporated into metals and metal oxide catalysts. Viewed as a whole, the catalytic properties of supported catalysts are determined by the catalytically active component itself, its distribution on a support material, the type of support material and the addition of promoters. Supported catalysts are generally applied in the form of bulk or honeycomb catalysts.

Bulk catalysts of transition metal oxides are widely used in catalytic postcombustion (CPC) processes. Their high porosity is achieved by granulating and tempering. Thus, inner surfaces up to several hundred square meters per gram of catalyst can be generated. Raney nickel is a representative example. Raney nickel is produced by hydrolysis of a Ni/Na alloy. By removing the Na from the alloy, the nickel skeleton remains with a very high porosity and surface.

Catalysts are called monoliths when their activity is limited to the geometric surface. Both ceramic and metal monolithic supports are used, each of which has specific advantages. Ceramic monoliths are mainly produced by extrusion from synthetic cordierite (2 MgO·2 Al_2O_3·SiO_2). The main advantages are the high melting point (>1,300 °C), high mechanical strength and a low thermal expansion coefficient providing pronounced thermal shock resistance. Monoliths that are made of nonporous metal (metal foils) provide good thermal properties and show mechanical and geometrical advantages [208]. The geometric surface areas of ceramic and metallic support structures are in the range of 2–4 m^2 dm^{-3} support volume, which is regarded too low for adequate performance of catalytic pollutant conversion in the off-gas. This limitation is overcome by coating the support with a thin layer of a mixture of inorganic oxides with a high internal surface area. The mixture is termed "washcoat." Due to the washcoat the surface of the catalytic structure is increased to between 10,000 and 40,000 m^2 dm^{-3} support volume. The catalytically active components, mainly precious metals, are then deposited onto the washcoat. The washcoat components generally support the catalytic function. For ceramic supports with square-shaped channels, a typical washcoat layer has a thickness of about 10–30 µm on the walls of the support and of about 100–150 µm in the corners [208].

Figure 7.5 depicts the cross-sectional profile of two monolithic catalysts: a surface-coated folded metal catalyst and a homogeneously doped honeycomb catalyst.

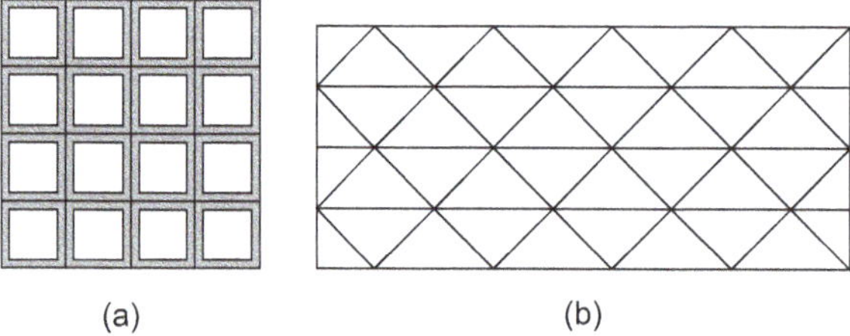

Fig. 7.5: Cross-sectional profile (cross-sectional area) of two monolithic catalysts: (a) homogeneously doped honeycomb catalyst and (b) surface-coated folded metal catalyst.

Monolithic catalysts consist of a single body with a multitude of parallel channels, where the catalytically active material is deposited along the walls of the channels. Parallel arrangement of many small channels made of a support material that is coated or doped with a catalyst – the washcoat – gives access to a large surface area of these catalysts. They can be produced in various forms differing in their dimensions, cell shape and density, and wall thickness. Most have square-shaped channels or hexagonal-shaped channels. Typical applications are Pt- or Pd-coated catalysts for CPC processes or the postoxidation of off-gas. A representative example are honeycomb catalysts with more than 200 cpsi (= channels per square inch). Monolithic catalysts are also applied in NO_x removal with SCR, for instance, in automobile exhaust control. These catalysts are manufactured by kneading V_2O_5 homogeneously into the support material TiO_2. Figure 7.6 shows a schematic sketch of honeycomb catalysts for off-as purification with a hexagonal-shaped and a square-shaped cross-sectional area. A term often used with honeycomb and metal plate catalysts is “catalyst pitch.” A pitch (p) refers to the width of a catalyst channel (catalyst cell) plus the cell wall thickness. A wider pitch results in lower interstitial gas velocities for a given flow rate. Appropriate choice of the catalyst pitch is crucial to prevent plugging of the channels, whereby the effective surface area is decreased through decrease of the number of catalytically active available sites.

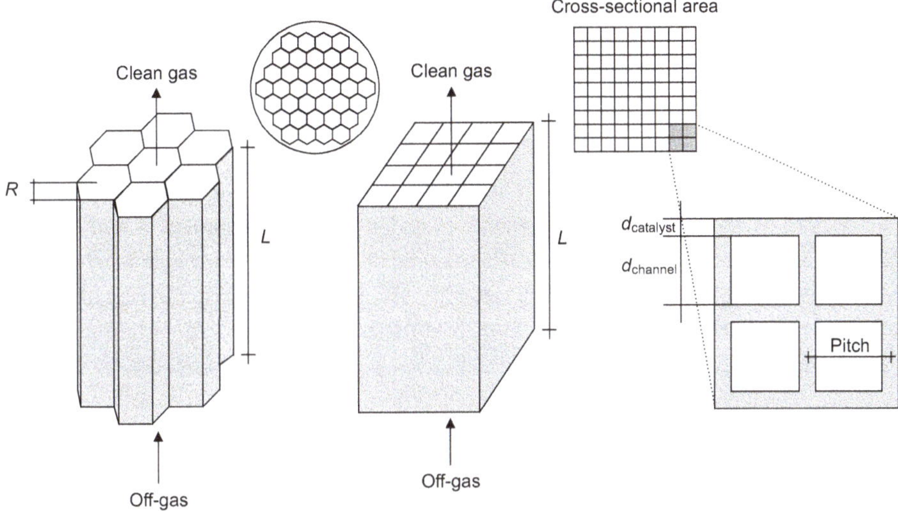

Fig. 7.6: Schematic sketch of honeycomb catalysts for off-gas purification: left: hexagonal-shaped cross-sectional area; right: square-shaped cross-sectional area, with channel length L (= length of the catalyst, sometimes also height of catalyst H), radius of the hexagonal-shaped channels R, channel side length $d_{channel}$, channel thickness $d_{catalyst}$ and pitch p.

A simple but effective way of increasing the selectivity of a heterogeneous catalyst is to embed the catalytically active component within a cage with a defined opening radius. Thus, only molecules, which are smaller than the cross section of the aperture, are able to enter the cage and to reach the catalyst. Reaction products can only leave the cage, if they are smaller than the cross section. The most popular representatives of this group of catalysts are zeolites (aluminosilicate minerals [209–211]). About 30 different natural zeolites are known. On industrial scale, zeolites are produced by swelling inside an immersion tube burner. Zeolites can be industrially produced with specified adsorption properties. The crystal structure is a tetrahedron with Si as the central atom. Several basic elements form the secondary structure from which the regular crystal structure is derived. The crystal structure is responsible for the adsorption properties of zeolites.

Substituting Si by Al results in a deficit of positive electric charge, which can be compensated by complementary cations. By varying the size of the complementary cation, the desired property of the zeolites can be achieved. Predominant applications of zeolites are drying, N_2/O_2 separation, ion exchange in water purification as well as deodorizing. The adsorptive removal of ammonia with clinoptilolite, a natural zeolite, finds application in manure deodorizing. Selective adsorption properties can be achieved by adjusting the grid of the cage with a suitable complementary cation. The specification of zeolites is based on the aperture ratio, which is determined by the complementary cation. The 4A-types are Na-substituted zeolites with an opening radius of 0.4 nm and 3A-types are K-substituted with the opening radius of 0.3 nm. Therefore, only H_2O and NH_3 are able to penetrate these zeolites. Ca-substitution leads to formation of 5A-types. X-types, such as 13X-zeolite, show the opening ratio of 12 rings. In off-gas purification, zeolite-based catalysts are used for various applications; both catalytic oxidation [212, 213] and catalytic reduction [214, 215].

7.2.4.3 Catalyst requirements

Catalysts must meet certain requirements in order to enable economical and reliable off-gas purification. Above all, catalysts should be characterized by a high level of catalytic activity. The activity of a catalyst describes its ability to increase the rate of a reaction under defined operation conditions. It is therefore specific to a certain chemical reaction at specified process parameters and not a fundamental property. In addition to being selective toward the respective reaction product, a catalyst should be robust and stable toward deactivation (= decrease of catalytic activity). This means that its properties should not change for the longest possible lifetime guaranteeing long-term stability. However, even though a catalyst is by definition not consumed in the chemical reaction itself, a change of properties during operation conditions is often inevitable. This leads to gradual decrease of catalytic activity, which may be reversible (e.g., fouling) or irreversible (e.g., poisoning). In the worst case, an instantaneous deactivation takes place by acute catalyst poisoning. When a catalyst is selected, special attention has to be paid to its chemical stability, its mechanical stability and the temperature range for application. This means that it has to be guaranteed that the catalyst is resistant to other substances that may be present even in traces in the off-gas (chemical resistance), that the catalyst is resistant to abrasion and fracture by vibrations and stresses that occur during operation (mechanical resistance), and that the catalyst does not undergo any structural changes at operation temperature (thermal resistance).

! The three main reasons for catalyst deactivation are fouling, poisoning and aging. Fouling is a partial reversible blocking of the active surface by decomposition products or matrix components. Poisoning is a quite critical type of activity loss since it may reduce the catalyst lifetime from years to a few weeks. It occurs when the off-gas contains substances (e.g., sulfur) that react with the catalyst support or the active component. Aging, on the contrary, is mainly caused by changes in crystal structure due to temperature effects. Therefore, upper temperature limits are specified for industrial catalysts. For metal oxide catalysts the upper temperature limits for industrial operation lie around 500 °C (for bulk catalysts) and 700 °C (for supported catalysts). For noble metal catalysts, they range from 700 °C (e.g., aluminum-oxide supported catalysts) to 1,000 °C (stabilized aluminosilicate-supported catalysts). Rapid temperature changes may result in thermal stress leading to fracture of a solid catalyst. Measures to prevent catalyst overheating during periods of intense release of heat due to high heat of conversion at high pollutant concentrations include the admixture of cold air and the reduction of preheating [216].

Mechanical damaging of a catalyst may occur due to abrasion, vibration of the plant, pressure surges, in case of loosening of the catalyst bed and due to solid particles (abrasive dusts). Elastic internal seals help prevent mechanical stress in reactors that are filled with honeycomb catalysts. A vertical inlet flow from the top to the bottom of the reactor helps avoiding catalyst bed fluidization. In any case, formation of funnel-like bed surfaces through high inlet flow rates or swirling incident flows must be avoided. Selecting appropriate geometries and catalyst types (e.g., honeycomb) as well as choosing the appropriate catalyst bed height helps influencing the pressure drop in catalytic off-gas purification plants. Typical values of specific pressure drop given per unit bed height are for honeycombs 1 kPa m^{-1} and for bulk catalysts 10 kPa m^{-1} [216].

7.2.4.4 Summary

To sum up, the following criteria must be met for successful application of off-gas purification catalysts:
- high catalytic activity for the desired reaction(s) at the lowest possible temperature to reduce the required residence time and thus reactor size;
- high selectivity toward the desired reaction(s), at best side reactions are suppressed;
- heat resistance to allow for adequate reaction temperatures and temperature insensitivity especially with changing off-gas compositions and temperatures;
- favorable surface/volume ratio to provide a high surface area;
- chemical resistance to prevent catalyst deactivation due to poisoning;
- mechanical stability to provide sufficient resistance to pressure and abrasion;
- low pressure drop as this lowers the performance;
- long life-time;
- low price; and
- cost-efficient recyclability.

7.2.5 Reactor design

Conventional chemical reactors comprise discontinuous (batch) and continuous tank and tubular reactors. In large-scale industries, mostly continuous reactors are used and in catalytic off-gas purification mainly packed-bed (= fixed bed) tubular reactors come into operation. In ideal tubular reactors, so-called plug flow reactors (PFR), the reactor content is fully mixed radially over the cross section (perpendicular to the flow direction), but not axially. Ideally the reaction broth is moving forward like a plug (thus, plug flow). Axial dispersion would result in limited conversion. The composition is changing continuously over the length. Hence, the material balance has to be set up for an infinitesimal segment meaning a differential volumetric element dV (Fig. 7.7).

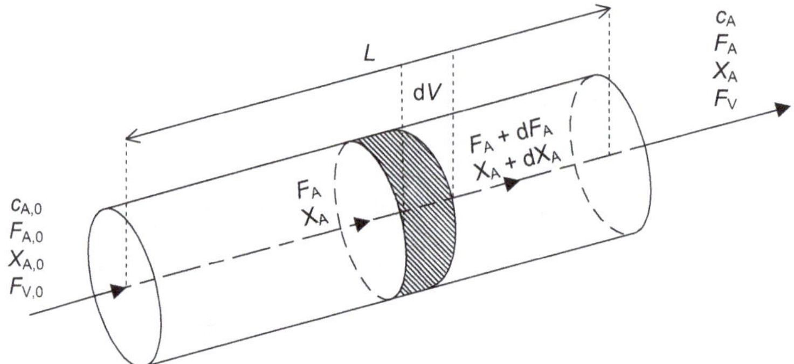

Fig. 7.7: Representative sketch of an ideal plug flow reactor for developing the reactor design balance.

7.2.5.1 Material balance for an ideal reactor

The general material balance (mole balance) of an ideal reactor in words is:

input = output + disappearance (conversion of reactant) by reaction + accumulation

This simple balance is probably the most important balance in chemical engineering and particular in off-gas treatment. We draw advantages from this balance in ESP design (Chapter 6), of course in chemical reaction as well as absorption technology (Chapter 8) and adsorption technology (Chapter 9).

For reactant A, the most prominent chemical species in CRE, this means:

Amount of reactant A fed ($F_{A,0}$) = amount of reactant discharged (F_A) + amount of reactant transformed into products by chemical reaction ($-r_A \cdot dV$) + amount of reactant accumulated in the reactor (dN_A/dt).

This can be accounted for through the following equation:

$$F_{A,0} = F_A + \int_0^V (-r_A) \cdot dV + \frac{dN_A}{dt} \tag{7.45}$$

From eq. (7.45), the design equation of an ideal tubular reactor (without accumulation) can be obtained:

$$V = F_{A,0} \cdot \int_0^{X_A} \frac{dX_A}{-r_A} = F_{V,0} \cdot c_{A,0} \cdot \int_0^{X_A} \frac{dX_A}{-r_A} \tag{7.46}$$

For homogeneous reactions, the reactor volume V_R of tubular reactors is often set equal to the volume of the reaction mixture V. Care must be taken as with internals, catalyst packings and so on the reactor volume differs from the volume of the reaction mixture.

The residence time τ is defined as the ratio of the reactor volume V_R to the volumetric (inflow!) flow rate $F_{V,0}$ (eq. (7.47)). The inverse residence time gives the SV, which can be referred to the reactor volume (eq. (7.48)) or the catalyst volume (eq. (7.44)). In addition to the SV, the area velocity AV can be used, which is defined as the volumetric flow rate per total active surface area of the catalyst (eq. (7.49)). For various applications, especially heterogeneous catalytic reactions, application of the SV or the area velocity is favored over residence time:

$$\tau = \frac{V_R}{F_{V,0}} = \frac{1}{SV} \tag{7.47}$$

$$SV = \frac{F_{V,0}}{V_R} \tag{7.48}$$

$$AV = \frac{F_{V,0}}{S_{cat}} \tag{7.49}$$

7.2.5.2 Design of ideal tubular reactors (PFR)

Under ideal (steady-state operation) conditions, accumulation is not an issue.

Thus:

input = output + conversion of reactant by reaction.

Consequently, for the differential volumetric element dV the following applies:

$$F_A = F_A + dF_A + (-r_A) \cdot dV \tag{7.50}$$

since

$$F_A = F_{A,0} \cdot (1 - X_A), \tag{7.51}$$

$$\frac{dF_A}{dX_A} = \frac{d(F_{A,0} \cdot (1 - X_A))}{dX_A} = -F_{A,0} \tag{7.52}$$

and thus,

$$F_{A,0} \cdot dX_A = (-r_A) \cdot dV \tag{7.53}$$

After separating variables and integration between $V = 0$ and V and $X_A = 0$ and X_A (eq. (7.54)), we get the design equation for an ideal tubular reactor (eqs. (7.55) and (7.56)):

$$\int_0^V \frac{dV}{F_{A,0}} = \int_0^{X_A} \frac{dX_A}{-r_A} \tag{7.54}$$

$$\frac{V}{F_{A,0}} = \frac{V}{(F_{V,0} \cdot c_{A,0})} = \frac{\tau}{c_{A,0}} = \int_0^{X_A} \frac{dX_A}{-r_A} \tag{7.55}$$

and

$$\tau = c_{A,0} \cdot \int_0^{X_A} \frac{dX_A}{-r_A} \tag{7.56}$$

Figure 7.8 shows the graphical representation of the design equation of an ideal tubular reactor.

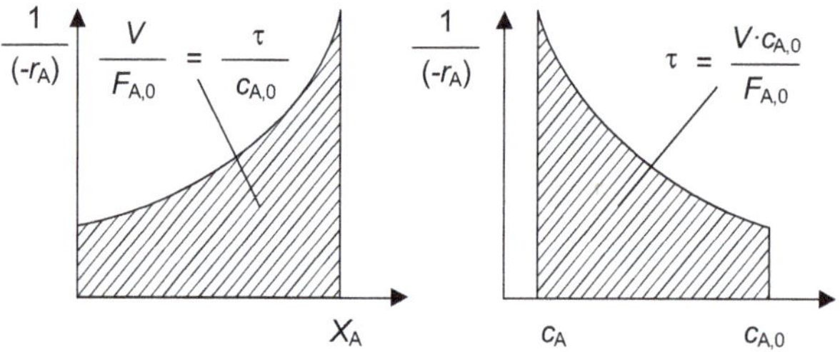

Fig. 7.8: Graphical representation of the design equation of an ideal tubular reactor.

Exercise 7.10: Design of an ideal tubular reactor – first-order reaction with varying volume
Derive the design equation of an ideal tubular reactor for an irreversible first-order reaction with varying volume ($\varepsilon_A \neq 0$).

Solver:
First, we state the general rate law of an irreversible first-order reaction of reactant A:

$$-r_A = k \cdot c_A$$

As the reaction volume changes during the course of reaction, we need to consider the change in reaction volume when stating the concentration of reactant A after a certain relative conversion. Provided that the reaction volume changes linearly with the relative conversion, the reaction volume after a certain relative conversion can be determined with

$$V = V_0 \cdot (1 + \varepsilon_A \cdot X_A)$$

Consequently, the concentration of reactant A after a certain relative conversion is obtained with

$$c_A = \frac{N_A}{V} = \frac{N_{A,0} \cdot (1 - X_A)}{V_0 \cdot (1 + \varepsilon_A \cdot X_A)} = c_{A,0} \cdot \frac{1 - X_A}{1 + \varepsilon_A \cdot X_A}$$

We can then state the rate law for a varying-volume reaction of first order in dependence of the relative conversion:

$$-r_A = k \cdot c_A = k \cdot c_{A,0} \cdot \frac{1 - X_A}{1 + \varepsilon_A \cdot X_A}$$

The integral form of the design equation of an ideal tubular reactor is given with

$$V = F_{A,0} \cdot \int_0^{X_A} \frac{dX_A}{-r_A}$$

Next, we insert the above-stated rate law into the design equation:

$$V = F_{A,0} \cdot \int_0^{X_A} \frac{dX_A}{-r_A} = F_{A,0} \cdot \int_0^{X_A} \frac{dX_A}{k \cdot c_{A,0} \cdot \frac{1-X_A}{1+\varepsilon_A \cdot X_A}} = \frac{F_{A,0}}{k \cdot c_{A,0}} \cdot \int_0^{X_A} \frac{dX_A}{\frac{1-X_A}{1+\varepsilon_A \cdot X_A}}$$

After integration between $X_A = 0$ and X_A, we get the following expression for the reaction volume that equals the reactor volume:

$$V = \frac{F_{A,0}}{k \cdot c_{A,0}} \cdot [-\varepsilon_A \cdot X_A + (-1 - \varepsilon_A) \cdot \ln(1 - X_A)]$$

Note: when the reaction volume does not change ($\varepsilon_A = 0$) during the course of the reaction, the volume can be obtained with

$$V = \frac{F_{A,0}}{k \cdot c_{A,0}} \cdot (-\ln(1 - X_A))$$

Deviations from ideal flow patterns

In the discussion of reactor design, we have assumed ideal flow patterns, meaning ideal plug flow in the tubular reactor. However, the flow characteristics of real reactors almost always deviate from this ideal. Even though reactor design can be based on ideal assumptions and good approximations can be achieved, in some cases it may be necessary to consider nonideal flow behavior. At this point, we would like to refer to the topic of residence time distributions in the relevant textbooks [180, 181].

7.2.6 Process selection

The process technology for off-gas purification based on chemical reactions in the gas phase is primarily selected on the basis of the appropriate type of reaction – oxidation versus reduction, thermal versus catalytic – in order to effectively convert hazardous pollutants into nontoxic species under the prevailing off-gas conditions; temperature, pressure and the composition of the off-gas (components and concentration of components) and required pollutant conversion. Economic considerations govern the choice of process technology. These are investment costs for the off-gas purification plant, operating costs regarding energy supply (try to avoid catalysts to become a significant amount of operating costs), waste heat utilization and pressure drop; the expected life-time of components (heat-exchangers, valves and catalysts), formation of unwanted byproducts and waste disposal of, for instance, spent catalysts [216].

According to [216], the basic operation of off-gas purification processes based on chemical reactions in the gas phase follows a series of process steps:

(1) the off-gas is transferred by suction;
(2) exhaust air is transported mechanically (e.g., by ventilators and compressors);
(3) interfering components are separated (e.g., by filters, adsorbers and absorbers);
(4) the off-gas is preheated by heat-exchangers;
(5) external energy is supplied (e.g., by burners and electrical resistance heating);
(6) the off-gas is mixed with oxidizing or reducing agents (e.g., oxygen, air and ammonia);
(7) pollutants are converted (e.g., by flame, thermal or catalytic reaction);
(8) waste heat is utilized (e.g., to heat steam or heat-transfer oil);
(9) the reaction products are separated; and
(10) the purified gas is discharged.

! In off-gas purification by oxidation in the gas phase, the off-gas purification can be integrated into the overall production process through waste heat utilization.

7.3 Off-gas purification by oxidation in the gas phase: combustion

Oxidation processes may be applied to convert combustible, hazardous pollutants with reducing properties into innocuous products. Usually the pollutants' concentrations in the off-gas are so low that recovery of these substances, for example, by adsorption, would not be economically justifiable. This group of pollutants primarily consists of organic compounds, but also odorous substances such as amines and inorganic gases, for example, hydrogen sulfide (H_2S) or ammonia (NH_3), can be treated. As an example, solvent-based pollutants overwhelmingly consist of carbon, hydrogen and oxygen ($C_mH_nO_o$). These substances may be completely oxidized (combusted) to carbon dioxide and water.

In this regard, an important group of pollutants are so-called volatile organic compounds (VOCs). The term "VOC" refers to gaseous and vaporous substances of organic origin including, for example, hydrocarbons, alcohols, aldehydes and organic acids. Numerous solvents, liquid fuels and synthetically produced substances occur as VOCs. In addition to that many organic compounds that arise from biological processes represent VOCs. Many hundreds of different compounds are known [217]. Off-gas purification for removal of VOCs is mainly accomplished by combustion. However, a substance-specific selection of the operating parameters is of fundamental importance; especially since many of them may contain halogens.

When the off-gas contains inorganic substances in addition to hydrocarbons, the process engineering effort required for purification by oxidation increases considerably, since the resulting inorganic compounds are toxic components according to [218] or catalyst poisoning components. After combustion, the purified off-gas must be subjected to further operational steps to remove the oxidized inorganic compounds by a downstream off-gas purification system, as discussed in Chapters 6 and 8.

7.3.1 Combustion fundamentals

Combustion (also burning, incineration) is a fast exothermic redox reaction in which a combustible (a fuel or pollutant) is oxidized with oxygen. The liberated energy from combustion can be used for various purposes. In off-gas purification it is mainly used to preheat the off-gas.

7.3.1.1 Complete versus incomplete combustion

Oxidation of hydrocarbons C_mH_n by combustion is based on the reaction in eq. (7.57). After complete combustion, carbon dioxide and water are the final reaction products. When a compound contains carbon, hydrogen and oxygen, the reaction equation for combustion changes to eq. (7.58):

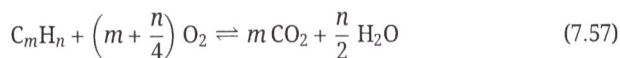

$$C_mH_n + \left(m + \frac{n}{4}\right) O_2 \rightleftharpoons m\, CO_2 + \frac{n}{2}\, H_2O \tag{7.57}$$

$$C_mH_nO_o + \left(m + \frac{n}{4} - \frac{o}{2}\right) O_2 \rightleftharpoons m\, CO_2 + \frac{n}{2}\, H_2O \tag{7.58}$$

Further constituents in the off-gas broaden the product range (e.g., SO_x, HF and NO_x), since inorganic substances are oxidized along with the hydrocarbons. For instance, sulfur oxidizes according to eq. (7.59) to SO_2. If chlorine is present in the off-gas, it may be converted to hydrogen chloride (HCl, eq. (7.60)) and subsequently to chlorine gas (Cl_2, eq. (7.61)). Accordingly, hydrofluoric acid (HF) may form in fluoride-laden gas (eq. (7.62)). If the off-gas contains phosphorus and arsenic, these compounds are converted to phosphorus oxide (P_2O_5, eq. (7.63)) and arsenic oxide (As_2O_5, eq. (7.64)), respectively. NO_x forms from nitrogen and nitrogen-containing constituents (= fuel NO_x). Since the avoidance of NO_x formation during combustion is one of the basic preventive measures for emission reduction, it is discussed in detail in Section 7.4:

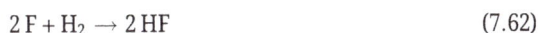

$$S + O_2 \rightarrow SO_2 \tag{7.59}$$

$$2\,Cl + H_2 \rightarrow 2\,HCl \tag{7.60}$$

$$4\,HCl + O_2 \rightleftharpoons 2\,Cl_2 + 2\,H_2O \tag{7.61}$$

$$2\,F + H_2 \rightarrow 2\,HF \tag{7.62}$$

$$4\,P + 5\,O_2 \longrightarrow 2\,P_2O_5 \tag{7.63}$$

$$4\,As + 5\,O_2 \longrightarrow 2\,As_2O_5 \tag{7.64}$$

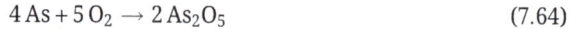

For combustion reactions, the minimum oxygen requirement and the percent excess air are often specified.

Minimum oxygen requirement: The minimum oxygen requirement ($F_{O_2,\,min}$) for complete combustion is derived from the reaction equation (stoichiometric equation) of the combustion process for stoichiometric oxygen supply. For a hydrocarbon C_mH_n, this yields $(m + (n/4))$ moles of oxygen per mole pollutant C_mH_n (see eq. (7.53)). In off-gas purification, the oxidant oxygen is generally supplied by air (= combustion air). The minimum air requirement ($F_{air,\,min}$) for complete combustion is obtained with the composition of dry air (approximately 21 vol% O_2 and 79 vol% N_2, eq. (7.65)):

$$F_{air,\,min} = \frac{F_{O_2,\,min}}{0.21} \tag{7.65}$$

Exercise 7.11: Minimum oxygen requirement – complete combustion of oxygen- and sulfur-containing hydrocarbons

Determine the minimum oxygen requirement for the pollutant $C_mH_nO_oS_p$.

Solver:

Set up the stoichiometric equation for combustion of $C_mH_nO_oS_p$. For simplification, the total reaction equation can be split into individual combustion reactions of C, H and S:

(1) $m\,C + m\,O_2 \longrightarrow m\,CO_2$

(2) $n\,H + \frac{n}{4}\,O_2 \longrightarrow \frac{n}{2}\,H_2O$

(3) $p\,S + p\,O_2 \longrightarrow p\,SO_2$

The minimum amount of oxygen arises from the individual combustion reactions (1)–(3) minus the amount of oxygen (O_2) already present in $C_mH_nO_oS_p$ $(o/2)$.

Thus, the minimum amount of oxygen is $m + \frac{n}{4} - \frac{o}{2} + p$.

The stoichiometric equation for combustion of $C_mH_nO_oS_p$ is

$$C_mH_nO_oS_p + \left(m + \frac{n}{4} - \frac{o}{2} + p\right)O_2 \longrightarrow m\,CO_2 + p\,SO_2 + \frac{n}{2}\,H_2O$$

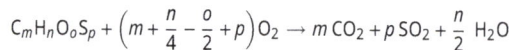

Percent excess air: To ensure complete combustion, combustion processes are generally carried out with excess oxygen (excess air). A useful term to express this is “percent excess air” (%EA):

$$\%EA = \left(\frac{\text{actual amount of air}}{\text{theoretical amount of air required for complete combustion}} - 1\right) \cdot 100\% \tag{7.66}$$

Another useful parameter is λ, which is specified as the ratio of the amount of air (or oxygen) available for combustion and the theoretical amount of air (or oxygen) required for complete combustion, as follows:

$$\lambda = \frac{\text{actual amount of air}}{\text{theoretical amount of air required for complete combustion}} \tag{7.67}$$

Incomplete combustion of hydrocarbons results in generation of carbon monoxide (CO, eq. (7.68)). Further possible products of incomplete combustion are methane, aldehydes (e.g., formaldehyde), organic acids or carbon. Even under oxygen excess, combustion of hydrocarbons proceeds via carbon monoxide formation. Accordingly, considerable amounts of CO may be present in the purified gas stream (clean gas) at low combustion temperatures or insufficient residence times in the combustion chamber. The permissible residual content of organic pollutants in the clean gas is usually reached easier than the permissible residual content of CO. In the purified gas stream from thermal or catalytic combustion equipment, emissions of organic substances shall not exceed 20 mg m^{-3}, expressed as total carbon. At the same time, emissions of CO shall not exceed the mass concentration of 0.10 g m^{-3} [218]. With regard to the residual CO concentration, it is therefore often necessary to carry out the combustion at higher temperatures or with a longer residence time.

As explained in the VDI guideline [219], the residual concentration of organic compounds is already greatly reduced at combustion temperatures below 760 °C. This means that the conversion of hydrocarbons takes place at relatively low temperatures (indicated by the ignition temperature). The oxidation of the resulting CO, on the other hand, requires higher temperatures. Consequently, the follow-up reaction step from CO to CO_2 is crucial for complete combustion of organic compounds. Temperatures of over 800 °C are therefore recommended:

$$C_mH_n + \left(\frac{n}{2} + \frac{m}{4}\right) O_2 \rightleftharpoons n\,CO + \frac{m}{2}\,H_2O \tag{7.68}$$

The effectiveness of a combustion process is expressed by the combustion efficiency (CE). It quantifies the ability to completely oxidize an organic substance at specified combustion conditions. It is defined as the ratio of the concentration of carbon dioxide to the sum of carbon dioxide and residual carbon monoxide concentrations in the clean gas (eq. (7.69)). An effective combustion process should be capable of achieving efficiencies >0.99. Veselý et al. [220] described the performance of an afterburner chamber equipped with a natural gas burner and reported a strong relation of the combustion efficiency to the overall temperature regime, the concentration of oxygen and the gas residence time:

$$\mathrm{CE} = \frac{c_{CO_2}}{c_{CO_2} + c_{CO}} \tag{7.69}$$

7.3.1.2 Specification of combustibles

For the design of thermal and catalytic off-gas purification systems, a series of substance properties of the combustible pollutant are crucial.

Calorific values: Above all, the amount of heat released during the combustion process – given by the calorific value – is of importance. The lower calorific value (LCV, also lower heating value, net calorific value) of a gaseous pollutant corresponds to the standard enthalpy of combustion $\Delta_c H^0$(= negative $\Delta_R H^0$ of the combustion reaction), also heat of combustion, with water in gaseous state. It represents the heat released in the combustion process per mole of substance to be combusted at constant pressure. Instead of one mole, the calorific value in technology is usually related to one kg of substance. The values always apply to complete combustion at constant pressure (in general P = 101.3 kPa) and the same initial and final temperature of the substances involved (0 or 25 °C). Note that for liquids (e.g., liquid fuels), the enthalpy of vaporization needs to be subtracted from the heat of combustion and the standard enthalpy of reaction has to be calculated from the heat of formation of the substance in liquid state. With the help of the molar mass of the individual substances and the molar volume (1 mol = 22.4 $\mathrm{dm^{-3}}_{\mathrm{STP}}$ at 0 °C and 101.3 kPa), the calorific value per liter can be calculated (mainly for liquids).

Sometimes confusion of the LCV and the higher calorific value (HCV, also higher heating value, gross calorific value) leads to misunderstandings. If the reaction product water is considered gaseous, it is referred to as LCV; if it is considered liquid, it is called higher calorific value. For determining the higher calorific value, it is assumed that all water vapor produced during the combustion process is fully condensed. Thus, the higher calorific value is the sum of the LCV and the heat of condensation of the water vapor (eqs. (7.70) and (7.71)). This is due to the fact that the condensation of water releases additional heat. In practice, the water leaves the combustion chamber in gaseous state, so that only the lower heating value can be made technically usable.

$$\mathrm{HCV} = \mathrm{LCV} + |\Delta_{\mathrm{cond}} H^0_{\mathrm{H_2O}}| \tag{7.70}$$

with

$$\Delta_{\mathrm{cond}} H^0_{\mathrm{H_2O}} = -43.99\ \mathrm{kJ\,mol^{-1}}\ (\mathrm{at}\ \theta = 25\,°\mathrm{C}) \tag{7.71}$$

Exercise 7.12: Lower and higher calorific value of hydrogen
Calculate the lower and the higher calorific value of hydrogen.

Substance data:

$$\Delta_f H^0_{\mathrm{H_2O_{(g)}}} = -241.8\ \mathrm{kJ\,mol^{-1}}$$

$$\Delta_f H^0_{\mathrm{H_2O_{(l)}}} = -285.8\ \mathrm{kJ\,mol^{-1}}$$

$$\Delta_{\mathrm{vap}} H^0_{\mathrm{H_2O}} = 43.99\ \mathrm{kJ\,mol^{-1}}$$

Solver:
The reaction equation for the combustion of hydrogen is

$$H_2 + 0.5\,O_2 \rightarrow H_2O$$

With the heat of formation of water, the standard enthalpy of reaction of hydrogen combustion can be calculated. The heats of formation of hydrogen and oxygen equal zero. Note: The heat of formation of an element is always zero. With the heat of formation of gaseous water $\Delta_f H^0_{H_2O_{(g)}}$, the LCV is obtained:

$$\text{LCV}: -\Delta_R H^0 = -(\nu_{H_2O} \cdot \Delta_f H^0_{H_2O_{(g)}}) = 241.8 \quad \text{kJ mol}^{-1}$$

With the heat of formation of liquid water $\Delta_f H^0_{H_2O_{(l)}}$, the higher calorific value is obtained:

$$\text{HCV}: -\Delta_R H^0 = -(\nu_{H_2O} \cdot \Delta_f H^0_{H_2O_{(l)}}) = 285.8 \quad \text{kJ mol}^{-1}$$

Alternatively, the negative higher calorific value can be obtained from the negative LCV when the enthalpy of evaporation is subtracted (or the enthalpy of condensation is added). Note: Be careful with the negative sign.

$$\text{HCV} = \text{LCV} + |\Delta_{cond} H^0_{H_2O}| = 241.8 \text{ kJ mol}^{-1} + 43.99 \quad \text{kJ mol}^{-1} = 285.8 \quad \text{kJ mol}^{-1}$$

Exercise 7.13: Combustion of methanol – $\Delta_c H^0$, LCV, and HCV of methanol

Methanol (CH_3OH) is to be combusted. Determine the heat of combustion and the lower and higher calorific value of methanol.

Solver:

First, set up the reaction equation (stoichiometric equation) for complete combustion of methanol to carbon dioxide and water. Take care that the reaction equation is stoichiometrically balanced, meaning that the number of moles of an element on the educts' side equals the number of moles of this element on the products' side:

$$CH_3OH + 1.5\,O_2 \rightarrow CO_2 + 2\,H_2O$$

From the reaction equation we see that in order to combust one mole of methanol, 1.5 moles of oxygen are required. As products one mole CO_2 and two moles water are formed.

From literature, the following substance data are known for 25 °C and 100 kPa [221]. Note that under these conditions, methanol and water are in liquid state. The heat of formation of oxygen is zero:

$$\Delta_f H^0_{CH_3OH_{(g)}} = -201.0 \quad \text{kJ mol}^{-1}$$

$$\Delta_f H^0_{CO_{2(g)}} = -393.2 \quad \text{kJ mol}^{-1}$$

$$\Delta_f H^0_{H_2O_{(l)}} = -285.8 \quad \text{kJ mol}^{-1}$$

$$\Delta_{vap} H^0_{H_2O} = 43.99 \quad \text{kJ mol}^{-1}$$

$$\Delta_{vap} H^0_{CH_3OH} = 37.6 \quad \text{kJ mol}^{-1}$$

The enthalpy of reaction can be obtained from the heats of formation of the products and educts:

$$\Delta_R H^0 = (\nu_{CO_2} \cdot \Delta_f H^0_{CO_{2(g)}} + \nu_{H_2O} \cdot \Delta_f H^0_{H_2O_{(g)}}) - \nu_{CH_3OH} \cdot \Delta_f H^0_{CH_3OH_{(l)}}$$

Since we only know the heat of formation of gaseous methanol, we have to calculate the heat of formation of liquid methanol with the enthalpy of evaporation of methanol:

$$\Delta_f H^0_{CH_3OH_{(l)}} = \Delta_f H^0_{CH_3OH_{(g)}} + \Delta_{vap} H^0_{CH_3OH}$$

Thus:

$$\Delta_R H^0 = (\nu_{CO_2} \cdot \Delta_f H^0_{CO_2(g)} + \nu_{H_2O} \cdot \Delta_f H^0_{H_2O_{(l)}}) - \nu_{CH_3OH} \cdot \left(\Delta_f H^0_{CH_3OH_{(g)}} + \Delta_{vap} H^0_{CH_3OH}\right)$$

$$= (2 \cdot (-285.8) + 1 \cdot (-393.2) - 1 \cdot (-201.0 - 37.6) = -726.2 \text{ kJ mol}^{-1}$$

$$\Delta_c H^0 = -\Delta_R H^0 = 726.2 \text{ kJ mol}^{-1}$$

Check: In the literature [186], you can find data for the heat of combustion:

$$\Delta_c H^0 = 726.1 \text{ kJ mol}^{-1}$$

In order to calculate the LCV, we need to add the enthalpy of evaporation of water to the negative heat of combustion (= enthalpy of reaction), since for determining the heat of combustion water was considered in liquid state:

$$\text{LCV: } \left|(-\Delta_c H^0) + \nu_{H_2O} \cdot \Delta_{vap} H^0_{H_2O}\right| = \left|\Delta_R H^0 + \nu_{H_2O} \cdot \Delta_{vap} H^0_{H_2O}\right| = |(-726.2) + 2 \cdot 43.99| = 638.2 \text{ kJ mol}^{-1}$$

$$\text{LCV} = \frac{638.2 \text{ kJ mol}^{-1}}{32.04 \text{ g mol}^{-1}} = 19.9 \text{ kJ g}^{-1} = 19.9 \text{ MJ kg}^{-1}$$

And:

$$\text{HCV: } \Delta_c H^0 = 726.1 \text{ kJ mol}^{-1}$$

Explosion level: Another important property of combustibles is their flammable (explosive) range in air; given by the lower explosive level (LEL) and the upper explosive level (UEL). The lower explosive level defines the lowest concentration of a combustible in air that is capable of producing a flash of fire when met with an ignition source; whereas the upper explosive level defines the highest concentration in this regard. Combustible concentrations below the LEL do not fuel and do not continue an explosion; they are considered too "lean." Concentrations above the UEL do not fuel and do not continue an explosion; they are considered too "rich." Table 7.6 lists combustion reactions, ignition temperatures, flammable ranges in air and LCVs for common substances.

7.3.1.3 Combustion mechanism and procedure

Concerning the reaction mechanism, combustion can be initiated by autoignition or by induced ignition. When combustibles (gases, vapors or mixtures thereof) ignite spontaneously without an external ignition source after a certain temperature has been reached, this is termed autoignition. Thus, the ignition temperature (also autoignition temperature) is the lowest temperature at which spontaneous combustion of a material occurs at specified conditions. In the absence of an external ignition source, this temperature is

Tab. 7.6: Combustion reactions, ignition temperatures [186], flammable (explosive) ranges in air in percent by volume (Fl. Lim.) [186], heats of combustion [186] and lower calorific values (25 °C, 101.3 kPa) [189] of common substances – all reactions stated as irreversible reactions.

Substance	Ignition temperature (°C)	Combustion reaction	Fl. Lim.	$\Delta_c H^0$* (kJ mol^{-1})	LCV** (kJ mol^{-1})	LCV (MJ kg^{-1})
Hydrogen (H_2)	560	$H_2 + 0.5\ O_2 \rightarrow H_2O$	4–74%	285.8	241.8	119.9
Carbon monoxide (CO)	609	$CO + 0.5\ O_2 \rightarrow CO_2$	12.5–74%	283.0	283.0	10.10
Methane (CH_4)	537	$CH_4 + 2\ O_2 \rightarrow CO_2 + 2\ H_2O$	5–15%	890.8	802.6	50.05
Ethane (C_2H_6)	472	$C_2H_6 + 3.5\ O_2 \rightarrow 2\ CO_2 + 3\ H_2O$	3.0–12.0%	1,560.7	1,428.8	47.52
Propane (C_3H_8)	450	$C_3H_8 + 5\ O_2 \rightarrow 3\ CO_2 + 4\ H_2O$	2.1–9.5%	2,219.2	2,043.1	46.33
Methanol (CH_3OH)	464	$CH_3OH + 1.5\ O_2 \rightarrow CO_2 + 2\ H_2O$	6–36%	726.1	638.12	19.91
Acetone (C_3H_6O)	465	$C_3H_6O + 4\ O_2 \rightarrow 3\ CO_2 + 3\ H_2O$	3–13%	1,789.9	1,657.9	28.55
Benzene (C_6H_6)	498	$C_6H_6 + 7.5\ O_2 \rightarrow 6\ CO_2 + 3\ H_2O$	1–8%	3,267.6	3,135.6	40.14
Toluene (C_7H_8)	480	$C_7H_8 + 9\ O_2 \rightarrow 7\ CO_2 + 4\ H_2O$	1–7%	3,910.3	3,734.3	40.53

*The heat of combustion $\Delta_c H^0$ refers to the state of aggregation of the respective substance (reactant or product) at 25 °C, e.g., gaseous CO, liquid H_2O and liquid methanol.
**LCV is obtained from $\left|(-\Delta_c H^0) + \nu_{H_2O} \cdot \Delta_{vap} H^0_{H_2O}\right|$.

required for self-sustained combustion as it is the lowest temperature where the rate of heat evolved from combustion increases beyond the rate of heat loss to the surroundings [186, 222]. Several parameters such as pressure, reactor shape and volume, presence of contaminants, flow rate, gravity and reactant concentration affect the autoignition temperature of a mixture of gases or vapors. The autoignition temperature is required to provide the activation energy for combustion. It reduces with increasing pressure and increasing concentration of oxygen [223].

Ignition can also be induced by the addition of heat, an electrical or laser spark or by radical addition. These ignition sources supply the minimum ignition energy.

! Combustion thus begins when:
- the amount of oxygen required for the combustion reaction (= minimum oxygen requirement) is present, and
- the substance to be combusted (pollutant or fuel) has reached the ignition temperature.

Upon ignition chain branching mechanisms occur, which are decisive for the rapid rate of a combustion reaction. This is in difference to a purely thermal, Arrhenius-type reaction [223]. For hydrocarbons, high- and low-temperature combustion mechanisms have been stated. Whereas for $\theta > 700$ °C, radical-chain mechanisms dominate, partially oxidized intermediates such as aldehydes, ketones, alcohols, peroxides and carboxylic acids form at temperatures $\theta < 700$ °C [216].

To achieve complete combustion, sufficient oxygen, good mixing of the off-gas and combustion air, adequate residence time in the combustion chamber, and a temperature that is high enough to ignite the pollutants must be guaranteed. This means that the oxygen must be brought into intimate contact with the combustible substance at a sufficient temperature and residence time for the combustion reaction to be completed. The rate of the combustion reaction is increased by inducing turbulence. In addition to reliable ignition and an adequate rate of reaction, flame-holding is crucial for efficient thermal combustion.

Oxidation of combustibles can be carried out by thermal combustion or by catalytic combustion. Thermal combustion is characterized by the presence of a flame during combustion. Catalytic combustion is a rapid flameless process, promoted by metallic or metal oxide-based catalysts. The apparatuses – the reactors – in which the pollutants are combusted are called combustors, combustion chambers, (after)burners or incinerators.

From a plant engineering and safety aspect, it is important to know that combustion cannot happen in tubes below a certain diameter (process safety draws many advantages from that by making use of flame retarders). This is due to the fact that radicals, which form during combustion, are quenched at the walls. This reduces the temperature below the autoignition temperature. The minimum diameter of a tube that defines the so-called quenching distance is set by the stoichiometry. It is around 1–4 mm for hydrocarbons. Catalytic combustion is an exception in this regard [223].

For further, more detailed information on combustion, refer to classic textbooks on fundamentals of combustion [224, 225] or technical aspects of combustion [226].

7.3.1.4 Technical considerations for pollutant combustion

Technically, pollutant oxidation in the gas phase can be accomplished by various methods. The selection is made depending on the composition of the off-gas and the requirements for the purity of the clean gas. In general, three types are distinguished (Fig. 7.9):

- thermal combustion (= thermal off-gas purification); in a combustion chamber or flare
- regenerative thermal combustion
- catalytic (flameless) combustion (= catalytic off-gas purification) with metals or metal oxides as catalysts

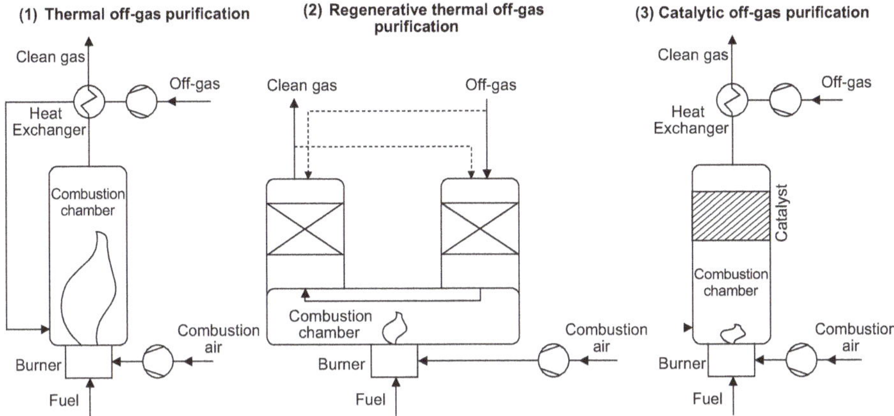

Fig. 7.9: Principal flow diagrams of (1) classical thermal, (2) regenerative thermal and (3) catalytic off-gas purification.

In the following discussion some considerations for selection of combustion methods are discussed [207]:

In thermal combustion, generally temperatures of more than 800 °C shall be used for complete pollutant combustion. If the off-gas does not contain sufficient oxidizable components through whose combustion heat is released, additional fuel (= auxiliary fuel) must be added to the combustion chamber to ensure a sufficiently high combustion temperature. The added fuel must have an adequately high lower calorific value. Moreover, the off-gas generally needs to be preheated. Combustion without off-gas preheating is only possible with very low volumetric off-gas flows in case that oxidation can take place via a flare. For larger volumetric flows, it is economical to increase the off-gas temperature via heat exchangers.

To preheat the off-gas, two basic concepts for heat exchange have been established industrially. These are classified as recuperative heat exchange and regenerative heat exchange. Firstly, in classical thermal combustion, the off-gas stream is heated via an indirect heat exchanger through cooling of the hot clean gas (= purified off-gas). According to Schultes [207], the preheating is successful up to 40–60% of the combustion chamber temperature and is, therefore, relatively incomplete. This means that the off-gas that flows into the combustion chamber is still undercooled and auxiliary fuel is required, whereby the amount of required fuel is relatively high. Secondly, regenerative thermal combustion can be applied which allows the addition of auxiliary fuel to be reduced as far as possible. Regenerative thermal combustion is equipped with at least two parallel regenerative heat exchangers with heat storage beds. The off-gas first flows through a heat accumulator (heat-storage material), which was previously heated by the hot combustion gases. As a result of the heat exchange, the off-gas temperature increases to up to 80–100% of the combustion chamber temperature [207]. Only if the calorific value of the pollutants in the off-gas is not sufficient for complete oxidation, additional fuel needs to be added to the combustion chamber. The hot clean gas then leaves the reactor through a second bed, through which the cold off-gas has previously flown through and cools down. The clean gas finally leaves the plant. The heat storage beds enable heat exchange between the cold off-gas and the hot clean gas to occur by switching the gas flow between off-gas and clean gas. However, a disadvantage of the concept with only two beds is that, when the gas flow path is switched from the off-gas side to the clean gas side, leakage flow may occur. Consequently, a third regenerator is required, which is briefly flushed with fresh air (purge air) between the heating-phase of the off-gas and the cooling-phase of the clean gas. The purge air may then be fed to the combustion chamber.

In catalytic combustion, the high combustion temperature required for complete oxidation by thermal combustion can be lowered by the help of noble metal catalysts. The presence of the catalyst reduces the activation energy of combustion, so that the pollutant conversion rate increases at lower temperatures.

!

Regenerative versus recuperative combustion

It is important not to confuse regenerative and recuperative combustion. Regenerative combustion involves a heat-storage material. The heat-storage material can either be applied as a fixed-bed, rotating device or moving bed. The fixed-bed regenerator is characterized by alternating flow direction and considering the temperature of the storage material, position- and time-dependent profiles. In moving-bed regenerators, the storage material moves and the thermal oxidation is generally carried out in the bed of the heat-storage material. Recuperative combustion uses gas–gas heat exchangers to preheat the off-gas with the hot clean gas. Common heat exchangers (recuperators) may be selected from the catalogue of the multitude of heat exchangers (whether plate type or tube type).

The decision as to whether combustion should be carried out thermally or with the use of a catalyst depends essentially on the composition of the off-gas and the pollutant properties. If the off-gas contains substances such as halogens, sulfur or phosphorus compounds that poison the catalyst, thermal combustion will be preferred. Due to the lower temperatures, catalytic combustion is preferred when waste heat recovery is not possible due to low pollutant contents in the off-gas. If off-gas, liquid and solid wastes are to be combusted at the same time, thermal combustors will most probably be chosen.

7.3.1.5 Summary

Remember some basic aspects regarding off-gas purification by combustion. First, for combustion to occur, the following conditions must be met:

- a minimum amount of oxygen must be available; usually it is taken from the available air;
- the required ignition temperature must be reached to start the combustion;
- in case of gases and vapors, the mixing ratio with oxygen must lie within the upper and lower ignition limits so that the heat released by the combustion reaction is sufficient to maintain a minimum temperature;
- the pollutant (or fuel) must be well mixed with the combustion air; and
- for complete combustion to occur, sufficient residence time in the reactor must be guaranteed.

Second, from the above-listed criteria, three combustion control parameters arise:

- sufficient oxygen supply;
- sufficient residence time in the reactor; and
- temperature > temperature of ignition.

Third, the choice of combustion technology (thermal, regenerative thermal or catalytic) depends on the composition of the off-gas. In general, off-gas purification by combustion is a simple and low-susceptibility technology. However, if pollutant concentrations are too low for autothermal combustion, a lot of auxiliary fuel is required to heat up the off-gas to the required combustion temperature of approximately 800 °C in thermal combustion technologies.

7.3.2 Thermal combustion

Thermal combustion is used to oxidize pollutants over a wide range of concentrations. Its application is nearly unlimited regarding off-gas composition and concentration. However, pollutant concentrations must be well below the lower limit of explosion. To ensure a sufficiently high rate of oxidation (r_{OX}), thermal combustion is conducted at elevated temperatures, well above the ignition temperature of the pollutant. Moreover, the off-gas feed has to be preheated with auxiliary energy to achieve this temperature. To prevent flashbacks during operation, the off-gas velocity v_{gas} has to lie between 3 and 15 m s^{-1}.

Air and auxiliary fuel are delivered continuously to the burner where the fuel is combusted. The burner may use the off-gas as combustion air for the auxiliary fuel or a separate source of outside air (= combustion air).

Indicated by the low concentration of the respective combustible pollutants (Tab. 7.7), the main disadvantage of thermal combustion are the high operating costs. In general, off-gas can support its own combustion when it has a heating value of at least

7,500 kJ m^{-3}. Direct recovery of the energy liberated by combustion decreases the amount of auxiliary fuel. Regenerative combustion may even autothermally process off-gas with a pollutant load of less than 10 g m^{-3}.

Tab. 7.7: Representative pollutants and their mean concentrations in thermal off-gas purification.

Pollutant		Mean concentration (g m^{-3}_{STP})
Solvents	Esters Alcohols Ethers Styrene Hydrocarbons	0.5–15
Odors	Amines Acids Aldehydes	Very low, traces

For complete combustion, the following operation conditions must be met for hydrocarbons and for organohalides (halogen-substituted hydrocarbons), respectively (Tab. 7.8).

Tab. 7.8: Required combustion temperatures and residence times for thermal off-gas purification with hydrocarbon and organohalide pollutants.

Pollutant	θ (°C)	τ (s)
Hydrocarbons	>750	0.5
Organohalides	>1,200	2

To conclude, thermal off-gas purification can be applied nearly unlimitedly regarding composition and concentration of the off-gas. However, technical limits are:
- the rate of oxidation depends on the combustion temperature: $r_{OX} = f(k) = f(T)$
- $v_{gas} > 3\text{–}15$ m s^{-1} to prevent flashbacks
- pollutant concentrations < LEL

Depending on the application case, thermal combustion is accomplished in
- flares or
- combustion chambers.

7.3.2.1 Combustion in flares

Gas flares (also flare stacks) are mainly used in chemical plants, petroleum refineries, natural gas processing plants and at oil and gas production sites. They are an important safety feature in the case of process malfunction or power failure [227]. Their use is determined by the following process conditions:

- start-up and shutdown of the industrial plant;
- excess process plant gas venting; and
- disturbances with controllable (e.g., safety valves) and uncontrollable (e.g., failure of cooling water) operating conditions that require safety release.

In general, off-gas purification flare systems consist of an elevated stack equipped with devices to maintain burning at the top of the stack and to prevent flashback within the system. The flare system is always running with a small amount of auxiliary gas that is continuously burned, like a pilot light. The flow rate of combustibles in the off-gas determines the size and brightness of the flame. Consequently, the flame varies depending on the process conditions. The released off-gas is transported through piping systems (so-called flare headers) to the elevated vertical flare where it is burned when it exits the flare stack. Sometimes vapor–liquid separators are installed prior to the flare stack to remove liquid constituents from the off-gas. In order to minimize formation of black smoke, steam can be injected into the flame. Heat from the flares is not used. Figure 7.10 shows a typical flare system.

The radiant heat intensity that is generated by the flame determines the height of the flare stack and also the required distance to the industrial plant and surroundings. When the height of the flare stack is fixed, the distance can be calculated and vice versa. Usually the distance is given due to legal constraints, hence the stack height is determined. As a guideline, for stack heights <23 m, a distance of 91 m, and for stack heights >23 m, a distance of 61 m is recommended [228].

7.3.2.2 Thermal combustion in combustion chambers

For thermal combustion in combustion chambers (also thermal afterburner), combustion air and unreacted off-gas feed are intensively mixed and enter the reaction zone of the combustion chamber, where the pollutants react at elevated temperature in a temperature range of 750–850 °C (an exception is the combustion of organohalides that requires higher temperatures, see Tab. 7.8). For complete combustion, a residence time of 0.5–1 s is recommended [207]. The combustion products continuously exit the outlet of the combustion chamber (reactor). To prevent settling of particulates and minimize flashbacks, sufficiently high gas velocities are required. The average gas velocity lies in the range of 3–15 m s^{-1}. In contrast to combustion in flares, the heat of combustion is generally used when thermal combustion is carried out in combustion chambers. Both direct recovery in the process or indirect recovery by suitable external heat exchange are common. Because of the high operating temperature, the reactor material

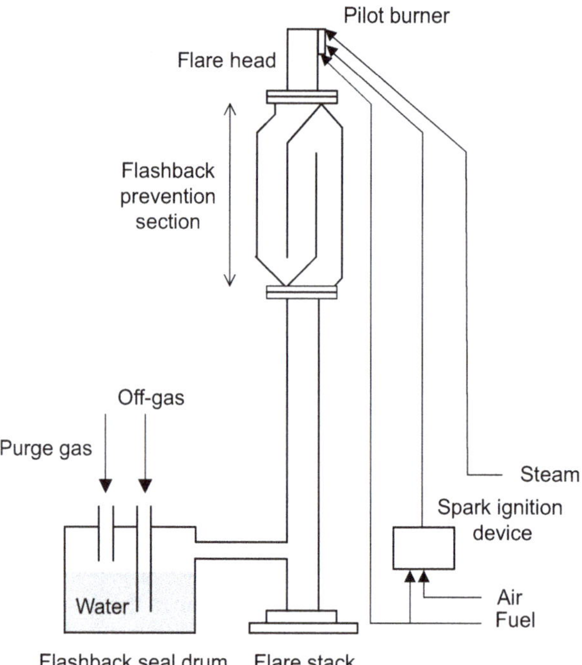

Fig. 7.10: Process scheme of thermal off-gas purification in a flare system.

must be capable of withstanding thermal stress. For this reason, combustion devices usually consist of an outer steel shell that is lined with refractory material. This is especially important when the off-gas purification plant has to be frequently shut down and restarted.

Figure 7.11 depicts the process flow scheme of a thermal combustion chamber with recuperative off-gas preheating. Such systems usually consist of a burner and a combustion chamber. The auxiliary fuel is burnt in the burner with either oxygen-containing off-gas or combustion air that is fed to the burner, when the oxygen content in the off-gas is too low. Typical burners are flat and lance burners [216]. It has to be noted that additional combustion air increases the mass flow rate and consequently the required energy. Burners and off-gas combustion chambers can also be combined in one unit (e.g., swirl combustion chamber). The standard auxiliary fuel is natural gas (LCV ≈ 33,500 kJ m^{-3}).

7.3.2.3 Design of combustion chambers for thermal combustion

The basic requirements for the design of thermal off-gas purification plants for complete off-gas specification as well as specified pollutant(s) properties including enthalpy of formation, lower explosion limit and temperature of ignition as well as the reaction equation of the combustion process are listed below.

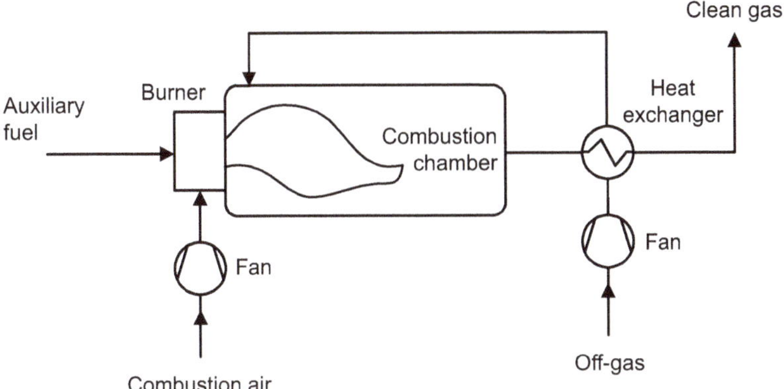

Fig. 7.11: Process scheme of thermal off-gas purification in a combustion chamber.

Reactor design can be carried out by evaluation of the:
- mass balance and energy balance of pollutant combustion;
- mass balance and energy balance of auxiliary fuel combustion;
- auxiliary fuel demand;
- overall mass balance; and
- overall energy balance.

In this context, we will get to know the adiabatic combustion temperature (also adiabatic flame temperature) θ_{ad} (or T_{ad}). For this theoretical temperature perfect combustion and the absence of any heat loss are assumed.

In the following we will discuss the calculation process based on the combustion of acetone (Exercise 7.14).

Exercise 7.14: Thermal combustion of acetone

Develop the calculation algorithm for thermal combustion on the example of acetone (C_3H_6O) combustion.

Formally, the off-gas is split into the pollutant (PO) acetone and a carrier gas (= off-gas without pollutant, OG). We start with considering only the pollutant; not the carrier gas.

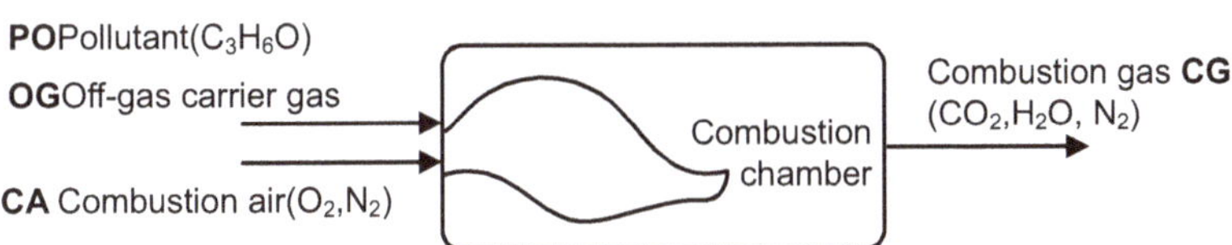

Fig. 7.12: Flow chart for the calculation algorithm of thermal combustion of acetone (Exercise 7.14).

Solver:
For reactor design, we evaluate:
(1) the mass balance of the pollutant-related combustion process;
(2) the energy balance of the pollutant-related combustion process;
(3) the overall energy balance; and
(4) the apparatus design.

With the stoichiometric equation of the combustion reaction we can determine:
- the stoichiometric (minimum) oxygen requirement;
- the stoichiometric (minimum) air requirement;
- the composition of the combustion gas (= clean gas from thermal off-gas purification);
- the adiabatic combustion temperature; and
- the amount of excess air regarding the required combustion temperature.

(1) Mass balance of the pollutant-related combustion process (pollutant combustion)

First step:
After specification of the amount of off-gas (flow rate), the temperature, the load of pollutant, the physicochemical properties of the pollutant and the theoretical minimum (= stoichiometric) oxygen requirement for complete combustion has to be determined. The calculation is based on the reaction equation of the combustion process of acetone:

$$C_3H_6O + 4\,O_2 \rightarrow 3\,CO_2 + 3\,H_2O$$

The minimum oxygen requirement ($O_{2,min}$) for the combustion of acetone is determined by its reaction stoichiometry (per mole acetone 4 moles O_2 are required, assumption that the ideal gas law is valid):

$$O_{2,\,min} = 4\,\text{mol mol}^{-1}\ C_3H_6O = 4\,\text{m}^3\ \text{m}^{-3}\ C_3H_6O$$

Second step:
The minimum amount of combustion air (CA_{min}) is calculated with eq. (7.65). Ambient air has an oxygen content of 21%, and therefore CA_{min} is

$$CA_{min} = \frac{4}{0.21}\ \text{mol mol}^{-1}\ C_3H_6O = 19\,\text{mol mol}^{-1}\ C_3H_6O = 19\,\text{m}^3\ \text{m}^{-3}\ C_3H_6O$$

Note: Humidity of the air is not considered.

Third step:
The amount of combustion gas (CG_{min}) for complete combustion is determined from the amount of oxygen supplied by the combustion air (= stoichiometric amount of combustion air minus minimum oxygen requirement) and the amount of combustion products CO_2 and water vapor:

$$\begin{aligned} CG_{min} = {} & CA_{min} - O_{2,\,min} + \text{amount of } CO_2\,(3\text{ moles per mole } C_3H_6O) \\ & + \text{amount of } H_2O\ (3\text{ moles per mole } C_3H_6O) \end{aligned} \tag{7.70}$$

$$\begin{aligned} CG_{min} = {} & 19\,\text{m}^3\ CA\,\text{m}^{-3}\ C_3H_6O - 4\,\text{m}^3\ O_2\ \text{m}^{-3}\ C_3H_6O + 3\,\text{m}^3\ CO_2\ \text{m}^{-3}\ C_3H_6O \\ & + 3\,\text{m}^3\ H_2O\,\text{m}^{-3}\ C_3H_6O = 21\,\text{m}^3\ \text{m}^{-3}\ C_3H_6O \end{aligned}$$

(2) Energy balance of the pollutant-related combustion process

Fourth step:
Next, the adiabatic temperature of combustion $T_{CG,ad}$ (or $\theta_{CG,ad}$) is determined for the case that only pollutant combustion is considered. The calculation is based on the overall energy balance of the combustion chamber.

With eq. (7.71), the overall energy input $\dot{Q}_{in}$ [kJ h^{-1}] can be obtained from the energy inputs of
- the pollutant (PO, eq. (7.72)),
- the off-gas (OG) without pollutant (=carrier gas, eq. (7.73)),
- the auxiliary fuel (AF, eq. (7.74)) and
- the combustion air (CA, eq. (7.75)); with F_m [kg h^{-1}], Cp [kJ kg^{-1} K^{-1}] and LCV [kJ kg^{-1}]:

$$\dot{Q}_{in} = \dot{Q}_{PO,in} + \dot{Q}_{OG,feed} + \dot{Q}_{AF} + \dot{Q}_{CA} \tag{7.71}$$

Pollutant (PO):

$$\dot{Q}_{PO,in} = F_{m,PO} \cdot (Cp_{PO} \cdot T_{PO} + LCV_{PO}) \tag{7.72}$$

Off-gas (OG) without pollutant (= carrier gas):

$$\dot{Q}_{OG,feed} = F_{m,OG,feed} \cdot \left(Cp_{OG,feed} \cdot T_{OG,feed}\right) \tag{7.73}$$

Auxiliary fuel (AF):

$$\dot{Q}_{AF} = F_{m,AF} \cdot (Cp_{AF} \cdot T_{AF} + LCV_{AF}) \tag{7.74}$$

Combustion air (CA) – when only pollutant combustion is considered:

$$\dot{Q}_{CA} = F_{m,CA,PO} \cdot (Cp_{CA} \cdot T_{CA}) \tag{7.75}$$

The energy output $\dot{Q}_{out}$ [kJ h^{-1}] of the combustion chamber is given by the energy content of the combustion gas:

$$\dot{Q}_{out} = \dot{Q}_{CG,out} = F_{m,CG} \cdot (Cp_{CG} \cdot T_{CG}) \tag{7.76}$$

Fifth step:
Based on the energy balance (eq. (7.77)), the adiabatic combustion temperature $T_{CG,ad}$ can be calculated. When the off-gas carrier gas is neglected and only combustion of the pollutant without any auxiliary fuel is considered, eq. (7.78) is obtained for the adiabatic combustion temperature $T_{CG,ad}$:

Energy balance:

$$\dot{Q}_{in} = \dot{Q}_{out} \tag{7.77}$$

$$T_{CG,ad} = \frac{F_{m,PO} \cdot (Cp_{PO} \cdot T_{PO} + LCV_{PO}) + F_{m,CA,PO} \cdot (Cp_{CA} \cdot T_{CA})}{F_{m,CG} \cdot (Cp_{CG} \cdot T_{CG})} \tag{7.78}$$

In general, the adiabatic temperature of combustion is not of interest as actually the temperature of combustion is specified by legislative requirements (e.g., θ_{CG} = 850 °C).

To achieve this required mean combustion temperature T_{CG}, the energy balance is extended with an unknown quantity of excess combustion air (ECA, $F_{m,ECA,PO}$, eq. (7.79)). After rearranging the energy balance, this unknown quantity can be determined according to the following equations:

$$F_{m,PO} \cdot (Cp_{PO} \cdot T_{PO} + LCV_{PO}) + (F_{m,CA,PO} + F_{m,ECA,PO}) \cdot (Cp_{CA} \cdot T_{CA}) = \left(F_{m,CG} \cdot C_{p,CG} + F_{m,ECA} \cdot Cp_{CA}\right) \cdot T_{CG} \tag{7.79}$$

$$F_{m,ECA,PO} \cdot (Cp_{CA} \cdot (T_{CG} - T_{CA}) = F_{m,PO}) \cdot (Cp_{PO} \cdot T_{PO} + LCV_{PO}) + F_{m,CA,PO} \cdot (Cp_{CA} \cdot T_{CA}) - (F_{m,CG} \cdot Cp_{CG}) \cdot T_{CG}$$

and

$$F_{m,ECA,PO} = \frac{F_{m,PO} \cdot (Cp_{PO} \cdot T_{PO} + LCV_{PO}) + F_{m,CA,PO} \cdot (Cp_{CA} \cdot T_{CA}) - F_{m,CG} \cdot (Cp_{CG} \cdot T_{CG})}{Cp_{CA} \cdot (T_{CG} - T_{CA})} \tag{7.80}$$

(3) Overall energy balance

The discussion of the energy balance in steps 1–5 did not consider the carrier of the pollutant, the off-gas carrier gas. As the concentration of the pollutant in the off-gas is low or even negligible in case of odorous substances, the energy demand for elevating the temperature of combustion of the off-gas needs to be evaluated. It has to be considered that some off-gas ($F_{m,ECA,PO}$ + $F_{m,CA,PO}$) has already been consumed by the combustion process of the pollutant. Thus, the amount of off-gas $F_{m,OG}$ to be heated corresponds to eq. (7.81):

$$F_{m,OG} = F_{m,feed} - \left(F_{m,ECA,PO} + F_{m,CA,PO}\right) \tag{7.81}$$

When repeating steps 1–5 after having specified the auxiliary fuel and its LCV_{AF}, and after having performed the mass balance for the combustion process of the auxiliary fuel, the amount of auxiliary fuel $F_{m,AF}$ can be evaluated (after rearranging the energy balance of step 5).

Sixth step:

Next, the amount of combustion gas for the auxiliary fuel $F_{m,CG,AF}$ per unit mass of auxiliary fuel is calculated. Then, the amount of excess combustion air $F_{m,ECA,AF}$ for a specified temperature of combustion T_{CG} can be deduced:

$$F_{m,AF} \cdot \left(Cp_{AF} \cdot T_{AF} + LCV_{AF}\right) + \left(F_{m,CA,AF} + F_{m,ECA,AF}\right) \cdot \left(Cp_{CA} \cdot T_{CA}\right) = \left(F_{m,CG,AF} \cdot Cp_{CG,AF} + F_{m,ECA,AF} \cdot Cp_{CG,AF}\right) \cdot T_{CG}$$

$$F_{m,ECA,AF} = \frac{F_{m,AF} \cdot \left(Cp_{AF} \cdot T_{AF} + LCV_{AF}\right) + F_{m,CG,AF} \cdot \left(Cp_{CA} \cdot T_{CA}\right) - \left(F_{m,CG,AF} \cdot Cp_{CG,AF}\right) \cdot T_{CG}}{Cp_{CA} \cdot \left(T_{CG} - T_{CA}\right)} \tag{7.82}$$

The total amount of auxiliary fuel $F_{m,AF,total}$ is calculated by dividing the net flow rate of off-gas $F_{m,OG}$ through $\left(F_{m,CA,AF} + F_{m,ECA,AF}\right)$ as shown in Equation 7.83.

Hint: some off-gas has already been consumed by the combustion process of the pollutant:

$$F_{m,AF,total} = \frac{F_{m,OG}}{F_{m,CA,AF} + F_{m,ECA,AF}} \tag{7.83}$$

Seventh step:

Finally, the off-gas of both the pollutant combustion and the combustion of the auxiliary fuel is summed up (eq. (7.84)):

$$F_{m,OG,total} = F_{m,OG,PO} + F_{m,OG,AF} \tag{7.84}$$

(4) Apparatus design

Target of design work is sizing of the volume V_R of the combustion chamber (= combustion reactor). Based on the mass balance in step 7, the combustion gas flow rate $F_{V,CG}$ in $m^3\ s^{-1}$ can be specified. Apart from the combustion of halogen-substituted hydrocarbons, the residence time τ of the combustion gas in the combustion reactor is recommended to be $\tau = 0.5$ s. In the latter case of halogen-substituted hydrocarbons, the temperature of combustion has to be 1,200 °C and the residence time 2 s.

The volume of the reactor is calculated according to the following equation:

$$V_R = \tau \cdot F_{V,CG} \tag{7.85}$$

! Reactor geometry is of cylindrical shape. In general, the ratio of the length of the reactor L to the diameter D is recommended to be between $L/D = 2.5$ and $L/D = 5$.

Exercise 7.15: Design of thermal off-gas combustion with recuperative off-gas preheating – acetone-laden off-gas

Design a thermal off-gas combustion process based on the specification given in Fig. 7.13 for the combustion of the pollutant acetone.

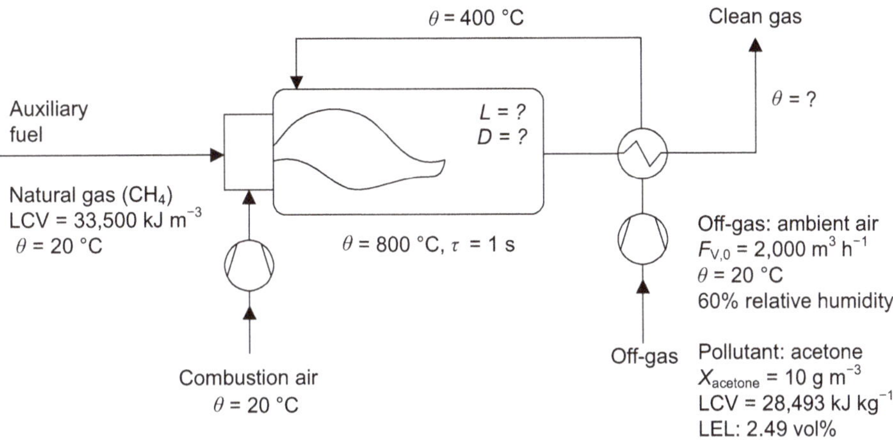

Fig. 7.13: Specification of the off-gas purification plant in Exercise 7.15 for thermal combustion of acetone with recuperative off-gas preheating.

Calculation values:

$$\rho_{air,20\,°C} = 1.2041\ \text{kg m}^{-3} \approx 1.20\ \text{kg m}^{-3};\ Cp_{air} \approx 1\ \text{kJ kg}^{-1}\ \text{K}^{-1}$$
$$Cp_{H_2O} \approx 2\ \text{kJ kg}^{-1}\ \text{K}^{-1}$$

Natural gas (NG):

$$LCV_{NG} \approx 33{,}500\ \text{kJ m}^{-3}$$

Acetone (C_3H_6O):

$$MM_{acetone} = 58\ \text{kg kmol}^{-1}$$
$$LCV_{acetone} = 28{,}493\ \text{kJ m}^{-3}$$

Note: It is common rule that the auxiliary fuel burner is not fed with off-gas. Usually, the burner is operated with fresh air from the ambient air (to avoid blockage of the burner by off-gas laden with solids).

Solver:
We work out the solution of this example in two parts using several approximations.

Part 1:

First approximation:

In the first approximation, we assume that no pollutant is contained in the off-gas and that the off-gas is not preheated:

- $\dot{Q}_{comb} = 0$
- no preheating

With $\dot{Q} = F_m \cdot c_P \cdot \Delta T$ and $\rho = \frac{F_m}{F_V}$, the required heat flow that is necessary to increase the temperature of the off-gas from 20 to 800 °C (off-gas without pollutant) can be determined:

$$\dot{Q}_{air} = F_{V,\text{off-gas}} \cdot \rho_{air,20\,°C} \cdot c_{P,air} \cdot (\theta_{comb.\,chamber} - \theta_{in})$$

$$= 2{,}000\,[m^3\,h^{-1}] \cdot 1.20\ [kg\,m^{-3}] \cdot 1\ [kJ\,kg^{-1}\,K^{-1}] \cdot (800-20)\,[K] = 1.87 \cdot 10^6\ kJ\ h^{-1} = 520\,kJ\ s^{-1} = 520\,kW$$

Note: $\Delta T = \Delta\theta$

Calculation of the volumetric flow rate of natural gas (NG) that is needed to enable this heat flow to be achieved results in:

$$F_{V,NG} = \frac{\dot{Q}_{air}}{LCV_{NG}} = \frac{1.87 \cdot 10^6\ [kJ\,h^{-1}]}{33{,}500\ [kJ\,m^{-3}]} = 55.9\ m^3\,h^{-1}$$

Second approximation:
We still assume that no pollutant is contained in the off-gas, but this time we preheat the off-gas:
- $\dot{Q}_{comb} = 0$
- preheating

15Mo3 (DIN171715) is a popular alloy steel that is used for heat exchangers. However, it limits the temperature to $\theta \leq 400$ °C.

Thus, the off-gas inlet temperature into the combustion chamber θ_{in} is set to 400 °C.

Heat flow:

$$\dot{Q}_{air,\,preheated} = F_{V,\text{off-gas}} \cdot \rho_{air,20\,°C} \cdot c_{P\,air} \cdot (\theta_{comb.\,chamber} - \theta_{in})$$

$$= 2{,}000\,[m^3\,h^{-1}] \cdot 1.20\,[kg\,m^{-3}] \cdot 1\ [kJ\,kg^{-1}\,K^{-1}] \cdot (800-400)\,[K] = 9.60 \cdot 10^5\ kJ\,h^{-1} = 267\ kW$$

Required quantity of natural gas:

$$\dot{F}_{V,NG} = \frac{\dot{Q}_{air,preheated}}{LCV_{NG}} = \frac{9.60 \cdot 10^5\ [kJ\,h^{-1}]}{33{,}500\ [kJ\,m^{-3}]} = 28.7\ m^3\,h^{-1}$$

Third approximation:
This time, we consider combustion of the pollutant acetone. The off-gas is preheated to 400 °C:
- $\dot{Q}_{comb} \neq 0$ and
- preheating

Combustion of the pollutant acetone releases heat, which reduces the required heat flow to increase the off-gas temperature from 400 to 800 °C:

$$\dot{Q}_{\text{off-gas, preheated}} = F_{V,\text{off-gas}} \cdot \rho_{air,20\,°C} \cdot c_{P,air} \cdot (\theta_{comb.\,chamber} - \theta_{in}) - \dot{Q}_{comb,\,acetone}$$

with

$$\dot{Q}_{comb,\,acetone} = \dot{F}_{m,acetone} \cdot LCV_{acetone}$$

Thus,

$$\dot{Q}_{\text{off-gas, preheated}} = 2{,}000\,[m^3\,h^{-1}] \cdot 1.20\,[kg\,m^{-3}] \cdot 1\ [kJ\,kg^{-1}\,K^{-1}] \cdot (800-400)\,[K] - 20\,[kg\,h^{-1}] \cdot 28{,}493\,[kJ\,kg^{-1}]$$

$$= 3.90 \cdot 10^5\ kJ\,h^{-1} = 108\,kW$$

Required quantity of natural gas:

$$\dot{F}_{V,NG} = \frac{\dot{Q}_{off-gas,preheated}}{LCV_{NG}} = \frac{3.90 \cdot 10^5 \left[kJ\,h^{-1}\right]}{33,500 \left[kJ\,m^{-3}\right]} = 11.6\ m^3\,h^{-1}$$

As you can see in the process scheme diagram (Fig. 7.13), the auxiliary burner does not use off-gas. It is operated with fresh air. This stream is termed "combustion air (CA)."

Detailed mass and energy balance:

Now, let us calculate the amount of substances involved and check with the mass balance.

Reaction equation for the combustion of acetone: $C_3H_6O + 4\ O_2 \rightarrow 3\ CO_2 + 3\ H_2O$.

First, the minimum amount of air that is required for complete combustion is determined. With $F_n = \frac{F_m}{MM}$ and the reaction stoichiometry of the combustion reaction, the minimum mass flow rate of air $(F_{m,O_2,min})$ is obtained:

$$F_{m,O_2,min} = F_{n,O_2,min} \cdot MM_{O_2} = F_{n,acetone} \cdot \nu_{O_2} \cdot MM_{O_2}$$

$$F_{m,O_2,min} = \frac{F_{m,acetone} \cdot \nu_{O_2} \cdot MM_{O_2}}{MM_{acetone}} = \frac{20 \cdot 4 \cdot 32}{58} = 44.14\ kg\,h^{-1}$$

With the molar volume MV at 20 °C ($MV_{20\,°C} = 24.05\ m^3\ kmol^{-1}$), we get the volumetric flow of oxygen that is required for complete combustion. The molar mass of oxygen is 32 kg $kmol^{-1}$:

$$F_{V,O_2,min} = \frac{F_{m,O_2,min} \cdot MV}{MM_{O_2}} = \frac{44.14 \cdot 24.05}{32} = 33.17\ m^3\,h^{-1}$$

The relative humidity of air is neglected:

$$F_{V,air,min} = \frac{F_{V,O_2,min}}{y_{O_2}} = \frac{33.17}{0.21} = 157.96\ m^3\,h^{-1}$$

$$F_{V,N_2} = F_{V,air,min} \cdot y_{N_2} = 157.96 \cdot 0.79 = 124.79\ m^3\,h^{-1}$$

Furthermore with $\rho = \frac{MM}{MV}$:

$$\rho_{N_2} = \frac{MM_{N_2}}{MV} = \frac{28}{24.05} = 1.16\ kg\,m^{-3}$$

$$\rho_{O_2} = \frac{MM_{O_2}}{MV} = \frac{32}{24.05} = 1.33\ kg\,m^{-3}$$

$$F_{m,N_2} = F_{V,N_2} \cdot \rho_{N_2} = 124.79 \cdot 1.16 = 145.29\ kg\,h^{-1}$$

The following mass and volumetric flow rates are obtained for the combustion products CO_2 and water:

$$F_{m,CO_2} = \frac{F_{m,acetone} \cdot \nu_{CO_2} \cdot MM_{CO_2}}{MM_{acetone}} = \frac{20 \cdot 3 \cdot 44}{58} = 45.52\ kg\,h^{-1}$$

$$F_{V,CO_2} = \frac{F_{m,CO_2} \cdot MV}{MM_{CO_2}} = \frac{45.52 \cdot 24.05}{44} = 24.88\ m^3\,h^{-1}$$

$$F_{m,H_2O} = \frac{F_{m,acetone} \cdot \nu_{H_2O} \cdot MM_{H_2O}}{MM_{acetone}} = \frac{20 \cdot 3 \cdot 18}{58} = 18.62\ kg\,h^{-1}$$

$$F_{V,H_2O} = \frac{F_{m,H_2O} \cdot MV}{MM_{H_2O}} = \frac{18.62 \cdot 24.05}{18} = 24.88 \text{ m}^3\text{ h}^{-1}$$

Check:
Mass balance:

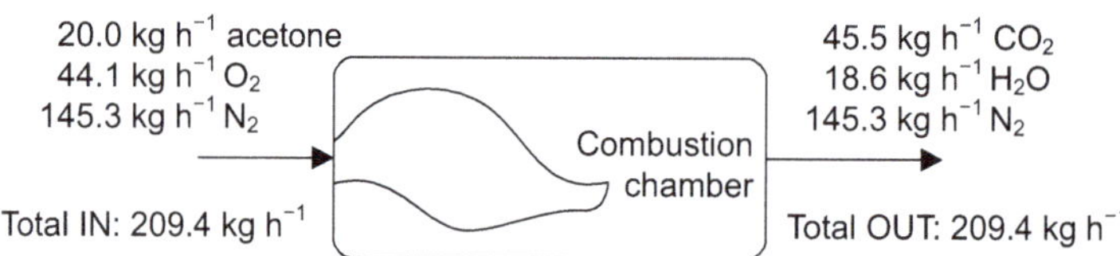

Fig. 7.14: Mass balance of the combustion chamber in Exercise 7.15.

$$\sum F_{m,\,in} = \sum F_{m,\,out}$$

Energy balance:

$$\sum \dot{Q}_{in} = \sum \dot{Q}_{out}$$

$$\dot{Q}_{in,total} = \dot{Q}_{air} + \dot{Q}_{comb} = \left(F_{m,O_2} + F_{m,N_2}\right) \cdot Cp_{air} \cdot (\theta_{comb.\,chamber} - \theta_{in}) + \dot{Q}_{comb,\,acetone}$$

$$(44.14 + 145.28) \cdot 1 \cdot (800 - 400) + 20 \cdot 28,493 = 645,630 \text{ kJ h}^{-1}$$

$$\dot{Q}_{out,total} = \sum (F_{m,i} \cdot Cp_i) \cdot \Delta T_{out} = \left[\left(F_{m,CO_2} + F_{m,N_2}\right) \cdot Cp_{air} + F_{m,H_2O} \cdot Cp_{H_2O}\right] \cdot$$

$$\Delta T_{out} = [(45.52 + 145.28) \cdot 1 + 18.62 \cdot 2] \cdot \Delta T_{out}$$

With $\dot{Q}_{in,total} = \dot{Q}_{out,total}$, ΔT_{out} can be determined.

$$\Delta T_{out} = \frac{\dot{Q}_{in,total}}{\sum (F_{m,i} \cdot CP_i)} = \frac{645,630}{[(45.52 + 145.28) \cdot 1 + 18.62 \cdot 2]} = 2,831.1 \text{ K}$$

$$\dot{Q}_{off-gas,\,800\,°C} = [(45.52 + 145.28) \cdot 1 + 18.62 \cdot 2] \cdot 800 = 182,437 \text{ kJ h}^{-1}$$

$$\Delta \dot{Q} = \dot{Q}_{in} - \dot{Q}_{off\text{-}gas,\,800\,°C} = 645,630 - 182,437 = 463,193 \text{ kJ h}^{-1}$$

$\Delta \dot{Q} = (F_{m,air} \cdot Cp_{air}) \cdot \Delta T$ and thus,

$$F_{m,air} = \frac{\Delta \dot{Q}}{Cp_{air} \cdot \Delta T} = \frac{463,193}{1 \cdot 400} = 1,158.0 \text{ kg}^{-1}$$

$$\rho_{air,\,20\,°C} = 1.20 \text{ kg m}^{-3}$$

$$F_{V,air} = \frac{1,158.0}{1.20} = 965.0 \text{ m}^3\text{ h}^{-1}$$

$$F_{V,air,total} = 157.96 + 965.0 = 1,123.0 \text{ m}^3\text{ h}^{-1}$$

Differential air flow:

$$\Delta F_{V,air} = 2,000 - 1,123 = 877 \text{ m}^3\text{ h}^{-1}$$

When 2,000 $m^3\ h^{-1}$ air are supplied, 877 $m^3\ h^{-1}$ need to be heated separately.

Part 2:
Now, we determine the heat that needs to be supplied by the auxiliary fuel natural gas. The entire energy content of the natural gas only enters through the auxiliary fuel burner (Fig. 7.15):

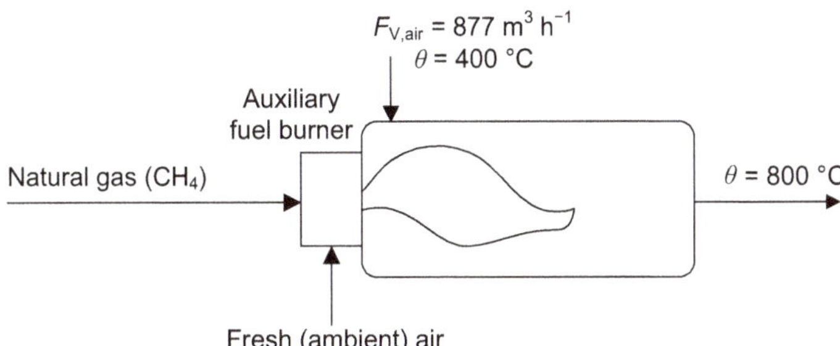

Fig. 7.15: Flow chart for supply of the auxiliary fuel natural gas in Exercise 7.15.

$$\dot{Q}_{air}(400-800\,°C) = \Delta F_{V,air} \cdot Cp_{air} \cdot (\theta_{comb.\,chamber} - \theta_{in}) = 877 \cdot 1.20 \cdot 1 \cdot (800-400) = 420,984\ \text{kJ h}^{-1}$$

This amount of energy needs to be provided by natural gas (methane CH_4).

Note: The numerical values in the individual calculation steps are given as rounded values. The results are based on calculated values without rounding.

The combustion of methane is represented by the following reaction equation:

$$CH_4 + 2\,O_2 \rightarrow CO_2 + 2\,H_2O$$

Now, we calculate the oxygen demand and the amount of combustion products CO_2 and H_2O for methane combustion (per 1 m^3 of methane).

Minimum oxygen demand: $O_{2,min} = 2\ m^3\ O_2$ per 1 m^3 CH_4:

$$\text{Air}_{min} = \frac{2}{0.21} = 9.52\ m^3\ m^{-3}\ \text{NG}$$

$$H_2O\text{:}\ m_{H_2O} = \frac{2 \cdot MM_{H_2O}}{MV} = \frac{2 \cdot 18}{24.05} = 1.50\ \text{kg}\ H_2O\ m^{-3}\ \text{NG}$$

$$N_2\text{:}\ m_{N_2} = \frac{(9.52 \cdot 0.79) \cdot MM_{N_2}}{MV} = \frac{7.52 \cdot 28}{24.05} = 8.76\ \text{kg}\ N_2\ m^{-3}\ \text{NG}$$

$$CO_2\text{:}\ m_{CO_2} = \frac{1 \cdot MM_{CO_2}}{MV} = \frac{1 \cdot 44}{24.05} = 1.83\ \text{kg}\ N_2\ m^{-3}\ \text{NG}$$

$$1\ m^3\ \text{NG}\ (CH_4)\ \text{equals}\ m_{CH_4} = \frac{V_{CH_4} \cdot MM_{CH_4}}{MV} = \frac{1 \cdot 16}{24.05} = 0.67\ \text{kg}$$

The mass of air is calculated with $m_{air} = \text{Air}_{min} \cdot \rho_{air} = 9.52 \cdot 1.20 = 11.43$ kg

For the energy content of the burner exhaust-gas we get:

$$Q_{\text{exhaust gas, }800\,°C} = \left[\left(m_{N_2} + m_{CO_2}\right) \cdot Cp_{air} + m_{H_2O} \cdot Cp_{H_2O}\right] \cdot \theta = [(8.76 + 1.83) \cdot 1 + 1.50 \cdot 2] \cdot 800 = 10,863\ \text{kJ}\ m^{-3}$$

Note that the burner is loaded with the off-gas to be purified.

When $Q_{air,in}$ is negligible:

$$Q_{in} = 1 \cdot LCV_{CH_4} = 33{,}500 \text{ kJ m}^{-3}$$

$$\Delta Q = Q_{in} - Q_{exhaust\ gas,\ 800\,°C} = 33{,}500 - 10{,}863 = 22{,}637 \text{ kJ m}^{-3}$$

This is the amount of heat that can be used to preheat the off-gas.

Requirement:

From that the volumetric flow rate of natural gas that is required for the burner is obtained:

$$F_{V,NG} = \frac{\dot{Q}_{air}}{\Delta Q} = \frac{420{,}984\ [\text{kJ h}^{-1}]}{22{,}637\ [\text{kJ m}^{-3}]} = 18.60 \text{ m}^3 \text{ h}^{-1}$$

As last step, we check the mass balance of the auxiliary fuel burner (Fig. 7.16). Compare the outcome with the approximations.

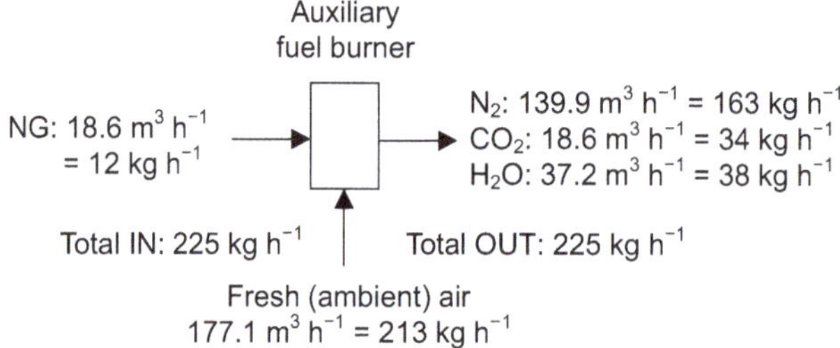

Fig. 7.16: Mass balance of the auxiliary fuel burner in Exercise 7.15.

Figure 7.17 serves as a guide for the calculation procedure of thermal off-gas purification processes.

For the sake of completeness, the electrically powered combustion process should also be mentioned. It can be applied to exhaust gas streams that are loaded with high concentrations of organic matter but low oxygen contents. By applying an electric field, the combustion process can be operated close to the minimum oxygen requirement due to the formation of oxygen radicals. The combustion temperature can be slightly lowered; but relatively high voltages are required [207]. Zigan [229] gives an overview about electric field applications in energy and process engineering, among others for off-gas purification by combustion. A comprehensive review on plasma-assisted ignition and combustion is given by Starikovskiy and Aleksandrov [230].

7.3.3 Catalytic combustion

Catalytic (flameless) combustion processes are a supplement to thermal combustion processes. Due to the activity of a catalyst the rate of reaction (= rate of combustion) is

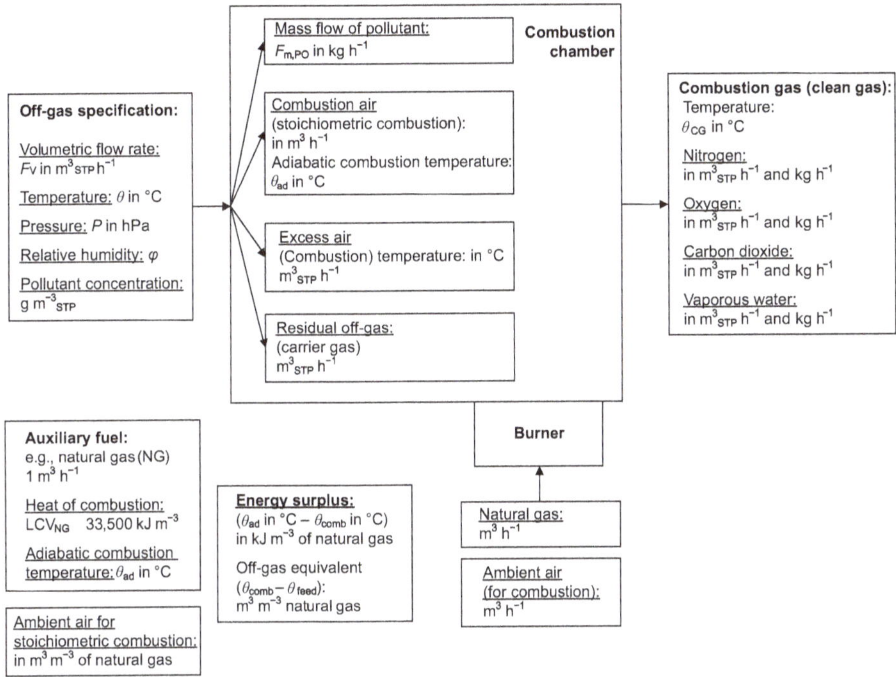

Fig. 7.17: Calculation guide for thermal off-gas purification processes.

increased, the reaction is permitted to occur at a lower temperature and the reactor volume is reduced. As heterogeneous catalysis is a surface phenomenon, the physical property of a large surface area is of decisive necessity. Catalysts may be porous pellets of cylindrical or spherical shape. Other common shapes are honeycombs, ribbons and wire mesh.

Figure 7.18 depicts the process flow scheme of a catalytic combustion chamber with recuperative off-gas preheating. The off-gas is continuously fed to the reactor at a gas velocity of 3–15 m s^{-1} and a temperature of 300–400 °C. The off-gas passes through the catalyst bed of the reactor and the combustion products exit the combustion chamber continuously. Heat recovery can support energy savings.

Metals of the platinum group (platinum, palladium) are recognized for their ability to promote combustion at low temperature. Other common catalysts include oxides of copper, chromium, vanadium, nickel and cobalt. However, these catalysts are subject to poisoning, particularly with halogens and sulfur compounds, zinc, arsenic, lead, mercury and particulates. Catalyst poisoning is a severe issue as it may shorten a catalyst's lifetime drastically.

The primary purpose of flameless combustion is to reduce the overall energy demand and emissions from byproduct formation due to high combustion temperatures (e.g., nitrogen oxides), while retaining thermal efficiencies in combustion systems. Flameless combustion is therefore characterized by some distinguished features. In

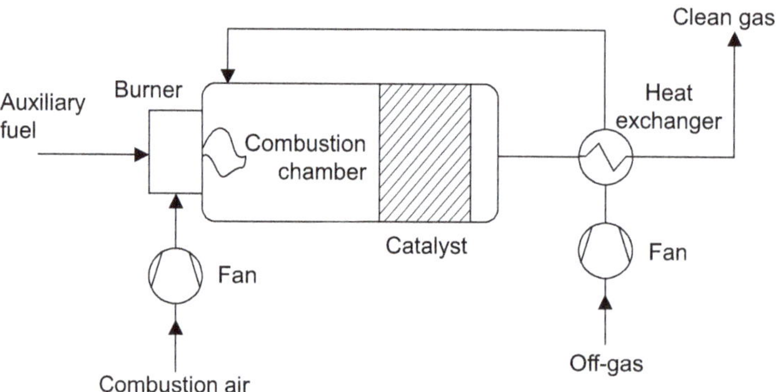

Fig. 7.18: Process scheme of catalytic off-gas purification.

addition to suppressed pollutant emission, these are homogeneous temperature distribution, reduced noise and thermal stress for burners and less restriction on fuels (since flame stability is not required).

In catalytic combustion, the upper operation temperature is limited due to thermal stability, but not a lower temperature. However, a low temperature results in a low reaction rate and therefore, a large catalyst volume is required. The thermal stability of the catalyst limits the upper temperature. Thermal catalyst deactivation may result in severe loss of activity.

7.3.3.1 Reactor design

Due to the similarity of catalytic oxidation and catalytic reduction, the calculation algorithm was already discussed in Section 7.2 in detail.

Tab. 7.9: Representative catalyst specification: type of catalyst (copper oxide pellets, $d = 4$ mm; catalytic combustion reaction).

Space velocity ($m^3_{STP}\ m^{-3}\ h^{-1}$)	10,000	20,000	30,000
Arrhenius frequency factor ($m^3_{STP}\ m^{-3}\ h^{-1}$)	$6.8 \cdot 10^{30}$	$1.2 \cdot 10^{16}$	$4.7 \cdot 10^{11}$
Energy of activation ($kJ\ mol^{-1}$)	263.8	121.9	79.0

Exercise 7.16: Design of a catalytic off-gas combustion process

Estimate the temperature needed for 90% (or 95%, 99%) conversion and different space velocities for the catalytic combustion of pollutant A in the catalytic systems given in Tab. 7.9.

Solver:

The catalytic combustion process is carried out in a tubular reactor. Assuming ideal behavior (no deviations from ideal flow behavior), the design equation of a PFR can be used. For a first-order reaction ($-r_A = k \cdot c_A$), the following design equation is obtained from the mass (mole) balance of a PFR at steady-state operation ($F_{A,0} = F_A + \int_0^V (-r_A) \cdot dV$) for constant-volume reaction ($\varepsilon_A = 0$; note: $\varepsilon_A \neq 0$ will not cause a severe mistake at low pollutant load in the off-gas):

$$\frac{V}{F_{A,0}} = \frac{V}{(F_{V,0} \cdot c_{A,0})} = \frac{\tau}{c_{A,0}} = \int_0^{X_A} \frac{dX_A}{-r_A} = \int_0^{X_A} \frac{dX_A}{k \cdot c_A} = \int_0^{X_A} \frac{dX_A}{k \cdot c_A \cdot (1 - X_A)}$$

In catalytic off-gas purification, it is more practical to relate the reaction rate constant to the catalyst volume (k''', also k_{SV}). After integration we get:

$$-\ln(1 - X_A) = k \cdot \tau = k \cdot \frac{V}{F_{V,0}} = k''' \cdot \frac{V_{cat}}{F_{V,0}} = \frac{k'''}{SV}$$

Note: Actually, it is an adiabatic process. As approximation, we consider it being isothermal.

First, we determine the catalyst volume-related reaction rate constant k''' for various operation temperatures:

$$k_{SV} = k''' = A''' \cdot e^{-\frac{E_A}{R \cdot T}}$$

With known k''', we can calculate the relative conversion X_A in the reactor:

$$X_A = 1 - e^{-\frac{k'''}{SV}}$$

Tab. 7.10: Calculated results for Exercise 7.16.

Space velocity		$10{,}000\ m^3_{STP}\ m^{-3}\ h^{-1}$		$20{,}000\ m^3_{STP}\ m^{-3}\ h^{-1}$		$30{,}000\ m^3_{STP}\ m^{-3}\ h^{-1}$	
Arrhenius frequency factor		$6.8 \cdot 10^{30}\ m^3_{STP}\ m^{-3}\ h^{-1}$		$1.2 \cdot 10^{16}\ m^3_{STP}\ m^{-3}\ h^{-1}$		$4.7 \cdot 10^{11}\ m^3_{STP}\ m^{-3}\ h^{-1}$	
Activation energy			$263.8\ kJ\ mol^{-1}$		$121.9\ kJ\ mol^{-1}$		$79.0\ kJ\ mol^{-1}$
θ (°C)	**T (K)**	**k''' ($m^3_{STP}\ m^{-3}\ h^{-1}$)**	**X_A**	**k''' ($m^3_{STP}\ m^{-3}\ h^{-1}$)**	**X_A**	**k''' ($m^3_{STP}\ m^{-3}\ h^{-1}$)**	**X_A**
100	373.15	$8.01 \cdot 10^{-7}$	$8.01 \cdot 10^{-11}$	$1.03 \cdot 10^{-1}$	$5.17 \cdot 10^{-6}$	4.10	$1.37 \cdot 10^{-4}$
200	473.15	$5.11 \cdot 10^{1}$	$5.10 \cdot 10^{-3}$	$4.18 \cdot 10^{2}$	$2.07 \cdot 10^{-2}$	$8.92 \cdot 10^{2}$	$2.93 \cdot 10^{-2}$
250	523.15	$3.11 \cdot 10^{4}$	0.955	$8.08 \cdot 10^{3}$	0.332	$6.08 \cdot 10^{3}$	0.138
300	573.15	$6.17 \cdot 10^{6}$	1	$9.32 \cdot 10^{4}$	0.991	$2.97 \cdot 10^{4}$	0.628
350	623.15	$5.24 \cdot 10^{8}$	1	$7.26 \cdot 10^{5}$	1	$1.12 \cdot 10^{5}$	0.976
400	673.15	$2.30 \cdot 10^{10}$	1	$4.17 \cdot 10^{6}$	1	$3.48 \cdot 10^{5}$	1

In Fig. 7.19 the relative conversion is shown as a function of temperature for the three different space velocities (10,000, 20,000 and 30,000 $m^3_{STP}\ m^{-3}\ h^{-1}$). It can be deduced from the results that the operation temperature must be >300 °C in order to achieve considerable relative conversion of the pollutant. Moreover, as expected, the relative pollutant conversion increases with decreasing SV due to the longer residence time in the reactor.

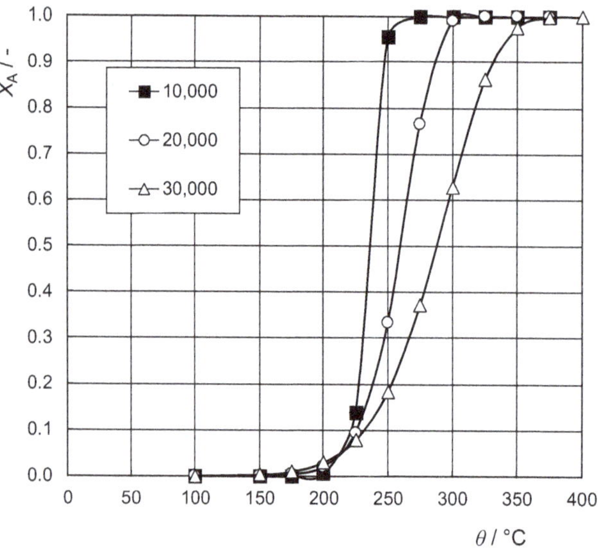

Fig. 7.19: Relative conversion of pollutant A combustion during catalytic off-gas purification at different temperatures for various space velocities (SV) (10,000, 20,000 and 30,000 m^3_{STP} m^{-3} h^{-1}).

In literature, different reactor concepts are proposed. Maier and Schlangen [231], for instance, suggested electrically heated wires with catalytically active surfaces as a novel application of catalytic combustion in off-gas purification.

7.3.3.2 Summary

To sum of, during catalytic combustion a heterogeneous catalyst (e.g., Pt, Cu, Cr, Va, Ni and Co)

- increases the rate of reaction,
- enables the reaction to proceed at low(er) temperatures (in general, at 300–400 °C),
- thus smaller reactors are needed.

Attention has to be paid to the off-gas composition as several species may act as catalyst poisons (e.g., halogens, S, Zn, As, Pb, Hg and particulate substances).

7.3.4 Combination of thermal and catalytic combustion

Thermal and catalytic combustion technologies can also be combined. The overall purpose of combination of these two technologies is to combust hydrocarbons in the gas phase first and to convert carbon monoxide and unburnt hydrocarbons in a consecutive step on the catalyst.

In case of unfavorable conditions such as varying off-gas flow rate and composition, inappropriate combustion temperature or residence time (e.g., during capacity changes), and secondary pollutants (e.g., CO or aldehydes) may be emitted from the thermal combustion chamber. A suitable countermeasure would be to increase the combustion temperature. However, this is often not possible due to equipment limitations. Moreover, increasing temperature results in increasing NO_x emissions, which have to be prevented. Through a subsequent catalytic combustion step secondary pollutant emissions can be avoided. Thus, thermal combustion can be run at a lower combustion temperature while ensuring emission limits [216].

7.4 Off-gas purification by reduction in the gas phase

Off-gas purification based on pollutant reduction in the gas phase is especially relevant for nitrogen oxides (catalytic denoxing). For this reason, denoxing will be discussed in particular in the following topic. Sulfur oxides (SO_x) can also be removed from off-gas via gas-phase reduction. The same principles of reaction kinetics, heterogeneous catalytic acceleration of the rate of reaction and reactor design apply as for catalytic oxidation.

7.4.1 Nitrogen oxides

Nitrogen oxides (nitrous oxides, NO_x) represent a special group of pollutants and provide an illustrative example in reductive off-gas purification. Since they are formed, among others, from atmospheric oxygen during combustion reactions, emission mitigation measures not only start with NO_x elimination in the off-gas but also before/at their formation. A distinction is therefore made between:

- primary measures – prevention of NO_x formation
- secondary measures – off-gas purification through NO_x reduction

In the following section, we address both approaches in more detail; NO_x emission prevention and emission control.

7.4.1.1 Formation of nitrogen oxides

NO_x emissions mainly originate in fuel composition and the combustion process. According to its formation, we need to distinguish between three types of NO_x:

- fuel NO_x: formed from nitrogen-containing fuel via CN intermediates
- thermal NO_x: formed by oxidation of nitrogen with oxygen in the combustion air at elevated temperature (eq. (7.86))
- prompt NO_x: formed from CN radicals in the fuel and nitrogen of the combustion air at high temperatures (eq. (7.87))

Thermal NO_x:

$$\begin{aligned}&(1)\ 2\,N_2 + O_2 \longrightarrow 2\,NO + 2N\\&(2)\ N + O_2 \longrightarrow NO + O\\&(1) + (2)\ N_2 + O_2 \longrightarrow 2\,NO\end{aligned} \tag{7.86}$$

Prompt NO_x:

$$\begin{aligned}&(1)\ CH + N_2 \longrightarrow HCN + N\\&(2)\ C + N_2 \longrightarrow CN + N\\&(3)\ N + OH \longrightarrow NO + H\end{aligned} \tag{7.87}$$

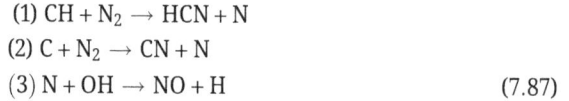

Mechanisms of formation and fractional NO_x amount depend on the conditions of combustion. Conversion of fuel into NO_x (fuel NO_x) may achieve a fractional amount of 50% for nitrogen contents in the fuel as low as 0.2 wt%. NO_x is preferably emitted gaseous.

The direct formation of NO_x from combustion air (thermal NO_x, eq. (7.86)) is controlled by the temperature and by the amount of oxygen and nitrogen according to the equilibrium condition in the following equation (with y in vol%):

$$K = \frac{y_{NO}{}^2}{y_{N_2} \cdot y_{O_2}} \tag{7.88}$$

The temperature dependence of the equilibrium constant is given by the correlation in the following equation:

$$K = 10 \cdot e^{\frac{-9{,}460}{T} + 1.074} \tag{7.89}$$

The above-mentioned equilibrium correlation can be used for estimating the maximum amount of thermal NO_x formed by chemical interaction of nitrogen and oxygen. Above an operating temperature of 600 °C, NO is the only stable nitrogen oxide. When decreasing the temperature, NO is oxidized with oxygen in excess to form nitrogen dioxide (NO_2, eq. (7.90)) and several nitrogen oxides in parallel and consecutive reactions. Nitrogen dioxide is of brownish occurrence and becomes visible when exceeding an amount of 500 mg m^{-3}:

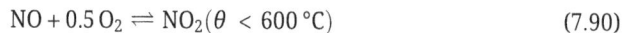

$$NO + 0.5\,O_2 \rightleftharpoons NO_2(\theta < 600\,°C) \tag{7.90}$$

The rate of NO oxidation depends on the temperature and the amount of NO. Estimation of conversion for NO contents <1 vol% and within the temperature limits of 0 and 80 °C can be based on the following algorithm:

$$\text{Amount of NO converted} = X_{\text{NO}} = \frac{k\left[\text{s}^{-1}\right] \cdot \tau\ [\text{s}] \cdot \text{NO content}\ [\text{vol\%}]}{1 + k\left[\text{s}^{-1}\right] \cdot \text{NO content}\ [\text{vol\%}]} \tag{7.91}$$

The temperature dependence of k is given in the following equation:

$$k = 0.062 \cdot e^{-0.011 \cdot T} \tag{7.92}$$

Besides nitrogen dioxide (N_2O) and nitrogen monoxide (NO), several other nitrogen oxides can be formed, for example, dinitrogen monoxide (N_2O), dinitrogen tetroxide (N_2O_4), dinitrogen trioxide (N_2O_3), dinitrogen pentoxide (N_2O_5), nitrous acid (HNO_2) and nitric acid (HNO_3). The amount of oxides and acids has to be considered in material selection of off-gas purification equipment.

7.4.1.2 NO_x emission mitigation measures

7.4.1.2.1 Primary measures – prevention of NO_x formation (emission prevention)

Emission prevention is the primary goal, and nitrogen oxides provide an excellent example for this approach. As the primary cause of NO_x formation is the simultaneous presence of (1) nitrogen and (2) oxygen at (3) high temperatures, these three parameters can be addressed to minimize NO_x formation [232, 233].

(1) Source of nitrogen
Many fuels only contain a limited amount of nitrogen and thus do not present a remarkable nitrogen source. In combustion processes, may it be fuel combustion or off-gas purification through combustion, the primary nitrogen source is the combustion air, which is normally used to provide the oxygen. Certainly, pure oxygen could be used for combustion, but this is not a viable option as the cost of pure oxygen in the required amounts is prohibitive.

(2) Amount of oxygen
As oxygen is the oxidant in oxidation reactions, its presence is unavoidable. To ensure complete combustion, oxygen is generally used in excess, meaning in higher quantities than the required stoichiometric quantity. As discussed in Sections 7.2 and 2.4, incomplete combustion may lead to unburnt or partially oxidized hydrocarbon species in the off-gas, above all to the formation of carbon monoxide. These species represent another source of serious concern. Optimized combustion chambers (low-NO_x burner) and improvements in combustible-air mixers and burners may reduce the amount of excess oxygen. Still, the use of excess oxygen cannot be eliminated completely.

Staged combustion (dual-step or two-stage combustion (TSC)) is a very effective measure for limiting fuel NO_x formation. During staged combustion, the combustion is initially run at oxygen deficit conditions (fuel-rich conditions). This has two beneficial effects. First, the maximum temperatures in the combustible-air mixing region are lowered; and second, the amount of available oxygen is reduced. Consequently, the concentration of oxygen in the hottest section of the combustion chamber is lowered. After heat has been removed from the system, additional air is supplied to the combustion process to ensure complete combustion. Such air-staging measures lower the formation of thermal NO_x due to lower peak temperatures during combustion. Combustion air is provided in two or more stages. For instance, first substoichiometric amounts of air are supplied as primary air, then secondary air is added, and finally tertiary air is supplied for complete combustion [223].

(3) Reaction temperature
The reaction temperature can be controlled in different ways. The first approach is to ensure good mixing to prevent hot spots. Sophisticated design of burners has brought insufficient mixing characteristics under control. The second approach is to dilute the combustion mixture. In the discussion of the adiabatic combustion temperature we saw that the maximum temperature in a combustion chamber largely depends on the gas mixture and volume, meaning the amount of off-gas (or fuel) and air since this quantity needs to be heated with the released heat of reaction. This effect is strongly pronounced for NO_x formation and counteracts the effect of excess air. On the one hand, NO_x formation increases when the supply of air is increased and even exceeds the stoichiometric quantity. This can be dedicated to the fact that more oxygen is available. On the other hand, this effect is annulled when the amount of excess air is further increased due to the decrease of combustion temperature because of the large gas amount related to the energy provided by combustion of fuels. This effect can also be accomplished by the supply of inerts. Drawback of this approach is that the overall efficiency of the combustion process suffers tremendously. The third approach, which is not as bad to the overall combustion efficiency as dilution with inerts or excess air, is flue gas recirculation (FGR). In FGR, recycled exhaust gas acts as diluent but only slightly reduces efficiencies as the only losses are minor losses in the recycle loop.

These methods enable NO_x formation to drop by a factor of two to five. However, for further reduction of NO_x emissions, reactive off-gas purification technologies must be installed [233].

To sum up, primary measures are based on the upgrading of fuel. Corresponding to the equilibrium condition, described by the law of mass action, formation of NO due to elevated temperature can be limited by limiting the temperature of combustion and by limiting either the oxygen content or the nitrogen content of the combustion air. Formation of NO passes a maximum for nitrogen/oxygen ratios of 0.5. High ratios and low ratios of nitrogen/oxygen decrease the formation of NO. Combustion with pure oxygen and substoichiometric combustion in connection with dual-step combustion are preferably applied. Similar results are achieved by FGR.

7.4.1.2.2 Secondary measures – off-gas purification through NO_x reduction (emission control)

Secondary measures are based on the reduction of NO_x with several additives with reducing properties. Fuel-based reduction is obtained by in-furnace reduction (IFR). The reduction process is nonselective. Nonselective reduction is also obtained by adding carbon monoxide (eq. (7.93) for NO) or hydrocarbons (C_nH_m, eq. (7.94) for NO) to the off-gas by either substoichiometric combustion or separate addition of the reducing agents. These processes are called nonselective reduction processes (so-called NSR processes). The processes are either carried out at a temperature of 850–950 °C or with catalytic support (NSCR) at reaction temperatures beyond 400 °C. In both applications the reaction temperature acts via the energy of activation (Arrhenius):

$$NO + CO \longrightarrow N_2 + CO_2 \tag{7.93}$$

$$\left(2n + \frac{m}{2}\right) NO + C_nH_m \longrightarrow \left(n + \frac{m}{4}\right) N_2 + n\,CO_2 + \frac{m}{2}\,H_2O \tag{7.94}$$

Selective reduction of nitrogen oxides (selective nitrogen reduction) is obtained by adding ammonia (NH_3) to the off-gas. Again, the process can be carried out at elevated temperature, necessary to overcome the activation energy, or it is carried out with catalytic support at an operation temperature of about 400 °C or even less. The first technique is a selective noncatalytic reduction process (SNCR) while latter process is a SCR process. The latter process is preferably applied in industrial denoxing.

Basically, three types of SCR reactions with NH_3 are distinguished. These are the standard SCR (eq. (7.95) for NO), the fast SCR (eq. 7.96) and the so-called slow SCR (eq. (7.97) for NO_2). The scientific literature on this topic is manifold (e.g., [234–236]).

Standard SCR of NO:

$$4\,NO + 4\,NH_3 + O_2 \longrightarrow 4\,N_2 + 6\,H_2O \tag{7.95}$$

Fast SCR of NO and NO_2:

$$2\,NO + 2\,NO_2 + 4\,NH_3 \longrightarrow 4\,N_2 + 6\,H_2O \tag{7.96}$$

Slow SCR with NO_2:

$$3\,NO_2 + 4\,NH_3 \longrightarrow 3.5\,N_2 + 6\,H_2O \tag{7.97}$$

Table 7.11 compares the thermodynamics – free standard enthalpy of reaction $\Delta_R G^0_{300\,°C}$ and standard enthalpy of reaction $\Delta_R H^0_{300\,°C}$ – of the reactions involved in nitrogen oxide (NO and NO_2) reduction with ammonia at 300 °C. All reactions are exothermic and irreversible at 300 °C.

Tab. 7.11: Thermodynamics (free standard enthalpy of reaction $\Delta_R G^0_{300\,°C}$ and standard enthalpy of reaction $\Delta_R H^0_{300\,°C}$) of nitrogen oxide (NO and NO_2) reduction with ammonia at 300 °C.

Reaction	$\Delta_R G^0_{300\,°C}$ (kJ mol^{-1} NO_x)	$\Delta_R H^0_{300\,°C}$ (kJ mol^{-1} NO_x)
$4\,NO + 4\,NH_3 + O_2 \rightarrow 4\,N_2 + 6\,H_2O$	−419.1	−406.1
$3\,NO + 2\,NH_3 \rightarrow 2.5\,N_2 + 3\,H_2O$	−307.1	−300.1
$3\,NO_2 + 4\,NH_3 \rightarrow 3.5\,N_2 + 6\,H_2O$	−516.3	−453.1
$4\,NO + NH_3 \rightarrow 2.5\,N_2O + 1.5\,H_2O$	−89.1	−118.2
$4\,NH_3 + 5\,O_2 \rightarrow 4\,NO + 6\,H_2O$	−252.7*	−225.5*

*Per mole NH_3.

Gaseous ammonia, liquid ammonia and aqueous ammonia (25%) are standard reducing additives. With regard to the toxic properties of ammonia, urea ($CO(NH_2)_2$) is the preferred reducing additive nowadays. Urea is stored and transported in solid bulk lots. It is dissolved in water prior to metering. At elevated temperature (>200 °C), urea is quantitatively transformed into ammonia and carbon dioxide according to the reaction equation in eq. (7.98). Via this route, ammonia is formed inside the reactor and will not form a potential source of toxic harm. Reducing additives are preferably distributed in the off-gas via dual-phase nozzles:

$$CO(NH_2)_2 + H_2O \rightarrow 2\,NH_3 + CO_2 \tag{7.98}$$

Technical measures for NO_x emission mitigation

Basically, a distinction is made between emission prevention (primary measures) and emission control (secondary measures), with preference always given to the former. The following technical measures are applied in NO_x emission mitigation (Fig. 7.20):

Primary measures:
- fuel upgrading
- low-NO_x burner (LNB)
- two-stage combustion (TSC)
- flue gas recirculation (FGR)
- in-furnace reduction (IFR)

Secondary measures:
- selective noncatalytic reduction (SNCR)
- selective catalytic reduction (SCR)
- nonselective catalytic reduction (NSCR)

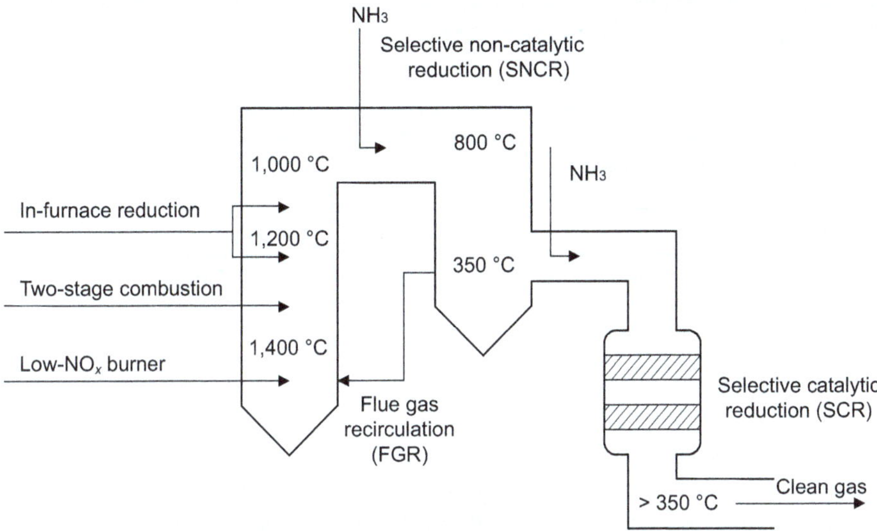

Fig. 7.20: Technical measures for NO_x emission mitigation.

7.4.1.3 Catalytic reduction of NO_x

7.4.1.3.1 Catalyst specification

Several heavy metal oxides have catalytic properties and could be used for catalytic denoxing. However, catalysts used in industrial applications are more or less exclusively based on the activity of vanadium pentoxide (V_2O_5) homogeneously mixed with the monolithic backbone titanium dioxide (TiO_2). Honeycomb structure is the preferred geometry of this type of catalyst due to the large catalyst surface of several hundred to >1,000 $m^2\ m^{-3}$. The optimum content of vanadium pentoxide is about 2%. In case of elevated SO_2 concentrations in the off-gas, SCR processes need the installation of so-called mixed catalysts of vanadium pentoxide and tungsten trioxide (WO_3) to limit the catalytic activity of vanadium pentoxide in converting SO_2 into sulfur trioxide.

Industrial catalytic denoxing is preferably based on the application of honeycomb catalysts. The standard cross-sectional dimension of honeycomb catalysts is 150 mm x 150 mm. The specific area is given by the number of channels of the catalyst cross section. Standard design is based on 40 channels per length of the cross-sectional area, resulting in a specific catalyst area of 850 $m^2\ m^{-3}$. Both catalysts with larger and smaller specific area are available. Based on V_2O_5 activation, the optimum temperature of operation is between 350 and 400 °C (but significant catalytic activity is already observed at much lower temperatures). The maximum temperature of operation must not exceed 450 °C. Therefore, the heat balance has to be considered due to the necessity of sectional gas cooling. The thermodynamics of NH_3 SCR on V_2O_5 catalysts and catalyst poisoning of V_2O_5 with NaCl are presented in Tab. 7.12.

Tab. 7.12: Thermodynamics (free standard enthalpy of reaction $\Delta_R G^0_{300\,°C}$ and standard enthalpy of reaction $\Delta_R H^0_{300\,°C}$) of reactions and catalyst poisoning during NO reduction with ammonia over V_2O_5 catalysts at 300 °C.

Reaction	$\Delta_R G^0_{300\,°C}$ (kJ mol^{-1})*	$\Delta_R H^0_{300\,°G}$ (kJ mol^{-1})*
$3\,V_2O_5 + 2\,NH_3 \rightarrow 3\,V_2O_4 + N_2 + 3\,H_2O$	−449.5	−235.7
$2\,V_2O_4 + 2\,NO \rightarrow 2\,V_2O_5 + N_2$	−314.5	−444.6
$V_2O_5 + 2\,NaCl + H_2O \rightarrow 3\,NaVO_2 + 2\,HCl$	56.9	132.1

*Formula conversion.

As mentioned, SCR denoxing can be based on first-order reaction kinetics. The rate of reaction and/or mass transfer may be rate-controlling. Actually turbulent diffusion is rate controlling within the optimum temperature range of operation. The rate constant changes with changing gas velocity. Representative design data are given by the following catalyst specification in Tab. 7.13.

Tab. 7.13: Representative catalyst specification for SCR denoxing.

Cross-sectional dimension	**150 mm × 150 mm**
Number of channels	40 per cross-sectional length
Standard length of the catalyst $L_{catalyst}$	650 mm
Specific area	850 m^2 m^{-3}
V_2O_5 content	2 wt%
Optimum temperature	350–400 °C
ΔP^* **(Pa m^{-1})**	$100 \cdot v$ (estimative for: $4\ m\ s^{-1} < v < 8\ m\ s^{-1}$)

*This algorithm is a simplification and shall therefore not be applied except for training purposes.

7.4.1.3.2 Application

Several modes of installation have been tested and applied. Reactors may be installed prior to dedusting, in the so-called high-dust mode; behind dedusting, in the so-called low-dust mode; and behind absorptive off-gas purification, the so-called clean-gas mode. The decision is based on detailed off-gas analysis, as several pollutants as well as high temperatures may rapidly poison and destroy the catalyst.

The design of denoxing equipment has to base on the proper performance of the reactor and the mixing chamber. The design of $DeNO_x$ reactors is based on the performance of PFRs.

The design strategy is as follows:

(1) specification of the wanted (needed) relative conversion $X_{pollutant}$
(2) specification of the start (feed) temperature

(3) calculation of the mass and heat balance
(4) specification of the gas velocity v_{gas}
(5) determination of k_{AV}-value (or k_{SV}-value)
(6) calculation of AV or SV
(7) calculation of the needed catalyst area and/or catalyst volume
(8) reactor-sizing

7.4.1.3.3 SCR of NO_x: reactor design

In selective catalytic reduction of NO_x, the rate of reaction of NO conversion ($-r_{NO}$) is in accordance with a first-order reaction kinetics ($-r_{NO} = k \cdot c_{NO}$). The type of reactor used in SCR equipment is a continuous tubular reactor (PFR under ideal conditions). The design algorithm is given by the following equation (7.99):

$$V = F_{NO,0} \cdot \int_0^{X_{NO}} \frac{dX_{NO}}{k \cdot c_{NO,0} \cdot (1 - X_{NO})} = F_{V,0} \cdot \int_0^{X_{NO}} \frac{dX_{NO}}{k \cdot (1 - X_{NO})} \tag{7.99}$$

After integration, the following equation is obtained:

$$k \cdot \tau = -\ln(1 - X_{NO}) \tag{7.100}$$

The unit of the reaction rate constant k is time^{-1} (e.g., h^{-1}). The residence time τ, derived from the ratio of the volume of reactor (strictly speaking reaction volume) V to the volumetric flow rate $F_{V,0}$, may also be related to the total surface area of catalyst S_{cat}. The design equation is then rearranged to result in an area-related design equation. The unit of the area-related reaction rate constant k'' is m h^{-1}. The inverse value of the area-related residence time is called area velocity (AV) or simply AV value. The area-related rate constant k'' can also be specified with the suffix AV. After rearranging, the design equation is given by the following algorithm:

$$\frac{k'' \cdot S_{cat}}{F_{V,0}} = -\ln(1 - X_{NO}) = \frac{k_{AV}}{AV} \tag{7.101}$$

Now, the design equation enables the evaluation of the total area of catalyst needed, and the reactor design can be carried out.

In case of granular catalysts (e.g. pellets), the residence time is mainly related to the volume of catalyst ($m^3_{catalyst}\ m^{-3}_{reactor}\ h^{-1}$). The residence time is substituted by the SV, and the rate constant k_{SV} (k''') is marked with the subscript SV. It has to be mentioned again that k-values are sometimes related to the volumetric inflow flow rate of the off-gas at standard conditions (STP).

Exercise 7.17: SCR lab training

During the CRE lab training, the following SCR experiment had to be done by the students: air was mixed with NO and preheated to the operation temperature. $NH_{3,aqu}$ was added to the mixture (where it was vaporized) with a metering pump in stoichiometric ratio to NO. After the heating section, the gas stream was led across a V_2O_5-doped SCR catalyst in the reactor. Gas samples were taken before and after the catalyst, and the NO content was analyzed:

$$4\,NO + 4\,NH_3 + O_2 \longrightarrow 4\,N_2 + 6\,H_2O$$

The following data are known for the catalyst.

Catalyst specification: 36 channels, channel cross section: 3 × 3 mm, catalyst length: 0.58 m.

Properties and additional information:

Temperature dependence of dynamic viscosity of air (T in K):

$$\mu[\mathrm{Pa \cdot s}] = 2 \cdot 10^{-14} \cdot T^3 - 5 \cdot 10^{-11} \cdot T^2 + 7 \cdot 10^{-8} \cdot T + 4 \cdot 10^{-7}$$

Temperature dependence of the diffusion coefficient of NO D_{NO} in air (T in K):

$$D_{NO}\ \left[\mathrm{m^2\,s^{-1}}\right] = 9 \cdot 10^{-10} \cdot T^{1.71}$$

The experimental results obtained are given in Tab. 7.14. Discuss them.

Tab. 7.14: Experimentally determined relative NO conversion X_{NO} for different reaction temperatures at different gas velocities v_{gas} (Exercise 7.17).

$\boldsymbol{\theta}$ **(°C)**	$\boldsymbol{X_{NO}}$ **(–)**	$\boldsymbol{v_{gas}}$ **(m s^{-1})**
260	0.90	9.51
250	0.86	9.38
240	0.84	9.25
230	0.80	9.12
220	0.76	8.99
210	0.73	8.86
200	0.67	8.73
190	0.61	8.60
180	0.54	8.47
170	0.46	8.34
160	0.38	8.21
150	0.31	8.08
140	0.24	7.95

In Fig. 7.21 the temperature-dependent NO conversion X_{NO} is plotted.

Which recommendations would you give for the design of SCR reactors based on these experimental results?

Solver:

Inflow area (= cross-sectional area) of the catalyst:

(note: assumption of square-shaped channels)

$A_{cross\ section}$ = channel side length^2 · number of channels = $(3 \cdot 10^{-3})^2 \cdot 36 = 3.24 \cdot 10^{-4}\ \mathrm{m^2}$

Surface of the catalyst:

S_{cat} = channel side length · channel length · 4 · number of channels = $(3 \cdot 10^{-3}) \cdot 0.58 \cdot 36 = 0.25\ \mathrm{m^2}$

Volumetric flow rate $F_V = F_{V,0}$:

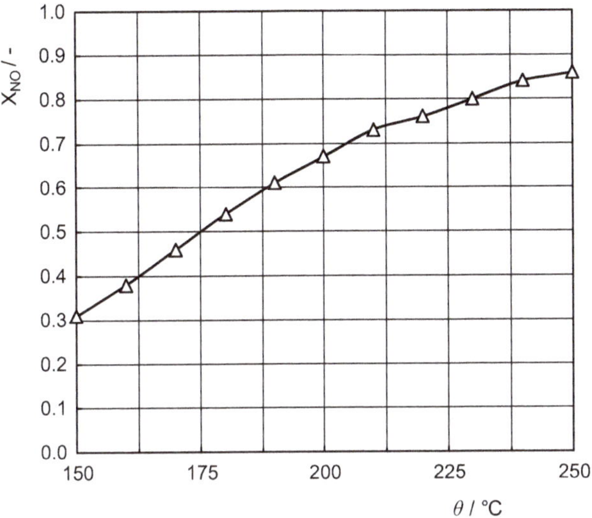

Fig. 7.21: Effect of temperature on the conversion of NO with NH_3 in SCR setup (Exercise 7.17).

Note: Strictly speaking, the reduction of NO is a reaction that takes place under volume increase ($\varepsilon_{NO} > 0$). Due to the low NO concentration; however, the volume change is negligible ($\varepsilon_{NO} \approx 0$).

$$F_V\,[\text{am}^3\,\text{h}^{-1}] = v_g \cdot 3{,}600 \cdot A_{\text{cross section}} = v_g\,[\text{m s}^{-1}] \cdot 3{,}600\,[\text{s h}^{-1}] \cdot 3.24 \cdot 10^{-4}$$

$$F_V\left[\text{m}^3{}_{\text{STP}}\,\text{h}^{-1}\right] = F_V\left[\text{am}^3\,\text{h}^{-1}\right] \cdot \frac{T_0}{T}$$

with $T_0 = 273.15\text{ K} = 0\text{ °C}$

Next, we set up the balance for a differential volumetric element of an ideal tubular reactor (reference base: surface of catalyst):

$$F_{NO} = F_{NO} + dF_{NO} + (-r''_{NO})dF_{NO}$$

$$dF_{NO} = r''_{NO}\,dS_{cat}$$

$$F_{NO} = F_{NO,0} \cdot (1 - X_{NO})$$

$$dF_{NO} = -F_{NO,0}\,dX_{NO}$$

From the previous discussion on reaction kinetics, we know that NO conversion can be well modeled with first-order reaction kinetics:

$$r''_{NO} = k'' \cdot c_{NO,0} \cdot (1 - X_{NO})$$

$$-F_{NO,0}dX = k'' \cdot c_{NO,0} \cdot (1 - X_{NO})dS_{cat}$$

$$-\frac{F_{V,0}}{k''} \cdot \frac{dX_{NO}}{(1 - X_{NO})} = dS_{cat}$$

$$-\ln(1 - X_{NO}) = k'' \cdot \frac{S_{cat}}{F_{V,0}} = k'' \cdot \tau''$$

Surface area-deduced residence time:

$$\tau'' = \frac{S_{cat}}{F_{V,0}} = \frac{1}{AV}$$

Now, we can determine the rate constant k'' (= k''_{exp}):

$$k'' = \frac{F_{V,0}}{S_{cat}} \cdot (-\ln(1 - X_{NO})) = k''_{exp}$$

The temperature dependence of the dynamic viscosity of air is calculated with

$$\mu[\text{Pa s}] = 2 \cdot 10^{-14} \cdot T^3 - 5 \cdot 10^{-11} \cdot T^2 + 7 \cdot 10^{-8} \cdot T + 4 \cdot 10^{-7}$$

The temperature-dependent diffusion coefficient of NO in air is calculated with

$$D_{NO}\left[\text{m}^2\,\text{s}^{-1}\right] = 9 \cdot 10^{-10} \cdot T^{1.71}$$

From the Arrhenius plot (Fig. 7.22) the activation energy E_A and the frequency factor A'' can be obtained:

$$k'' = A'' \cdot \exp\left(-\frac{E_A}{RT}\right)$$

$$\ln k'' = \ln A'' + \left(-\frac{E_A}{RT}\right)$$

The calculated results are given in Tab. 7.15.

Tab. 7.15: Calculated results for the calculation procedure in Exercise 7.17 (part 1).

θ (°C)	T (K)	X_{NO} (–)	v_{gas} (m s^{-1})	F_V (am^3 h^{-1})	F_V (m^3_{STP} h^{-1})	k''_{exp} (m h^{-1})	ln k''	T^{-1} (K^{-1})	μ (Pa s)	D_{NO} (m^2 s^{-1})
260	533.15	0.9	9.51	11.09	5.68	101.94	4.62	0.0019	$2.65 \cdot 10^{-5}$	$4.14 \cdot 10^{-5}$
250	523.15	0.86	9.38	10.94	5.71	85.85	4.45	0.0019	$2.62 \cdot 10^{-5}$	$4.01 \cdot 10^{-5}$
240	513.15	0.84	9.25	10.79	5.74	78.91	4.37	0.0019	$2.59 \cdot 10^{-5}$	$3.88 \cdot 10^{-5}$
230	503.15	0.8	9.12	10.64	5.77	68.33	4.22	0.0020	$2.55 \cdot 10^{-5}$	$3.75 \cdot 10^{-5}$
220	493.15	0.76	8.99	10.49	5.81	59.72	4.09	0.0020	$2.52 \cdot 10^{-5}$	$3.62 \cdot 10^{-5}$
210	483.15	0.73	8.86	10.33	5.84	54.00	3.99	0.0021	$2.48 \cdot 10^{-5}$	$3.50 \cdot 10^{-5}$
200	473.15	0.67	8.73	10.18	5.88	45.06	3.81	0.0021	$2.44 \cdot 10^{-5}$	$3.38 \cdot 10^{-5}$
190	463.15	0.61	8.6	10.03	5.92	37.70	3.63	0.0022	$2.41 \cdot 10^{-5}$	$3.26 \cdot 10^{-5}$
180	453.15	0.54	8.47	9.88	5.96	30.62	3.42	0.0022	$2.37 \cdot 10^{-5}$	$3.14 \cdot 10^{-5}$
170	443.15	0.46	8.34	9.73	6.00	23.92	3.17	0.0023	$2.33 \cdot 10^{-5}$	$3.02 \cdot 10^{-5}$
160	433.15	0.38	8.21	9.58	6.04	18.27	2.91	0.0023	$2.30 \cdot 10^{-5}$	$2.90 \cdot 10^{-5}$
150	423.15	0.31	8.08	9.42	6.08	13.96	2.64	0.0024	$2.26 \cdot 10^{-5}$	$2.79 \cdot 10^{-5}$
140	413.15	0.24	7.95	9.27	6.13	10.16	2.32	0.0024	$2.22 \cdot 10^{-5}$	$2.68 \cdot 10^{-5}$

Slope of the Arrhenius plot: $\frac{-E_A}{R} = -4{,}118.6$

$E_A = 34{,}242.04\ \text{J mol}^{-1} = 34.24\ \text{kJ mol}^{-1}$ (indicating a significant contribution of mass transfer)

Axis intercept: $\ln A'' = 12.424$

$$A'' = 248{,}699.35\ \text{m h}^{-1}$$

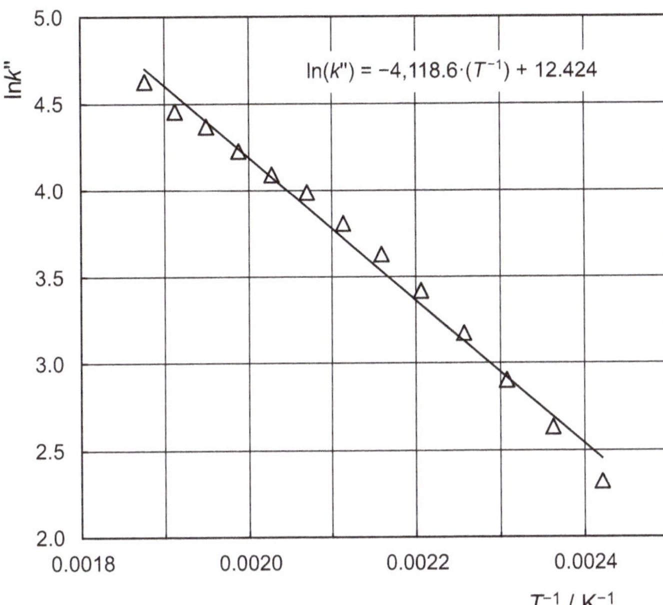

Fig. 7.22: Arrhenius plot of NO conversion with NH_3 in SCR setup (Exercise 7.17).

Next, we calculate the hydraulic diameter d_H:

$$d_H = \frac{4 \cdot \text{cross-sectional area}}{\text{perimeter}} = \frac{4 \cdot (3 \cdot 10^{-3})^2}{4 \cdot 3 \cdot 10^{-3}} = 3 \cdot 10^{-3}\ \text{m}$$

Furthermore, we know:

$$\text{Re} = \frac{d_H \cdot v_{gas}}{\nu} \text{ and } \text{Sc} = \frac{\nu}{D}$$

Thus,

$$\text{Re} \cdot \text{Sc} = \frac{d_H \cdot v_{gas}}{D_{NO}} = \frac{3 \cdot 10^{-3}[\text{m}] \cdot v_{gas}[\text{m s}^{-1}]}{D_{NO}[\text{m}^2\ \text{s}^{-1}]}$$

With

$$\text{Gz} = \text{Re} \cdot \text{Sc} \cdot \frac{d_H}{L_{cat}} = \text{Re} \cdot \text{Sc} \cdot \frac{3 \cdot 10^{-3}}{0.58} < 16$$

we do know how to calculate the Sherwood number Sh:

$$\text{Sh} = \frac{\beta_g \cdot d_H}{D_{NO}} = 3.66 + \frac{0.085 \cdot \text{Gz}}{1 + 0.047 \cdot \text{Gz}^{0.667}}$$

With known Sherwood number, we can determine the mass transfer coefficient β_g:

$$\beta_g[\text{m s}^{-1}] = \frac{\text{Sh} \cdot D_{NO}}{d_H} = \frac{\text{Sh} \cdot D_{NO}\ [\text{m}^2\ \text{s}^{-1}]}{3 \cdot 10^{-3}\ [\text{m}]}$$

The results are summed up in Tab. 7.16.

Tab. 7.16: Calculated results for the calculation procedure in Exercise 7.17 (part 2).

θ (°C)	T (K)	X_{NO} (–)	v_{gas} (m s^{-1})	μ (Pa s)	D_{NO} (m^2 s^{-1})	Re · Sc	Gz	Sh	β_g (m s^{-1})
260	533.15	0.90	9.51	$2.65 \cdot 10^{-5}$	$4.14 \cdot 10^{-5}$	688.89	3.56	3.93	0.054
250	523.15	0.86	9.38	$2.62 \cdot 10^{-5}$	$4.01 \cdot 10^{-5}$	701.83	3.63	3.94	0.053
240	513.15	0.84	9.25	$2.59 \cdot 10^{-5}$	$3.88 \cdot 10^{-5}$	715.32	3.70	3.94	0.051
230	503.15	0.80	9.12	$2.55 \cdot 10^{-5}$	$3.75 \cdot 10^{-5}$	729.41	3.77	3.95	0.049
220	493.15	0.76	8.99	$2.52 \cdot 10^{-5}$	$3.62 \cdot 10^{-5}$	744.12	3.85	3.95	0.048
210	483.15	0.73	8.86	$2.48 \cdot 10^{-5}$	$3.50 \cdot 10^{-5}$	759.51	3.93	3.96	0.046
200	473.15	0.67	8.73	$2.44 \cdot 10^{-5}$	$3.38 \cdot 10^{-5}$	775.61	4.01	3.96	0.045
190	463.15	0.61	8.6	$2.41 \cdot 10^{-5}$	$3.26 \cdot 10^{-5}$	792.49	4.10	3.97	0.043
180	453.15	0.54	8.47	$2.37 \cdot 10^{-5}$	$3.14 \cdot 10^{-5}$	810.19	4.19	3.98	0.042
170	443.15	0.46	8.34	$2.33 \cdot 10^{-5}$	$3.02 \cdot 10^{-5}$	828.79	4.29	3.98	0.040
160	433.15	0.38	8.21	$2.30 \cdot 10^{-5}$	$2.90 \cdot 10^{-5}$	848.34	4.39	3.99	0.039
150	423.15	0.31	8.08	$2.26 \cdot 10^{-5}$	$2.79 \cdot 10^{-5}$	868.93	4.49	4.00	0.037
140	413.15	0.24	7.95	$2.22 \cdot 10^{-5}$	$2.68 \cdot 10^{-5}$	890.64	4.61	4.01	0.036

Mass transfer resistance $1/\beta_g$ and reaction resistance $1/k_S$ may be added, but by assuming that the mass transfer and reaction rates act in series; the reaction is first order and it is related to the surface area of the catalyst area:

$$\frac{1}{k''_{exp}} = \frac{1}{k''_{total}} = \frac{1}{\beta_g} + \frac{1}{k_S}$$

Rearrangement of the above-stated equation gives the expression for determination of the reaction rate constant k_S at the catalyst surface:

$$k_S = \frac{k''_{exp} \cdot \beta_g}{\beta_g - k''_{exp}}$$

Tab. 7.17: Mass transfer coefficients β_g and rate constants k_S for Exercise 7.17.

k''_{exp} (m h^{-1})	β_g (m s^{-1})	β_g (m h^{-1})	k_S (m h^{-1})
101.94	0.054	195.46	213.05
85.85	0.053	189.46	156.99
78.91	0.051	183.54	138.43
68.33	0.049	177.70	111.02
59.72	0.048	171.94	91.51
54.00	0.046	166.26	79.98
45.06	0.045	160.66	62.62
37.70	0.043	155.13	49.80
30.62	0.042	149.69	38.49
23.92	0.040	144.33	28.68
18.27	0.039	139.05	21.03
13.96	0.037	133.86	15.58
10.16	0.036	128.74	11.03

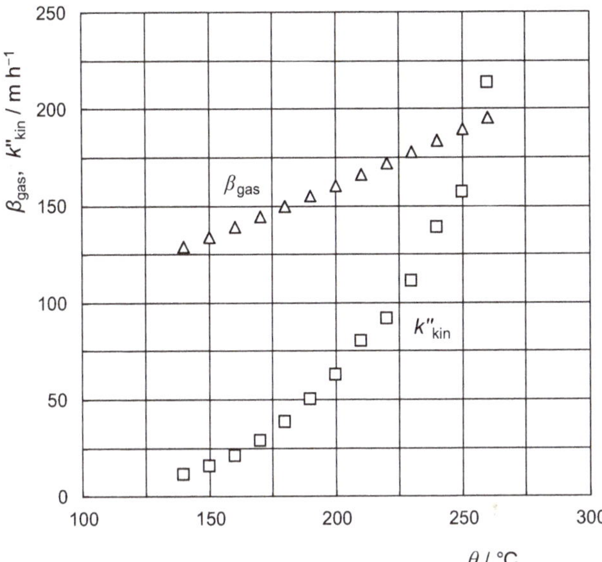

Fig. 7.23: Mass transfer coefficients in the gas phase β_g and rate constants k_S (k''_{kin}) for catalytic conversion of NO with NH_3 in SCR setup (Exercise 7.17).

At the intersection of the curves, the contribution of both resistances is equal. When temperature is increased, catalysis becomes increasingly mass transport-controlled, and when temperature is decreased, it becomes reaction-controlled.

8 Absorptive off-gas purification

When after World War II welfare began to spread, the Western-world people pretty soon registered that this welfare is poisoned; poisoned with smoke and with several gaseous pollutants. Looking back in history, pollution control more or less had to respond on the worst pollutants. In the early twentieth century off-gas dedusting became a subject of technology development. Suffering from a lack of high efficiency and pollutant-resistant materials, cyclones, electrostatic precipitators and Venturi scrubbers were preferably applied in pollution control, still leaving high dust load in the clean gas. In the 1980s and 1990s fabric filters established in pollution control, giving access to clean gas dust load of (much) less than 10 mg $m^{-3}{}_{STP.}$ We have discussed off-gas dedusting in Chapter 6. However, air quality still was not as expected. The welfare we addressed above was mainly based on incinerating coal, and with some delay oil and also municipal waste. These energy carriers provided ambient air with several pollutants of devastating impact. SO_2, HCl and HF, especially in combination with inversion layers, accelerated wood decline and severe damaging of artificial structures, while NO_x led to acid rain and eutrophication of soil and water bodies. Research and engineering had to work hard on getting these problems with appropriate technologies under control. It is truly fascinating to track the development of SO_x control technologies within the last decades. In parallel to progress in SO_x control the nitrogen oxides embarrassed the environment in synergism with hydrocarbon pollution, ending up in smog problems. As a consequence, we had to develop high-efficiency technologies such as selective catalytic reduction, as addressed in Chapter 7. At present, we still have by far not obtained the clean air quality level of a healthy environment (see the statistics in Chapter 1). The greenhouse problem is seemingly attaining a tsunami-like level, although we still struggle with the ambient air quality level referring to SO_x, NO_x and their colleagues.

If you ever have to deal with any of these topics, do not hesitate to start with a literature search (keep the recommendation of Henry Ford in mind that spending a few hours in the library might save months of hard work in the lab).

The intention behind this chapter on absorptive off-gas purification is not to compete the "mass transfer unit operations" community, but to address the specific needs of pollutant absorption. If you are interested in detailed discussion of "mass transfer unit operations," do not hesitate to pick up the specific literature, such as [237–240].

If you want to make yourself acquaint with the chemistry of a specific pollutant, start with the *Gmelin Handbook of Inorganic and Organometallic Chemistry*, e.g., [241, 242]. But start with studying the Gmelin system for not getting lost in the periodic system of elements. (If, for example, you want to search for SO_2, you may stay with volume 9; if you want to search for NO_x, you have to visit volume 4, nitrogen.) You will find comprehensive information about technologies in Ullmann's *Encyclopedia of Industrial Chemistry* (e.g., [243]). If you need detailed information about

https://doi.org/10.1515/9783110763928-008

pollution control technologies, you will find a great source of data and information in [244]. The discussion of greenhouse aspects is dating back to the nineteenth century [245]. Be careful with the greenhouse gas tsunami. You are recommended to start with an updated review [246]. You may also start with [247].

Although dating back to the year 2010, the report "On innovations in carbon capture and storage (CCS) and carbon capture and utilization (CCU)" is a recommended treatise on this topic [248]. Many CCU processes have been proposed; however, this topic is still subject of intensive investigations. We may, for example, make use of it for hydrogen storing purposes or synthesis of several bulk chemicals [249, 250], but not for free. CO_2 is the carbon-based constituent with the lowest energy level.

Highly attracted by the challenges we face, we are willing to solve some of these problems by applying absorption technologies in a step-by-step approach in this chapter with a focus on the basics and the engineering of absorption technologies. An introductory promise: This will be a fascinating trip.

In Tab. 1.1, we have specified a really bad off-gas. We have had to get a very high dust load with a distinct amount of PM2.5 under control. We had a focus on that in Chapter 6. Then we had to construct compliance technologies for NO_X-control in Chapter 7. What has been left is an engineering miracle cure for the pollutants TCDD (2,3,7,8-tetrachlorodibenzo-*p*-dioxin), SO_X (300 ppm), HCl (30 ppm), HF (10 ppm) and even CO_2 (10 vol%). Except TCDD, these guys like water. In aqueous solutions, HCl will readily form the "strong" hydrochloric acid. HF also has a preference for water, but not that distinct. For sure you are familiar with the properties of CO_2 in many beverages. And finally, SO_X, forming sulfurous acid with water, is indeed a huge challenge in industrial off-gas purification. Have in mind, who is hiding behind SO_X (we have discussed it in Section 2.5.5).

Let us convince these friends to jump from off-gas into appropriate scrubbing liquids, although we always have to keep in mind that the separation of a pollutant (being the solute in absorptive off-gas purification) from the gaseous carrier is just the transfer of an off-gas problem into a wastewater problem. As a consequence, process design and operation have to solve the off-gas problem as well as the wastewater problem.

Absorptive off-gas treatment processes are applied in off-gas purification as well as in process gas treatment. Absorptive separation of acidic pollutants such as HCl, HF, SO_2 (NO_X and CO_2) from off-gas is the main application of absorption processes in pollution control

Target of absorption processes is the selective separation (precipitation) of specified solutes from the gaseous carrier with a scrubbing liquid. The scrubbing liquid or solvent is called absorbent due to the principle mode of action. The substance to be absorbed is called solute. Selectivity and efficiency of separation depend on the phase interaction of the gaseous solute and the scrubbing liquid or solvent. Process design is based on the discussion of physical–chemical interactions between the solute, which has to be separated from the feed gas, and the absorbent, which has to suffice the requirements of separation and the complete process. Our engineering activity will focus on the choice of absorbent and the technology. Mass transfer and the hydraulic process conditions, the ratio of the solvent phase to the gas phase, the flow rate of both phases, the pressure of operation and the temperature determine the design of absorption processes.

From this brief introduction you may conclude that we shall spend plenty of time with discussing the physio-chemical basics and constraints in a first step, and that for sure we will have the process design on screen. Therefore, we have to discuss phase contact patterns and perform balances in a next step, and we must not ignore mass transfer and the hydraulics of equipment. Yes, that is the plan.

8.1 Physiochemical basics

Based on the interaction between the solute, which has to be separated from the gaseous carrier phase, and the absorbent we have to distinguish between two types of absorption processes.

8.1.1 Physiosorption

There is no chemical interaction between the solute to be separated from the gaseous phase and the solvent. The chemical nature of the separated solute does not change. Absorption of oxygen with water from air, shown in eq. (8.1), is a representative example for physiosorption (also termed physisorption):

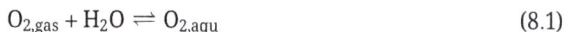

$$O_{2,gas} + H_2O \rightleftharpoons O_{2,aqu} \tag{8.1}$$

Solubility of oxygen in water is only controlled by the pressure of operation and by the temperature. The chemical nature of oxygen does not change. Due to the dependency of the solubility of oxygen in water on pressure and temperature, absorption of oxygen from air in water (= transfer of oxygen from air into water at high pressure and/or low temperature) and desorption of oxygen from water (= transfer of oxygen from water into the gas phase at low pressure and/or high temperature) can be carried out by pressure swing (we change the pressure) and/or by temperature swing (we change the temperature) operation. Figure 8.1 shows a representative process for SO_2 separation from off-gas by physiosorption and desorption by temperature swing.

The process shown in Fig. 8.1 has exemplarily been applied in the "Linde desulfurization process of synthesis gas" [251]. In an earlier version of this Linde process, the focus was on separation of SO_2 from off-gas by physiosorption with tetraethylene glycol dimethyl ether [252]. The rich gas may be fed to a sulfuric acid plant. Sulfur is a major contaminant of crude oil. Having in mind the huge capacity of present crude

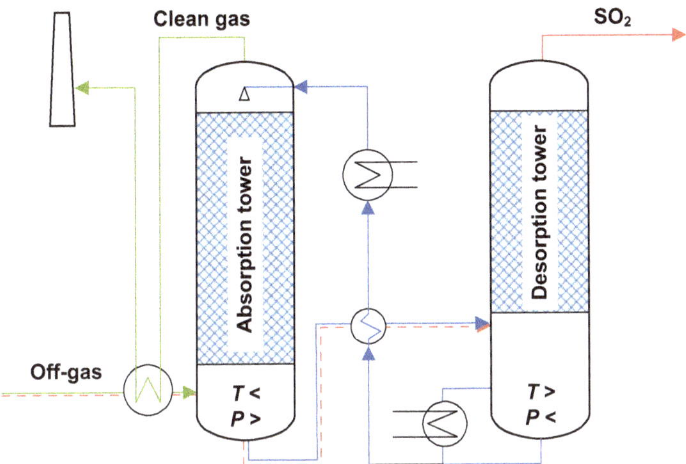

Fig. 8.1: Absorption for separation of SO_2 from off-gas by physiosorption and desorption of SO_2 from the absorbent by temperature swing.

oil refining industry you can imagine that it is impossible to get rid of the sulfur by disposing of or by incineration. As early as in 1883, Carl Friedrich Claus invented the famous Claus process [253]. In a modified version of the original Claus process tens of million tons of elemental sulfur are produced from crude oil. This process is a brilliant combination of the oxidation of hydrogen sulfide (H_2S) with sulfur dioxide (SO_2), also demonstrating the redox properties of sulfur dioxide. Several CCU/CCS processes make use of the acid–base interaction of CO_2 and Na_2CO_3 or K_2CO_3 [254, 255].

8.1.2 Chemisorption

Chemical interaction of the solute to be separated from the gaseous carrier phase and the scrubbing liquid is boosting absorption performance. Separation is based on chemical reaction of the gaseous solute and the solvent with or without admixture of reactive ingredients. Absorption of the gaseous solute is limited by either the amount of additive in the solvent or the chemical interaction of the solute with the absorbent.

The absorption of hydrochloric acid (HCl) from gaseous carrier phase with water is a representative example for absorption processes with chemical reaction.

According to eqs. (8.2) and (8.3), hydrochloric acid dissociates immediately when dissolved in aqueous solution. The dissociation constant K_{Diss} is about $K_{Diss} = 1{,}000$ at ambient temperature [256]. The total concentration of hydrochloric acid, including dissociated hydrochloric acid and the nondissociated amount in the absorbent water, will rise during absorption until the nondissociated amount of hydrochloric acid in the aqueous mixture interacts with the corresponding gaseous carrier phase.

Due to (nearly) quantitative dissociation, the gas–liquid partition of hydrochloric acid is controlled by a negligible partial pressure of the nondissociated amount of hydrochloric acid in the aqueous phase:

$$HCl_{gas} + H_2O \rightleftharpoons HCl_{aqu} \tag{8.2}$$

$$HCl_{aqu} \rightleftharpoons H^+ + Cl^- \quad K_{Diss} \sim 10^3 \tag{8.3}$$

Increasing the pH value of the aqueous phase (= consuming protons), for example, by metering sodium hydroxide to the absorbent, as shown in eq. (8.4), will immediately decrease the amount of nondissociated acid. Absorption of hydrochloric acid will be complete due to instant neutralization of the protons formed by dissociation, as long as sodium hydroxide is added to the liquid carrier phase:

$$H^+ + Cl^- + Na^+ + OH^- \rightleftharpoons H_2O + Na^+ + Cl^- \tag{8.4}$$

Absorption of hydrochloric acid in water as well as absorption of hydrochloric acid in caustic aqueous solution is representatives for absorption with chemical reaction or chemisorption.

Temperature dependency and pressure dependency of chemisorption processes is much less distinct than in physiosorption. Therefore, separation of the dissolved solute from the liquid carrier phase by pressure swing or by temperature swing operation is not very efficient or even negligible. Regeneration of the solvent phase must be carried out by other unit operations (e.g., distillation, ion exchange and precipitation). Figure 8.2 shows the comparison of physiosorption and chemisorption.

Physiosorption

Gas phase

$O_{2,g}$

$O_{2,aqu}$

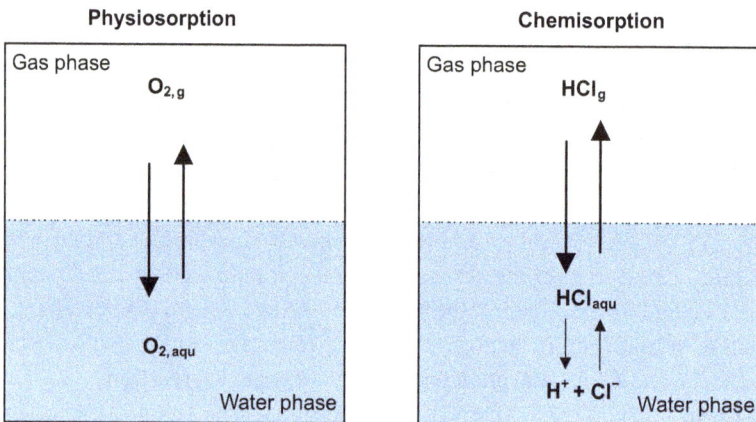

Water phase

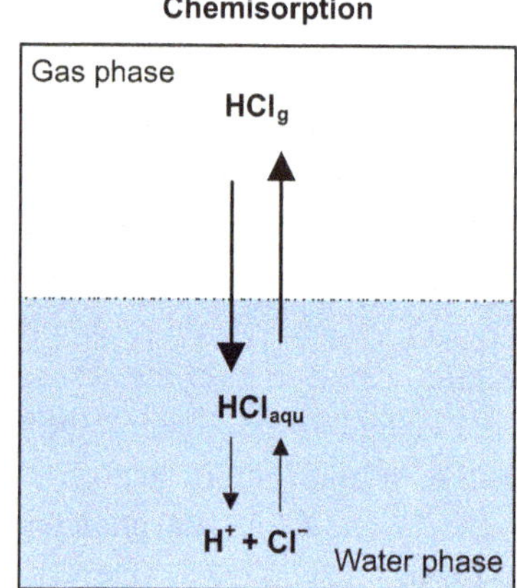

Fig. 8.2: Comparison of physiosorption and chemisorption.

8.1.3 Absorbent selection

In physical absorption as well as absorption with chemical reaction, the absorbent has to suffice the following specification (if you conclude from this specification that the scrubbing liquid water or aqueous solutions might do a great job, you are very right):

- high solubility of the solute in the solvent,
- high selectivity for the solute to be absorbed,
- simple regeneration,
- complete separability of absorbent and solute,
- low volatility and low viscosity of the absorbent,
- nontoxic and noncorrosive properties of the absorbent and
- low cost of investment and operation.

!

The solubility of the gaseous solute in the absorbent shall be high, thus increasing the rate of absorption and decreasing the solvent requirement. Generally, solvents of chemical nature similar to that of the solute to be absorbed will provide good solubility. The absorbent shall be inexpensive to avoid costly losses, and it shall be readily available. Low viscosity is preferred for reasons of rapid absorption, improved flooding characteristics in absorption towers, low pressure drop on pumping and good heat transfer characteristics. In summary we mainly talk about water.

Moreover, the material of construction required for the equipment shall not be unusual or expensive.

8.1.4 Gas–liquid partition

Modeling of gas–liquid partition is dating back to an empirical correlation, observed by W. Henry. From experimentally obtained gas–liquid equilibrium data he concluded that the solubility of a gaseous solute in an absorbent correlates with the operation pressure when carrying out the experiment at constant temperature.

Although derived from empirical observation, Henry's law is in accordance with the Gibbs phase rule for specifying multiphase/multicomponent systems. According to the Gibbs phase rule a three-component (C) two-phase (P) system is quantitatively described by three (F) independent parameters ($F = C - P + 2$). A gas–liquid equilibrium is therefore completely specified by the temperature, the pressure and the content of the solute.

According to Gibbs, the gas–liquid equilibrium is obtained when the chemical potential $\mu_{A,g}$ of a solute A in the gaseous phase is in balance with the chemical potential of the solute A in the absorbent phase $\mu_{A,l}$, according to eq. (8.5) [257].

For

$$\mu_{A,g} = \mu_{A,g,0} + RT \ln p_A \text{ and } \mu_{A,l} = \mu_{A,l,0} + RT \ln a_A \tag{8.5}$$

at equilibrium, we get $\mu_{A,g} = \mu_{A,l}$ and therefore, $\mu_{A,g,0} + RT \ln p_A = \mu_{A,l,0} + RT \ln a_A$.

After separation of variables we obtain $\mu_{A,g,0} - \mu_{A,l,0} = \Delta\mu = RT \ln a_A - RT \ln p_A$, and after rearrangement we end up with eq. (8.6). We (incidentally) sum up this expression after rearrangement with the abbreviation H_A/P:

$$\frac{y_A}{a_A} = e^{-\frac{\Delta\mu}{R \cdot T}} = \frac{H_A}{P} \tag{8.6}$$

Eq. (8.6) demands activity a_A of the solute A in the absorbent phase, and actually fugacity in the gas phase. For convenience, we make use of the following simplification:

For $a_A \to x$ at $x \to 0$ and the help of the Dalton law $p_A = y_A \cdot P$, we simplify eq. (8.6) to end up with the handsome eq. (8.7), as also deduced by Henry from experiments:

$$y_A = \frac{H_A}{P} \cdot x_A \tag{8.7}$$

Let us briefly discuss the assumption $a \to x$ at $x \to 0$ with the sketch in Fig. 8.3 for the system ethanol (EtOH)/water. The sketch shows the two boundaries $x_{EtOH} \to 0$ and $x_{EtOH} \to 1$. At $x_{EtOH} \to 0$, we will not fail when deducing the inclination of the vapor pressure curve by the linear correlation of eq. (8.7). At $x_{EtOH} \to 1$, the vapor pressure of ethanol is well approached by the Raoult correlation $p_{EtOH} = p^0_{EtOH} \cdot x_{EtOH}$. In absorption, we talk about the left-hand side boundary of Fig. 8.3.

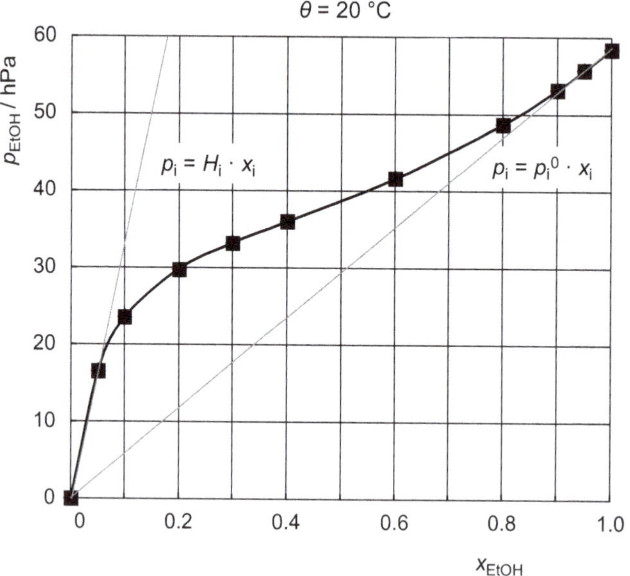

Fig. 8.3: Comparison of the Henry approach ($p_i = H_i \cdot x_i$) and the Raoult approach ($p_i = p_i^0 \cdot x_i$) for the system ethanol (EtOH)/water at 20 °C; subscript i stands for ethanol.

For the given system pressure P, eq. (8.7) may be simplified to the following equation:

$$y_A = m \cdot x_A \text{ with } \frac{H_A}{P} = m \tag{8.8}$$

The empirically developed correlation of Henry is seemingly in accordance with the thermodynamic approach. Thermodynamics explain the physio-chemical background of the equilibrium constant H_A (Henry constant or Henry coefficient of solute A), which implies the standard chemical potential of the mass transfer process.

The solubility of every substance of multi-component mixtures of the gaseous phase is specified by the temperature, the Henry constant H_A and the partial pressure $p_A = y_A \cdot P$ of the solutes. The ratio of the Henry constant of two substances indicates the selectivity of partition. The Henry constant of many gaseous substances is published in several handbooks. Table 8.1 is a representative selection of Henry constants of several gaseous solutes and the absorbent water at a temperature of 293 K.

Be careful with the unit of the Henry constant. !

Tab. 8.1: Henry constant H_A for the absorption of selected gaseous solutes (A) in water at 293 K [258].

Substance	$H_{A,\,293\,K}$ (MPa)
Hydrogen	6,830
Nitrogen	8,040
Air	6,630
CO	5,360
Oxygen	4,000
NO	2,640
CO_2	142
Cl_2	53
SO_2	3.5
HCl	0.28

Table 8.2 shows an empirical approach of the temperature dependency of the Henry constant of oxygen and hydrochloric acid.

Tab. 8.2: Temperature dependency of the Henry constants of oxygen and hydrochloric acid [258].

Solute A	Temperature range (°C)	H_A correlation
O_2	0–80	H_{O_2} [MPa] = $2{,}480.16 + 86.47 \cdot \theta - 0.39 \cdot \theta^2$
HCl	0–60	H_{HCl} [MPa] = $0.242 + 0.00185 \cdot \theta - 1.04 \cdot 10^{-5} \cdot \theta^2$

Practical application has to consider the type of the equilibrium correlation. Several adjustments to individual necessity have been published [259]. For example, liquid phase concentrations c_A may be correlated with the molar fraction in the gas phase y^*, as follows:

$$y_A^* = k' \cdot c_A \tag{8.9}$$

The partition constant k' is then specified by the unit pressure per concentration. Simplification of Henry correlations may also be based on mass load. For very dilute conditions, the molar fraction y^* and the specific molar load Y^* do not deviate significantly, resulting in the simplification $y^* = Y^*$, shown in eq. (8.10) with A being the solute and I being the inert carrier gas:

$$y_A = \frac{n_A}{n_A + n_I}; \quad Y_A = \frac{n_A}{n_I} = \frac{y_A}{1 - y_A} \tag{8.10}$$

Many absorption processes (e.g., in pollution control) deal with dilute concentrations of the solute to be separated from the gaseous phase: $n_A \ll n_I$, hence: $y_A \cong Y_A$.

Based on the correlation between molar fraction and specific molar load, as deduced from eq. (8.10), $Y_A = \frac{n_A}{n_I} = \frac{y_A}{1-y_A}$, the gas–liquid equilibrium may be expressed according to the following equation:

$$\frac{Y_A^*}{1 + Y_A^*} = \frac{H_A \cdot X_A}{P \cdot (1 + X_A)} \tag{8.11}$$

Due to $\frac{n_A}{n_I} = \frac{p_A}{p_I}$, and the specific mass load $Y_A = \frac{m_A}{m_I}$ in kg A kg^{-1}, we can easily correlate:

$$\frac{p_A}{p_I} = \frac{p_A}{P - p_A} = \frac{n_A}{n_I} = \frac{\frac{m_A}{MM_A}}{\frac{m_I}{MM_I}} = Y_A \cdot \frac{MM_I}{MM_A}$$

and derive the specific mass load from the partial pressure, as already applied when discussing the general basics (for reference see eq. (2.11) with $Y_A = \frac{MM_A \cdot p_A}{MM_I \cdot (P - p_A)}$.

Gas solubility may also be specified by the Bunsen absorption coefficient a. The Bunsen absorption coefficient is derived from the volume of dissolved pure solute ($p_A = 1$ bar) in the absorbent, as follows:

$$a_A = \frac{m^3_{A,STP}}{m^3 \cdot bar} \tag{8.12}$$

The Bunsen absorption coefficient correlates with the Henry constant H_A, as shown in eq. (8.13), where MV is the molar volume of the perfect gas in m^3 mol^{-1}, ρ_l the density of the absorbent (note: subscript l for liquid) in kg m^{-3}, MM_l the molar mass of the absorbent in kg mol^{-1}, and H_A the Henry constant in bar:

$$\alpha = \frac{\mathrm{MV} \cdot \rho_{\mathrm{l}}}{\mathrm{MM}_{\mathrm{l}} \cdot H_{\mathrm{A}}} \tag{8.13}$$

Table 8.3 shows a-values of selected gases in water at a temperature of 273 K [258].

Tab. 8.3: Bunsen coefficient for selected gases (A) in water: $T = 273$ K.

Substance	α ($m^3_{A,STP}$ m^{-3} bar^{-1})
Oxygen	0.04825
Air	0.02847
NO	0.03491
H_2S	4.609
CO_2	1.691

Exercise 8.1: Physiosorption and the melting pot of lover's oaths

Graz City, the beautiful capital of the Austrian province Styria, houses about 300,000 inhabitants, including approximately 80,000 students (and, therefore, plenty of great restaurants and beer gardens). Mur river, meanwhile a high-quality waterbody again, crosses the city with plenty of bridges connecting the riversides, and the Erzherzog Johann bridge linking the main square with the Kunsthaus. The bridge railings of Erzherzog Johann bridge are definitely a hot spot of lover's oaths with thousands of lockers decorating the bridge railing, commemorating eternal love oaths. City administration let the civil engineers already check the statics of the bridge.

On a bright morning in May, at this time Mur river has a mean temperature of 8 °C because of the snowmelt in the alps, a female sixth-semester undergraduate student of Chemical Engineering of the Graz University of Technology passes the bridge, when suddenly registering a young gentleman with both hands fixing the railing top and his head already beyond the railing, in a seemingly desperate condition. The well-trained young lady fixes the forearm of the poor boy and tries to get the problem under control with the kind statement: Do not do it. The young gentleman turns his head toward the student, and with a smile on his face he asks the student: "As an environmentalist I would really like to have an idea about present oxygen content of the Mur river water." The student unbags her cellphone, activates the hand calculator, asks the gentleman for the barometric pressure, which he replies to have been $P = 756$ mm Hg (equivalent $P = 1{,}008$ hPa) at the main square meteorological station just a few minutes ago, and after a few seconds she tells him: "Well, based on my calculations the mean oxygen content of Mur river is about 11.8 mg kg^{-1}." Fully excited of the brilliant and quick answer the gentleman asks the student to kindly explain calculations and the basics. So they walk to the ground-floor restaurant nearby Kunsthaus to discuss calculations. Having the Gibbs phase rule in mind she did probably even tell him that oxygen solubility in water is fixed at constant pressure, molar fraction in air and temperature (we cannot witness as we did not participate the meeting). She did perform calculations for $\theta = 8$ °C at $P = 1{,}008$ hPa and $\theta = 20$ °C, 40 °C and 60 °C at $P = 1{,}013$ hPa with the Henry constants from Tab. 8.2 (H_{O_2} [MPa] = $2{,}480.16 + 86.47 \cdot \theta - 0.39 \cdot \theta^2$ and θ in °C). For comparison of results, she prepared a summary on the back side of a beer coaster, shown in Tab. 8.4. It is not reported whether they have married meanwhile, but how did she perform calculations, and what day of the week was it?

Solver:

a) Let us start with the day of the week.

It for sure must have been Sunday, because during the week the Chemical Engineering students spend the whole day either in the lecture halls, in the libraries or in the lab (and sometimes even in the beer gardens).

b) How did she perform calculations?

She did start with calculating the Henry constant (chemical engineering students do have these data available on their cellphone). Being careful, she had in mind that the specification of the pressure in this example is hPa. With eq. (8.7) (Henry's law) she immediately obtained the molar fraction of oxygen in water, since the molar fraction of oxygen in air is $y_{O_2} = 0.21$, and, by keeping in mind that we talk about a very dilute system, hence $x_{O_2} \cong X_{O_2}$ (the molar fraction of oxygen in water does not deviate much from the molar load of oxygen in terms of mol O_2 mol^{-1} H_2O. After multiplying the nominator with the molar mass of oxygen ($MM_{O_2} = 32$ g mol^{-1}) and the denominator with the molar mass of water ($MM_{H_2O} = 18$ g mol^{-1}), she got the oxygen load of water in terms of g O_2 g^{-1} H_2O. Not very happy with the outcome, she converted this result into mg O_2 kg^{-1} H_2O by multiplying with 10^6 to end up with $X_{O_2} = 11.8$ mg O_2 kg^{-1} H_2O. Let us now do calculations together as exemplarily shown for 8 °C and 1,008 hPa.

$$H_{O_2,8\,°C} = 2{,}480.16 + 86.47 \cdot 8 - 0.39 \cdot 8^2 = 3{,}461 \text{ MPa} = 3{,}461 \cdot 10 \cdot 10{,}13 = 35{,}066{,}413 \text{ hPa}$$

$$x_{O_2} = \frac{P}{H_{O_2}} \cdot y_{O_2} \approx X_{O_2} = \frac{1{,}008 \cdot 0.21}{35{,}066{,}413} \approx 6.04 \cdot 10^{-6} \text{ mol mol}^{-1}$$

$$6.04 \cdot 10^{-6} \cdot \frac{32}{18} \cdot 10^6 = 11.8 \text{ mg } O_2 \text{ kg}^{-1} \text{ } H_2O$$

Table 8.4 shows the calculated results for $\theta = 20$ °C, 40 °C and 60 °C at $P = 1{,}013$ hPa.

Tab. 8.4: Temperature dependency of oxygen absorption from air into water for $P = 1{,}013$ hPa.

θ (°C)	H_{O_2} (MPa)	H_{O_2} (hPa)	x_{O_2} (equivalent X_{O_2}) (mol mol^{-1})	X_{O_2} (mg kg^{-1})
20	4,054	41,062,563	$5.18 \cdot 10^{-6}$	9.2
40	5,315	53,840,545	$3.95 \cdot 10^{-6}$	7.0
60	6,264	63,457,967	$3.35 \cdot 10^{-6}$	6.0

!

Discussion: Oxygen saturation of water from air shows a distinct temperature and pressure dependency. How about the numbers? Would you kindly compare the outcome with literature data [260]. Before you complain the simplification $x_{O_2} \cong X_{O_2}$ (for very dilute systems), we rather make a comparison. We start with the famous Henry approach of eq. (8.7), but rearrange it:

$$\frac{P \cdot y_{O_2}}{H_{O_2}} = x_{O_2} = \frac{n_{O_2}}{n_{O_2} + n_{H_2O}} \tag{8.7}$$

After inverting and rearranging again we end up with the following equations:

$$\frac{n_{O_2}}{n_{O_2} + n_{H_2O}} = 1 + \frac{1}{X_{O_2}} = \frac{X_{O_2} + 1}{X_{O_2}} = \frac{H_{O_2}}{P \cdot y_{O_2}}$$

and

$$X_{O_2}\left[\text{mol mol}^{-1}\right] = \frac{1}{\left(\frac{H_{O_2}}{P \cdot y_{O_2}} - 1\right)} \tag{8.14}$$

If you now compare the outcome with the results of Tab. 8.4, you will probably agree with the simplification $x \cong X$ for very dilute systems.

Exercise 8.2: Does the Henry constant H_{O_2} correlate with the Bunsen absorption coefficient α for oxygen absorption in water?

Solver:
We rearrange eq. (8.13) for the Henry constant and check for comparability with the Bunsen coefficient from α Tab. 8.3 (keep care of the units); $T = 273$ K:

$$H_{O_2} = \frac{\text{MV} \cdot \rho_l}{\alpha \cdot \text{MM}_l} = \frac{22.4 \cdot 1{,}000 \cdot 1{,}000}{1{,}000 \cdot 0.04825 \cdot 18 \cdot 10} = 2{,}579\ \text{MPa} \tag{8.13}$$

Then we compare the result with the Henry-correlation for oxygen in Tab. 8.2 to obtain $H_{O_2} = 2{,}480$ MPa. When we now compare the oxygen load X_{O_2}, as discussed in Exercise 8.1, we will get $X_{O_2} = 15$ mg kg^{-1} with the Henry constant from Tab. 8.2 and $X_{O_2} = 14.5$ mg kg^{-1} via Bunsen coefficient α of oxygen from Tab. 8.3. We may accept the difference.

Exercise 8.3: Next step is chemisorption

Determine the gas–liquid equilibrium for the system HCl/air/H_2O at 20 °C in terms of Y in mg kg^{-1} for the gaseous phase as well as for the water phase X in mg kg^{-1}; $pK_{HCl} = -3$ ($K_{Diss} = 1{,}000$) [256]. Before we develop the solver strategy, we collect some physical properties of aqueous hydrochloric acid, shown in Tab. 8.5.

Tab. 8.5: Physical properties of aqueous hydrochloric acid at 293 K [261].

w_{HCl} (wt%)	X_{HCl} (mol kg^{-1})	c_{HCl} (mol L^{-1})	ρ_{HCl} (kg L^{-1})	μ_{HCl} (mPa s)
0.5	0.138	0.137	1.0007	1.008
1.0	0.277	0.275	1.0031	1.015
2.0	0.56	0.533	1.0081	1.029
3.0	0.848	0.833	1.013	1.044
4.0	1.143	1.117	1.0179	1.059
5.0	1.444	1.403	1.0228	1.075
6.0	1.751	1.691	1.0278	1.091
7.0	2.064	1.983	1.0327	1.108
8.0	2.385	2.277	1.0377	1.125
9.0	2.713	2.547	1.0426	1.143
10.0	3.047	2.873	1.0476	1.161
12.0	3.74	3.481	1.0576	1.199

Solver:
From Tab. 8.2 we get the Henry constant H_{HCl} in MPa $= 0.242 + 0.00185 \cdot \theta - 1.04 \cdot 10^{-5} \cdot \theta^2$ and for $\theta = 20$ °C, $H_{HCl} = 0.275$ MPa.

First step:
According to Fig. 8.2 we must now extract the nondissociated amount of hydrochloric acid for the concentration data or load data of Tab. 8.5, before we address the gas–liquid equilibrium. For demonstration, we pick out the data set for 5 wt% of HCl in the aqueous phase from Tab. 8.5. The corresponding molar load is $X_{HCl,0} = 1.444$ mol HCl kg^{-1} H_2O, equivalent $1.444 \cdot 36.45 \cdot 1{,}000 = 52{,}634$ mg HCl kg^{-1} H_2O. Now we let HCl dissociate.

$$HCl_{aqu} \rightleftharpoons H^+_{aqu} + Cl^-_{aqu}$$

Equilibrium:

$$X_{HCl,0} \cdot (1-\alpha) = X_{HCl,0} \cdot \alpha + X_{HCl,0} \cdot \alpha$$

$$\frac{(X_{HCl,0} \cdot \alpha)^2}{X_{HCl,0} \cdot (1-\alpha)} = K_{Diss} = \frac{(1.444 \cdot \alpha)^2}{1.444 \cdot (1-\alpha)} = 1{,}000$$

$$X_{HCl,0} \cdot \alpha^2 + K_{Diss} \cdot \alpha - K_{Diss} = 0$$

$$\alpha_{1,2} = \frac{-K_{Diss} \pm \sqrt{K_{Diss}^2 + 4 \cdot K_{Diss} \cdot X_{HCl,0}}}{2 \cdot X_{HCl,0}} = \frac{-1{,}000 \pm \sqrt{1{,}000^2 + 4 \cdot 1{,}000 \cdot 1.444}}{2 \cdot 1.444} = 0.998 \tag{8.15}$$

From this, we get:

molar load of non-dissociated acid: $X_{HCl,0} \cdot (1-\alpha) = 2.08 \cdot 10^{-3}$ mol HCL $kg^{-1} H_2O$
molar load of dissociated acid: $X_{HCl,0} \cdot \alpha$

Second step:
Then we transfer the mass-based load data of the nondissociated hydrochloric acid into mole-based load data, and in a further step, we extract the molar fraction of hydrochloric acid in water from mole-based load data:

$$\frac{X_{HCl,0} \cdot (1-\alpha)\ [\text{mol HCl kg}^{-1}\text{H}_2\text{O}]}{\left(\frac{1{,}000}{18}\right) [\text{mol H}_2\text{O kg}^{-1}\text{H}_2\text{O}]} = X_{HCl,\text{non-diss}} \left[\text{mol HCl mol}^{-1}\text{H}_2\text{O}\right] = \frac{2.08 \cdot 10^{-3}}{\left(\frac{1{,}000}{18}\right)} = 3.74 \cdot 10^{-5}$$

and

$$\frac{X_{HCl,\text{non-diss}}}{1 + X_{HCl,\text{non-diss}}} = x_{HCl,\text{non-diss}} = 3.74 \cdot 10^{-5} \tag{8.16}$$

(Do not get confused now, but remember the discussion of load and molar fraction for very dilute systems.)

Third step:
We let the Henry law act on the molar fraction of nondissociated hydrochloric acid in the aqueous phase to obtain the molar fraction of hydrochloric acid in air:

$$y_{HCl} = x_{HCl,\text{non-diss}} \cdot \frac{H_{HCl}}{P} = \frac{3.7 \cdot 10^{-5} \cdot 0.2748 \cdot 10 \cdot 1{,}013}{1{,}013} = 1.03 \cdot 10^{-4} \tag{8.17}$$

Fourth step:
For construction of the gas–liquid equilibrium in terms of mg kg^{-1}, the molar fraction of hydrochloric acid in the gas phase is retransferred into molar load and finally mass-based load of the gas phase:

$$Y_{HCl,air}\left[\text{mol HCl mol}^{-1}\text{air}\right] = \frac{Y_{HCl}}{1-Y_{HCl}} = \frac{1.03\cdot 10^{-4}}{1-1.03\cdot 10^{-4}} = 1.03\cdot 10^{-4}$$

$$Y_{HCl,air}\left[\text{mol HCl mol}^{-1}\text{air}\right]\cdot\frac{1{,}000}{28.84} = Y_{HCl,air}\left[\text{mol HCl kg}^{-1}\text{air}\right]$$

$$= 1.03\cdot 10^{-4}\cdot\frac{1{,}000}{28.84} = 3.57\cdot 10^{-3}\ \text{mol HCl kg}^{-1}\text{air}$$

and

$$Y_{HCl,air}\left[\text{mol HCl kg}^{-1}\text{air}\right]\cdot MM_{HCl}\left[\text{g mol}^{-1}\right]\cdot 1{,}000\left[\text{mg g}^{-1}\right] = Y_{HCl,air}\left[\text{mg HCl kg}^{-1}\text{air}\right]$$

$$= 3.57\cdot 10^{-3}\cdot 36.45\cdot 1{,}000 = 1.03\cdot 10^{2}\ \text{mg HCl kg}^{-1}\text{air} \tag{8.17}$$

The HCl load in the gas phase $Y_{HCl,air} = 1.03\cdot 10^{2}$ mg HCl kg^{-1} air is in equilibrium with the total mass load of $X_{HCl} = 52{,}634$ mg HCl kg^{-1} air.

After having done the calculations for all the data of Tab. 8.5, we draw a sketch of the gas–liquid equilibrium HCl/air/water at $T = 293$ K and $P = 1{,}013$ hPa, shown in Fig. 8.4.

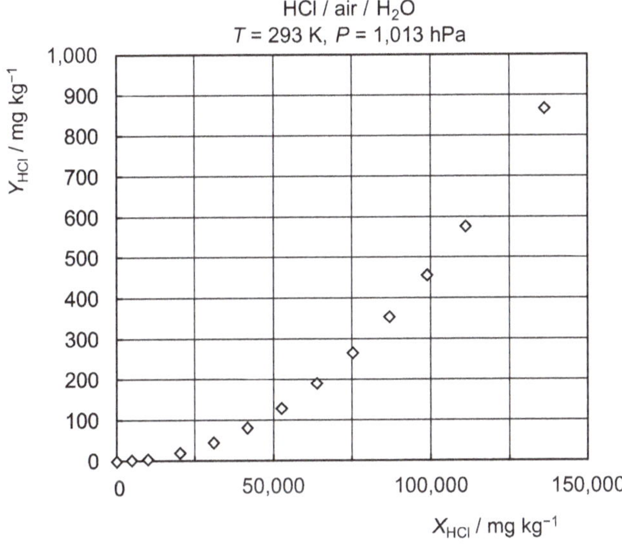

Fig. 8.4: Gas–liquid equilibrium of the system HCl/air/water: $T = 293$ K, $P = 1{,}013$ hPa.

Discussion: This type of presentation of gas–liquid equilibrium data offers many advantages in performing mass balances. Since HCl is a strong acid, the pH value can easily be deduced from the concentration. According to Tab. 8.5 a concentration of 5 wt% of hydrochloric acid is equivalent to a concentration of 1.4 mol L^{-1}, and it is readily dissociated. Then the concentration of protons (H^+) is $C_{H^+} = 1.4$ mol L^{-1}, too. In the Chemistry basics they told us that the pH is the negative decadic logarithm of the H^+ concentration, and $-\lg_{10}(1.4) = -0.14$. If you try to validate that with a pH electrode you will fail. You may probably be curious of getting familiar with absorption of hydrochloric acid at constant pH value. So we have to discuss that in Exercise 8.4.

Exercise 8.4: Gas–liquid equilibrium for the system HCl/air/H_2O at constant pH value

We want to absorb hydrochloric acid in aqueous scrubbing liquid at constant pH value of pH = 2 by, for example, metering NaOH according to eq. (8.4) to the scrubbing liquid under pH control.

Before we develop the solver, we must discuss the chemical background. We start with protolysis of the acid HA as follows:

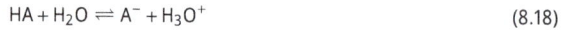

$$HA + H_2O \rightleftharpoons A^- + H_3O^+ \tag{8.18}$$

We let acid HA undergo protolysis by withdrawing some H^+ by neutralization. As a consequence, the concentration of the anion A^- must increase by the same amount of H^+ withdrawn from HA:

$$(X_{HA} - X_{H^+}) = (X_{A^-} + X_{H^+}) + X_{H^+} \tag{8.19}$$

Then we apply the law of mass action and rearrange for the amount of acid HA:

$$\frac{(X_{A^-} + X_{H^+}) \cdot X_{H^+}}{(X_{HA} - X_{H^+})} = K_{Diss}$$

$$\frac{(X_{A^-} + X_{H^+}) \cdot X_{H^+}}{K_{Diss}} = (X_{HA} - X_{H^+}) \tag{8.20}$$

Solver:

We have fixed the pH value at pH = 2. We select a series of mass load data for Cl^- (= X_{A^-}) from 0 g kg^{-1} to 30 g kg^{-1}, determine the molar load of (X_{HCl} – X_{H^+}) according to eq. (8.20) at constant $X_{H^+} = 10^{-pH} = 10^{-2} =$ 0.01 mol kg^{-1}. (Since the molar load of free acid (X_{HCl} – X_{H^+}) will be very low, we can easily transfer into molar fraction.) At this point you are already familiar with the procedure. We again apply the Henry approach to obtain the molar fraction of y_{HCl} in the gas phase and convert it into the corresponding mass load Y_{HCl}. But, let us do calculations for $X_{Cl^-} = 5$ g kg^{-1} step by step.

First step:

$$\frac{(X_{A^-} + X_{H^+}) \cdot X_{H^+}}{K_{Diss}} = (X_{HA} - X_{H^+}) \; \frac{(0.14 + 0.01) \cdot 0.01}{1{,}000} = 1.5 \cdot 10^{-6} \left[\text{mol HCl kg}^{-1} H_2O\right]$$

$$= \frac{1.5 \cdot 10^{-6}}{\frac{1{,}000}{18}} = 2.7 \cdot 10^{-8} \text{mol HCl mol}^{-1} \; H_2O = x_{HCl}$$

Second step:

$$y_{HCl} = x_{HCl,\text{non-diss}} \cdot \frac{H_{HCl}}{P} = \frac{2.7 \cdot 10^{-8} \cdot 0.2748 \cdot 10 \cdot 1{,}013}{1{,}013}$$

$$= 7.5 \cdot 10^{-8} \text{ mol HCl kg}^{-1} \; H_2O = Y_{HCl} \; \left[\text{mol HCl kg}^{-1} \; H_2O\right]$$

Third step:

$$Y_{HCl,air} \left[\text{mol HCl mol}^{-1} \text{ air}\right] = \frac{y_{HCl}}{1 - y_{HCl}} = \frac{7.5 \cdot 10^{-8}}{1 - 7.5 \cdot 10^{-4}8} \approx 7.5 \cdot 10^{-8}$$

$$Y_{HCl,air} \left[\text{mol HCl mol}^{-1} \text{ air}\right] \cdot \frac{1{,}000}{28.84} = Y_{HCl,air} \left[\text{mol HCl kg}^{-1} \text{ air}\right] = 7.5 \cdot 10^{-8} \cdot \frac{1{,}000}{28.84} = 2.6 \cdot 10^{-6}$$

$$Y_{HCl,air} \left[\text{mol HCl kg}^{-1} \text{ air}\right] \cdot MM_{HCl} \left[\text{g mol}^{-1}\right] \cdot 1{,}000 \left[\text{mg g}^{-1}\right] = Y_{HCl,air} \left[\text{mg HCl kg}^{-1} \text{ air}\right]$$

$$= 2.6 \cdot 10^{-6} \cdot 36.45 \cdot 1{,}000 = 9.4 \cdot 10^{-2} \text{ mg HCl kg}^{-1} \text{ air}$$

For comparison with absorption of hydrochloric acid without pH control, shown in Fig. 8.4, we construct the gas–liquid equilibrium sketch for HCl absorption at constant pH of pH = 2, shown in Fig. 8.5. Note that pH control has a significant effect on the gas–liquid equilibrium.

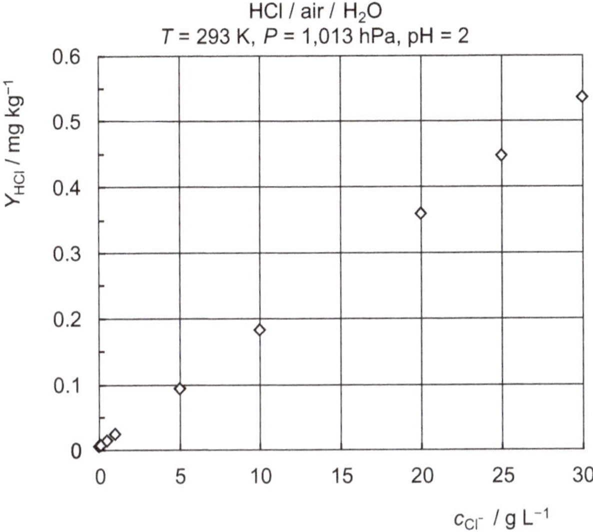

Fig. 8.5: Gas–liquid equilibrium for the system HCl/air/H_2O at constant pH of 2.

Discussion: When you compare Fig. 8.5 and Fig. 8.4 you will register that at pH = 2 the gas phase load $Y_{HCl,air}$ is two orders of magnitude less than for absorption of hydrochloric acid with pure water. You may conclude from this finding that hydrochloric acid absorption at elevated pH value will be an ease, and you are right.

We also have to address the gas–liquid equilibrium basics for the pollutants SO_2 and CO_2 to complete the first part of our engineering job. We will rather focus on SO_2, because it really fixed attention of the environmentalists and the engineering community for decades. We will briefly address CO_2 and discuss its interaction with SO_2 absorption. Let us start with the chemical basics. With water, SO_2 as well as CO_2 form divalent acids. Depending on the pH value different species of protolysis domain the aqueous solution. We can readily forecast the composition and fraction of species for given pH value with the law of mass action, shown in eqs. (8.21)–(8.23) [262].

With the two protolysis steps (with K_1 and K_2 being the dissociation constants of the first and second steps of protolysis) for the acid H_2A

$$\frac{[\mathrm{H}^+]\cdot[\mathrm{HA}^-]}{[\mathrm{H_2A}]} = K_1 \text{ and } \frac{[\mathrm{H}^+]\cdot[\mathrm{A}^{2-}]}{[\mathrm{HA}^-]} = K_2$$

$$\text{with } \frac{[\mathrm{H}^+]\cdot[\mathrm{A}^{2-}]}{K_2} = [\mathrm{HA}^-] \text{ you get } \frac{[\mathrm{H}^+]^2\cdot[\mathrm{A}^{2-}]}{[\mathrm{H_2A}]\cdot K_2} = K_1$$

$$\text{and } [A^{2-}] = \frac{[H_2A] \cdot K_2 \cdot K_1}{[H^+]^2} \text{ and } [HA^-] = \frac{[H^+] \cdot [H_2A] \cdot K_2 \cdot K_1}{[H^+]^2 \cdot K_2}$$

we solve for the amount of nondissociated acid in eq. (8.21) (e.g., sulfurous acid H_2SO_3),

$$\frac{[H^+]^2}{\left([H^+]^2 + [H^+] \cdot K_1 + K_1 \cdot K_2\right)} = [H_2A] \tag{8.21}$$

the amount of dissociated acid HA^- after the first step of protolysis in the following equation (e.g., hydrogen sulfite HSO_3^-):

$$\frac{[H^+] \cdot K_1}{\left([H^+]^2 + [H^+] \cdot K_1 + K_1 \cdot K_2\right)} = [HA^-] \tag{8.22}$$

and the amount of dissociated acid A^{2-} after the second step of protolysis in the following equation (e.g., sulfite, SO_3^{2-}):

$$\frac{K_1 \cdot K_2}{\left([H^+]^2 + [H^+] \cdot K_1 + K_1 \cdot K_2\right)} = [A^{2-}] \tag{8.23}$$

! Comment: Do not panic because of the square brackets. The chemists, for example, often specify concentrations with square brackets ($c_A = [A]$). The thermodynamic people sometimes specify activity with square brackets, and we specify molar load with square brackets (when advantageous). If we apply thermodynamics seriously, these square brackets contain the activity of the corresponding species. When lazily assuming the activity coefficient $\gamma \approx 1$ the square brackets may also specify mol kg^{-1} or mol L^{-1}.

You may wonder about missing initial concentration or load of species in eqs. (8.21)–(8.23). Do not worry. At this level you may assume the total load of species $X_0 = 1$ mol kg^{-1}.

From literature [263], we collect data for the dissociation constants K_1 and K_2 with T in K and the temperature range of 20 °C (293.15 K) to 70 °C (343.15 K), shown in the following equations:

$$K_{1,\,H_2SO_3} = 7.7 \cdot e^{-0.0207 \cdot T} \tag{8.24}$$

$$K_{2,\,H_2SO_3} = 9.6 \cdot 10^{-6} \cdot e^{-0.0168 \cdot T} \tag{8.25}$$

$$K_{1,\,H_2CO_3} = -1.84 \cdot 10^{-6} + 8.152 \cdot 10^{-9} \cdot T \tag{8.26}$$

$$K_{2,\,H_2CO_3} = -6.17 \cdot 10^{-15} \cdot T^2 + 4.43 \cdot 10^{-12} \cdot T - 7.265 \cdot 10^{-10} \tag{8.27}$$

We may also extract handsome correlations for the temperature-dependent Henry constant of SO_2 and of CO_2 shown in eqs. (8.28) ($H^*_{SO_2}$) and (8.29) ($H^*_{CO_2}$) (carefully consider the units of H^* in eqs. (8.28) and (8.29)) [258]:

$$H^{*}_{SO_2}(T)\,[\text{hPa kg mol}^{-1}] = 1.013 \cdot 10^{7} \cdot e^{\left(-\frac{2,838}{T}\right)} \tag{8.28}$$

$$H^{*}_{CO_2}(T)\,[\text{bar kg mol}^{-1}] = 14.193 \cdot e^{(0.0287 \cdot T)} \tag{8.29}$$

Do not worry about the different algorithms for specifying the dissociation constants K_1 and K_2 in eqs. (8.24)–(8.27). They are just fit equations based on literature data [263]. Do not make use of these Henry correlations beyond the specified temperature range. Incidentally (referring to Exercise 2.10) we perform calculations for $\theta = 58$ °C and $P = 980$ hPa. For going straightforward in calculations, you construct a list with the independent variable pH value. Then you transfer the pH value into X_{H^+} ($X_{H^+} = 10^{-pH}$). With eqs. (8.21–8.23) you then calculate the fraction of H_2A, HA^- and A^{2-}. Then, you sum up the fractions to the denominator (sum = $H_2A + HA^- + A^{2-}$). Afterward you form the ratio of the species fractions and the denominator and multiply it with 100 to obtain the relative amount of species in %. Figure 8.6 shows the outcome of these calculations. For comparison, Figure 8.6 shows the protolysis of sulfurous acid (H_2SO_3) and carbonic acid (H_2CO_3).

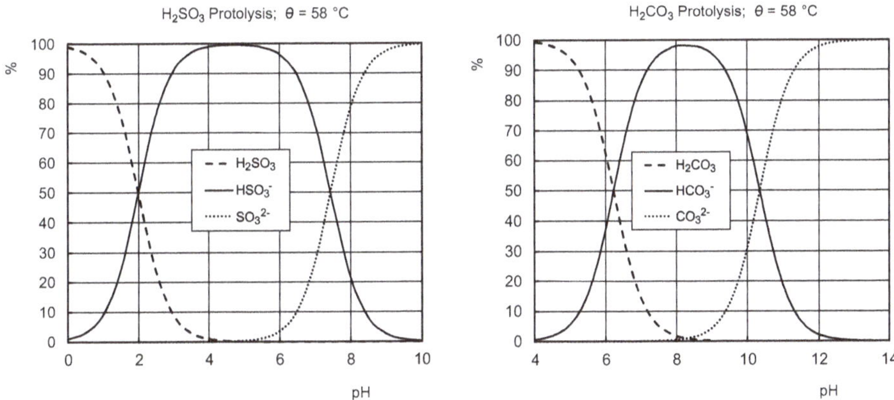

Fig. 8.6: pH dependency of sulfurous acid (H_2SO_3) protolysis and carbonic acid protolysis (H_2CO_3) at 58 °C.

Figure 8.6 is a very important chart for discussing absorption of SO_2 and CO_2 and the interaction of these constituents. But how to interpret the message of this chart? Let us go back to the definition of the Henry Law (eq. (8.7)) by referring to Fig. 8.2. The Henry constant correlates the molar fraction of the nondissociated species in the aqueous phase with the molar fraction of this species in the gaseous phase. Seemingly we may nearly neglect the amount of nondissociated H_2SO_3 at pH values beyond pH = 4. We may assume that we will readily absorb SO_2 at pH values beyond pH > 4. On the other hand, H_2CO_3 is nearly 100% nondissociated at pH = 4. Absorption of CO_2 at pH = 4 will be very low. For off-gas, such as the off-gas we constructed in Tab. 1.1, this difference in absorption properties is of enormous importance. Just have in mind that we talk about 300 ppm of SO_2, but 10 vol% of CO_2. If we operate the absorption at an elevated pH value of pH = 7, we for sure will readily absorb SO_2, but most of the caustic neutralizer (e.g., NaOH) fed to the scrubbing liquid will be consumed by CO_2. If we operate the absorption process at pH = 5, SO_2 will readily absorb, leaving CO_2 nearly unaffected in the gas phase. Let us compare this outcome with the absorption properties of HCl. As discussed in Exercises 8.3 and 8.4, HCl will readily absorb at very low pH values of pH < 3. Seemingly we are able to separate the pollutants HCl, SO_2 and CO_2 more or less selectively by appropriate pH control of the scrubbing liquid. This is a very important boundary in absorption process design. Finally, having in mind the procedure from discussing HCl absorption we may construct the gas–liquid equilibrium for SO_2 absorption.

Exercise 8.5: The system SO_2/off-gas/water at pH = 5

We recall the off-gas data of Tab. 1.1 and Exercise 2.10:

$MM_{\text{inert gas}} = 30.08\ \text{g mol}^{-1}$

$c_{SO_2} = 300$ ppm

$\theta = 58\ °C$

$P = 980$ hPa

$H^*_{SO_2, 58\,°C} = 1{,}921\ \text{hPa kg mol}^{-1}$ (8.28)

$K_{1,\ H_2SO_3} = 1.02 \cdot 10^{-2}$

$K_{2,\ H_2SO_3} = 3.68 \cdot 10^{-8}$

With this data, we determine the molar fraction of H_2SO_3 with eq. (8.21) through (8.23) for $X_{SO_2,0} = 1$ mol SO_2 kg^{-1} H_2O.

First step:

Equilibrium composition of H_2SO_3 at pH = 5 and $X_{SO_2,0} = 1$ mol SO_2 $kg^{-1} H_2O$:

$$\frac{[H^+]^2 \cdot X_{SO_2,0}}{\left([H^+]^2 + [H^+] \cdot K_1 + K_1 \cdot K_2\right)} = \frac{\left[10^{-5}\right]^2 \cdot X_{SO_2,0}}{\left(\left[10^{-5}\right]^2 + \left[10^{-5}\right] \cdot 1.02 \cdot 10^{-2} + 1.02 \cdot 10^{-2} \cdot 3.68 \cdot 10^{-8}\right)} = 5.72 \cdot 10^{-4} \quad (8.21)$$

$$\frac{[H^+] \cdot K_1 \cdot X_{SO_2,0}}{\left([H^+]^2 + [H^+] \cdot K_1 + K_1 \cdot K_2\right)} = \frac{\left[10^{-5}\right] \cdot 1.02 \cdot 10^{-2} \cdot X_{SO_2,0}}{\left(\left[10^{-5}\right]^2 + \left[10^{-5}\right] \cdot 1.02 \cdot 10^{-2}{}_1 + 1.02 \cdot 10^{-2} \cdot 3.68 \cdot 10^{-8}\right)} = 5.86 \cdot 10^{-1} \quad (8.22)$$

$$\frac{K_1 \cdot K_2 \cdot X_{SO_2,0}}{\left([H^+]^2 + [H^+] \cdot K_1 + K_1 \cdot K_2\right)} = \frac{1.02 \cdot 10^{-2} \cdot 3.68 \cdot 10^{-8} \cdot X_{SO_2,0}}{\left(\left[10^{-5}\right]^2 + \left[10^{-5}\right] \cdot 1.02 \cdot 10^{-2} + 1.02 \cdot 10^{-2} \cdot 3.68 \cdot 10^{-8}\right)} = 2.16 \cdot 10^{-3}$$

Second step:

Calculation of the molar fraction of H_2SO_3 from the data of step 1:

$$x_{H_2SO_3} = \frac{5.72 \cdot 10^{-4}}{5.72 + 10^{-4} + 5.86 \cdot 10^{-1} + 2.16 \cdot 10^{-3}} = 9.72 \cdot 10^{-4}$$

Third step:

Calculation of $X_{H_2SO_3}$

$$X_{H_2SO_3} = x_{H_2SO_3} \cdot X_{SO_2,0}$$

To avoid misunderstanding: The molar load of nondissociated acid $X_{H_2SO_3}$ in the aqueous phase is isolated by multiplying the molar fraction of sulfurous acid in the aqueous phase $x_{H_2SO_3}$ with selected values of freely chosen initial load in mol kg^{-1} of SO_2 in the aqueous phase. (The molar fraction $x_{H_2SO_3}$ in the aqueous phase is independent of the total load $X_{SO_2,0}$.) How to find appropriate values of $X_{SO_2,0}$? You may either guess or test for different values to finally comply with the maximum SO_2 load in terms of Y_{SO_2} in g SO_2 kg^{-1} gas in the gas phase.

Fourth step:
Determination of y_{SO_2} with the Henry correlation of eq. (8.28) and the system pressure $P = 980$ hPa:

$$H^*_{SO_2}(T) = 1.013 \cdot 10^7 \cdot e^{\left(-\frac{2,838}{T}\right)} = 1,921.531 \text{ hPa kg mol}^{-1} \text{ and } y_{SO_2} = \frac{H_{SO_2}}{P} \cdot x_{H_2SO_3}$$

We may again simplify: $y_{SO_2} = Y_{SO_2}$

Fifth step:
Calculation of the mass load Y_{SO_2} in g kg^{-1} of SO_2 in the off-gas according to eq. (8.30), with the molar mass of SO_2 $MM_{SO_2} = 64$ g mol^{-1} and the mean molar mass of the inert gas ($MM_{\text{inert gas}}$) = 30.08 g mol^{-1}, resulting in 1,000/30.08 = 33.25 mol off-gas kg^{-1} off-gas:

$$Y_{SO_2}\left[\text{mol SO}_2 \text{ mol}^{-1} \text{ off-gas}\right] \cdot \left[\text{mol off-gas kg}^{-1} \text{ off-gas}\right] \cdot MM_{SO_2} = Y_{SO_2}\left[\text{g kg}^{-1}\right] \tag{8.30}$$

We perform the calculations of steps 3–5 for several $X_{SO_2,0}$ values. Table 8.6 shows the data for selected $X_{SO_2,0}$ values, and Fig. 8.7 shows the corresponding equilibrium graph.

Tab. 8.6: Gas–liquid equilibrium of the system SO_2/off-gas/water: $\theta = 58$ °C, $P = 980$ hPa.

$X_{SO_2,0}$ (g kg^{-1})	$X_{SO_2,0}$ (mol kg^{-1})	$x_{H_2SO_3}$	$X_{H_2SO_3}$ (mol kg^{-1} nondiss)	$y_{H_2SO_3} = Y_{H_2SO_3}$ (mol mol^{-1})	$Y_{H_2SO_3}$ (g kg^{-1})
0	0	0	0	0	0
1	0.016	$9.72 \cdot 10^{-4}$	$1.95 \cdot 10^{-5}$	$2.98 \cdot 10^{-5}$	0.063
4	0.063	$9.72 \cdot 10^{-4}$	$6.08 \cdot 10^{-5}$	$1.19 \cdot 10^{-4}$	0.254
8	0.125	$9.72 \cdot 10^{-4}$	$1.22 \cdot 10^{-4}$	$2.38 \cdot 10^{-4}$	0.507
11	0.172	$9.72 \cdot 10^{-4}$	$1.67 \cdot 10^{-4}$	$3.28 \cdot 10^{-4}$	0.697

At pH = 5, the mass load of SO_2 in the gas phase Y_{SO_2} is assumingly very low for given X_{SO_2} in the aqueous phase (e.g., $Y_{SO_2} = 0.507$ g kg^{-1} at $X_{SO_2} = 8$ g kg^{-1}). The reason is very simple. According to Figure 8.6, the fraction of nondissociated sulfurous acid in the aqueous phase at pH = 5 is very low, but must not be neglected.

We will need the graph of Fig. 8.7 when we discuss SO_2 absorption technologies, and we need it for the scrubber design.

Attention: Do not mix up the different specifications $x_{species}$, $X_{species}$, $X_{species,0}$ and also y and Y, when you do these calculations for the first time. We must also outline the specific expression for the Henry constant in terms of $H^*_{species}$ in hPa kg mol^{-1}. This is for sure a very handsome expression, but do not make use of it without having the specific dimensions in mind. If you ignore that you will definitely end up in a disaster. Perhaps we may keep in mind that our off-gas will also contain some residual hydrochloric acid. This constituent will be readily absorbed at pH = 5.

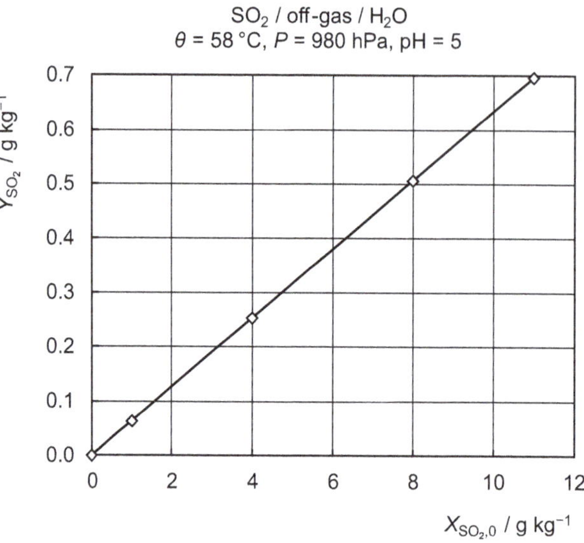

Fig. 8.7: Equilibrium graph for the system SO_2/off-gas/water at 58 °C and 980 hPa.

Exercise 8.6: The system CO_2/off-gas/water at pH = 8.6

We again recall the off-gas data of Tab. 1.1 and Exercise 2.10, and we figure out the difference between the acidic constituent SO_2 and the acidic properties of CO_2. When comparing the protolysis data of sulfurous acid and carbonic acid in Figure 8.6, we register that HCO_3^- is the governing species at pH = 8.6. Different to discussing the system SO_2/off-gas/water we must redefine the specification of the inert gas in Tab. 1.1, which now just consists of oxygen and nitrogen.

New off-gas specification:
Flow rate $F_V = 22{,}104.5\ m^3_{STP}\ h^{-1}$
$MM_{inert\ gas} = 28.5\ g\ mol^{-1}$
O_2: 13.3 vol%
N_2: 86.7 vol%
$\rho_{inert\ gas} = 1.273\ kg\ m^{-3}_{STP}$
$F_{V,CO_2} = 2{,}456.1\ m^3_{STP}\ h^{-1}$
$Y_{CO_2} = 171.34\ g\ kg^{-1}$ inert gas
$P = 980$ hPa
By applying a magic (e.g., a huge heat exchanger), we lower the off-gas temperature to $\theta = 25$ °C.

From eq. (8.29) we get

$$H^*_{CO_2}(T) = 14.193 \cdot e^{(0.0287 \cdot 25[°C])} = 29.09\ \text{bar kg mol}^{-1} = 29.5 \cdot 10^5\ \text{hPa kg mol}^{-1}$$

From eq. (8.26) we get

$$K_{1,H_2CO_3} = -1.84 \cdot 10^{-6} + 8.152 \cdot 10^{-9} \cdot 25 = 5.91 \cdot 10^{-7}$$

From eq. (8.27) we get:

$$K_{2,H_2CO_3} = -6.17 \cdot 10^{-15} \cdot 58^2 + 4.43 \cdot 10^{-12} \cdot 58 - 7.265 \cdot 10^{-10} = 4.58 \cdot 10^{-11}$$

By applying eqs. (8.21)–(8.23) we obtain the molar fraction x of H_2CO_3 at pH = 8.6 of $x_{H_2CO_3} = 4.16 \cdot 10^{-3}$ (see Fig. 8.6). Finally, we may apply the solver.

Solver:
We do calculations by deducing the load of CO_2 in the gas phase Y_{CO_2} from several data for CO_2 load in the aqueous phase X_{CO_2}, shown in Tab. 8.7. We guess a series of X_{CO_2} data, for example from $X_{CO_2} = 0$ to $X_{CO_2} = 4\ \text{g kg}^{-1}$. Again, we do calculations according to the algorithms of the following equations:

$$X_{H_2CO_3} = X_{CO_2} \cdot x_{H_2CO_3} = X_{CO_2} \cdot 4.16 \cdot 10^{-3}$$

$$y_{CO_2} = \frac{H_{CO_2}}{P} \cdot X_{H_2CO_3} = \frac{29.5 \cdot 10^5}{980} \cdot X_{H_2CO_3}$$

and

$$Y_{CO_2} = \frac{y_{CO_2}}{1 - y_{CO_2}}$$

$$Y_{CO_2}\left[\text{mol } CO_2 \text{ mol}^{-1} \text{ off-gas}\right] \cdot \left[\text{mol off-gas kg}^{-1} \text{ off-gas}\right] \cdot MM_{CO_2} = Y_{CO_2}\left[\text{g kg}^{-1}\right]$$

Tab. 8.7: Gas–liquid equilibrium of the system CO_2/off-gas/water: $\theta = 25$ °C, $P = 980$ hPa.

X_{CO_2} (g kg^{-1})	X_{CO_2} (mol kg^{-1})	$X_{H_2CO_3}$ (mol kg^{-1})	Y_{CO_2} (mol mol^{-1})	Y_{CO_2} (mol g^{-1})	Y_{CO_2} (g kg^{-1})
0	0	0	0	0	0
0.25	0.006	$2.36 \cdot 10^{-5}$	$7.16 \cdot 10^{-3}$	$2.38 \cdot 10^{-4}$	10.47
0.50	0.011	$4.73 \cdot 10^{-5}$	$1.44 \cdot 10^{-2}$	$4.79 \cdot 10^{-4}$	21.09
0.75	0.017	$7.09 \cdot 10^{-5}$	$2.18 \cdot 10^{-2}$	$7.24 \cdot 10^{-4}$	31.87
1.00	0.023	$9.45 \cdot 10^{-5}$	$2.93 \cdot 10^{-2}$	$9.73 \cdot 10^{-4}$	42.80
1.50	0.034	$1.42 \cdot 10^{-4}$	$4.45 \cdot 10^{-2}$	$1.48 \cdot 10^{-3}$	65.15
2.00	0.045	$1.89 \cdot 10^{-4}$	$6.03 \cdot 10^{-2}$	$2.00 \cdot 10^{-3}$	88.17
2.50	0.057	$2.36 \cdot 10^{-4}$	$7.65 \cdot 10^{-2}$	$2.54 \cdot 10^{-3}$	111.90
3.00	0.068	$2.84 \cdot 10^{-4}$	$9.32 \cdot 10^{-2}$	$3.10 \cdot 10^{-3}$	136.37
4.00	0.091	$3.78 \cdot 10^{-4}$	$1.28 \cdot 10^{-1}$	$4.26 \cdot 10^{-3}$	187.66

With the data of Tab. 8.7, we can now draw a graph for X_{CO_2} and Y_{CO_2}, shown in Fig. 8.8. We also consider the feed load of CO_2 in the off-gas $Y_{CO_2} = 171.34\ \text{g kg}^{-1}$.

!

Discussion: You may conclude from this results that water, even at elevated pH value, is not a great choice for CO_2 absorption from incineration off-gas. You may compare the gas–liquid equilibrium for pH = 7.5 and pH = 9.5, and you will register a tremendous difference in the outcome. If you would perform absorption of all acidic off-gas constituents HCl, SO_2 and CO_2, you would nearly quantitatively remove HCl and SO_2 and a fraction of CO_2 when performing the absorption process at pH = 8.6. This outcome probably does not comply with your economic intention, but the discussion of this series of examples demonstrates the whole variety of process configurations you have to consider in discussing absorptive off-gas purification. The discussion of this topic hopefully did not scare you but encourage you in facing these challenges.

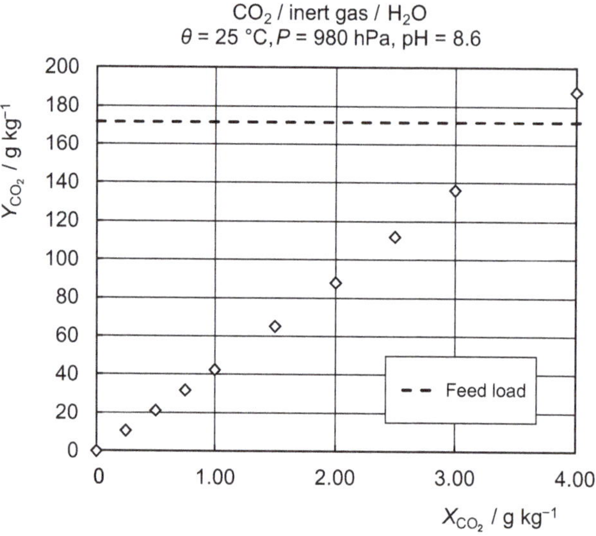

Fig. 8.8: Equilibrium graph for the system CO_2/off-gas/water at 25 °C and 980 hPa.

Exercise 8.7: How about solutes with caustic properties?

NH_3 is a very noxious substance, but has a variety of different industrial applications. You have to develop an algorithm for the determination of the gas–liquid equilibrium for the system air/NH_3/water with consideration of pH dependency, pressure dependency and temperature dependency. You may assume an NH_3 load of the off-gas of as much as 1 vol% (carrier: nitrogen). Construct the gas–liquid equilibrium graph. Assume an emission limit for NH_3 of 10 ppm (= 0.01 vol%).

Data collection:

We start with the most important information we need, the Henry constant for absorption of NH_3 in water. You may find data in [239, 264, 265], and elsewhere. We make use of the data from Perry, but adjust them to our needs by transferring them into a (ugly but helpful) power law approach, shown as follows:

$$H_{NH_3}(T)[\text{bar}] = 1.013 \cdot 10^{\left(-0.430219 + 0.0157205 \cdot \theta + 3.9369 \cdot 10^{-5} \cdot \theta^2\right)} \tag{8.31}$$

You may find the physical properties of NH_3/water mixtures in the *Handbook of Chemistry and Physics* [261] and in [258, 260]. Table 8.8 shows selected physical properties of aqueous solutions of ammonia. Kindly check the melting point (mp) data. You may make use of this advantageous temperature span in NH_3 absorption.

Tab. 8.8: Physical properties of NH_3/H_2O mixtures.

w_{NH_3} (wt%)	X_{NH_3} (mol kg^{-1})	ρ_{NH_3} (kg L^{-1})	mp (°C)	μ (mPa s)
0	0	0.9998	0	1.009
1	0.593	0.994	−1.14	1.015
2	1.198	0.99	−2.32	1.029

Tab. 8.8 (continued)

w_{NH_3} (wt%)	X_{NH_3} (mol kg^{-1})	ρ_{NH_3} (kg L^{-1})	mp (°C)	μ (mPa s)
5	3.09	0.977	−6.08	1.071
10	6.524	0.958	−13.55	1.141
16	11.184	0.936	−25.63	1.218
20	14.679	0.923	−36.42	1.254
24	18.542	0.91	−51.38	1.28

From [266] you may collect vapor pressure data for NH_3. In [260], you will find vapor pressure data of NH_3 over aqueous mixtures. For your convenience, these data have been adapted in an empirical approach as follows:

$$p_{NH_3}[\text{hPa}] = x_{NH_3} \cdot 1{,}013 \cdot 10^{\left(6.442 - \frac{1.922}{T}\right)} \tag{8.32}$$

Aqueous ammonia ($NH_{3,aqu}$) dissociates into ammonium cation ($NH^+_{4,aqu}$) and hydroxide (OH^-). This reaction is preferably expressed in the version of eq. (8.3):

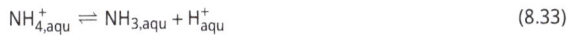

$$NH^+_{4,aqu} \rightleftharpoons NH_{3,aqu} + H^+_{aqu} \tag{8.33}$$

From [263] we can collect the temperature-dependent dissociation constant for aqueous ammonia, adapted for our purposes in an empirical approach, shown as follows:

$$K_{Diss,NH_4^+} = 9.727 \cdot 10^{-11} \cdot e^{6.559 \cdot 10^{-2} \cdot \theta} \tag{8.34}$$

With K_{Diss} we can deduce the protolysis equilibrium from the Law of Mass Action for aqueous ammonium and aqueous ammonia between pH = 6 and pH = 12 according to the following equations [262]:

$$x_{NH_4^+} = \frac{1}{1 + \frac{K_{Diss,NH_4^+}}{10^{-pH}}} \tag{8.35}$$

$$x_{NH_3} = \frac{1}{1 + \frac{10^{-pH}}{K_{Diss,NH_4^+}}} \tag{8.36}$$

Figure 8.9 shows the pH-dependent molar fraction of ammonium (NH_4^+) and ammonia (NH_3) at 293 K.

You may conclude from Fig. 8.9 that you may readily remove ammonia from off-gas by absorption at pH values of the absorbent of pH < 7 (you could even have co-absorption with SO_2 in mind, when you compare the prevailing species at pH < 7), while ammonia will undergo physiosorption at pH values of pH > 12. Therefore, we will now discuss NH_3 absorption in water by physiosorption from air at T = 293 K.

Solver:
From previous examples you probably do have an idea how to construct the gas–liquid equilibrium graph for the system NH_3/air/water. We assume a series of mass-based data for aqueous NH_3 solutions in wt%. We transfer this data into molar fractions of NH_3 in water with eq. (8.37). The molar mass of NH_3 is MM_{NH_3} = 17 g mol^{-1}, the molar mass of air is MM_{air} = 28.84 g mol^{-1} and the molar mass of H_2O is MM_{H_2O} = 18 g mol^{-1}:

$$x_{NH_3} = \frac{\frac{NH_3[\text{wt\%}]}{MM_{NH_3}}}{\frac{NH_3[\text{wt\%}]}{MM_{NH_3}} + \frac{H_2O[\text{wt\%}]}{MM_{H_2O}}} \tag{8.37}$$

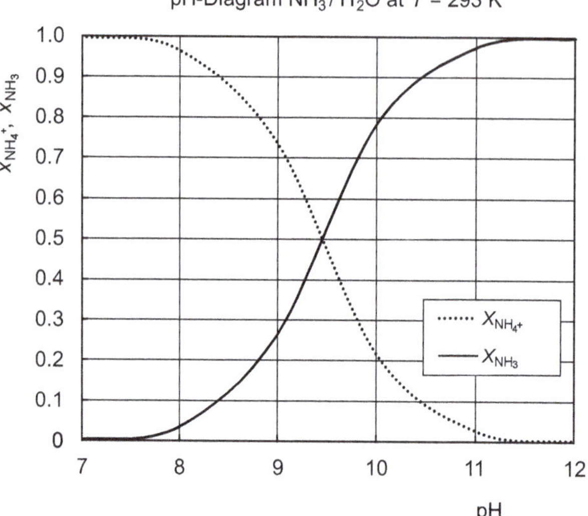

Fig. 8.9: pH-dependent protolysis equilibrium of aqueous ammonia at 293 K.

Next, we apply the Henry law (8.7) to get the molar fraction of NH_3 in air. Then, we transfer the mass-based NH_3 content of the aqueous phase into mass-based load X_{NH_3} in mg kg^{-1} with eq. (8.38) and finally transfer the molar fraction of NH_3 in air into the mass load of NH_3 Y_{NH_3} in mg kg^{-1} with eq. (8.39) (you will check the equations). We collect the data in Tab. 8.9:

$$y_{NH_3} = \frac{H_{NH_3}}{P} \cdot x_{NH_3} \tag{8.37}$$

$$X_{NH_3}\left[\text{mg kg}^{-1}\right] = \frac{NH_3[\text{wt\%}] \cdot 10^6}{(100 - NH_3[\text{wt\%}])} \tag{8.38}$$

Tab. 8.9: Gas–liquid equilibrium data for the system NH_3/air/H_2O at θ = 20 °C and P = 1,013 hPa.

w_{NH_3} (wt%)	x_{NH_3}	X_{NH_3} (mg kg^{-1})	y_{NH_3}	Y_{NH_3} (mg kg^{-1})
0	0		0	0
0.001	0	10	0	5
0.01	0.0001	100	0.0001	50
0.05	0.0005	500	0.0004	251
0.1	0.0011	1,001	0.0009	503
0.2	0.0021	2,004	0.0017	1,006
0.3	0.0032	3,009	0.0026	1,510
0.40	0.0042	4,016	0.0034	2,015
0.5	0.0053	5,025	0.0043	2,521
1	0.0106	10,101	0.0085	5,062
1.183	0.0125	11,972	0.0101	5,997

$$Y_{NH_3}\left[\text{mg kg}^{-1}\text{ air}\right]=\frac{NH_3[\text{wt\%}]\cdot 10^6}{(100-NH_3[\text{wt\%}])}=\frac{y_{NH_3}\cdot MM_{NH_3}\cdot 10^6}{\left(1-y_{NH_3}\right)\cdot MM_{air}} \quad (8.39)$$

Finally, we draw a sketch of the gas–liquid equilibrium for the system NH_3/air/H_2O in Fig. 8.10.

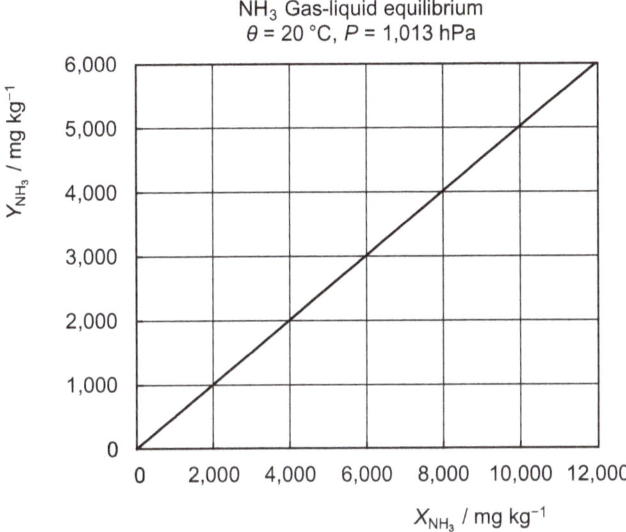

Fig. 8.10: Gas–liquid equilibrium of the system NH_3/air/H_2O at $\theta = 20$ °C and $P = 1{,}013$ hPa.

Discussion: In this series of examples we have discussed different algorithms to demonstrate how you can do the job. The idea behind this discussion was not to confuse you, but to demonstrate the different strategies you may make use of. Try to check the outcome of this final Exercise 8.7 by varying temperature and pressure of operation, and whether the basic needs for absorption are met (what will happen, when you raise the pressure or temperature or vice versa; do the results comply with the basics of absorption?).

8.2 Technologies, processes, their thermodynamics and kinetics

We have discussed the basics of absorption in Section 8.1 so far. Except an excursion into technologies and processes in Fig. 8.1 we have not discussed the practical part of our job. We will go through this topic with the example of SO_2 absorption. As discussed the absorption process of SO_2 starts with the transfer of the solute from the gaseous carrier phase into the (liquid) aqueous absorbent phase according to the following equation:

$$SO_{2,gas} \rightleftharpoons SO_{2,aqu} \rightleftharpoons H_2SO_{3,aqu} \tag{8.40}$$

Aqueous sulfurous acid undergoes immediate dissociation in two steps, shown in eqs. (8.41) and (8.42), quantifiable via dissociation constant K_{1,H_2SO_3} and K_{2,H_2SO_3} (eqs. (8.24) and (8.25)):

$$H_2SO_{3,aqu} \rightleftharpoons H^+ + HSO_3^- \tag{8.41}$$

$$HSO_3^- \rightleftharpoons H^+ + SO_3^{2-} \tag{8.42}$$

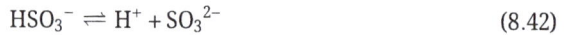

If we admix a caustic reactant to the aqueous solution of H_2SO_3 we will consume the protons H^+ and shift the equilibrium composition of constituents (irreversibly) to the right-hand side of eqs. (8.40)–(8.42). Figure 8.6 shows the composition of aqueous constituents depending on the amount of caustic reactant fed to the aqueous solution of H_2SO_3 under pH control. Figure 8.11 shows the basic setup of an absorption plant

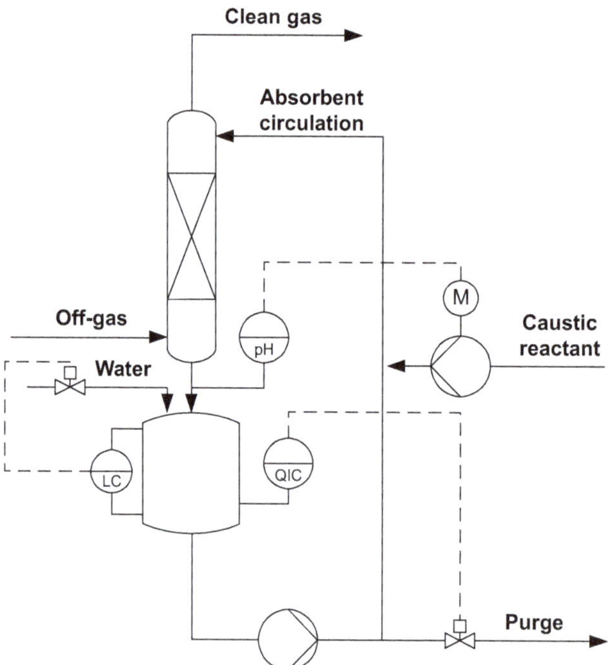

Fig. 8.11: Basic setup of an absorption plant.

According to Fig. 8.11 we need an absorption tower, a circulation liquid tank, some pipework and control equipment, and of course the off-gas and the absorbent. In case of absorption of acidic solutes, we provide pH control via admixture auf caustic reactant, a quality control (e.g., conductivity or density of the scrubbing liquid) and a level control. Figure 8.11 suggests a packed tower. In case of SO_2 absorption, packed towers are only recommended when NaOH, KOH or NH_3 is used as caustic reactant. See the location of the pH-measurement. Always locate the pH-electrode in the scrubbing liquid effluent of the column. You may also locate it in the circulation tank. Figure 8.11 also indicates that pH of the scrubbing liquid is highest on top of the scrubber. This may cause instantaneous sulfite precipitation and carbonate precipitation. Sulfite precipitation will limit the quality of the effluent. Carbonate precipitation limits the process economics of SO_2 scrubbing processes. Figure 8.11 also suggests a voluminous tank for the circulation liquid. It is recommended to offer the scrubbing liquid sufficient residence time in the circulation tank of half an hour or even more, because of a simple reason. The scrubbing liquid is definitely not equilibrated when it enters the circulation tank. It may contain unwanted solid sulfites and carbonates. High residence time in the circulation tank will offer an opportunity of minimizing the amount of these constituents by redissolving; a very slow reaction. The circulation tank may also be equipped with an oxidation air sparger (not shown in Fig. 8.11). The size of the circulation tank must also consider degassing of the scrubbing liquid to prevent the circulation pump from cavitation. (Process economics will not suffer from a huge circulation tank, but draw many advantages from this wise decision.)

We may make use of any kind of tower and tower internals, as shown in Figure 6.22 and presented in Section 6.3.4.3, but the choice of equipment very much depends on the process and the caustic reactant. For caustic reactants except NaOH, KOH or NH_4OH, you are strongly recommended to install single stage or multistage spray scrubbers to avoid blocking, fouling and scaling of internals by solids. Grid trays and structured packings can stand solids load to some extent. You may consider bubble cap trays or valve trays for absorbing strong acidic solutes such as HCl, but just without solids. In SO_2 absorption you basically may consider any chemical substance with caustic properties, such as NaOH, KOH, NH_4OH and Na_2CO_3. We talk about transition metal oxides such as ZnO, FeO or NiO and the oxides of the group of earth alkaline metals, such as MgO, CaO and their carbonates. Whether you believe it or not, discussing technologies and processes is just thermodynamics at a glance. We will go through the main steps of absorption with the caustic reactants mentioned above.

8.2.1 NaOH

The final product of SO_2 absorption with NaOH is (expected to be) $Na_2SO_{4,aqu}$. In extension of eqs. (8.41) and (8.42) we will form $NaHSO_3$ at $pH < 6.5$ and an increasing fraction Na_2SO_3 at $pH > 6.5$ (see Figure 8.6) in the absorption tower. Depending on the phase contact pattern and the oxygen content of the off-gas we will partially transfer the S-species from oxidation state S(IV) to S(VI). If we want to produce high-grade aqueous Na_2SO_4 we must finish oxidation of S(IV) in the scrubbing liquid circulation tank. The kinetics of oxidation will be appraised separately. Sodium sulfate has a very high water solubility. Figure 8.12 shows the temperature dependency of sodium sulfate solubility.

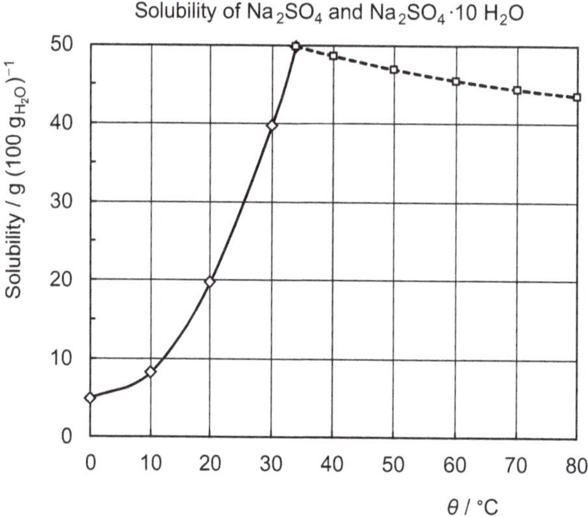

Fig. 8.12: Temperature dependency of Na_2SO_4 solubility in water [260].

Table 8.10 lists relevant physical properties of aqueous Na_2SO_4 solutions.

Tab. 8.10: Selected physical properties of aqueous Na_2SO_4 solutions at $T = 298$ K [261].

$w_{Na_2SO_4}$ (wt%)	$\rho_{H_2SO_4}$ (kg m^{-3})	$c_{Na_2SO_4}$ (g L^{-1})	$\mu_{Na_2SO_4}/\mu_{H_2O}$ (-)
0	999.97	0	1
1	1,008.9	10.3	1.024
3	1,025.2	30.8	1.089
5	1,043.6	52.2	1.161
10	1,090.5	109	1.387
16	1,150.3	184	1.808
20	1,192.8	238.1	2.223

According to the data of Tab. 8.10 the density might help in quality control of the process (QIC in Fig. 8.11). Table 8.10 also shows that viscosity rises with increasing Na_2SO_4 content. Latter property may have a negative effect on mass transfer in the liquid phase.

Oxidation of aqueous sodium sulfite should be an ease, but does not make any sense when the off-gas also contains CO_2, because in latter case you will preferably absorb the CO_2 when you run the process at pH > 8 (see Figure 8.6). Oxidation of

$NaHSO_3$, on the other hand, is an operation challenge. According to eq. (8.43) you cannot avoid SO_2 liberation during oxidation, except when you control the pH in the oxidation reactor (e.g., circulation liquid tank). If you operate scrubber and oxidation reactor (e.g., circulation liquid tank) with a hydraulic connection, you probably will even not register that. When you have equipped the oxidation reactor with a separate venting hose, you may slip into a compliance problem due to (unwanted) secondary SO_2 emission because of the following equation:

$$2\,HSO_3^- + 0.5\,O_2 \rightleftharpoons SO_4^{2-} + SO_2 + H_2O \tag{8.43}$$

8.2.2 MgO

MgO is prepared by calcining magnesite. The calcining temperature has a decisive effect on the reactivity of magnesium oxide. Calcining of magnesite is already complete at temperatures of 800 °C. At higher temperature, reactivity drops to finally break down due to sintering and dead burning of MgO.

MgO has a very low water solubility, while water solubility of the product of desulfurization, $MgSO_{4,aqu}$ is beyond 300 g L^{-1} [261]. When crystallizing, $MgSO_4$ may either form dihydrate or heptahydrate, depending on the temperature of crystallization.

8.2.3 ZnO, NiO and FeO

Water solubility of $NiSO_4$ and $FeSO_4$ is about 400 g L^{-1}, water solubility of $ZnSO_4$ is beyond 300 g L^{-1} [261]. When crystallizing, the transition metal sulfates form multihydrates. The type of hydrate again depends on the temperature. Application of transition metal oxides in off-gas desulfurization is rather limited to integrated processes in the metallurgical industry.

8.2.4 $CaCO_3$

Limestone has mainly been used in flue gas desulfurization of coal fired power plants. Usage of limestone is limited to very low operation pH, because of the multistep reaction path. Lime stone can be applied in slurries due to the negligible water solubility. After having dissolved in the scrubbing liquid the aqueous solute SO_2 will attack $CaCO_3$ to form dissolved hydrogen sulfite and gaseous CO_2. This is a very slow reaction. It needs huge residence time in the circulation tank, and oxidation of hydrogen sulfite at low pH is also slow, except when accelerated with catalysts.

8.2.5 $Ca(OH)_2$ (slaked lime)

Slaked lime is prepared by slaking calcined limestone. Its reactivity depends on the hydration time (the longer the better). It is widely used in desulfurization because of the product $CaSO_4$ or $CaSO_4 \cdot 2\,H_2O$ (again depending on the operation temperature of the scrubber and the circulation tank), which can be admixed to construction material when quality (sulfite content, contamination with Cl^- and heavy metal ions) permit application.

Usage of slaked lime does force engineering and operation of desulfurization plants into discussion and consideration of a multicomponent system, including the neutralizing reactant $Ca(OH)_2$, Ca^{2+} ions HSO_3^- ions, SO_3^{2-} ions, HSO_4^- ions, SO_4^{2-} ions as well as the solid products $CaCO_3$, $CaSO_3$ and $CaSO_4$ and the chemical interaction of these constituents. The challenge starts with the wetting of the neutralizing reactant $Ca(OH)_2$. The lime particle shall be penetrated with water before being fed to the scrubbing liquid. Nonsteady-state diffusion of liquids in solids is very slow. According to Crank [267], nonsteady-state diffusion of a solute with constant surface concentration C_0 into a spherical particle may be approached by the following equation:

$$\frac{C_t}{C_0} = 1 + 2 \cdot \frac{r_P}{r(t)} \cdot \sum_{n=1}^{\infty} \frac{(-1)^n}{n} \cdot \sin \frac{n \cdot \pi \cdot r(t)}{r_P} \cdot e^{-\frac{D \cdot n^2 \cdot \pi^2 \cdot t}{r_P^2}} \tag{8.44}$$

Figure 8.13 shows the water penetration depth of a lime particle with a particle diameter $d_P = 30\ \mu m$ after 24 h and after 48 h ($n = 10$ and $D = 1 \cdot 10^{-16}\ m^2\ s^{-1}$).

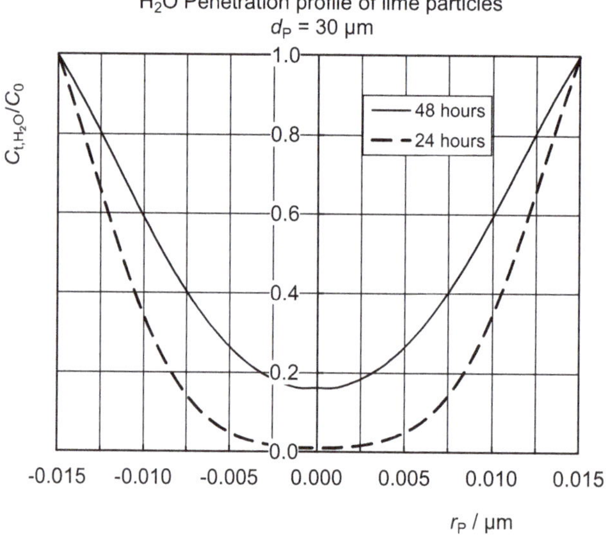

Fig. 8.13: Water penetration depth of a lime particle with $d_P = 30\ \mu m$ after 24 h and after 48 h ($n = 10$ and $D = 1 \cdot 10^{-16}\ m^2\ s^{-1}$).

The reactivity of lime particles is, depending on the aging of the slurry, more or less limited to the particle surface. The problem becomes worse when particles roll up to large clumps.

The effect of particle size on precipitation efficiency in dry off-gas desulfurization has been reported [268]. The effect of droplet size on desulfurization efficiency in absorption processes has also been reported [269]. Referring to Fig. 8.13, this performance limitation may be caused by limited accessibility of the lime particle in the droplet. It may also be caused by insulation of the droplet surface with $CaSO_3$. We have to keep in mind that the reactant $Ca(OH)_2$ and the primary products of desulfurization do have a very low water solubility. Table 8.11 lists the thermodynamic equilibrium constants for the major constituents of absorptive desulfurization with lime and their interaction with water [263]. Seriously conclude from the data of Tab. 8.11 that the target product $CaSO_4$ or gypsum ($CaSO_4 \cdot 2H_2O$) has the highest water solubility, while the bad guys $CaSO_3$ and $CaCO_3$ because of their low water solubility just have in mind to block all your pipework.

Tab. 8.11: Thermodynamic equilibrium constant K at 293 K for governing reactions in absorptive desulfurization with lime.

Reaction equation	K_{293}
$Ca(OH)_2 \rightleftharpoons Ca^{2+} + 2OH^-$	$4.376 \cdot 10^{-6}$
$CaSO_4 \rightleftharpoons Ca^{2+} + SO_4{}^{2-}$	$5.845 \cdot 10^{-5}$
$CaSO_3 \rightleftharpoons Ca^{2+} + SO_3{}^{2-}$	$4.467 \cdot 10^{-7}$
$CaCO_3 \rightleftharpoons Ca^{2+} + CO^{2-}$	$5.136 \cdot 10^{-9}$

From the first equation we may deduce the solubility of $Ca(OH)_2$ and the pH value of lime slurries according to the following equations:

$$C_{OH^-} = 2 \cdot \sqrt[3]{K_{293}} = 2 \cdot \sqrt[3]{4.376 \cdot 10^{-6}} = 0.033 \tag{8.45}$$

According to eq. (8.46) the corresponding pH value is

$$pH_{Ca(OH)_2} = 14 - \log_{10}(0.033) = 12.5 \tag{8.46}$$

When comparing the equilibrium constant of the products you will register for the solubility of constituents: $CaSO_4 > CaSO_3 > CaCO_3$. These properties may help explain the limited efficiency of lime-powered desulfurization processes, because the constituent with the largest amount in the off-gas, $CaCO_3$, has the lowest water solubility. The constituent formed at elevated pH, $CaSO_3$, stupidly has a much lower water solubility than the wanted product $CaSO_4$, thereby limiting and retarding the oxidation in the slurry due to its solid surface insulating nature, and $CaSO_3$ will readily contribute to blocking the pipework of your absorption tower by scaling.

8.2.6 Activity: the link between energy and the ability to "offer work"

At this point we have to make a quick side step, since we have mentioned thermodynamic equilibrium constants several times (also see Chapter 7).

When somewhere down the line between 1876 and 1878 Josiah Willard Gibbs published his treatise on the relationship between energy and the ability to provide work (how come, we talk about the same dimension) it truly was the sunrise in chemical engineering after the medieval darkness of alchimism. His gift and message to the off-gas absorption family includes a warning: Whenever you make use of thermodynamic data, for example in Tab. 8.11, you have to work with activities a_i of the species involved, with $a_i = X_i \cdot \gamma_\pm$. In the previous examples we lazily assumed $\gamma_\pm = 1$ (we hopefully will not be blamed for that). Never neglect the activity coefficient $\gamma_\pm$ for ionic strength $I > 0.0001$ mol kg^{-1}.

Let us briefly go through the algorithm, before we discuss the need of considering the activity coefficient in absorption calculations with an example. For details see the specific literature [270–273].

In dilute aqueous solutions the mean activity coefficient $\gamma_\pm$ is well approached by the Debye–Hückel limiting law, depicted in eqs. (8.48)–(8.52), with the mean activity coefficient $\gamma_\pm$ being calculated with the following equation:

$$\gamma_\pm = \sqrt[(m+n)]{\gamma_+^m \cdot \gamma_-^n} \tag{8.47}$$

The decadic logarithms of the ion specific activity coefficients γ_+ and γ_- are deduced from the prefix $A = 0.51$ (at $\theta = 25$ °C; for details see [270]), the oxidation state of the corresponding ion z_+ and z_- and the ionic strength I.

For $I < 0.001$ mol kg^{-1} we may apply the following equation:

$$\lg(\gamma_+) = -A \cdot z_+^{\,2} \cdot \sqrt{I} \tag{8.48}$$

$$\lg(\gamma_-) = -A \cdot z_-^{\,2} \cdot \sqrt{I} \tag{8.49}$$

The ionic strength is calculated via eq. (8.50). Always check your system for the ionic strength. It is not a hard job, but very helpful:

$$I = 0.5 \sum \left(z_i^2 X_i \right) \tag{8.50}$$

For $0.001 < I < 0.1$ mol kg^{-1} you may apply the following equations:

$$\lg(\gamma_+) = -A \cdot z_+^{\,2} \cdot \frac{\sqrt{I}}{1 + \sqrt{I}} \tag{8.51}$$

$$\lg(\gamma_-) = -A \cdot z_-^{\,2} \cdot \frac{\sqrt{I}}{1 + \sqrt{I}} \tag{8.52}$$

For $I > 0.1$ mol L^{-1}, kindly consider that at higher ionic strength of $I > 0.1$ mol kg^{-1}, the Debye–Hückel approach will fail, except when, for example, corrected with the Pitzer approach, the Guggenheim approach or the Meisner approach. We will make use of the empirical approach, as proposed by Bromley [274], shown in the following equation:

$$\lg(\gamma_\pm) = -A \cdot z_+ \cdot z_- \cdot \frac{\sqrt{I}}{1+\sqrt{I}} + (0.06 + 0.6 \cdot B_{MX}) \cdot z_+ \cdot z_- \cdot \frac{I}{\left(1 + \frac{1.5 \cdot I}{z_+ \cdot z_-}\right)^2} + B_{MX} \cdot I \qquad (8.53)$$

with the interaction parameter B_{MX}

$$B_{MX} = B_M + B_X + \delta_M \cdot \delta_X \qquad (8.53a)$$

Table 8.12 shows selected Bromley ion parameters B_M and B_X and the interaction parameters δ_M and δ_X for cations and anions.

Tab. 8.12: Bromley ion parameters B_M and B_X and interaction parameters δ_M and δ_X.

Cation	B_M	δ_M	Anion	B_X	δ_X
H^+	0.088	0.103	F^-	0.030	−0.930
Na^+	0.000	0.028	Cl^-	0.064	−0.067
K^+	−0.452	−0.079	OH^-	0.076	−1.000
Mg^{2+}	0.057	0.157	SO_4^{2-}	0.000	−0.400
Ca^{2+}	0.037	0.119	CO_3^{2-}	0.028	−0.670

Exercise 8.8: Do we really need this?

In the introduction of this chapter we agreed upon the problem that absorption may primarily transfer an off-gas problem into a wastewater problem. We may get the wastewater problem under control by separating the absorbed pollutant from circulation water by precipitation. We may, for example, dispatch SO_2 after dissolving in aqueous solution of lime ($Ca(OH)_2$), successful oxidation of the dissolved sulfurous acid species (HSO_3^- or SO_3^{2-}) with oxygen from air to obtain SO_4^{2-} in the scrubbing liquid, and precipitate SO_4^{2-} with Ca^{2+} to form $CaSO_4$ or $CaSO_4 \cdot 2H_2O$ (depending on the operation temperature). We ask thermodynamics for information about $CaSO_4$ solubility in water, since higher amounts of dissolved Ca^{2+} may cause severe scaling problems with SO_3^{2-}, and exemplarily we want to find out the effect of high Na_2SO_4-load (let us assume 1 mol kg^{-1} of the scrubbing liquid on $CaSO_4$ solubility). We do the job at 25 °C.

Solver:

We start with pure $CaSO_4$. With the law of mass action and the data of Tab. 8.11 we get with eq. (8.54), by assuming ideal solubility of $CaSO_4$, in the first step:

$$X_{CaSO_4}\left[\text{g kg}^{-1}\right] = \sqrt{K} \cdot MM_{CaSO_4} = \sqrt{5.845 \cdot 10^{-5}} \cdot 136 = 1.04 \text{ g kg}^{-1} \qquad (8.54)$$

The reading of [272] says: $X_{CaSO_4} = 2.09$ g kg^{-1}, equivalent to $7.46 \cdot 10^{-3}$ mol kg^{-1}. Seemingly we are just 100% off the experimental result.

Following the discussion above, we determine the ionic strength with eq. (8.5) in the second step with $z_+ = z_- = 2$:

$$I = 0.5 \cdot \Sigma X_i \cdot z_i^2 = 0.5 \cdot \left(7.46 \cdot 10^{-3} \cdot 2^2 + 7.46 \cdot 10^{-3} \cdot 2^2\right) = 3.06 \cdot 10^{-2}\ \text{mol kg}^{-1}$$

suggesting eqs. (8.51) and (8.52) for the determination of $\gamma_\pm$. With ($z_+ = z_- = 2$) we may simplify

$$\lg\gamma_+ = \lg\gamma_- = -0.51 \cdot z_+^2 \cdot \frac{\sqrt{I}}{1+\sqrt{I}} = -0.51 \cdot 2^2 \cdot \frac{\sqrt{3.06 \cdot 10^{-2}}}{1+\sqrt{3.06 \cdot 10^{-2}}} = -3.04 \cdot 10^{-1},$$

and after determining

$$\gamma_\pm = \sqrt[m+n]{\gamma_+^m \cdot \gamma_-^n} = \sqrt[2+2]{0.497^2 \cdot 0.497^2} = 0.497 \text{ with eq. (8.47)}$$

and reinserting the results in the law of mass action according to eq. (8.54) we obtain

$$X_{CaSO_4}\left[\text{g kg}^{-1}\right] = \sqrt{\frac{K}{(\gamma_\pm)^2}} \cdot MM_{CaSO_4} = \sqrt{\frac{5.845 \cdot 10^{-5}}{(0.497)^2}} \cdot 136 = 2.09\ \text{g kg}^{-1}, \tag{8.55}$$

and correspondingly the solubility of Ca^{2+}

$$X_{Ca^{2+}} = \frac{2.09 \cdot 40}{136} = 0.61\ \text{g kg}^{-1}$$

We just had to apply the Debye–Hückel theory for $0.001 < I < 0.1$ to meet the experimentally determined water solubility of $CaSO_4$ at $\theta = 25$ °C. Seemingly we may trust in these algorithms.

In a next step we want to discuss, whether Na_2SO_4 has an effect on Ca^{2+} solubility in water at 25 °C. We assume a Na_2SO_4-load of 1 mol kg^{-1}. Consider that $z_+ = 1$ and $z_- = 2$.

First step:
We assume that the ionic strength of Ca^{2+} and SO_4^{2-} is not significant compared to the ionic strength of Na_2SO_4:

$$I = 0.5 \cdot \Sigma X_i \cdot z_i^2 = 0.5 \cdot \left(2 \cdot 1 \cdot 1^2 + 1 \cdot 2^2\right) = 3\ \text{mol kg}^{-1}$$

Second step:
From Tab. 8.12 we get the data for determining B_{MX}:

$$B_{MX} = B_M + B_X + \delta_M \cdot \delta_X = 0.000 + 0.000 + 0.028 \cdot (-0.400) = -0.0112$$

Third step:
With eq. (8.53) we may now determine $\log\gamma_\pm$, but be careful, z_+ of $Ca^{2+} = 2$ and z_- of $SO_4^{2-} = 2$. Na_2SO_4 controls the ionic strength, but the activity acts on the target constituent $CaSO_4$. Do not mix that up:

$$\lg(\gamma_\pm) = -0.51 \cdot 2 \cdot 2 \cdot \frac{\sqrt{3}}{1+\sqrt{3}} + (0.06 + 0.6 \cdot (-0.112) \cdot 2 \cdot 2 \cdot \frac{3}{\left(1+\frac{1.5 \cdot 3}{2 \cdot 2}\right)^2} + (-0.112) \cdot 3 = -1.185$$

and $\gamma_\pm = 0.065$

The solubility of Ca^{2+} is then

$$X_{Ca^{2+}} = \frac{K}{X_{SO_4^{2-}} \cdot \gamma_\pm^2} = \frac{5.845 \cdot 10^{-5}}{1 \cdot 0.065^2} = 1.37 \cdot 10^{-2}\ \text{mol kg}^{-1} = 1.37 \cdot 10^{-2} \cdot 40 = 0.55\ \text{g } Ca^{2+}\ \text{kg}^{-1}$$

The experimental value for $X_{Ca^{2+}} = 0.63$ g kg^{-1}.

Discussion: At the chosen Na_2SO_4-load of $X_{NaSO_4} = 1$ mol kg^{-1} the load of dissolved calcium $X_{Ca^{2+}}$ is comparable with the results of $CaSO_4$ solubility in pure water.

If we would do this calculation for $\gamma_\pm = 1$, we would get

$$X_{Ca^{2+}} = \frac{K}{X_{SO_4{}^{2-}}} = \frac{5.845 \cdot 10^{-5}}{1} = 5.85 \cdot 10^{-5} \text{mol kg}^{-1} = 5.85 \cdot 10^{-5} \cdot 40 = 2.34 \cdot 10^{-3}\text{ g Ca}^{2+}\text{ kg}^{-1}$$

The difference of this result to the experimentally observed value does not need much of an explanation. It clearly shows why it is much recommended to operate with activity of constituents in absorption technology. You may check that for $I = 0.1$. The outcome of this experiment will tell you that solubility of $CaSO_4$ is at minimum for $I = 0.1$.

8.2.7 Double-alkali processes

To avoid the disadvantages of lime-based desulfurization, such as $CaSO_3$-scaling and $CaCO_3$-scaling, double-alkali processes have been intensively investigated in the 1970s and 1980s. According to eq. (8.55) the process principle is seemingly simple and invitingly:

$$\left(Na_2SO_4 + Ca(OH)_2 \rightleftharpoons CaSO_4 + 2\,NaOH\right)$$

$$SO_4{}^{2-} + Ca(OH)_2 \rightleftharpoons CaSO_4 + 2\,OH^- \tag{8.55}$$

Following this technology, SO_2 is absorbed in aqueous sodium hydroxide. After oxidation sodium sulfate, the effluent undergoes precipitation with lime to form $CaSO_4$ and the regenerated absorbent. $CaSO_4$ is separated from the regenerated absorbent by sedimentation and solid–liquid separation [244]. From a thermodynamics point of view, we have to combine the reactions of eqs. (8.56) and (8.57) to characterize this process. Thermodynamics will tell you that eq. (8.55) is an equilibrium reaction with a preference of the educts:

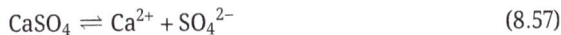

$$Ca(OH)_2 \rightleftharpoons Ca^{2+} + 2\,OH^- \tag{8.56}$$

$$CaSO_4 \rightleftharpoons Ca^{2+} + SO_4{}^{2-} \tag{8.57}$$

By substituting Ca^{2+} from eq. (8.56) into eq. (8.57) via the law of mass action you may isolate the OH^--load, ending up in an OH^--load of $X_{OH^-} < 0.2$ mol kg^{-1} [242], not a very attractive outcome, limiting application to low sulfur load of the off-gas.

8.2.8 Oxidation of sulfite and hydrogen sulfite

According to 1272/2008/EG sodium sulfite is not classified as water pollutant. The fish toxicity level LC50 is <460 mg L^{-1} and water hazard class is 1 (Germany, [275]).

Eurogypsum (http://www.eurogypsum.org) recommends limitation of calcium sulfite hemihydrate in FGD (flue gas desulfurization) gypsum to less than 0.5 wt%. FGD gypsum quality parameters are determined in accordance with the VGB Instruction Sheet "Analysis of FGD Gypsum" [276]. Therefore, wastewater from desulfurization must undergo oxidation, either performed by sparging air in the circulation tank (Fig. 8.11) or in a separate oxidizing reactor. As already mentioned we shall operate the desulfurization process in the hydrogen sulfite domain. According to eq. (8.43) oxidation of hydrogen sulfite will cause liberation of SO_2, demanding pH control to avoid secondary SO_2 emission. The Gmelin [241] already reports the loss of SO_2 during oxidation of aqueous $NaHSO_3$ by sparging air in the reaction broth [277]. From the data of this paper we may deduce the conclusion that hydrogen sulfite has attracted research more than hundred years ago. In these early days, researchers already registered that Fe^{2+}, Cu^{2+}, Co^{2+} and Mn^{2+} have a catalytic effect on oxidation of S(IV). They also registered that Fe^{2+} preferably catalyzed oxidation of hydrogen sulfite, and that the concentration of Fe^{2+} should not exceed 10^{-4} mol L^{-1}. At the edge of the 19th to the twentieth century, Fe^{2+}-catalyzed reactions in acidic environment were also investigated by John Horstman Fenton, after whom the so-called Fenton reactions are named nowadays. It is fascinating that the optimum Fe^{2+} concentration was identified on an empirical scale. We nowadays do have the data on the stability of $Fe(OH)^{n+}$ complexes available, and it is state-of-the-art knowledge that $Fe(OH)_3$ is readily precipitated at pH < 4, causing an immediate loss of catalytic activity in hydrogen sulfite oxidation, the main reason, why Fe^{2+}-catalyzed oxidation of hydrogen sulfite just needs a few milligrams of the catalyst per liter. Many lime-based neutralizers contain Fe^{2+} or Fe^{3+} in traces. Even hematite (Fe_2O_3) dissolves in aqueous hydrogen sulfite atmosphere according to eq. (8.58) to provide catalytic activity in hydrogen sulfite oxidation, probably even not registered by the staff of FGD plants:

$$Fe_2O_3 + HSO_3^- + 3H^+ \rightleftharpoons 2Fe^{2+} + SO_4^{2-} + 2H_2O \quad \Delta_R G^0_{293} = -53.8\,\text{kJ}\,\text{mol}^{-1}\,\text{Fe} \tag{8.58}$$

Catalyzed oxidation reactions of HSO_3^- are heterogeneous reactions, covering the whole span of reaction regimes between kinetic control and mass transfer control, as discussed in Chapter 7. In [278] uncatalyzed oxidation of Ca-hydrogensulfite in the kinetic regime is reported. Lancia et al. [279] investigated Mn^{2+}-catalyzed Ca-hydrogensulfite oxidation in a limestone slurry. They suggested a parallel oxidation rate of catalyzed and uncatalyzed hydrogen sulfite oxidation. S. Bengtssons and Bjerleb [280] investigated the rate of Co^{2+}-catalyzed sulfite oxidation in the kinetic regime. They proposed a rate law with reaction order 1.5 for sulfite and 0.5 for the catalyst, and they reported increasing rate with increasing pH value. These outline is intended to mention the broad field of investigation of S(IV) oxidation in FGD effluents. It is by far not representative, but will hopefully encourage you to screen the literature before stressing the lab. Anyhow, it is possible to completely oxidize S(IV) in the hydrogensulfite domain. Successful oxidation of S(IV) in the sulfite domain very much depends on the chemical environment.

Sodium sulfite scrubbing liquids perform well in caustic sulfite oxidation. It is not recommended to apply oxidation at elevated pH in lime-powered or limestone-powered FGD processes. You may end up with impressive $CaSO_3$ scaling. $CaSO_3$ is indeed a bad guy. It may even form in hydrogen sulfite environment to release SO_2 (you will check that with thermodynamics for $Ca^{2+} + 2HSO_3^- \rightleftharpoons CaSO_3 + H_2O + SO_2$).

Do you have an idea about the difference of "kinetically controlled" oxidation and "mass transfer-controlled" oxidation of S(IV) species? If not, it is much less complicated than it sounds in the very first moment. You can easily check that in the lab. You just need sufficient samples of fresh effluent (e.g., 1 L per experiment) in a glass beaker and an appropriate air sparger. You start sparging the broth with air, and you monitor it by taking samples after distinct time sequences, and you let the samples analyze for the chemical oxygen demand (the lab staff does know how to do the job). You finally get a concentration vs. time sketch for each experiment. Now it is your turn to draw the correct conclusions from the outcome by comparing the inclination of the concentration/time graph (actually you interpret the rate of conversion, as explained in detail in Chapter 7). The steeper the inclination is, the faster S(IV) was oxidized. You best start with a base experiment with sparging air with a specific flow rate. In the next experiment, you raise the air flow rate and compare the outcome with the first experiment. If you register a higher rate of conversion in the second experiment (steeper inclination of your concentration/time graph) your process is gladly mass transfer-controlled, and you may order the plant operators to speed up sparging. If you do not register any difference, you run the process in the kinetically controlled regime. Now it is your turn to think about admixing appropriate catalysts, such as Co^{2+} or Fe^{2+}. But be careful with dosing these miracle cure ingredients. We talk about traces (just a few mg L^{-1}). (Do not hesitate to go through Chapter 7 for more details about the magic of catalysis.)

8.3 Design of absorption processes

Design of absorption processes covers:
- the mass balance (and the energy balance),
- the choice of mass transfer equipment,
- the sizing of the mass transfer equipment,
- the specification of peripheral equipment and
- cost estimation and plant optimization.

We will have a focus on the mass balance and the choice and sizing of equipment. For convenience we assume isothermal operation (e.g., we do not consider neutralization enthalpy).

Sizing can be based on the procedure listed:
- complete specification of the off-gas problem,
- selection of the absorbent,

- specification of the process data (gaseous feed, clean gas, absorbent feed),
- determination of the gas–liquid equilibrium,
- specification of mass (and energy) balance,
- phase ratio,
- phase contact pattern,
- selection of equipment,
- choice of the correct design concept (theory of separation stages, HTU-NTU concept) and
- sizing of equipment.

8.3.1 Mass balances

Let us introductorily agree upon the necessity of clear specification of the units, because the mass balance simply demands that the total amount of mass entering the absorption tower will leave the absorber (no accumulation within the absorber). Mass may be transferred from the gas phase to the liquid phase and vice versa, but you cannot make mass vanish. Volumes and volumetric flows, especially of the gas phase, may change. Molar flows of constituents may change due to the magic of chemical conversion, which has to be considered (see Chapter 7).

For your convenience we list and explain the symbols before we start with the balances. For not getting lost in specifying the type of flow (mass, volumetric or molar), the phase and the base (mass, molar or volumetric) we simplify the symbols for the specific needs in absorption technology (the equivalent standard symbols are listed in brackets). Let us start with the rules of the game, symbols and abbreviations and their equivalents to the standard list of symbols and abbreviations.

Symbols and equivalent symbols:

G ($F_{\mathrm{m,g}}$ or $F_{\mathrm{n,g}}$)	Gas flow rate in kg h^{-1} or kmol h^{-1}
G ($F_{\mathrm{V,g}}$)	Volumetric gas flow rate in $\mathrm{m}^3\ \mathrm{h}^{-1}$
L ($F_{\mathrm{m,l}}$ or $F_{\mathrm{n,l}}$)	Absorbent flow rate in kg h^{-1} or kmol h^{-1}
L ($F_{\mathrm{V,l}}$)	Volumetric liquid phase flow rate $\mathrm{m}^3\ \mathrm{h}^{-1}$
x,y	Molar fraction or volume fraction or mass fraction (x is reserved for the liquid phase and y for the gas phase)
X,Y	Specific molar load in mol mol^{-1} or mol kg^{-1}; and specific mass load (X is reserved for the liquid phase and Y for the gas phase)

To explain the reason for this simplification, let us just try to specify the mass flow of the gaseous solute on top of the column: We just specify it with G, in standard notation we would have to name it $F_{\mathrm{m,A,g,T}}$ (a little bit much of subscripts).

Subscripts:
A substance to be absorbed (solute)
I carrier gas (= inert gas)
T top of the absorption tower
B bottom of the absorption tower

Specification of the gaseous phase:

$$G_T = G_I + G_{A,T} \quad \text{with} \quad G_{A,T} = G_T \cdot y_{A,T} \tag{8.59}$$

All the gas entering the absorption tower on top T consists of the inert gas I and the solute A. The amount of solute flow $G_{A,T}$ correlates with the fraction y_A of the total gas flow G_T eq. (8.59):

$$G_B = G_I + G_{A,B} \quad \text{with} \quad G_{A,B} = G_B \cdot y_{A,B} \tag{8.60}$$

All the gas leaving the absorption tower at the bottom B consists of the inert gas I and the solute A. The amount of solute flow $G_{A,B}$ correlates with the fraction $y_{A,B}$ of the total gas flow G_B (eq. (8.60)).

Specification of the absorbent phase:

$$L_T = L_I + L_{A,T} \quad \text{with} \quad L_{A,T} = L_T \cdot x_{A,T} \tag{8.61}$$

All the liquid phase (absorbent) entering the absorption tower on top T consists of the carrier liquid I and the solute A. The amount of solute flow $L_{A,T}$ correlates with the fraction $x_{A,T}$ of the total liquid phase flow L_T (eq. (8.61)):

$$L_B = L_I + L_{A,B} \quad \text{with} \quad L_{A,B} = L_B \cdot x_{A,B} \tag{8.62}$$

All the liquid phase leaving the absorption tower at the bottom B consists of the carrier liquid I and the solute A. The amount of solute flow $L_{A,B}$ correlates with the fraction $x_{A,B}$ of the total liquid flow L_B (eq. (8.62)).

With the mass flow rate or molar flow rate of the solute A to be absorbed (eq. (8.63)):

$$G_T \cdot y_{A,T} = \frac{G_T \cdot G_{A,T}}{G_T} = G_{A,T} = \frac{G_{A,T} \cdot G_I}{G_I} = G_I \cdot Y_{A,T} \tag{8.63}$$

and eqs. (8.64), (8.65) and (8.66), respectively:

$$G_B \cdot y_{A,B} = G_I \cdot Y_{A,B} \tag{8.64}$$

$$L_T \cdot x_{A,T} = L_I \cdot X_{A,T} \tag{8.65}$$

$$L_B \cdot x_{A,B} = L_I \cdot X_{A,B} \tag{8.66}$$

we have specified all the entering constituents and all the exiting constituents.

Now the mass balance for the species A to be absorbed, based on the specific load of the solute, can be generated, but we have to distinguish between co-current flow

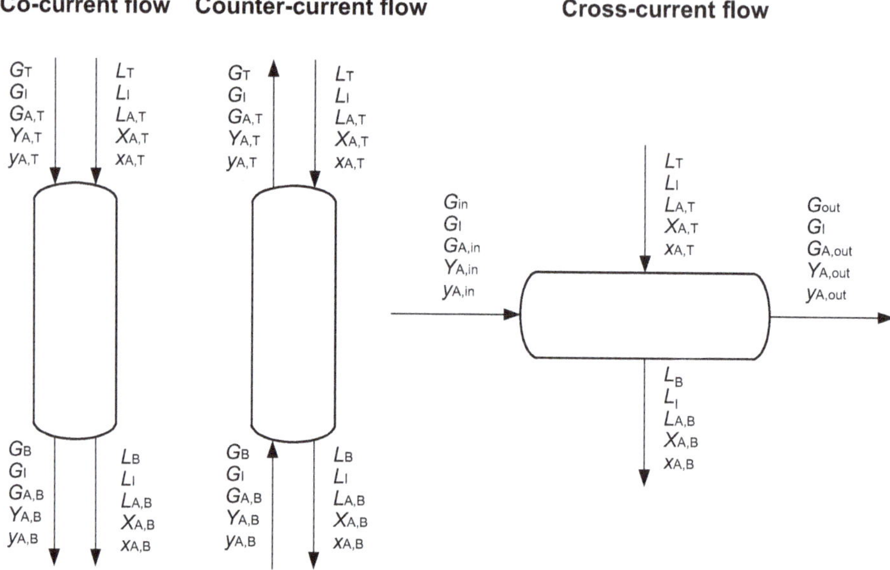

Fig. 8.14: Contact patterns and balance data.

pattern and counter-current flow pattern, shown in the sketch of Fig. 8.14. Figure 8.14 shows the industrially applied flow patterns. We will limit the discussion to co-current and counter-current flow.

For co-current operation (both phases are fed on top of the column and they leave the column at the bottom) we correspondingly obtain the overall mass balance:

$$G_T + L_T = G_B + L_B \tag{8.67}$$

and the mass balance of the solute to be transferred from the gas phase into the liquid phase:

$$Y_{A,T} \cdot G_I + X_{A,T} \cdot L_I = Y_{A,B} \cdot G_I + X_{A,B} \cdot L_I \tag{8.68}$$

After separation of variables and rearrangement for L/G, we get the inclination of the operation line:

$$G_I \cdot (Y_{A,T} - Y_{A,B}) = L_I \cdot (X_{A,T} - X_{A,B}) \tag{8.69}$$

$$-\frac{L}{G} = \frac{Y_{A,T} - Y_{A,B}}{X_{A,T} - X_{A,B}} \tag{8.70}$$

For counter-current operation, the mass balance is given by the following equation:

$$G_B + L_T = G_T + L_B \tag{8.71}$$

and respectively the constituent balance by the following equation:

$$Y_{A,T} \cdot G_I + X_{A,B} \cdot L_I = Y_{A,B} \cdot G_I + X_{A,T} \cdot L_I \quad (8.72)$$

resulting in the corresponding inclination of the operation line:

$$\frac{L}{G} = \frac{Y_{A,B} - Y_{A,T}}{X_{A,B} - X_{A,T}} \quad (8.73)$$

Figure 8.15 shows the load specification of the solute A for cross-current and for counter-current phase contact pattern. Figure 8.15 also shows the operation line with the phase contact pattern specific *L*/*G* ratio. Have a look at the scale of abscissa and ordinate. Since the scale is specified in terms of g kg^{-1}, the inclination shows the *L*/*G* ratio in kg absorbent kg^{-1} inert gas. Also compare the difference in load for counter-current and co-current operation.

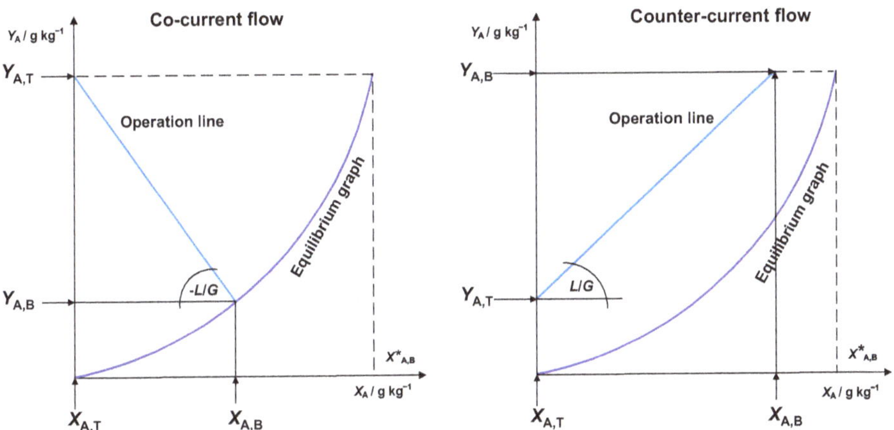

Fig. 8.15: Comparison of balance data, the operation line and the inclination of the operation line for co-current and for counter-current operation of absorption processes.

In case of co-current operation, the amount of transferred solute decreases with decreasing phase ratio of the liquid to the gas phase (−*L*/*G* becomes smaller because of less absorbent fed to the absorption tower for same flow rate of gas), while the concentration of the transferred solute in the liquid phase increases. Extrapolated to *L*/*G* = 0 the feed concentration of the solute to be transferred is in equilibrium with the maximum load of the transferred solute in the liquid phase $X^*_{A,B}$ (dashed lines in Fig. 8.15). In case of counter-current operation the operation line will intersect the equilibrium graph at minimum *L*/*G*. Again the gas phase load $Y_{A,B}$ is in equilibrium with the liquid phase load $X^*_{A,B}$ (dashed lines in Fig. 8.15 for counter-current operation).

Exercise 8.9: How to make use of these balance equations?

The fridge of a food market is powered with the coolant NH_3. In case of emergency due to leaking coolant pipework the venting air $F_{V,air} = 1{,}000\ m^3_{STP}\ h^{-1}$ may contain an ammonia load of 400 mg kg^{-1}. It must be detoxified to a maximum load of $Y_{NH_3,max} = 10$ mg kg^{-1} in the exiting air by absorbing the ammonia in water in an ejector spray scrubber (see Fig. 6.23). The contaminated water is drained to a municipal wastewater facility. Temperature of the water saturated venting air is $\theta = 12$ °C. You have to determine the minimum amount of water to be fed to the spray scrubber.

Solver:

We talk about a co-currently operated scrubber, the balance setup of which is shown in Fig. 8.15. With eq. (8.31) we do have access to the Henry constant, and with the algorithm discussed in Exercise 8.7 we are able to construct the gas–liquid equilibrium graph for the system NH_3/air/water. The equilibrium line can be transferred into the linear correlation $Y_{NH_3} = 0.3657 \cdot X_{NH_3}$. Via the algorithm discussed in Exercise 2.1, we can transfer $F_{V,air}$ into G ($F_{V,air} \cdot \rho_{STP} = G = 1{,}000 \cdot 1.286 = 1{,}286$ kg air h^{-1}).

For $Y_{NH_3,B} = 0.3657 \cdot X_{NH_3,B}$, we get: $10 = 0.3657 \cdot X_{NH_3,B}$ and $X_{NH_3,B} = 27.34$ mg NH_3 kg^{-1} H_2O

Now we have to apply eq. (8.70):

$$-\frac{L}{G} = \frac{Y_{NH_3,T} - Y_{NH_3,B}}{X_{NH_3,T} - X_{NH_3,B}} = \frac{400 - 10}{0 - 27.3} = \frac{390}{-27.3} = -14.26 \text{ kg } H_2O \text{ kg}^{-1}\text{air (with } L/G = +14.26 \text{ kg kg}^{-1})$$

to finally get L: L [kg H_2O h^{-1}] = 14.26 [kg H_2O kg^{-1} air] · G [kg air h^{-1}] = 1,286 · 14.26 = 18,348 kg H_2O h^{-1}

Discussion: Just correctly apply algorithms and do not get confused by eq. (8.70).

In counter-current operation (see Fig. 8.15) the number of mass transfer stages increases with decreasing phase ratio.

When the operation line and the equilibrium function (line) intersect, the load of the transferred substance in the absorbent ($X_{A,B}$) will be at maximum ($X_{A,B} = X^*_{A,B}$), which is the equilibrium load in the absorbent phase. $X^*_{A,B}$ will correspond with the gaseous feed load $Y_{A,T}$, but the number of transfer stages will be infinite. The minimum amount of absorbent is derived from this (hypothetic) state of operation. For practical application the amount of absorbent shall be 1.5 times to 2 times the minimum amount $(L/G)_{eff} = 1.5 \cdot (L/G)_{min}$ to $2 \cdot (L/G)_{min}$. Absorption with chemical reaction may need different values of (L/G) due to specific process conditions and different gas–liquid equilibrium characteristic.

Exercise 8.10: Curious of counter-current operations?

Being fascinated by the ease of process balancing you now want to find out, what would happen if you would operate your emergency absorption black box of Exercise 8.9 in counter-current mode according to the sketch in Fig. 8.14. You also want to figure out, whether counter-current operation would provide any savings, and how an actual phase ratio of $(L/G)_{actual} = 2 \cdot (L/G)_{min}$ acts on load X.

Solver:

a) Comparison of operation for same L/G-ratio

We start with eq. (8.73) by considering the different balance setup for counter-current operation. We have fixed the L/G -ratio with $L/G = 14.26$, $Y_{NH_3,T} = 10$ mg kg^{-1}, $Y_{NH_3,B} = 400$ mg kg^{-1} and $X_{NH_3,T} = 0$ (hopefully the fresh water is not poisoned with NH_3), leaving the unknown $X_{NH_3,B}$ to become calculated from eq. (8.73):

$$\frac{L}{G} = \frac{Y_{NH_3,B} - Y_{NH_3,T}}{X_{NH_3,B} - X_{NH_3,T}}\ 14.26 = \frac{400-10}{X_{NH_3,B}-0}, \text{ and } X_{NH_3,0} = 27.34 \text{ mg } NH_3 \text{ kg}^{-1} H_2O$$

Discussion: Any difference? No.

b) Determination of $(L/G)_{min}$
We let the L/G ratio drop for fixed = 10 mg NH_3 kg^{-1} H_2O and $Y_{NH_3,B}$ = 400 mg NH_3 kg^{-1} H_2O until the operation line will intersect the equilibrium line, with the equilibrium line being specified by $Y_{NH_3,B} = 0.3657\ X_{NH_3,B}$: ·
$Y_{NH_3,B} = 0.3657 \cdot X_{NH_3,0}$ and $400 = 0.3657 \cdot X_{NH_3,0}$with $X_{NH_3,0}$ = 1,093 mg NH_3 kg^{-1} H_2O, and

$$\left(\frac{L}{G}\right)_{min} = \frac{Y_{NH_3,B} - Y_{NH_3,T}}{X_{NH_3,B} - X_{NH_3,T}} = \frac{400-10}{1{,}093.8-0}$$

As a consequence, we will need much less water at minimum L/G ratio, with

$$L = 0.36 \cdot G = 0.36 \cdot 1{,}286 = 459 \text{ kg } H_2O \text{ h}^{-1}$$

Discussion: Isn't the significant difference of $X_{NH_3,B}$ = 1,093 mg NH_3 kg^{-1} H_2O at $(L/G)_{min} = 0.36$ impressive compared to just 27.34 mg NH_3 kg^{-1} H_2O in co-current operation?

c) Determination of $X_{NH_3,B}$ for $(L/G)_{eff} = 2 \cdot (L/G)_{min}$
$(L/G)_{eff} = 2 \cdot (L/G)_{min} = 2 \cdot 0.36 = 0.72$, and

$$\left(\frac{L}{G}\right)_{eff} = \frac{Y_{NH_3,B} - Y_{NH_3,T}}{X_{NH_3,B} - X_{NH_3,T}} = 0.72 = \frac{400-10}{X_{NH_3,B}-0}, \text{ and } X_{NH_3,B} = 542 \text{ mg } NH_3 \text{ kg}^{-1} H_2O$$

Discussion: Still impressive compared to co-current operation.

8.3.2 Selection of equipment

Selection of equipment depends on the flow pattern and the hydraulic design boundaries of the total amount of absorbent and gaseous phase. Selection further has to distinguish between the type of phase contact and the mass transfer regime, as shown in Tab. 8.13.

Tab. 8.13: Equipment, phase contact pattern and flow regime.

Type of apparatus	Phase contact	Regime
Tray columns Bubble cap- Sieve- Fluidized bed	Discontinuous (stage-wise)	Bubble
Packed columns Packed bed Structured packings	Continuous	Film
Film absorber	Continuous	Film
Bubble column	Continuous	Bubble

Tab. 8.13 (continued)

Type of apparatus	Phase contact	Regime
Agitated vessel	Continuous	Bubble
Spray tower	Continuous	Droplet
Venturi scrubber	Continuous	Droplet

8.4 Apparatus design for absorption towers with stage-wise phase contact: apparatus diameter and apparatus height

Mass transfer equipment is specified by the type of phase contact which is either continuous or stage-wise (see Fig. 6.22). Figure 8.16 shows the hydraulics of a bubble cap column, representative for stage-wise phase contact. Figure 8.16 already suggests the need of well-designed gas flow as well as liquid phase flow to keep the bubble cap tray in an appropriate hydraulic operation range. Figure 8.17 shows a schematic sketch of

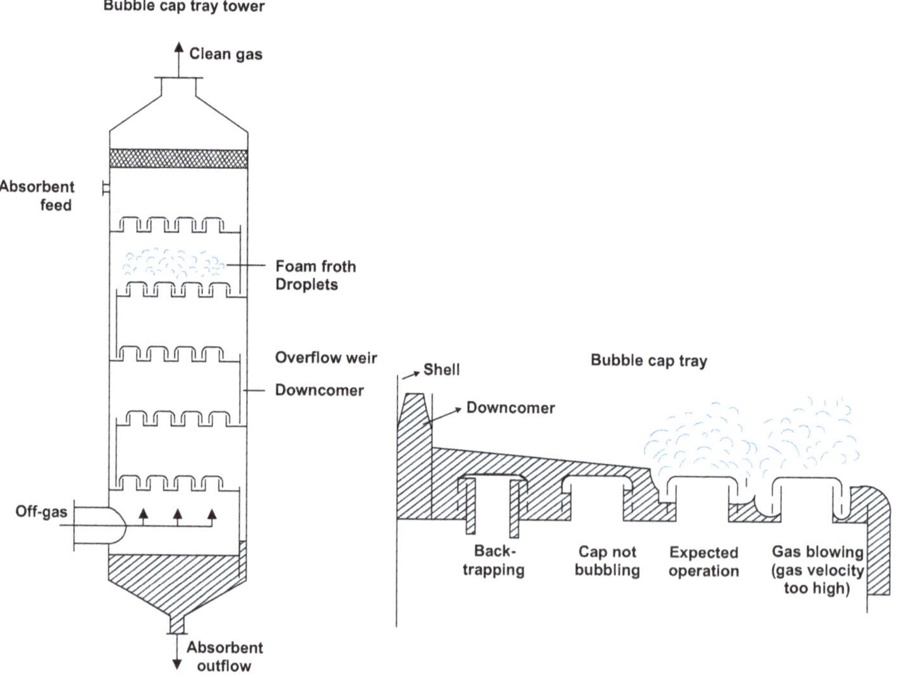

Fig. 8.16: Bubble cap tray column and bubble cap tray with different hydraulic loads.

the operation range of tray columns, specifically explained with the operation range of bubble cap trays. Apparatus design of course has to consider the specific needs of phase contact. The design of absorption towers with stage-wise phase contact is based on the theory of separation stages. In case of continuous phase contact the mass transfer concept is applied. Literature offers plenty of detailed design instructions. If ever you have to construct an absorption tower, trust the algorithms published in [237, 240]. To suffice our needs, we will go through short cut methods for the design of tray-type absorption towers for stage-wise phase contact and the design of packed towers for continuous phase contact, and we will keep an eye on spray contactors.

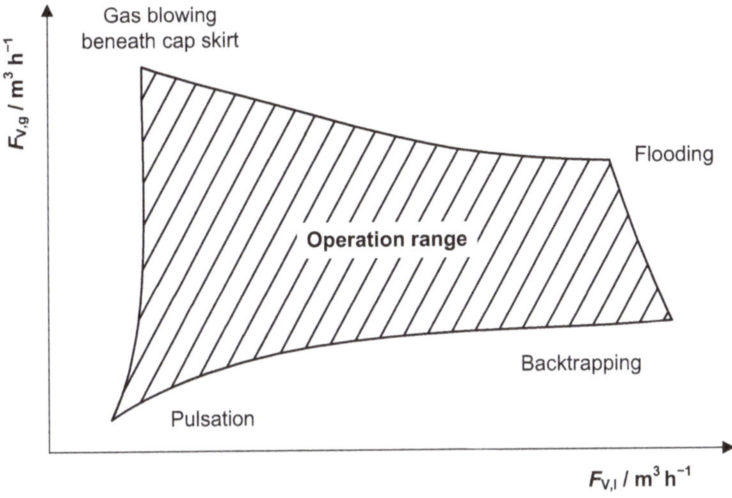

Fig. 8.17: Operation range of tray columns in general, specifically bubble cap columns.

8.4.1 The hydraulics of stage-wise phase contact

The hydraulic design of tray columns is based on comparing gravitational force F_G, buoyancy force F_B and flow resistance force F_D acting on a single droplet at turbulent flow conditions in the gas phase above the tray. At hydraulic equilibrium ($F_D = F_G - F_B$) a specific particle, for example a droplet, will remain in the gas phase at a specific level. When we increase the gas flow rate beyond this steady state velocity (the gas velocity v_g is higher than the settling velocity of our droplet v_{dr}) the droplet will move upward to the next tray and accumulate there to finally flood the tray. Figure 8.18 shows how to derive the force balance for a single droplet with diameter d_{dr} in the gas space at turbulent gas flow rate between two neighboring trays and therefore, $C_{w,turb} = 0.4$.

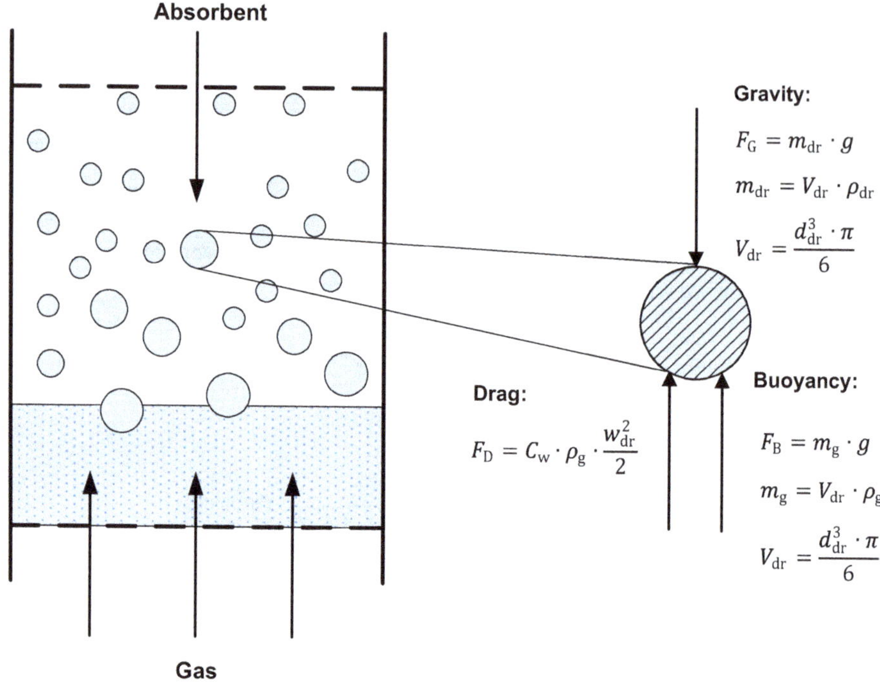

Fig. 8.18: Schematic sketch to set up the force balance for a single droplet with diameter d_{dr} in the gas space at turbulent gas flow rate between two neighboring trays and therefore, $C_{w,turb} = 0.4$.

The force balance can be derived as follows:

$$\frac{d_{dr} \cdot \pi \cdot \rho_g \cdot v_g^2 \cdot C_w}{8} = \frac{d_{dr} \cdot \pi \cdot g \cdot (\rho_l - \rho_g)}{6}$$

$$v_g = \sqrt{\frac{4 \cdot d_{dr} \cdot g \cdot (\rho_l - \rho_g)}{1.2 \cdot \rho_g}} = \frac{1}{\sqrt{\rho_g}} \sqrt{\frac{4 \cdot d_{dr} \cdot g}{1.2}} \cdot \sqrt{(\rho_l - \rho_g)}$$

From this force balance the hydraulic design of staged phase contactors such as bubble cap trays, valve trays or sieve trays has been deduced. Souders and Brown [281] are great names in tray hydraulics. We will trust their merits.

The last line of the derivation shows the important outcome of the force balance for turbulent gas flow, acting on our droplet, explicitly outlined in the following equation:

$$v_g = \frac{1}{\sqrt{\rho_g}} \sqrt{\frac{4 \cdot d_{dr} \cdot g}{1.2}} \cdot \sqrt{(\rho_l - \rho_g)} \tag{8.74}$$

We rearrange eq. (8.74) for the expression in eq. (8.75) and rename the right-hand side part of eq. (4.75) with F like capacity factor:

$$v_g \cdot \sqrt{\rho_g} = \sqrt{\frac{4 \cdot d_{dr} \cdot g}{1.2}} \cdot \sqrt{\left(\rho_l - \rho_g\right)} = F$$

or

$$F = v_g \cdot \sqrt{\rho_g} \tag{8.75}$$

For fixed gas velocity and fixed physical properties of the liquid and the gas phase our droplet with diameter d_{dr} will remain in the position of the gas phase between neighboring trays where we have placed it, when we suffice the balance of eqs. (8.74) or (8.75). In eq. (8.76) we just rename the left-hand side expression of eq. (8.75) with F_m like maximum capacity factor:

$$F_m = \sqrt{\frac{4 \cdot d_{dr} \cdot g}{1.2}} \cdot \sqrt{\left(\rho_l - \rho_g\right)} \tag{8.76}$$

For successful application in tray design we have got just one problem: We do not know the droplet diameter d_{dr}. At this crucial point, Souders and Brown come into play. To find a way out of this problem, they isolated the droplet related part of eq. (8.76) and renamed it with load factor k_V, shown in eq. (8.77):

$$k_V = \sqrt{\frac{4 \cdot d_{dr} \cdot g}{1.2}} \tag{8.77}$$

and they did perform a series of experiments (you may assume several hundreds of experiments) in test columns with the test system air/water to gain experimental correlations for the k_V value of different tray types with “unknown droplet size and droplet size distribution,” such as

– Bubble cap trays in eq. (8.78), with Δz being the tray distance in mm, z_g the height of the bubble cap in mm and d_G the diameter of the bubble cap in mm

$$k_V = 0.05 \cdot \frac{\sqrt{\left(\Delta z - z_g\right)}}{\sqrt[3]{d_G^2}} \tag{8.78}$$

– Sieve trays in eq. (8.79) with the hole diameter d_B

$$k_V = 0.0045 \cdot \frac{\sqrt{(\Delta z)}}{\left(d_B\right)^{0.2}} \tag{8.79}$$

– and Koch valve trays (tray distance: 400 mm) in the following equation:

$$k_V = \frac{4 \cdot V_g \cdot \sqrt{\left(\frac{\rho_g}{(\rho_l - \rho_g)}\right)}}{D^2 \cdot \pi} \tag{8.80}$$

This brilliant idea of hiding the unknown droplet size in the k_V value was a breakthrough in tray column design. With the appropriate k_V value at hand we just need to determine the column diameter via gas flow $F_{V,g}$ and the gas velocity according to the adjusted equation:

$$v_{gas} \cdot \sqrt{\rho_g} = k_V \cdot \sqrt{(\rho_l - \rho_g)} \tag{8.76a}$$

The column diameter D in m of tray-type absorption columns is then calculated from the actual maximum gas flow rate $F_{V,g,max}$ in $m^3\ s^{-1}$ and the effective (actual) gas velocity v_{eff} in $m\ s^{-1}$ according to the following equation:

$$D = \sqrt{\frac{4 \cdot F_{V,g,max}}{v_g \cdot \pi}} \tag{8.81}$$

Let us summarize the plan of action for successful tray design:

First step: Tray selection
Second step: Calculation of k_V with eqs. (8.77), (8.78), (8.79) or (8.80)
Third step: Calculation of F_m with eq. (8.76)
Fourth step: Calculation of v_g with eq. (8.75)
Fifth step: Calculation of D with eq. (8.81)

This brilliant job was already done by Souders and Brown [281] nearly one hundred years ago, and the algorithm still performs well. If you are interested in detailed design of tray columns, try to get access to the famous Hoppe-Mittelstrass [282]. They also worked hard on the subject. Several software packages on tray column design are based on the fundamental investigations of K. Hoppe and M. Mittelstrass. We must not forget to praise the merits of J. Stichlmair, who also did a great job on tray type hydraulics in his PhD thesis [261].

The Souders and Brown algorithm and the Stichelmair algorithm (not discussed in this chapter) may cover your needs in tray column design. For further details in tray column design see for example [237, 240].

Exercise 8.11: Let the Souders and Brown algorithm show up

The fridge of a food market powered with the coolant NH_3, already tried to heal in Exercises 8.9 and 8.10, still has got a problem. What is the problem? In case of emergency due to leaking coolant pipework the venting air $F_{V,air} = 1{,}000\ m^3_{STP}\ h^{-1}$ may contain an ammonia load of 400 mg kg^{-1}. It must be detoxified to a maximum load of $Y_{NH_3,max} = 10$ mg kg^{-1} by absorbing the ammonia in water in a tray column. The contaminated water is drained to a municipal wastewater facility. Temperature of the water saturated venting air is $\theta = 12$ °C and $P = 1{,}013$ hPa. Being the vendor of "best equipment supply" you have to give the customer an idea about the diameter of the column (the customer will also need the height of the tower to find an appropriate location for installing the equipment in the vicinity of the fridge). Density of water is $\rho_{H_2O} = 1{,}000$ kg m^{-3} and density of air ρ_{air} at 12 °C is $\rho_{air} = 1.23$ kg am^{-3} and $F_{V,max}\left[\text{am}^3\ \text{h}^{-1}\right] = \frac{1{,}000 \cdot (273.15 + 12)}{273.15}$.

Solver:
First step:
We decide for a sieve tray column and test eq. (8.79) for applicability for:

$$\Delta z = 300 \text{ mm and } d_B = 8 \text{ mm}$$

The "free cross-sectional area" is $A_{free} = 8\%$ of the tray area.

Second step:

$$k_V = 0.0045 \cdot \frac{\sqrt{\Delta z}}{(d_B)^{0.2}} = 0.0045 \cdot \frac{\sqrt{300}}{(8)^{0.2}} = 0.051$$

Third and fourth steps:
Now we may make eqs. (8.75) and (8.76) work:

$$v_g = k_V \cdot \frac{\sqrt{(\rho_l - \rho_g)}}{\sqrt{\rho_g}} = 0.051 \cdot \frac{\sqrt{1{,}000 - 1.23}}{1.23} = 1.46 \text{ m s}^{-1}$$

This is the gas velocity v_g over the whole cylindrical cross-sectional area of our column.

Fifth step:
We determine the diameter of our column with eq. (8.81):

$$D = \sqrt{\frac{4 \cdot F_{V,g,max}}{v_g \cdot \pi}} = \sqrt{\frac{4 \cdot 1{,}043.9}{1.46 \cdot \pi \cdot 3{,}600}} = 0.5 \text{ m}$$

Discussion: With 0.5 m in diameter the scrubber will not block much of the ground floor area. How do you think about the gas velocity? We have mentioned above that the free cross-sectional area is 8% of the tray area, resulting in a total area of holes of $A_{holes} = 0.016$ m^2. This area must be passed by both phases. Neglecting the liquid phase flow (we still do not know it) we talk about a minimum gas velocity of

$$v_{g,holes} = \frac{F_{V,max}}{A_{holes}} = \frac{1{,}043.9}{3{,}600 \cdot 0.016} = 18.3 \text{ m s}^{-1}, \text{ and that is a lot.}$$

8.4.2 The theory of separation stages for determining N_{th}

Now, you may probably miss instructions for the design of the height of tray towers. Do not worry, we have already done the hard graft with our mass balances in Section 8.3.1 and will now finish the job with Exercise 8.12 after having prepared appropriate (hopefully not confusing) instructions.

Apparatus design of absorption processes with stage-wise phase contact is based on the theory of separation stages. A separation stage is defined to be a mixing unit of limited size. Both phases enter this mixing unit. After mixing, both phases separate and leave the mass transfer stage at equilibrium. The number of mixing stages N_{th} times the height between the mixing stages results in the "net active height" of the absorption tower.

Design strategy:

1. Solvent selection
2. Determination of the gas–liquid equilibrium
3. Evaluation of the mass balance
4. Calculation of the minimum amount of solvent
5. Calculation of the actual amount of solvent
6. Determination of the theoretical number of trays N_{th}
7. Estimation (calculation) of the separation efficiency
8. Calculation of the recommended number of trays
9. Determination of the "net height" of the absorption tower

To explain "net height": We talk about the number of trays and the corresponding height between trays, but not about the feeder section for the absorbent inflow on top of the column, the demister section, the absorbent collector at the bottom and the clean gas outflow section. With your hand calculator at hand we want to get the number of "ideal" or "theoretical" separation stages N_{th} fixed in a simple step-by-step procedure after having fixed the process specification.

Specification:
$Y_{A,B}$, $Y_{A,T}$: The inflow load $Y_{A,B}$ and the outflow load $Y_{A,T}$ of the off-gas are specified, and the gas flow rate G (= $F_{V,g}$) is fixed. The gas–liquid equilibrium $Y^* = f(X)$ is specified. The inflow load of the absorbent phase $X_{A,T}$, is fixed (e.g., $X_{A,T} = 0$). In case of physiosorption, we make use of the Henry law in molar fraction y or molar load Y (Section 8.1.4):

$$y_A^* = \frac{H_A}{P} \cdot x_A \tag{8.8}$$

or

$$\frac{Y_A^*}{1+Y_A^*} = \frac{H_A \cdot X_A}{P \cdot (1+X_A)} \tag{8.11}$$

For convenience, you may also apply a power-law approach or a polynomial approach as long as these empirical approaches compare well with the equilibrium data set (but, be careful with polynomial approaches).

Solver:
First step:
For specified feed gas load $Y_{A,B}$ you have to determine the intersection with the equilibrium function $Y^* = f(X) \cdot X_{A,B}$

Second step:
With the specified gas load $Y_{A,B}$ and $Y_{A,T}$, and the specified absorbent load $X_{A,T}$ and the maximum absorbent load $X_{A,B}$, as calculated via equilibrium function you can determine the minimum liquid phase flow rate L_{min} with the following equation:

$$L_{min} = \frac{G \cdot (Y_{A,B} - Y_{A,T})}{X_{A,B} - X_{A,T}} \tag{8.73}$$

Third step:
With the selected or recommended multiple of the $(L/G)_{min}$ ratio (f = 1.5 to 2 . . .) you determine $(L/G)_{eff}$:

$$\left(\frac{L}{G}\right)_{eff} = \frac{L_{eff}}{G} = f \cdot \left(\frac{L}{G}\right)_{min} \tag{8.82}$$

Fourth step:
At this level we have specified $Y_{A,B}$ and $Y_{A,T}$, $X_{A,T}$ and $(L/G)_{eff}$, and we can calculate the outflow load of the absorbent $X_{A,B,1}$ by rearranging the balance for counter-current operation:

$$X_{A,B,1} = \frac{Y_{A,B} - Y_{A,T}}{\left(\frac{L}{G}\right)_{eff}} + X_{A,T} \tag{8.73}$$

With $X_{A,B,1}$ from eq. (8.73) and the equilibrium function $Y^* = f(X) \cdot X_{A,B}$ we calculate the equilibrium load $Y^*_{A,B,1}$ of the first stage.

With $Y^*_{A,B,1}$ and eq. (8.73) we calculate $X_{A,B,2}$.

We repeat this procedure (n-times) until $Y_{A,B,n} < Y_{A,T}$ and get from the number n of calculation loops the number of theoretical (= ideal) separation stages $n = N_{th}$. Confusing? Do not worry. In Exercise 8.12 we go through these instructions again.

Exercise 8.12: Much less complicated than it looks like

The fridge of a food market powered with the coolant NH_3, already tried to heal in Exercises 8.9 and 8.10, still has got a problem. What is the problem? In case of emergency due to leaking coolant pipework the venting air $F_{V,air} = 1{,}000\ m^3_{STP}\ h^{-1}$ may contain an ammonia load of 400 g kg^{-1}. It must be detoxified to a maximum load of $Y_{NH3,max} = 10$ mg kg^{-1} by absorbing the ammonia in water in a tray column. The contaminated water is drained to a municipal wastewater facility. Temperature of the water saturated venting air is $\theta = 12$ °C, and P = 1,013 hPa. Being the vendor of "best equipment supply" you have to give the customer an idea about the diameter of the column (the customer will also need the height of the tower to find an appropriate location for installing the equipment in the vicinity of the fridge). Density of water is $\rho_{H_2O} = 1{,}000$ kg m^{-3} and density of air at 12 °C is $\rho_{air} = 1.286$ kg m^{-3}_{STP}, the mass flow rate of air is $G = F_{air} = F_{V,air} \cdot \rho_{air} = 1{,}000 \cdot 1.286 = 1{,}286$ kg h^{-1}.

Process specification:

$$Y_{NH_3,B} = 400 \text{ mg kg}^{-1} \text{ and } Y_{NH_3,T=10} \text{ mg kg}^{-1} \text{ and } X_{NH_3,T} = 0 \text{ mg kg}^{-1}$$

$$Y^*_{NH_3} = 0.3657 \cdot X_{NH_3}$$

Solver:
First step:

$$Y^*_{NH_3} = 0.3657 \cdot X_{NH_3} \text{ and } X_{NH_3,B,max}\left[\text{mg kg}^{-1}\right] = \frac{Y^*_{NH_3}}{0.3657}$$

Second step:

$$\left(\frac{L}{G}\right)_{min} = \frac{Y_{NH_3,B} - Y_{NH_3,T}}{X_{NH_3,B,max} - 0} = \frac{400 - 10}{1{,}094} = 0.36 \text{ kg H}_2\text{O kg}^{-1} \text{ air}$$

$$L_{min} = 0.36 \cdot 1{,}286 = 486.7 \text{ kg H}_2\text{O h}^{-1}$$

Third step:

$$\left(\frac{L}{G}\right)_{eff}\left[\text{kg H}_2\text{O kg}^{-1} \text{ air}\right] = 1.5 \cdot \left(\frac{L}{G}\right)\frac{486.7}{1{,}286.4} \quad L_{eff} = \left(\frac{L}{G}\right)_{eff} \cdot G = 0.53 \cdot 1{,}286.4 = 688 \text{ kg H}_2\text{O h}^{-1}$$

Fourth step:

$$X_{NH_3,1} = \frac{Y_{NH_3,B} - Y_{NH_3,T}}{\left(\frac{L}{G}\right)_{eff}} + X_{NH_3,T} = \frac{400 - 10}{0.53} + 0 = 729.2 \text{ mg NH}_3 \text{ kg}^{-1} \text{ H}_2\text{O}$$

Fifth step:

$$Y^*_{NH_3,1} = 0.3657 \cdot X_{NH_{3,1}} = 0.3657 \cdot 729.2 = 267 \text{ mg NH}_3 \text{ kg}^{-1} \text{ air}$$

Sixth step:

$$X_{NH_3,2} = \frac{Y_{NH_3,1} - X_{NH_3,T}}{\left(\frac{L}{G}\right)_{eff}} + X_{NH_3,T} = \frac{267 - 10}{0.53} + 0 = 480 \text{ mg NH}_3 \text{ kg}^{-1} \text{ H}_2\text{O}$$

Seventh step:

$$Y^*_{NH_3,2} = 0.3657 \cdot X_{NH_3,T} = 0.3657 \cdot 480 = 175 \text{ mg NH}_3 \text{ kg}^{-1} \text{ air}$$

Being well trained now, you prepare a list with results in Tab. 8.14.

Tab. 8.14: Number of separation stages n and corresponding balance loads Y_{NH_3} and X_{NH_3} (and equilibrium load) data.

n	Y_{NH_3} (mg NH_3 kg^{-1} air)	n	X_{NH_3} (mg NH_3 kg^{-1} H_2O)
Feed	400	1	729
1	267	2	498
2	176	3	309

Tab. 8.14 (continued)

n	Y_{NH_3} (mg NH_3 kg^{-1} air)	n	X_{NH_3} (mg NH_3 kg^{-1} H_2O)
3	113	4	193
4	71	5	113
5	42	6	59
6	21	7	21
7	8		

With tray distance $z = 300$ mm we obtain a theoretical column height of

$$H_{th} = H_{ideal} = n \cdot z = 7 \cdot 0.3 = 2.1 \text{ m}$$

8.4.3 Overall mass transfer efficiency of staged phase contactors

Do you trust in the outcome of Exercise 8.12? Please, do not. Your results may easily be off reality by 50%, but why? Up to now we did not consider any deviation from ideal phase contact per separation stage, when actually both phases, the gas phase and the absorbent phase, do not leave the tray at equilibrium. We may deduce the stage separation efficiency from experiments, a very expensive adventure, or we me make use of reported experience. By definition the overall efficiency of separation is derived from the ratio of theoretical number of stages to the actually needed number of stages $E = N_{th}/N_{eff}$. O'Connell [283] (again a brand in the column design community) suggested an empirically evaluated algorithm, which considers the molar mass of the absorbent MM_l, the dynamic viscosity of the absorbent and the density of the absorbent μ_l/ρ_l (= kinematic viscosity, SI units), as shown in the following equation:

$$E = 0.60 - 24 \cdot \sqrt{\frac{m \cdot MM_l \cdot \mu_l}{\rho_l}} \tag{8.83}$$

$$m = \frac{H}{P} \tag{8.8}$$

Exercise 8.13:
We will let his proposal act on the results of Exercise 8.12 and find with eqs. (8.83) and (8.8), and MM_l = 1,000 kg m^{-3}, ρ_l = 1,000 kg m^{-3} and μ_l = 0.001 Pa s:

$$E = 0.6 - 24 \cdot \sqrt{\frac{m \cdot MM_l \cdot \mu_l}{\rho_l}} = 0.6 - 24 \cdot \sqrt{\frac{0.59 \cdot 18 \cdot 0.001}{1{,}000}} = 0.52$$

Correspondingly, $H_{column} = \frac{H_{th}}{E} = \frac{2.1}{0.52} = 4 \text{ m}$

Discussion of results of Exercises 8.11, 8.12 and 8.13: Our job was to design a counter-currently operating absorption tower. We decided for a sieve tray column. Following the Theory of Separation Stages, we determined N_{th} with $N_{th} = 7$. With the help of the Souders and Brown correlation for sieve trays we determined the column diameter to be $D = 0.5$ m. With a tray distance of $\Delta z = 300$ mm and an overall mass transfer efficiency after O'Connell of $E = 0.52$ we obtained a net apparatus height (without bottom and top sections and without demister) of $H = 4$ m. These results are sound. Going through this topic hopefully did encourage you not to fear apparatus design. We did not discuss the pressure drop determination, but do not fear to go into literature (do not fear the literature).

8.5 Apparatus design for absorption towers with continuous phase contact: apparatus diameter and apparatus height

We start with a brief discussion of the preconditions we need for convincing a solute to transfer from the gaseous carrier into the absorbent phase.

The driving force for mass transfer is either originating in a concentration gradient, a gradient of the partial pressure or density. These gradients induce mass transfer by molecular diffusion and convective diffusion.

Under quiescent conditions equimolar molecular diffusion is mass transfer-determining. Equimolar molecular diffusion is modeled by Fick's law. Corresponding to Fick's first law eq. (8.84), molar flux $F_{n,i}$ in mol time^{-1} of a substance perpendicular to a specified plane correlates with the diffusion coefficient D_i in m^2 s^{-1}, the area A in m^2 and the concentration gradient dc/dx:

$$F_{n,i} = \frac{D_i \cdot A \cdot dc_i}{dx} \tag{8.84}$$

With the partial pressure of a substance deduced from the perfect gas law we obtain the following equation:

$$F_{n,i} \cdot R \cdot T = \frac{-D_i \cdot A \cdot dp_i}{dx} \tag{8.85}$$

Equimolar diffusion is limited to independent diffusion of all constituents. When diffusion is selectively limited to one component by phase separation, diffusion is unidirectional. Modeling is then performed by Stefan's law eq. (8.86). The molar flux $F_{n,i}$ in mol time^{-1} of a substance i correlates with the concentration gradient dc/dx and the parameter diffusion coefficient D_i, the mass transfer area A, and the molar concentration c of all substances:

$$F_{n,i} \cdot (c - c_i) = \frac{-D_i \cdot A \cdot c \cdot dc_i}{dx} \tag{8.86}$$

Unidirectional diffusion $F_{n,i}$ of substance i is increased over equimolar diffusion by the factor $c/(c-c_i)$.

Convective diffusion is a diffusional transfer of molecule aggregates under free flow or forced flow eq. (8.87). Gas–liquid mass transfer processes are based on convective transport of the solute (convective mass transfer under steady-state conditions has similarity to convective heat transfer):

$$F_{n,i} = \beta_i \cdot A \cdot (c_{i,b} - c_{i,IF}) = \beta_i \cdot A \cdot \Delta c_i \tag{8.87}$$

where $c_{i,b}$ is the concentration of substance i in the fluid phase (bulk phase b) in mol volume^{-1}, $c_{i,IF}$ is the concentration of substance i at the interface (IF) in mol volume^{-1}, β_i is the mass transfer coefficient in m time^{-1}, Δc_i is the concentration difference of the substance i within the phase boundary of thickness δ, and $F_{n,i}$ is the molar flow in mol time^{-1}.

Within the phase boundary the flow pattern is laminary, and mass transfer is controlled by molecular diffusion:

$$n_i = \frac{D_i \cdot A \cdot \Delta c_i}{\delta} = \beta_i \cdot A \cdot \Delta c_i \tag{8.88}$$

with $$\beta_i = \frac{D_i}{\delta}$$

Analogous to the heat transfer coefficient α the mass transfer coefficient ß is controlled by the flow pattern, the geometric shape of the interface and by the physical properties of the fluid. The thickness δ of the boundary layer cannot be specified locally. It is a link between diffusion and mass transfer.

Mass transfer across an interface considers the transport of a substance i from a bulk phase I across the interface to a second bulk phase II. Modeling is again analogous to modeling of heat transfer, except the difference that both phases form the interface boundary without an additional physical discontinuum. The interface is specified by the properties of the involved substances and the kind of formation which depends on the design specification of the mass transfer equipment. Modeling is based on the assumption of equilibrium between both phases at the physical interface.

Opposite to heat transfer the separate determination of the individual mass transfer resistances is not possible. Experimental determination of the overall mass transfer resistance, which is assumed to result from the sum of the individual phase resistances, is possible. Our discussion is based on the assumption of negligible resistance of the interface. This assumption actually can be an approximate simplification only for several interface reactions and hydraulic activity such as interface turbulences or interfacial mass transfer retardation.

The film theory is the number one standard in modeling mass transfer processes with continuous phase contact. It is based on the following boundaries:

- The mass transfer resistance is limited to the phase boundary layers.
- Across the phase boundary layers, mass transfer is controlled by molecular diffusion.
- The interface does not cause any resistance.
- At the interface the transferring substance is at any time at equilibrium.
- Mass transfer within the bulk phases is controlled by turbulent convection.

8.5.1 The film theory

Considering the transferring substance i, which passes the interface from phase I (for example gaseous phase) to phase II (for example absorbent phase), we can specify the transfer process by the schematic sketch shown in Fig. 8.19.

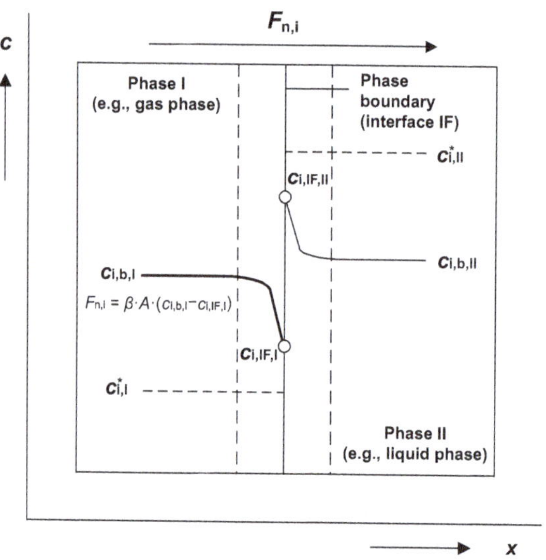

Fig. 8.19: Schematic sketch to derive the mass transfer of a solute i from the gas phase across the interface of a gas–liquid system into the absorbent phase.

The following symbols are used:

A	Interfacial area (m^2)
β	Mass transfer coefficient within a phase (also β_i); phase I (β_I) and phase II (β_{II}), respectively (m time^{-1})
$F_{n,i}$	Molar flow rate of component i (mol time^{-1})
$c_{i,IF,I}$	Concentration of i at the interface (IF) of phase I (mol m^{-3})
$c_{i,IF,II}$	Concentration of i at the interface (IF) of phase II (mol m^{-3})
$c_{i,b,I}$	Concentration of i in bulk phase (b) I (mol m^{-3})
$c_{i,b,II}$	Concentration of i in bulk phase (b) II (mol m^{-3})
$c^*_{i,I}$, $c^*_{i,II}$	Corresponding equilibrium concentrations (mol m^{-3})
K^*	Equilibrium coefficient (–)

To specify the transfer process, we can apply the following algorithm:
For transfer of component i from phase I to the interface, the following applies:

$$F_{n,i} = \beta_I \cdot A \cdot (c_{i,b,I} - c_{i,IF,I})$$

$$\frac{F_{\text{n,i}}}{\beta_{\text{I}} \cdot A} = (c_{\text{i,b,I}} - c_{\text{i,IF,I}})$$

Since substance i arriving at the interface is transported through it, the following applies to the transfer to phase II:

$$F_{\text{n,i}} = \beta_{\text{II}} \cdot A \cdot (c_{\text{i,IF,II}} - c_{\text{i,b,II}})$$

$$\frac{F_{\text{n,i}}}{\beta_{\text{II}} \cdot A} = (c_{\text{i,IF,II}} - c_{\text{i,b,II}})$$

By definition, substance i is in equilibrium across the interface:

$$c_{\text{i,IF,I}} = K^* \cdot c_{\text{i,IF,II}}$$

K^* is the equilibrium coefficient (e.g., $m = H/P$; but local value of y^*/x)

The bulk phases are in equilibrium according to

$$c_{\text{i,b,I}} = K^* \cdot c^*_{\text{i,b,II}}$$

with $c^*_{\text{i,b,I}}$ being the corresponding equilibrium concentration of i in bulk phase I to any concentration of i in bulk phase II ($c_{\text{i,b,II}}$), and $c^*_{\text{i,b,II}}$ as the corresponding equilibrium concentration of i in bulk phase II to any concentration of i in bulk phase I ($c_{\text{i,b,I}}$).

After introducing the equilibrium concentration $c^*_{\text{i,II}}$ (also $c^*_{\text{i,b,II}}$) in phase II

$$c_{\text{i,b,I}} = K^* \cdot c^*_{\text{i,II}}$$

we get:

$$\frac{F_{\text{n,i}}}{\beta_{\text{I}} \cdot A \cdot K^*} + \frac{F_{\text{n,i}}}{\beta_{\text{II}} \cdot A} = \left(c^*_{\text{i,II}} - c_{\text{i,IF,II}}\right) + (c_{\text{i,IF,II}} - c_{\text{i,b,II}})$$

$$\left(c^*_{\text{i,II}} - c_{\text{i,IF,II}}\right) + (c_{\text{i,IF,II}} - c_{\text{i,b,II}}) = \frac{F_{\text{n,i}}}{A} \cdot \left(\frac{1}{\beta_{\text{I}} \cdot K^*} + \frac{1}{\beta_{\text{II}}}\right)$$

Now we may adjust the mass transfer (eq. (8.89), extracted from Fig. 8.19)

$$\left(c^*_{\text{i,II}} - c_{\text{i,b,II}}\right) = \frac{F_{\text{n,i}}}{A} \cdot \left(\frac{1}{\beta_{\text{I}} \cdot K^*} + \frac{1}{\beta_{\text{II}}}\right) \tag{8.89}$$

to our needs by introducing the overall mass transfer coefficient k_{II} (or $k_{\text{o,l}}$), if phase resistances refer to phase II eq. (8.90), and eq. (8.91) (with k_{I} or $k_{\text{o,g}}$), if phase resistances refer to phase I,

$$\frac{1}{k_{\text{II}}} = \frac{1}{\beta_{\text{II}}} + \frac{1}{K^* \cdot \beta_{\text{I}}} = \frac{1}{k_{\text{o,l}}} \tag{8.90}$$

$$\frac{1}{k_{\text{I}}} = \frac{K^*}{\beta_{\text{II}}} + \frac{1}{\beta_{\text{I}}} = \frac{1}{k_{\text{o,g}}} \tag{8.91}$$

and with overall mass transfer then being related to eq. (8.91) for phase I, eq. (8.92), or to eq. (8.90) for phase II, eq. (8.93):

$$F_{n,i} = k_I \cdot A \cdot \Delta c_i = k_I \cdot A \cdot \left(c_{i,b,I} - c^*_{i,I}\right) \tag{8.92}$$

$$F_{n,i} = k_{II} \cdot A \cdot \Delta c_i = k_{II} \cdot A \cdot \left(c^*_{i,II} - c_{i,b,II}\right) \tag{8.93}$$

In absorption with chemical reaction, mass transfer of the liquid phase may be enhanced. Equation (8.91) is then extended with the enhancement factor E, as shown in the following equation:

$$\frac{1}{k_I} = \frac{K}{\beta_{II} \cdot E} + \frac{1}{\beta_I} = \frac{1}{k_{o,g}} \tag{8.91a}$$

After having discussed the mode of convincing our solute to transfer from the gas phase into the liquid phase, we have to offer a device for continuous mass transfer across the gas–liquid interface, and we have to work on the basics for an appropriate design of this device. We will find a helpful hand in the HTU-NTU concept.

8.5.2 The HTU-NTU concept

Modeling of gas–liquid mass transfer with continuous phase contact is based on the kinetic theory of counter-flow operation (Fig. 8.20). Continuous phase contact is applied in packed towers, film absorbers and spray scrubbers.

Modeling starts with the mass transfer balance for a specific volume V (or area A) of the mass transfer tower, a loan from Chapter 7, as follows:

$$dF_A = r_A \cdot dV$$

$$r_A \cdot V = r''_A \cdot A$$

$$r_A = \frac{A}{V} \cdot r''_A = a \cdot r''_A$$

$$-r''_A = -\frac{1}{A} \cdot \frac{dN_A}{dt} = k_{o,g} \cdot \left(c_g - c^*_g\right)$$

$$dF_A = a \cdot r''_A \cdot dV = G \cdot dy = -k_{o,g} \cdot \left(c_g - c^*_g\right) \cdot a \cdot dV$$

$$F \cdot dy = F_V \cdot C \cdot dy = -k_{o,g} \cdot C \cdot \left(y_g - y^*_g\right) \cdot a \cdot dV$$

$$F \cdot dy = F_V \cdot \frac{P}{R \cdot T} \cdot dy = -k_{o,g} \cdot \frac{P}{R \cdot T} \cdot \left(y_g - y^*_g\right) \cdot a \cdot dV$$

$$F_V \cdot \frac{P}{R \cdot T} \cdot dy = -k_{o,g} \cdot \frac{P}{R \cdot T} \cdot \left(y_g - y^*_g\right) \cdot a \cdot dV \text{ with } dV = CSA \cdot dh$$

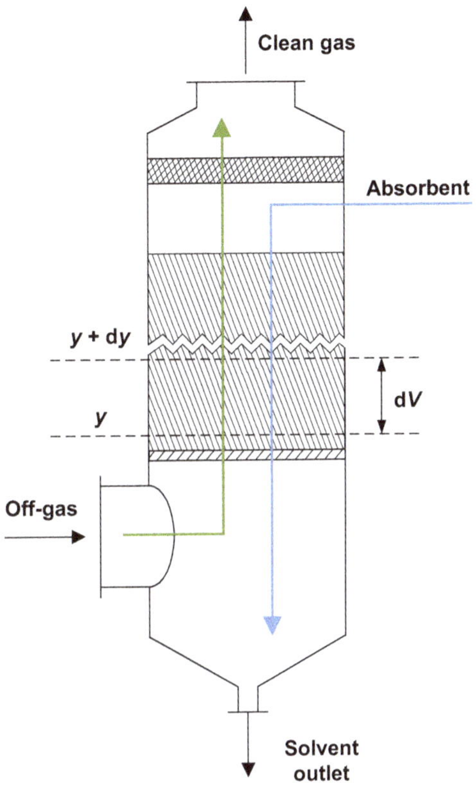

Fig. 8.20: Schematic sketch to derive the mass transfer balance for packed towers.

$$F_V \cdot \mathrm{d}y = -k_{o,g} \cdot \left(y_g - y_g^*\right) \cdot a \cdot \mathrm{CSA} \cdot \mathrm{d}h$$

$$\frac{F_V}{k_{o,g} \cdot a \cdot \mathrm{CSA}} \cdot \int_{y_{g,in}}^{y_{g,out}} - \frac{\mathrm{d}y}{\left(y_g - y_g^*\right)} = \int_0^h \mathrm{d}h$$

or

$$\frac{F_V}{k_{o,g} \cdot a \cdot \mathrm{CSA}} \cdot \int_{y_{g,out}}^{y_{g,in}} \frac{\mathrm{d}y}{\left(y_g - y_g^*\right)} = \int_0^h \mathrm{d}h$$

We outline the last line, shown in the following equation:

$$H = \int \mathrm{d}h = \frac{G}{k_{o,g} \cdot \mathrm{CSA} \cdot a} \int_{y_{g,out}}^{y_{g,in}} \frac{\mathrm{d}y}{\left(y_g - y_g^*\right)} = \frac{F_V}{k_{o,g} \cdot \mathrm{CSA} \cdot a} \int_{y_{g,out}}^{y_{g,in}} \frac{\mathrm{d}y}{\left(y_g - y_g^*\right)} \tag{8.94}$$

(Note: in the following discussion we simplify y_g and y_g^* with y and y^*, respectively.)

Chilton und Colburn [284], famous brands in the mass transfer community, have named the expression of the integral $\int \frac{dy}{y-y^*}$, which specifies the "number of transfer units" needed for separation, the so-called NTU-value. The expression $\frac{G}{k_{o,g} \cdot CSA \cdot a}$ is called the "height of a transfer unit" or HTU value.

The NTU value is the total driving force (y-y^*) of the substance i between the entrance (bottom) $y_{i,B}$ of the scrubber ($h = 0$) and the (net) top or outlet ($h = H$) of the scrubber $y_{i,T}$. Based on "overall mass transfer" (related to either phase) it is called NTU_{og} (overall gas), and the HTU-value is called HTU_{og}, as depicted in the following equation:

$$H = HTU_{og} \cdot NTU_{og} = \frac{G}{k_{o,g} \cdot CSA \cdot a} \cdot \int \frac{dy}{y-y^*} \tag{8.95}$$

with

$$\frac{1}{k_I} = \frac{1}{k_{o,g}} = \frac{m}{\beta_{II}} + \frac{1}{\beta_I} \text{ and } m = H/P \tag{8.91 and 8.8}$$

The specific mass transfer area a and the mass transfer coefficient $k_{o,g}$ must be determined prior to determining the HTU-value. The mass transfer coefficient is either determined experimentally or it is derived from dimensional analysis. The specific mass transfer area a depends on the type of contactor.

The height of a transfer unit HTU can also be deduced from separate mass transfer coefficients for gas phase and liquid phase and the fractional amounts of the gas phase height HTU_g and the liquid phase height HTU_l, shown in eqs. (8.96), (8.97) and (8.98), with G being the flow rate of the gas phase in terms of m^3 $time^{-1}$, L as the flow rate of the absorbent phase in terms of m^3 $time^{-1}$, and $m = H/P$. These algorithms probably suggest that you rather try to work with "overall mass transfer coefficients":

$$HTU = HTU_g + \frac{m \cdot G \cdot HTU_l}{L} \tag{8.96}$$

with

$$HTU_l = \frac{G}{\beta_I \cdot a \cdot CSA} \tag{8.97}$$

and

$$HTU_g = \frac{L}{\beta_g \cdot a \cdot CSA} \tag{8.98}$$

For doing the calculations you need the specific mass transfer area a from specification of the packings or the spray. We will soon address the determination of the mass transfer coefficient (either the $k_{o,g}$ or $k_{o,l}$ value or $ß_g$ (former $ß_I$ for phase I) and $ß_l$ (former $ß_{II}$ for phase II)).

Being at war with mathematics you may probably slip into a problem with getting the NTU-value determined, especially when discussing absorption with chemical reaction. It is then recommended to do these calculations by graphical integration by:

- performing a diagrammatic scheme of the gas–liquid equilibrium (x/y-scheme),
- determining the reciprocal of the driving force $1/(y - y^*)$ for distinct values of x,
- drawing a diagrammatic scheme of $1/(y - y^*)$ versus y between feed (y_B) and outlet (y_T) and
- determining the area below the graph between lower limit and upper limit, which represents the NTU value.

For parallel straight graphs of the equilibrium curve and the operation line, the NTU-value agrees with the number of theoretical stages N_{th}. The driving force $(y - y^*)$ may then be approximated by the logarithmic mean $(y - y^*)_{lm}$, with the NTU depicted in eq. (8.99) via eq. (8.100):

$$\text{NTU} = \int \frac{dy}{y - y^*} = \frac{y_B - y_T}{(y - y^*)_{lm}} \tag{8.99}$$

with

$$(y - y^*)_{lm} = \frac{(y - y^*)_{in} - (y - y^*)_{out}}{\ln\left(\frac{(y - y^*)_{in}}{(y - y^*)_{out}}\right)} \tag{8.100}$$

In case of great solubility of the solute in the absorbent, which may be assumed to be observed in absorption with chemical reaction, as for example absorption of HCl or SO_2 at pH = 6, the equilibrium concentration is nearly negligible ($y^* \cong 0$) and determination of the NTU value simplifies to the following equation:

$$\int \frac{dy}{y - y^*} = \int \frac{dy}{y} = \ln\left(\frac{y_{in}}{y_{out}}\right) \tag{8.101}$$

8.5.3 Mass transfer

When analyzing the HTU value we briefly mentioned the $k_{o,l}$ and $k_{o,g}$ value, and the corresponding *ß*-values, but just made a comment on how to get access to this important design data. We may definitely rely on the support of dimensional analysis for getting these data determined. Literature offers plenty of correlations [237, 240, 258, 264].

The data you need:

- diffusion coefficient of the solute in the gas phase and the liquid phase D in $m^2\ s^{-1}$
- characteristic length of the packing l in m; irregular packings: $l = 4 \cdot V/A$ with V as the volume and A as the surface area

- the velocity of both phases v in m s^{-1}
- the density of both phases ρ in kg m^{-3}
- the kinematic viscosity of both phases ν in m^2 s^{-1}

You may, for example, find these data in [266, 285] or [258, 264].

In a first step, the dimensionless groups:

- Reynolds number: $\mathrm{Re} = \frac{v \cdot l}{\nu}$
- Schmidt number: $\mathrm{Sc} = \frac{\nu}{D}$
- Galilei number: $\mathrm{Ga} = \frac{g \cdot l^3}{\nu^2}$
- Archimedes number: $\mathrm{Ar} = \frac{g \cdot d_\mathrm{p} \cdot \mu}{\nu} \cdot \left(\frac{p_\mathrm{s}}{p_\mathrm{f}} - 1\right)$

have to be calculated, to gain access to the Sherwood number: $\mathrm{Sh} = \frac{\beta \cdot l}{D}$.

In the second step, correlation of the dimensionless groups for correct specification of the Sherwood number is performed for specific mass transfer phase contact patterns:

For forced convective flow within a phase (e.g., packed tower), the following equation applies:

$$\mathrm{Sh} = C \cdot \mathrm{Re}^m \cdot \mathrm{Sc}^n \tag{8.102}$$

For convective flow of a liquid phase within a phase (e.g., packed towers, falling film) eq. (8.103) applies:

$$\mathrm{Sh} = C \cdot \mathrm{Re}^m \cdot \mathrm{Sc}^n \cdot \mathrm{Ga}^q \tag{8.103}$$

And for spherical particles (droplets, bubbles) eq. (8.104) is valid.:

$$\mathrm{Sh} = 2 + C \cdot \mathrm{Re}^m \cdot \mathrm{Sc}^n \tag{8.104}$$

Table 8.15 shows the parameters for the Sherwood correlations of eqs. (8.102–8.104).

Tab. 8.15: Parameters for the Sherwood correlations of eqs. (8.102–8.104).

Equipment	Dimensionless number	C	m	n	q
Packed towers	$\mathrm{Sh_g}$	0.69	0.59	0.33	
Packed towers	$\mathrm{Sh_l}$	0.32	0.59	0.5	0.17
Spray towers	$\mathrm{Sh_g}$	0.60	0.5	0.33	
Film absorbers	$\mathrm{Sh_l}$	1.39	0.33	0.5	0.167

You will also need the specification of the type of packings. These data are, for example, collected in [237]. You will also get product specifications for a huge variety of packings and structured packings from manufacturers. Tables 8.16 through 8.18 exemplarily list the data for Pall rings (plastic) [286].

Tab. 8.16: Pall rings (plastic): product specification.

Size (mm)	Wall thickness (mm)	Specific area ($m^2\ m^{-3}$)	Void volume (%)	Bulk density ($kg\ m^{-3}$)
15	1.0	350	88	120
25	1.3	220	90	085
35	1.5	160	90	078
50	1.5	110	92	072
90	2.0	066	91	057

Tab. 8.17: Pall rings (plastic): product specification.

Size (mm)	K_1	K_2	B_{max} ($m^3\ m^{-2}\ h^{-1}$)	K_3	K_4	B_{min} ($m^3\ m^{-2}\ h^{-1}$)
15	0.090	0.090	115	0.45	0.115	09.0
25	0.100	0.080	145	0.40	0.100	10.5
35	0.107	0.071	185	0.35	0.090	12.0
50	0.113	0.065	220	0.30	0.080	13.0
90	0.125	0.055	310	0.25	0.070	15.0

Tab. 8.18: Pall rings (plastic): product specification.

Size (mm)	K_5	K_6	K_7
15	16	0.07	0.012
25	09	0.07	0.0095
35	05.8	0.07	0.0080
50	03.8	0.07	0.0068
90	02.5	0.07	0.0050

With this background information we may perform the hydraulic design of packed bed absorption towers.

8.5.4 Hydraulic design of packed bed absorption towers

Present design rules for the design of packed absorption towers also date back to the hard work that R. Beck did on this topic [286]. It is easy to apply in a step-by-step concept.

First step:
Determination of the actual gas velocity v_g with eq. (8.105). Calculation starts with the estimation of the specific absorbent load B in $m^3\ m^{-2}\ h^{-1}$, which is recommended to be between the lower limit B_{min} and the upper limit B_{max}; $B_{min} < B < B_{max}$:

$$v_{eff}\ \left[m\,s^{-1}\right] = K_1 \cdot \sqrt{\frac{\rho_l}{\rho_g}} \cdot \left(1 - K_2 \cdot f(\mu_l) \cdot \sqrt{B}\right) \tag{8.105}$$

$$f(\mu_l) = 0.80 \cdot 0.225 \cdot \sqrt[4]{\mu_l}$$

We need v_{eff} to get access to the gas velocity v_g, shown in eq. (8.106), above the packing:

$$v_g = v_{eff} \cdot \left(1 + 7.5 \cdot \left(\frac{d}{D}\right)^2\right) \tag{8.106}$$

The following symbols are used:

K_1, K_2	Constants corresponding to the specification of packings (Tab. 8.17)
ρ_l, ρ_g	Density of the absorbent and density of the gaseous carrier ($kg\ m^{-3}$)
B	Specific absorbent load ($m^3\ m^{-2}\ h^{-1}$)
μ_l	Dynamic viscosity of the absorbent (cP) (note: 1 cP = 10^{-3} Pa s)
v_g	Actual gas velocity ($m\ s^{-1}$)

Second step:
Determination of the specific number of theoretical stages n_{th} per unit length of the selected packing with the following equation:

$$n_{th} = K_3 \cdot \sqrt{\frac{\rho_l}{\rho_g}} \cdot \left(1 - K_4 \cdot \sqrt{B}\right) \cdot \frac{f\left(\frac{d}{D}\right)}{v_g \cdot \sqrt[4]{H} \cdot (1 + D)} \tag{8.107}$$

with

$$f\left(\frac{d}{D}\right) = \frac{1.72}{1 + 7.5 \cdot \left(\frac{d}{D}\right)^2} - 0.72$$

with the symbols:

n_{th}	Specific number of theoretical plates per meter
K_3, K_4	Constants corresponding to the specification of packings
D	Diameter of the tower (m)
H	Height of the packing (m)
d	Diameter of the packing (m)

From analytical or graphical determination of the number of theoretical stages N_{th} and the specific number of theoretical stages per unit height n_{th} (eq. (8.107)) of the absorption tower, the total height of the column is determined. Alternatively, the height of the tower is deduced from the HTU-NTU concept.

Third step:
Determination of the specific pressure drop Δp in dPa m^{-1} with the following equation:

$$\Delta p\ \left[\text{dPa m}^{-1}\right] = K_5 \cdot \left(\frac{1{,}000 \cdot \rho_g}{\rho_l \cdot v_{eff}^2}\right)^n \cdot f(B, \mu_l) \tag{8.108}$$

with

$$f(B, \mu_l) = 10^{(K_6 + K_7 \cdot B) \cdot f(\mu_l)}$$

and $n = 1.25 - 0.002 \cdot B$

To avoid asymmetric absorbent flux over the cross-sectional area, the height of a single packing is limited to a maximum height H_x. Every packing in series has to be placed on a separate bearing plate. The absorbent is collected on the bottom of every packing and redistributed on top of the packing below. The size of the packing d, the optimum height of the packing H_x and the diameter of the tower D shall be within recommended limits, as follows:

$$\text{opt.}\ H_x = 120 \cdot d \text{ with } d \text{ in m} \tag{8.109}$$

$$H_x / D = 5 \text{ to } 10 \text{ (Pall rings)} \tag{8.110}$$

Our engineering toolbox, containing a model-based mass transfer concept (Section 8.5.1), an application concept (Section 8.5.2) with application oriented mass transfer correlations and a hydraulic design concept (Section 8.5.4), now has to prove applicability in Exercise 8.14.

Exercise 8.14: Does the fridge venting system accept continuous phase contact?
We already tried to "heal" the fridge of a food market, powered with the coolant NH_3 with the magic of a sieve tray absorption tower in Exercises 8.9 through 8.12. Now we want to check how a packed absorption tower might perform. We briefly summarize the problem:

In case of emergency due to leaking coolant pipework the venting air $F_{V,air} = 1{,}000$ m$^3{}_{STP}$ h^{-1} may contain an ammonia load of 400 g kg^{-1}. It must be detoxified to a maximum load of $Y_{NH_3,max} = 10$ g kg^{-1} by absorbing the ammonia in water in a packed bed absorption tower. The contaminated water is drained to a municipal wastewater facility. Temperature of the water saturated venting air is $\theta = 12$ °C, and $P = 1{,}013$ hPa. Being the vendor of "best equipment supply" you have to give the customer an idea about the diameter of the column (the customer will also need the height of the tower to find an appropriate location for installing the equipment in the vicinity of the fridge).

Process specification:
Density of water is $\rho_{H_2O} = 1{,}000$ kg m^{-3} and density of air is $\rho_{air} = 1.286$ kg m$^{-3}{}_{STP}$
and $\rho_{air,12\,°C} = 1.223$ kg am^{-3} at $\theta = 12$ °C

The mass flow rate of air is $G = F_{air} = F_{V,air} \cdot \rho_{air} = 1{,}000 \cdot 1.286 = 1{,}286$ kg h^{-1}

$$F_{V,g} = \frac{1{,}000 \cdot (273.15 + 12)}{273.15} \cdot \frac{1{,}013}{1{,}013} = 1{,}044 \text{ am}^3 \text{ h}^{-1}$$

$$\left(\frac{L}{G}\right)_{eff} [\text{kg } H_2O \text{ kg}^{-1} \text{ air}] = 1.5 \cdot \left(\frac{L}{G}\right) \frac{486.7}{1{,}286.4}$$

$$L_{eff} = \left(\frac{L}{G}\right)_{eff} \cdot G = 0.53 \cdot 1{,}286.4 = 688 \text{ kg } H_2O \text{ h}^{-1}$$

(see Exercise 8.12)

$$X_{NH_3,1} = \frac{Y_{NH_3,B} - Y_{NH_3,T}}{\left(\frac{L}{G}\right)_{eff}} + X_{NH_3,T} = \frac{400 - 10}{0.53} + 0 = 729.2 \text{ mg } NH_3 \text{ kg}^{-1} H_2O$$

(see Exercise 8.12)

$$Y_{NH_3,B} = 400 \text{ mg kg}^{-1} \text{ and } Y_{NH_3,T} = 10 \text{ mg kg}^{-1} \text{ and } X_{NH_3,T} = 0 \text{ mg kg}^{-1}$$

$$Y^*_{NH_3} = 0.3657 \cdot X_{NH_3}$$

Solver:
We decide for the Beck design concept and check for applicability of the packing type "Pall ring plastic, size 35 mm," specified in Tabs. 8.16–8.18.

First step (eq. (8.105)):

$$v_{eff} = K_1 \sqrt{\frac{\rho_{H_2O}}{\rho_{air}}} \left[1 - K_2 f(\mu_l) \cdot \sqrt{B}\right] = 0.107 \sqrt{\frac{1{,}000}{1.223}} \left[1 - 0.07 \cdot 1.025 \cdot \sqrt{11}\right] = 2.28 \text{ m s}^{-1}$$

$$f(\mu_l) = 0.80 + 0.225 \cdot \sqrt[4]{\mu_l} = 0.80 + 0.225 \cdot \sqrt[4]{1.0} = 1.025$$

With the gas velocity v_{eff} and the gas flow rate $F_{V,g} = 1{,}044$ am^3 h^{-1}, we have access to the cross-sectional area (CSA) = 0.112 m^2 and the diameter $D = 0.4$ m (just a reminder: $F_{V,g}/v_{eff}$ = CSA, and CSA = $D^2 \cdot \pi/4$).

Second step (eq. (8.106)):

$$v_g = v_{eff} \cdot \left[1 + 7.5 \cdot \left(\frac{d}{D}\right)^2\right] = 2.28 \cdot \left[1 + 7.5 \cdot \left(\frac{\frac{35}{1{,}000}}{0.4}\right)^2\right] = 2.42 \text{ m s}^{-1}$$

Third step (eq. (8.107)):

$$n_{th} = \frac{K_3 \cdot \sqrt{\frac{\rho_{H_2O}}{\rho_{air}}} \cdot (1 - K_4 \cdot \sqrt{B}) \cdot f\left(\frac{d}{D}\right)}{v_g \cdot \sqrt[4]{H} \cdot (1 + D)} = \frac{0.35 \cdot \sqrt{\frac{1{,}000}{1.223}} \cdot (1 - 0.09 \cdot \sqrt{12}) \cdot 0.9}{2.42 \cdot \sqrt[4]{2} \cdot (1 + 0.4)} = 1.7 \text{ m}^{-1}$$

$$f\left(\frac{d}{D}\right) = \frac{1.72}{1 + 7.5 \cdot \left(\frac{d}{D}\right)^2} - 0.72 = \frac{1.72}{1 + 7.5 \cdot \left(\frac{0.035}{0.4}\right)^2} - 0.72 = 0.9$$

From Exercise 8.12 we may use the number of separation stages $N_{th} = 7$ and estimate the height of the column according to

$$\frac{N_{th}}{n_{th}} = H = \frac{7}{1.7} = 4.2 \text{ m}$$

Fourth step (eq. (8.108)):

$$\Delta p = K_5 \cdot \left(1{,}000 \cdot \frac{\rho_{air}}{\rho_{H_2O}} \cdot v_g^2\right)^n \cdot f(B,\eta_l) = 5.8 \cdot 10 \cdot \left(1{,}000 \cdot \frac{1.23}{1{,}000} \cdot 2.46^2\right)^n \cdot 1.31 = 897 \text{ Pa m}^{-1}$$

$n = 1.25 - 0.003676 \cdot B = 1.25 - 0.003676 \cdot 12 = 1.23$
$f(B,\mu_l) = 10^{(K_6 + K_7 \cdot B) \cdot f(\mu_l)} = 10^{(0.076 + 0.008 \cdot 5.4) \cdot 1.025} = 1.31$
$f(\mu_l) = 0.80 + 0.225 \cdot \sqrt[4]{0.001} = 1.025$

Finally, we check for the recommendations of eqs. (8.109) and (8.110).
opt. $H_x = 120 \cdot d = 120 \cdot 0.035 = 4.2$ m (we are very close to this recommendation)
$H_x/D = 5$ to $10 = 4.2/0.4 = 10.5$ (we will survive this deviation from recommendation of eq. (8.110).

The actual specific liquid phase load B is outside the recommended value. Latter would need an absorbent flow rate of $G\ (F_m) = 1{,}500$ kg h^{-1}, far beyond the value we obtained from the mass balance to suffice the needs of our fridge emergency scrubbing problem. We do not want to dump more waste water than needed to suffice the application needs ($X_{NH_3,in} = 400$ mg kg^{-1} air and $X_{NH_3,out} < 10$ mg kg^{-1} air). Our scrubber will operate at the lower operation limit, but it will do its job.

Fifth step:
Determination of the NTU value

The balance line and the equilibrium line are straight lines. We may apply eqs. (8.99) and (8.100) for determining the NTU value:

$$Y^*_{NH_3,in} = 0.3657 \cdot Y_{NH_3,1} = 0.3657 \cdot 729.2 = 267 \text{ mg NH}_3 \text{ kg}^{-1} \text{ air (see Exercise 8.12)}$$

$$Y_{NH_3,in} = 400 \text{ g kg}^{-1}$$

$$Y_{NH_3,out} = 10 \text{ g kg}^{-1}$$

$$Y^*_{NH_3,out} = 0$$

$$(Y - Y^*)_{lm} = \frac{(Y - Y^*)_{in} - (Y - Y^*)_{out}}{\ln\left(\frac{(Y - Y^*)_{in}}{(Y - Y^*)_{out}}\right)} = \frac{(400 - 267)_{in} - (10 - 0)_{out}}{\ln\left(\frac{(400 - 267)_{in}}{(10 - 0)_{out}}\right)} = 47.5$$

$$\text{NTU} = \int \frac{dY}{Y - Y^*} = \frac{Y_B - Y_K}{(Y - Y^*)_{lm}} = \frac{400 - 10}{47.5} = 8.2 \qquad (8.212)$$

Comparison of the NTU-value with n (Tab. 8.14) is sound.

Sixth step:
Determination of the HTU value

We decide for eqs. (8.96) through (8.98) for $a = 185$ m^2 m^{-3} and $d_P = 0.035$ m and the dimensionless numbers

$$\text{Sh} = \frac{\beta \cdot d_P}{\nu},\ Sc = \frac{\nu}{D},\ \text{Re} = \frac{v \cdot d_P}{\nu} \text{ and Ga} = \frac{g \cdot d_P^3}{\nu^2}$$

For packed towers the Sherwood-correlations of eqs. (8.102) and (8.103) have to be applied:

$$\text{Sh}_g = C \cdot \text{Re}^m \cdot \text{Sc}^n$$

$$\text{Sh}_g = C \cdot \text{Re}^m \cdot \text{Sc}^n \cdot \text{Ga}^q$$

We collect the data in a list, shown in Tab. 8.19.

Tab. 8.19: Design data for NH_3 mass transfer by absorption in a packed column.

	ρ (kg m^{-3})	v (m s^{-1})	Ga	Re	Sc	Sh	β (m s^{-1})
Gas phase	1.22	$1.39 \cdot 10^{-5}$		5727	0.57	94.3	0.066
Liquid phase	1,000	$1.0 \cdot 10^{-6}$	$4.21 \cdot 10^{8}$	52.5	6.10	239.2	$1.12 \cdot 10^{-3}$

With the data of Tab. 8.19 and the specification of the liquid phase flow rate ($L = 688$ kg h^{-1} equivalent 0.688 m^3 h^{-1}) and the gas flow rate ($G = 1{,}044$ am^3 h^{-1}) we obtain for

$$\mathrm{HTU_g} = \frac{G}{\beta_g \cdot a \cdot \mathrm{CSA}} = \frac{1{,}040}{3{,}600 \cdot 0.066 \cdot 185 \cdot 0.13} = 0.19 \text{ (see eq. (8.97))}$$

$$\mathrm{HTU_l} = \frac{L}{\beta_l \cdot a \cdot \mathrm{CSA}} = \frac{0.688}{3{,}600 \cdot 0.001 \cdot 185 \cdot 0.13} = 0.19 \text{ (see eq. (8.98))}$$

$$\mathrm{HTU} = \mathrm{HTU_g} + \frac{m \cdot G \cdot \mathrm{HTU_l}}{L} = 0.19 + \frac{0.58 \cdot 0.035 \cdot 0.19}{0.85} = 0.19 \text{ (see eq. (8.96))}$$

and $H = \mathrm{HTU} \cdot \mathrm{NTU} = 0.19 \cdot 8.2 = 1.6$ m. This outcome strongly deviates from the outcome of eq. (8.107) in step 3. However, we must keep in mind that we operate our scrubber at the lower end of the operation range. With the outcome of Exercise 8.13 in mind, we rather trust a mass transfer efficiency of $E = 0.5$, resulting in $H_{eff} = H/E = 1.6/0.5 = 3.2$ m – still a significant difference off the results of eq. (8.107), but more reliable.

8.5.5 Design of spray scrubbers

Finally, we have to appraise spray scrubbers. We will limit the discussion to the performance of the Venturi scrubber in absorption. The design of Venturi scrubbers has already been discussed in Section 6.3.4.4. We may therefore just focus on the design procedure.

First step:
Estimation of v_{eff} is rather adjusted to the needs of dust precipitation.

Second step:
Estimation of the mean droplet diameter $d_{1,2}$ (in general, droplet diameter d_{dr}) according to eq. (6.56) (see Venturi scrubber design, Section 6.3.4.4):

$$d_{1,2} = 0.585 \cdot \sqrt{\frac{\sigma}{\rho_l \cdot v_g^2}} + 53.2 \cdot \left(\frac{F_{V,l}}{F_{V,g}}\right)^{1.5} \cdot \left(\frac{\mu_l}{\sqrt{\sigma \cdot \rho_l}}\right)^{0.45} = 5 \cdot 10^{-5}\,\mathrm{m} \qquad (6.56)$$

Third step:
Determination of the terminal velocity v_t with eq. (8.111) (not needed in cross-flow mode)

$$v_t = \sqrt{\frac{4 \cdot d_{dr} \cdot \left(\rho_l - \rho_g\right) \cdot g}{15 \cdot \rho_g}} \tag{8.111}$$

Fourth step:
Determination of the specific mass transfer area a with the following equation:

$$a = \frac{6 \cdot B}{\rho_l \cdot (v_t - v_{eff}) \cdot d_{dr}} \tag{8.112}$$

Fifth step:
Determination of the droplet Re number with eq. (8.113):

$$\text{Re} = \frac{v_t \cdot \pi \cdot d_{dr}}{2 \cdot \nu_g} \tag{8.113}$$

Sixth step:
Determination of β_g, following the instructions of eq. (8.104) and Tab. 8.15:

$$\beta_g = \frac{\text{Sh} \cdot 2 \cdot D}{d_{dr} \cdot \pi} \tag{8.104}$$

The symbols used are:

v_t	terminal falling velocity of the droplet (m s^{-1})
d_{dr}	diameter of the droplet (m)
ρ_l, ρ_g	density of the absorbent, density of the gaseous carrier (kg m^{-3})
g	acceleration due to gravity (m s^{-2})
B	specific cross-sectional absorbent load (kg m^{-2} s^{-1})
v_{eff}	effective gas velocity (m s^{-1})
ν_g	kinematic viscosity (m^2 s^{-1})
D	diffusion coefficient (m s^{-1})

We will check this algorithm for applicability in the final Exercise 8.15.

Exercise 8.15: Can we get HCl and SO_2 of our off-gas problem, as specified in Tab. 1.1, under control? 
For specification of our problem let us recall the absorption-related data of Tab. 1.1. We implement the off-gas data and clean gas data of Exercise 2.10.

Tab. 8.20: Absorption-relevant off-gas data.

Specification	Value	Unit
p_{stat}	980	hPa
θ	200	°C
$F_{V,STP,\ dry}$	24,561	$m^3_{STP,dry}$ h^{-1}
$F_{m,\ dry\ gas}$	32,982	kg h^{-1}

Tab. 8.20 (continued)

Specification	Value	Unit
N_2	78	vol%
O_2	12	vol%
ROV	3	vol%
CO_2	10	vol%
$X_{H_2O,\text{volume-based}}$	100	g $m^{-3}{}_{STP,dry}$
$h_{(1+x),200}$	416	kJ kg^{-1} dry gas
H_2O	11.08	vol%
p_{H_2O}	108.5	hPa
SO_x	300	ppm
SO_x off-gas	856	mg $m^{-3}{}_{STP,dry}$
SO_x clean gas	175	mg $m^{-3}{}_{STP,dry}$
HCl	30	ppm
HCl off-gas	49	mg $m^{-3}{}_{STP,dry}$
HCl clean gas	3	mg $m^{-3}{}_{STP,dry}$
MM_{mean}	30.1	kg $kmol^{-1}$
$\rho_{STP,dry}$	1.34	kg $m^{-3}{}_{STP,dry}$
μ_{gas}	$1.5 \cdot 10^{-5}$	Pa s
ρ_{actual}	0.852	kg am^{-3}
$v_{g,actual}$	$1.76 \cdot 10^{-5}$	m s^{-1}
$D_{HCl,g}$	$1.8 \cdot 10^{-5}$	m^2 s^{-1}
$D_{SO_2,g}$	$1.5 \cdot 10^{-5}$	m^2 s^{-1}

We investigate a combination of Venturi scrubber for absorption of HCl and packed bed absorption tower for absorption of SO_2 at pH = 5.5. In Exercise 6.10, a mass-based L/G ratio of $L/G = 1$ and a gas velocity in the Venturi throat of $v_g = 150$ m s^{-1}, resulting $\Delta p_{Venturi} = 143$ hPa in $d_{1,2} = 50$ µm was proposed for successful dedusting.

a) Absorption of HCl in the Venturi quench

We make use of off-gas dedusting in a Venturi quench to get some HCl removed from the off-gas. In Exercise 8.9 we have discussed single stage co-current absorption. In the present application we cannot afford high water consumption. So we have to limit the waste water effluent from the quench to 104 kg h^{-1} (equivalent an L/G ratio for balancing this step of $L/G = 0.003$ kg kg^{-1}. By applying the algorithm as discussed in Exercise 8.3 you will be able to change the HCl load from $Y_{HCl,in} = 36$ mg kg^{-1} to $Y_{HCl,out} =$ 5 mg kg^{-1}.

Solver:

First step:

Determination of the cooling limit temperature according to Section 2.5.4, Exercise 2.7:

$\theta_{g,out} = 54.9$ °C

Second step:

Determination of specific cross-sectional absorbent load B in kg m^{-2} s^{-1}.

According to Section 6.3.4.4, the throat diameter is $d_{throat} = 322$ mm and the cross-sectional area is $A_{throat} = 0.082$ m^2. For $L/G = 1$, the circulation mass flow rate of the scrubbing liquid is $F_{m,l} = 33{,}000$ kg h^{-1}, and $B = 112.3$ kg m^{-2} s^{-1}.

Third step:
Determination of the specific mass transfer area a for $v_{rel} = v_{throat} \cdot f$, and $f = 0.35$
$v_{rel} = 150 \cdot 0.35 = 52.5\ m\ s^{-1}$

$$a = \frac{6 \cdot B}{\rho_l \cdot v_{rel} \cdot d_{dr}} = \frac{6 \cdot 112.3}{1{,}000 \cdot 52.5 \cdot 50 \cdot 10^{-6}} = 256\ m^2\ m^{-2}$$

Fourth step:
Determination of the droplet Re number

$$Re = \frac{v_{rel} \cdot \pi \cdot d_{dr}}{2 \cdot v_g} = 234.8$$

Fifth step:
Determination of Sh_g with eq. (8.104) and the data of Tab. 8.15:

$$Sh = 2 + C \cdot Re^m \cdot Sc^n = 0.6 \cdot 234.8^{0.5} \cdot \left(1.76 \cdot 10^{-5}\right)^{0.33} = 11.7$$

$$\beta_g = \frac{Sh \cdot 2 \cdot D}{d_{dr} \cdot \pi} = \frac{11.7 \cdot 2 \cdot 1.5 \cdot 10^{-5}}{50 \cdot 10^{-5} \cdot \pi} = 2.28\ m\ s^{-1}$$

$$H = \frac{F_{v,g}}{\beta_g \cdot a \cdot A_{throat}} = \frac{44{,}054}{3{,}600 \cdot 2.28 \cdot 257 \cdot 0.082} = 0.26\ m$$

Also, the mass transfer height H in the downstream spray of the Venturi scrubber was determined for β_g only, this is a sound result. We shall be able to get the hydrochloric acid removed in the Venturi quench to an outflow load of $X_{HCl,out} = 5\ mg\ kg^{-1}$. We do install an SO_2 scrubber in the second stage, operated at pH = 5.5. We will, therefore, remove HCl quantitatively.

b) Absorption of SO_2 in a packed bed absorber.
We decide for Pall rings plastic, 90 mm packings. NaOH is fed to the circulation line of the scrubber, as shown in Fig. 8.11. Referring to the physical properties of Na_2SO_4 in Tab. 8.21 we could run the process with a very high Na_2SO_4 load in the scrubber effluent. We keep it at a smooth level of 10 wt% [261].

Tab. 8.21: Physical properties of Na_2SO_4.

$w_{Na_2SO_4}$ (wt%)	$\rho_{Na_2SO_4}$ (kg m^{-3})	$c_{Na_2SO_4}$ (g L^{-1})	$\mu_{Na_2SO_4}/\mu_{H_2O}$ (–)
10	1,090.5	109	1.387

We add NaOH with a mass-based concentration of 20 wt%. The physical properties of NaOH are shown in Tab. 8.22.

Tab. 8.22: Physical properties of NaOH [261].

w_{NaOH} (wt%)	ρ_{NaOH} (kg m^{-3})	c_{NaOH} (g L^{-1})	μ_{NaOH}/μ_{H_2O} (mg kg^{-1})
20	1,219.2	243.8	4.61

We repeat the procedure, as discussed in Exercise 8.14, and we will summarize the results in Tab. 8.23.

Tab. 8.23: Results of scrubber design for SO_2 absorption from off-gas with NaOH.

Hydraulics (eqs. (8.105)–(8.110))	Value
v_{eff} (m s^{-1})	3.44
D (m)	2
v_g (m s^{-1})	3.5
n_{th} (m^{-1})	0.9
Δp (Pa m^{-1})	517
NTU (eq. (8.99))	2.44
HTU_g (m)	0.58
$H = NTU \cdot HTU_g$ (m)	1.4
E	0.5
$H_{packing}$ (m)	2.8
NaOH consumption (L h^{-1})	89
Na_2SO_4 effluent (L h^{-1})	340

Discussion: Seemingly, getting SO_2 under control at pH = 5.5 is not a huge problem. We have assumed mass transfer to be limited to the gas phase resistance. For sure, we should have compared the outcome with a different size of packing. However, we are within the operation range of the chosen packing. Comparison of NaOH consumption and Na_2SO_4 effluent shows some flexibility in concentration of the neutralizer as well as the solute in the effluent.

8.5.5.1 Summary

The intention behind this chapter "absorption" was focused on the specific needs of off-gas purification. As introductory mentioned, we do not want to compete the absorption community. Our toolbox contains the discussion of physiosorption and chemisorption. We tried to prepare and underline the difference. We tried to find a link to the thermodynamics community. We demonstrated the reason, why we should care for a relaxed relationship to the thermodynamics community by discussing the difference between energy and the ability to provide or consume work. Then we appraised the technologies, processes and ingredients, before we had fun with the design of absorption processes. We started with balancing of processes for the common phase contact patterns. We tried to figure out the difference between stage-wise phase contact and continuous contact, and we developed the different solver approaches and strategies. We always took care of discussing the different topics with several examples, by considering the boundary of performing calculations with a hand calculator. For sure you will find great support from chemical engineering software packages. We do have several great simulation software available on the engineering market. We hopefully did prepare relevant literature for your support in the engineering of absorption processes in off-gas purification. And finally, we hopefully could encourage you to have fun with engineering in the off-gas purification business.

9 Adsorption

The mass transfer unit operation adsorption plays a major role in off-gas purification. Adsorption processes are preferably applied in solvent recovery from off-gas but also find application in the process for the removal of water vapor, odors or several other vapor-phase impurities [287]. Adsorption of CO_2, COS or H_2S on activated carbon is also investigated and applied in today's off-gas purification. H_2S adsorption with impregnated activated carbon is a representative example for adsorption with chemical reaction, when heavy metal-doped activated carbon is used. H_2S forms very stable heavy metal sulfides with ZnO, FeO, etc. This process is an irreversible adsorption process and regeneration of the adsorbent is not possible. Spent sulfide laden-activated carbon may be processed in roasting processes.

However, main application of adsorption is in water treatment because of today's stringent wastewater emission limits. To remove toxic compounds or compounds that may pose problems of taste or odor, nearly 100 million pounds of activated carbons are annually used in the USA alone [288]. Dry sorptive precipitation of acidic substances such as hydrochloric acid, hydrogen fluoride and sulfur dioxide (SO_2) with powdered lime and limestone is of increasing importance. Precipitation of dioxins has to be outlined. Be careful with SO_2 adsorption from oxygen-laden carrier gas. We have outlined the specific properties and problems of sulfuric acid in off-gas purification in Section 2.5.5. SO_2 may easily form SO_3 on the active adsorbent surface, and with a minimum off-gas humidity, you immediately end up with highly concentrated sulfuric acid with disastrous effects on safety and equipment [289].

9.1 Basics

Adsorption is a surface phenomenon. Constituent molecules selectively adhere and concentrate on the surface of a porous solid, called adsorbent. In adsorptive off-gas purification the target pollutant is the transferred component, also called adsorptive. The forces adhering the adsorbed molecules on the surface can be either of physical nature, as Van der Waals and electrostatic forces, or of chemical valence forces [290]. Former adsorption process is called physisorption and latter is called chemisorption. Capillary condensation is another phenomenon that occurs in micropores of adsorbents. In capillary condensation the residual pore volume remaining after multilayer adsorption is filled with vapor condensate. A meniscus is immediately formed at the vapor–liquid interface once condensation has started, allowing for equilibrium load of constituents below saturation vapor pressure of the pure liquid. Physisorption is a reversible process. The accumulation of molecules on the adsorbent surface can be reversed by energy (e.g., heat) in the desorption step. The energy demand for the reversion of chemisorption is higher as

https://doi.org/10.1515/9783110763928-009

compared to physisorption. As already mentioned, the downstream process of activated carbon from chemisorption may need thermal destruction of the adsorbent.

A typical process flow chart of adsorption application is pictured in Fig. 9.1. For continuous operation, two adsorption towers, packed with a specific adsorbent, are arranged in parallel. After one of the adsorbers becomes saturated with the adsorbate, which means that the capacity of the adsorbent is spent, desorption and constituent recovery, as well as adsorbent regeneration has to be performed. Therefore, purge gas or steam is preheated and fed to the "saturated" adsorber, whereby the adsorbate is released from the surface of the adsorbent. When the purge gas is subsequently cooled, the target pollutant or constituent can be recovered in the condenser. This step is often denoted as regeneration step. Cyclic regeneration can be accomplished by thermal swing, pressure swing, chemical reaction (e.g., elution or supercritical extraction), or inert purge or displacement purge. As already discussed in Chapter 8, adsorption shall be operated at high pressure and low temperature, and desorption prefers low pressure and high temperature. Adsorption columns for off-gas purification are generally designed in fixed-bed, moving-bed or fluidized-bed setup.

The basic concepts involved in the design of adsorption plants are similar to other separation processes. The off-gas is passed through a bed of the adsorbents at velocities that are reconcilable with pressure drop under process conditions to allow mass transfer of the target constituent from the carrier gas to the adsorbent [278]. Information about the equilibrium between the solid and the fluid phase are the basic needs in adsorption process design, giving access to the adsorption capacity of the adsorbent. The rate of adsorption, mainly controlled by the rate of diffusion, as well as material- and energy balances need to be performed.

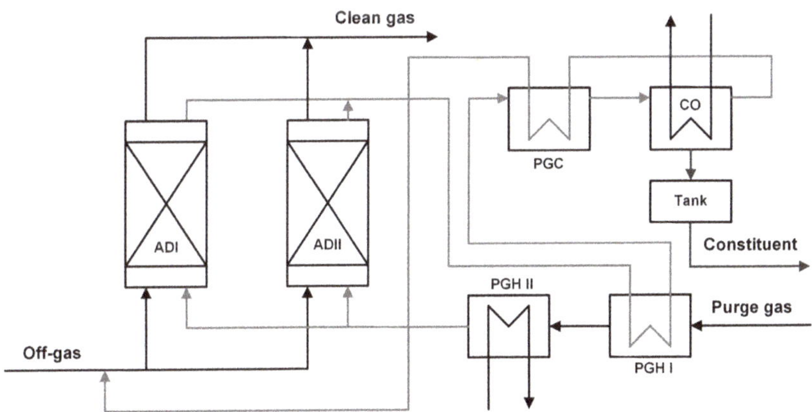

Fig. 9.1: Flow chart of adsorption process with two adsorption towers ADI and ADII, two purge gas preheaters PGH I and PGH II, a purge gas cooler PGC and a condenser CO.

! When you count the number of heat exchangers in Fig. 9.1 you may correctly conclude that you should make yourself familiar with heat exchange and the design of heat exchangers in adsorption process design.

Do not hesitate to ask the library for detailed basics [291–293]. Noll et al. [294] are still a nice introduction into adsorption technology. Since we talk about a mass transfer unit operation, you may also pick up their contributions to adsorption technology, for example [295].

9.2 Adsorbents

Adsorbents are characterized by their surface properties, in particular, the surface area and surface polarity.

The adsorption capacity is directly related to the adsorbents surface area, and it is the most important characteristic of an adsorbent. Commercial adsorbent materials are therefore prepared to have a very large surface area per unit mass, resulting in a large number of micropores below the adsorption surfaces. The size of micropores affects the accessibility of the molecules to the adsorption surface, as schematically shown in Fig. 9.5. Macropores have their origin in the texture of raw materials and result from granulation treatment. Macropores provide the diffusion paths for the molecules to migrate from the adsorbent particle surface to the micropores. According to the International Union of Pure and Applied Chemistry (IUPAC) [296] recommendation, the pore size of adsorbents is classified into (i) micropores that are ≤2 nm in size, (ii) mesopores that are between 2 and 50 nm in size, and (iii) macropores being ≥50 nm in size.

Another important property of adsorbents is their pore size distribution, characterizing the absorptivity of molecules. Surface polarity corresponds to affinity with polar substances. Polar adsorbents are hydrophilic as for example aluminosilicates such as zeolites, porous alumina, silica gel or silica-alumina. Nonpolar adsorbents are hydrophobic as for instance carbonaceous adsorbents or polymer adsorbents. These adsorbents have higher affinity with hydrocarbons.

The specification of industrially applied adsorbents is briefly described in the following section. With regard to metal oxides the adsorption capacity of magnesium oxide, titanium oxide, zirconium oxide and cerium oxide has to be pointed out. Magnesium oxide is preferably used in removing polar molecules such as pigments, sulfide compounds from gasoline and silica from water. Magnesium hydroxide is an appropriate adsorbent for phosphate removal in waste water treatment. Oxides of the four valence metals show selective adsorption properties in removing anions from aqueous feed. Hydrous titanium is a selective adsorbent for recovering uranium in seawater. The monohydrate of zirconium oxide is found to adsorb phosphate ions and cerium oxide is effective in fluoride ion adsorption in industrial waste water treatment [291].

9.2.1 Activated carbon

Activated carbons are microporous carbonaceous adsorbents. Commercially available activated carbons are prepared from carbon containing source materials such as lignite, coal, wood, nut shell, petroleum coke and synthetic polymers.

The base material is pyrolized and carbonized to remove the volatile fraction and low molecular weight products of pyrolysis. The residual carbonaceous material undergoes activation in oxidizing atmosphere (air, water at about 800 °C or carbon dioxide at higher temperature). During activation, micropores are formed. The yield of activated carbon from raw materials is less than 50%. Carbonization and activation can also be performed with inorganic chemicals such as zinc chloride or phosphoric acid. This method provides larger micropores, preferred for the adsorption of larger molecules [291].

Micropores are in the form of two-dimensional spaces between neighboring graphite-like layers. During activation and after activation the surface partially forms oxygen complexes. Surface oxides add a polar nature to activated carbons resulting in elevated hydrophilicity, acidity and negative ζ– potential [291].

Oxygen complexes exist in:
- strong carboxylic groups,
- weak carboxylic groups,
- phenolic groups and
- carbonyl groups.

Activated carbon partially contains ashes derived from starting materials. The amount of ash can range from 1% to 12%. Ashes mainly consist of alumina, silica, iron and alkaline earth metals. Ashes may increase hydrophilicity of activated carbon and catalytic activity [291].

In commercial use, activated carbon is of powder form (powdered activated carbon (PAC)) with an average size of 15–25 µm and granular or pelletized form (granular activated carbon (GAC)). PAC assures that intraparticle diffusion will not be rate-determining. Major industrial application of PAC is decolorization in food processes and water treatment. Due to the considerable dependency of the ζ-potential on the pH-value PAC positively affects coagulation, sedimentation and filtration.

GACs are either applied in crushed granules or in pelletized form. Size of granules differs depending on the application. In gas phase adsorption cylindrically extruded pellets of between 4 and 6 mm or crushed and sieved granules are preferably used. Main applications in gas phase adsorption are solvent recovery, air and gas purification, flue gas desulfurization and bulk gas separation. In liquid phase adsorption, smaller adsorbent particles are advantageous considering intraparticle diffusion. In most applications, spent GAC is regenerated by thermal methods.

Carbon molecular sieving (CMS) effects can be prepared by using proper starting materials and selecting conditions such as carbonizing temperature, activation temperature and time. Main applications of CMS are separation of oxygen and nitrogen on the basis of different diffusion properties. In liquid-phase application adjustment of fragrance in winery is representative. CMS is specified by uniform and narrow micropore size distribution.

Activated carbon fiber (ACF) is manufactured by pyrolizing and carefully activating synthetic fibers (phenolic resin, polyacrylic resin, viscose rayon). Mean diameter of ACF is about 7–15 μm, resulting in very fast intrafiber diffusion. Overall rate of adsorption is controlled by longitudinal diffusion in the bed. ACF is supplied as fiber mat, cloth and cut fibrous chip. Main application of ACF sheet adsorbent is solvent recovery [288, 291]. Table 9.1 shows the physical properties of several activated carbon adsorbents [291, 297].

Tab. 9.1: Physical properties of several activated carbon adsorbents.

Material	Shape	Porosity	Bulk density (kg m^{-3})	Pore radius (nm)	Surface area (m^2 g^{-1})
Shell-based	G	0.5–	433–	2	800–1,100
	P	0.65	575	2	1,500
Wood-based	C	0.7–	192–	2.2	750–1,450
	P	0.3–0.5	144–560	–	600–1,200
	G	0.6	320	2–4	600–1,200
Coal-based	G	0.65	320–480	2–3.8	500–1,200
	P	0.56–0.67	400–480	2	1,000–1,400

G, granular; P, powder; C, cylindrical pellets; S, spherical pellets.

9.2.2 Silica and alumina

9.2.2.1 Silica gel

Pure silica, SiO_2, is nonpolar and chemically inactive. Activation with hydroxyl functional groups will result in very polar and hydrophilic properties. Silica gel is prepared by coagulation of colloidal solution of silicic acid followed by controlled dehydration. Liquid sodium silicates are neutralized with sulfuric acid and then coagulated from hydrogel. Sodium sulfate, formed during neutralization, is removed by washing. The hydrogel is dried, crushed and sieved. Spherical silica gel particles are formed by spray drying of hydrogel. Micropore size distribution depends on the manufacturing process and is either 2.0–3.0 nm or about 7 nm. The amount of fixed water is about 40–60 mg g^{-1}. When elevating the temperature above $\theta = 350$ °C the adsorption capacity for water is lost due to the loss of fixed water. Main application of silica gel is dehumidification of gases and hydrocarbons [291]. Table 9.2 shows the physical properties of selected silica gel adsorbents [291, 297].

Tab. 9.2: Physical properties of selected silica gel adsorbents.

Shape	Porosity	Bulk density (kg m^{-3})	Pore radius (nm)	Surface area (m^2 g^{-1})
G	0.35–0.5	640–770	2–4	650–900
P	0.51	650–830	2–2.8	650–700
S	0.45	740–820	7	250

G, granular; P, powder; C, cylindrical pellets; S, spherical pellets.

9.2.2.2 Active alumina

Active alumina is mainly γ-alumina with a pore radius of 1.5–6 nm and a porosity of 40–75%. Porous alumina particles are produced by dehydration of alumina (tri-)hydrates ($Al_2O_3 \cdot 3H_2O$) to about 6% of moisture content. Active alumina is also used in dehumidification and for the removal of polar gases from hydrocarbons. Table 9.3 shows the physical properties of selected active alumina adsorbents [288, 291].

Tab. 9.3: Physical properties of selected active alumina adsorbents.

Material	Shape	Porosity	Bulk density (kg m^{-3})	Pore radius (nm)	Surface area (m^2 g^{-1})
Active alumina	G	0.25–0.35	800	3.5–4.5	230
	S	0.5–0.6	750–800	4–5	400

G, granular; P, powder; C, cylindrical pellets; S, spherical pellets.

9.2.3 Zeolite

Zeolite is an aluminosilicate mineral which swells under the blowpipe. About 30 natural zeolites have been identified so far. Many types can be synthesized industrially. Crystalline structures are composed of tetrahedral units with Si located in the center. Several units form secondary units, the arrangement of which forms regular crystalline structures. Regular crystalline structures provide unique adsorption characteristics. Replacement of Si by Al results in a deficit of positive valence, requiring the addition of cations such as alkaline earth ions. These cations are easily exchangeable, and the size and the properties of these ions modify adsorption characteristics of zeolites since they affect the size (aperture) of the window between the cells [291].

The main applications of zeolite as adsorbent are drying, deodorization, air separation, ion exchange for water purification especially for removing ammonium ion and heavy metal ions. Clinoptiliolite, a natural zeolite, is well-known for its adsorption selectivity of the ammonium ion [291].

The adsorption selectivity of synthetic zeolites can be adjusted by altering the exchangeable cation located near the window between neighboring cells. The specification of the zeolite, which is actually the specification of the aperture of the window, indicates the type of cation. So-called 4A-type zeolites contain Na ion at this site. The aperture of the window is 0.4 nm. The 3A-type zeolites contain K at this site, limiting the aperture to 0.3 nm only H_2O and NH_3 can penetrate through the window. Introducing Ca, which has two valences, will result in so-called 5A-type zeolite. X-type zeolite, made up of 12-membered rings, has much larger windows. It is usually called 13X zeolite [288, 291].

9.2.4 Adsorbent selection

Choice of adsorbents depends on the field of application. Prior to any technical discussion the complete specification of the gaseous or liquid carrier fluid, the specification of the quality of the recovered adsorbate is necessary. Demonstrated with the example of dehumidification, any of the above-described adsorbents shows adsorption capacity. But adsorption character and adsorption capacity are very different. Qualitatively Fig. 9.2 points out the difference. The sketch shows moisture isotherms for several desiccants.

Activated carbon, being hydrophobic, has low humidity adsorption capacity, particularly at low relative humidity. Silica gel has high adsorption capacity at high humidity, but modest water uptake at low moisture levels. Zeolites have the highest uptake at low humidity and moderate capacity at high humidity. Deep drying preferably demands zeolites (high capacity at low humidity) while silica gel is suitable for more modest dew point requirements.

The different shape of the isotherms of type I silica gel and type II silica gel demonstrates the influence of the manufacturing process on adsorption properties [298].

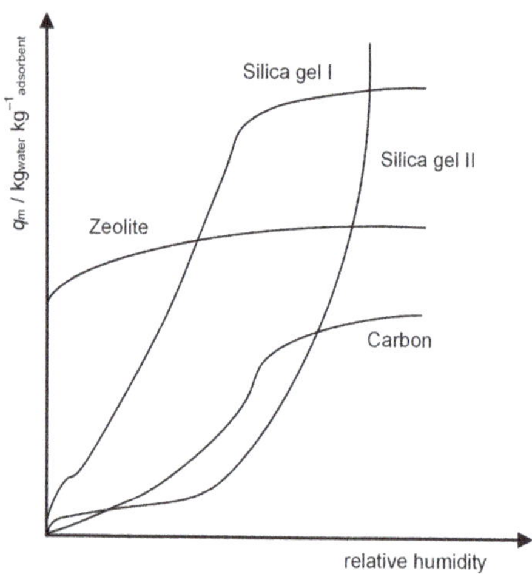

Fig. 9.2: Moisture isotherms for several desiccants.

9.2.5 Evaluation of pore-related properties

9.2.5.1 Porosity

The porosity of pores is determined by mercury penetration. Mercury (Hg) can penetrate pores with a radius larger than 7.5 µm. By comparing the mass of a glass pycnometer filled with mercury m_{Hg}, with the mass of the sample m_{sample} and with the mass of the sample filled with mercury after evacuation m_p, the particle density (= particle volume) ρ_p can calculated according to eq. (9.1), with ρ_{Hg} being the density of mercury [291]:

$$\rho_p = \frac{m_{sample}}{\frac{m_{Hg} - m_p + m_{sample}}{\rho_{Hg}}} \tag{9.1}$$

By replacing mercury with a liquid that is able to penetrate smaller pores (e.g., H_2O), the particle density can be approximated more accurately. Another method is to employ a displacement type pycnometer.

9.2.5.2 Pore size distribution and surface area

Pore size distribution is either evaluated by the mercury penetration, by the nitrogen adsorption or by the molecular probe method. The internal surface area is derived from BET measurements [291].

9.3 Adsorption equilibrium

The fundamental concept in adsorption is based on adsorption isotherms giving the equilibrium concentration of a certain substance in the fluid phase Y and the solid phase q for a constant temperature. Adsorption isotherms can be determined by bringing an adsorbent in contact with a gaseous or liquid containing the adsorbate. After a sufficiently long time, equilibrium will be reached and the amount of adsorbate is determined in this state. The maximum amount of the adsorbate taken up by the adsorbent, per unit mass (or volume) of the adsorbent, is also called adsorption capacity q_0 or loading. The adsorption capacity is crucial for the capital cost since it gives the amount of adsorbent needed and, by implication, the volume of the adsorber vessels. The adsorption equilibrium strongly depends on the temperature and system pressure. Lower temperatures and higher pressure favor the adsorption step while higher temperatures and lower pressure favor desorption.

Based on the Brunauer–Deming–Deming–Teller [299] classification, the IUPAC proposed a recommendation to classify physisorption isotherms into six types [296] shown in Fig. 9.3. The amount adsorbed q/q_0 is plotted against the equilibrium relative pressure p/p_0, where p_0 is the saturation pressure of the pure adsorptive. If the operation temperature is above the critical temperature of the adsorptive, it is recommended to plot the amount adsorbed solute against the pressure p. (Comment: You can also replace the partial pressure by c (= p/RT) and use the equilibrium concentration c/c_{Sat} as x-axis with c_{Sat} being the saturation concentration). The type of isotherm for a specific system depends on superficial properties, pore structure features the adsorbate and the adsorption itself. Either monolayer adsorption (single-molecule adsorption), multilayer adsorption (polymolecular adsorption) or capillary condensation occur and affect the shape of the adsorption isotherm. Horizontal isotherms indicate irreversible adsorption [296, 300].

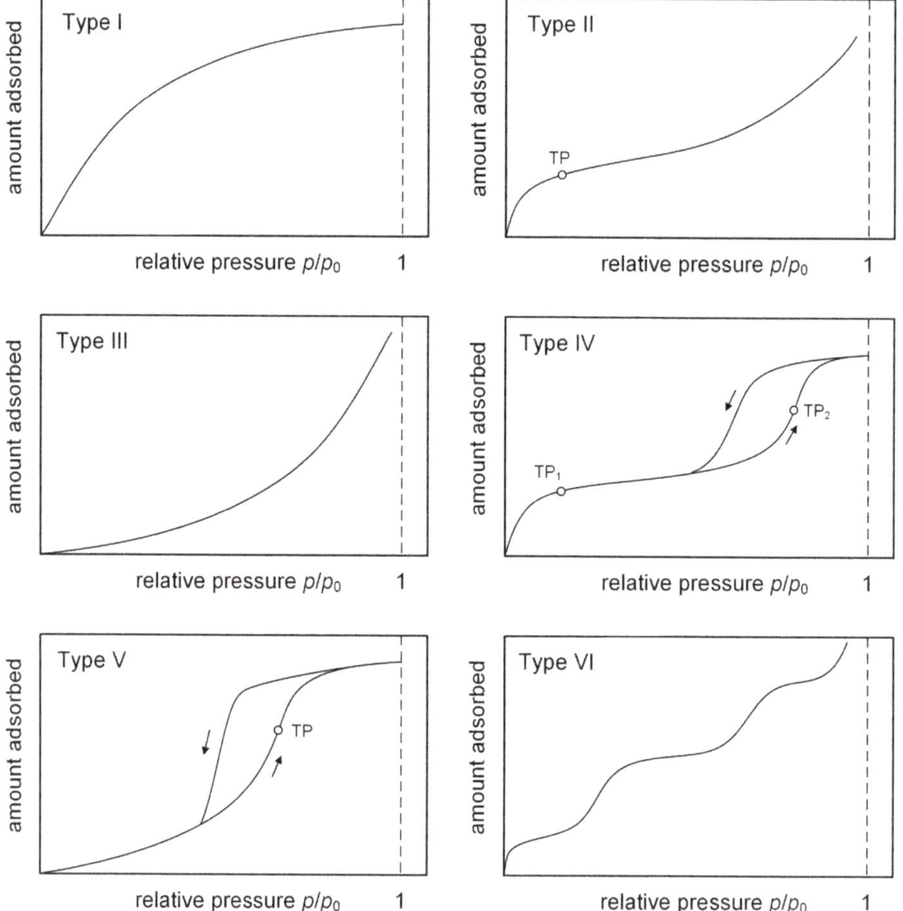

Fig. 9.3: Classification of adsorption isotherm for monoconstituent substance according to IUPAC [296, 301].

9.3.1 Isotherm classification

(1) Type I isotherms (we like this type of adsorption curves) describe single-molecule adsorption characterized by the continuous increase in the adsorption volume up to a specific relative pressure $p/p_0 \sim 1$. The shape of type I isotherms is convex (concave to the x-axis) throughout which is typically for microporous adsorbents having a relatively small external surface, for example, activated carbons [296]. Type I isotherms also occur when micropores are filled up by multilayer adsorption. For mathematical description of type I isotherms the Langmuir equation (eq. (9.2)) is best suitable. Type I isotherms are favorable for adsorption processes but unfavorable for the desorption step.

(2) Type II isotherms represent unrestricted monolayer and multilayered adsorption, capillary filling as well as capillary condensation. The middle section of type II isotherms, starting at the inflection (TP), is nearly linear and indicates the transition from monolayer adsorption to multilayer adsorption [296]. Type II isotherms are typical for nonporous or macroporous adsorbents and are generally used at lower adsorption temperature than the critical temperature of the adsorbate (the adsorption saturation point will not be reached). The monomolecular heat of adsorption is greater than the condensation heat.

(3) Type III isotherms exhibit a concave shape (convex to the x-axis) throughout and represent unrestricted multilayer adsorption. The pore allocation of type III adsorbents is the same as for type II but interaction between the adsorbent surface and the adsorbate is responsible for the different shape. The monomolecular heat of adsorption is lower than the condensation heat.

(4) Type IV isotherms exhibit two inflections and a characteristic hysteresis loop. Adsorption hysteresis arise when desorption follows a different path from adsorption, a result of capillary condensation in the mesopores. The initial part of type IV isotherms follows the same shape as type II isotherms, indicating unrestricted monolayer- and multilayered adsorption, but tends to level off after exceeding a specific p/p_0 value. The majority of industrial mesoporous adsorbents give type IV isotherms.

(5) Type V isotherms are related to type III isotherms but show an S-shaped course. The interactions between the adsorbent and adsorbate are rather weak but can be retained with certain porous adsorbents. Type V isotherms also exhibit a hysteresis loop.

(6) Type VI isotherms represent stepwise multilayer adsorption on a nonporous solid. The sharpness of the steps depends on the system and the temperature while the step-height gives the monolayer capacity for each adsorbed layer.

Modeling of adsorption isotherms is derived from the chemical potential. Adsorption isotherms can be described in several mathematical forms either based on simplified physical phenomena or empirically with two or four empirical parameters [291]. In the following sections, the most common isotherm equations are presented.

9.3.2 Isotherm equations

9.3.2.1 Langmuir isotherm

The most commonly used adsorption isotherms are classified Langmuir isotherms. Langmuir isotherms were already developed about 100 years ago but are still successful in fitting a wide variety of adsorption data quite well and easy [302]. The Langmuir equation [303] is a semiempirical approach having one parameter and can be derived from adsorption–desorption kinetics or from statistic thermodynamics. The Langmuir model is based on the assumption of monomolecular covering of the adsorbate surface without any interaction between the adsorbed molecules, leading to an energetically uniform surface. With the fractional coverage $\theta = q/q_0$ and the partial pressure of the adsorbate p (or the concentration $c = p/R\cdot T$), the adsorption rate is $k_a \cdot p \cdot (1-\theta)$ and the desorption rate is $k_d \cdot \theta$ when assuming first-order kinetics. k_a is the rate constant for adsorption and k_d is the rate constant for desorption. For equilibrium condition ($\mu_g = \mu_{Ad}$) the adsorption equilibrium constant $K = k_a/k_d$ and hence the Langmuir equation is derived according to eq. (9.2) [291, 304]:

$$\theta = \frac{q}{q_0} = \frac{K \cdot p}{1 + K \cdot p} \tag{9.2}$$

For application in adsorption from gas phase, conversion from molar into mass ratios is recommended and can be carried out by replacing the partial pressure p in the Langmuir equation via the ideal gas law, resulting in the following equation [304]:

$$q_m = \frac{P \cdot MM_{CG} \cdot K \cdot q_{0,m} \cdot Y}{1 + \left(\frac{K \cdot P \cdot MM_{CG}}{MM_{ads}}\right) \cdot Y} \tag{9.3}$$

In eq. (9.3), q_m is the mass load of adsorbate to the mass of adsorbent in kg kg^{-1}, $q_{0,m}$ is the mass load capacity of the adsorbent in kg kg^{-1}, MM_{CG} is the molar mass of the carrier gas in kg kmol^{-1}, MM_{ads} is the molar mass of the adsorbate in kg kmol^{-1}, P is the total pressure in hPa and Y is the mass load of adsorbate to the mass of the carrier gas in kg kg^{-1}.

Equation (9.3) can further be simplified by substituting $K \cdot q_{0,m} = a$ and $K = b$, resulting in the following equation:

$$q_m = \frac{P \cdot MM_{CG} \cdot a \cdot Y}{1 + \left(\frac{b \cdot P \cdot MM_{CG}}{MM_{ads}}\right) \cdot Y} \tag{9.4}$$

In several types of adsorption applications, eq. (9.4) may be specified by the parameter m and summarized with $q_{0,m}$, as follows:

$$q_{\mathrm{m}} = \frac{q_{0,\mathrm{m}} \cdot Y}{m + Y} \tag{9.5}$$

9.3.2.2 Henry isotherm

For low surface coverage, accompanied by sufficiently low amounts adsorbed q compared to the capacity of the adsorbent q_0, the adsorption isotherms reduce to a linear form. The degree of adsorption is proportional to the carrier fluid concentration (= constant slope of the isotherms) and can thus be described with Henry's law. The Langmuir equation is reduced according to eq. (9.6):

$$\theta = \frac{q}{q_0} = K_{\mathrm{H}} \cdot p \tag{9.6}$$

where K_{H} is the Henry coefficient and is analog to the adsorption equilibrium constant K. For highly concentrated carrier fluid ($p \gg 1/K$) the adsorption sites are saturated and $\theta = 1$ [291].

9.3.2.3 Freundlich isotherm

Besides the Langmuir isotherms, the Freundlich isotherms are commonly used although they are, thermodynamically considered, inconsistent. The Freundlich-type equation [305], as shown in eq. (9.7), does not give any limit to the adsorption capacity, whereby the amount adsorbed goes to infinity when concentration increases. It is only applicable below the saturation concentration (solubility or vapor pressure) where condensation prevails and adsorption phenomena are no more significant [291]. Although the Freundlich equation fails to depict the required linear behavior in the Henry's law region, it can generally be used to correlate data on adsorbents in a relatively wide range of concentration [297]. The adsorption equilibrium constant K_{F} as well as n_{F} needs to be determined empirically:

$$q = K_{\mathrm{F}} \cdot c^{\frac{1}{n_{\mathrm{F}}}} \tag{9.7}$$

A combination of the Henry-type equation and the Freundlich-type equation is proposed by Radke and Prausnitz (eq. (9.8)) [306]. This type of isotherm equation contains three empirical constants and is able to correlate adsorption data in a wide range of concentration:

$$q = \frac{1}{\frac{1}{K_{\mathrm{H}} \cdot p} + \frac{1}{K_{\mathrm{F}} \cdot p^{\frac{1}{n_{\mathrm{F}}}}}} \tag{9.8}$$

9.3.2.4 Brunauer–Emmet–Teller (BET) isotherm

The classical isotherm for multilayer adsorption on a homogeneous, flat surface is the BET isotherm [297]. Adsorption in multilayer is considered to be on different attractive forces compared to surface adsorption. Adsorption above monolayers equals condensation of adsorbate molecules, giving rise to the BET [299] equation:

$$\frac{q}{q_{\text{mon}}} = \frac{K_B \cdot \frac{p}{p_0}}{\left(1 - \frac{p}{p_0}\right) \cdot \left(1 - \frac{p}{p_0} + K_B \cdot \frac{p}{p_0}\right)} \tag{9.9}$$

where q_{mon} is the amount adsorbed by monomolecular coverage on the surface [291].

9.3.2.5 Fowler and Guggenheim isotherm

Considering interaction between adsorbing molecules, Fowler and Guggenheim [307] extended the Langmuir correlation, according to the following equation:

$$p = \frac{\theta}{K \cdot (1-\theta)} \cdot e^{\frac{2 \cdot u \cdot \theta}{k_B \cdot T}} \tag{9.10}$$

The interaction energy is positive in case of repulsion and negative in case of attraction and is represented by $2 \cdot u$ (k_B is the Boltzmann constant). When adsorbed molecules are free to move on the adsorbent surface (mobile adsorption), the Langmuir equation is modified to the following equation:

$$p = \frac{\theta}{K \cdot (1-\theta)} \cdot e^{\frac{\theta}{1-\theta}} \tag{9.11}$$

Considering mobile adsorption and interaction, the isotherm correlation corresponds to the following equation:

$$p = \frac{\theta}{K \cdot (1-\theta)} \cdot e^{\frac{\theta}{1-\theta} + \frac{2 \cdot u \cdot \theta}{k_B \cdot T}} \tag{9.12}$$

Exercise 9.1: Equilibrium of acetone/air/GAC at varying temperature

For a separation task you are asked to investigate the temperature dependence of the adsorption of acetone on GAC and determine the equilibrium data. What type of isotherm fits the equilibrium best?

After having spent sufficient time in the laboratory, you get all equilibrium data for the investigated temperature $\theta = 30$, 100 and 160 °C. You identify type I adsorption isotherm for the system acetone/air/GAC and therefore correctly conclude to use the Langmuir equation to fit your dataset. Next you transform the data to a linear form and estimate the parameter $q_{0,m}$ and m for the selected adsorption temperature, as given in Tab. 9.4 [292], graphically (or via linear regression).

Tab. 9.4: Langmuir parameter $q_{0,m}$ and m for the adsorption of acetone for selected adsorption temperature $\theta = 30$, 100 and 160 °C.

θ (°C)	$q_{0,m}$ (kg kg^{-1})	m
30	0.411	0.043
100	0.273	0.217
160	0.179	0.407

According to eq. (9.5):

$$q_{m,\ \theta=30\,°C} = \frac{0.411 \cdot Y}{0.043 + Y}$$

$$q_{m,\ \theta=100\,°C} = \frac{0.273 \cdot Y}{0.217 + Y}$$

$$q_{m,\ \theta=160\,°C} = \frac{0.179 \cdot Y}{0.407 + Y}$$

You can now plot the Langmuir isotherms for the needed range of concentration, as exemplarily shown in Fig. 9.4. The unfavorable effect of increasing temperature on adsorption is clearly pictured. The amount adsorbed decreases with increasing temperature.

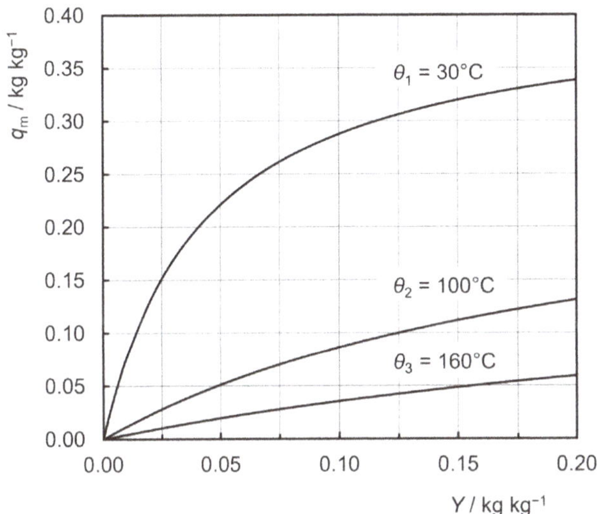

Fig. 9.4: Langmuir isotherms (type I) for the system acetone/air/GAC at the temperature $\theta = 30$, 100 and 160 °C.

A large collection of equilibrium data for common standard systems, can be found in [308]. Standard systems therefore do not necessarily have to be examined in the laboratory. You can also save time by contacting the adsorbent manufacturer. !

9.3.3 Micropore adsorption

In case of comparable size of micropores and adsorbate, adsorption takes place by attractive force from the wall surrounding the micropores. The molecules thus fill the micropores volumetrically, which is comparable to capillary condensation. By using the adsorption potential, the equilibrium relation for a given adsorbate and adsorbent combination can be given independently of the temperature, according to the following equation [291]:

$$W_{MP} = \frac{q}{\rho_{ads}} = W_{MP}(A) \tag{9.13}$$

In eq. (9.13), W_{MP} is the volume of micropores filled by the adsorbate, ρ_{ads} is the density of the adsorbed phase and A is the adsorption potential. $W_{MP}(A)$ is thus the adsorption characteristic curve. For operation conditions below the critical temperature, ρ_{ads} is the density of the liquid phase at the adsorption temperature. For adsorption operation conditions above the critical temperature, ρ_{ads} needs to be calculated according to eq. (9.14), where ρ_b is the density of the liquid at normal boiling point T_b and ρ_c is the density at the critical temperature T_c of the adsorbed phase:

$$\rho_{ads} = \rho_b - \frac{(\rho_b - \rho_c) \cdot (T - T_c)}{T_c - T_b} \tag{9.14}$$

The density at the critical temperate is given with $\rho_c = M_{ads}/b_{VdW}$, where M_{ads} is the molar mass of the adsorbate and b is the Van der Waals constant $b_{VdW} = (R \cdot T_c)/(8 \cdot p_c)$.

The adsorption potential A is the difference in free energy between the adsorbed phase and the saturated liquid and can be defined according to the following equation:

$$A = -R \cdot T \cdot \ln \frac{p}{p_S} \tag{9.15}$$

Assuming Gaussian-type distribution for the adsorption characteristic curve $W(A)$, the volume of micropore filled with the adsorbate can be described by the Dubinin-Radushkevich [309] equation:

$$W_{MP} = W_0 \cdot e^{-k \cdot A^2} \tag{9.16}$$

This equation was generalized by Dubinin and Astakhov [310] according to the following equation:

$$W_{MP} = W_{MP,0} \cdot e^{-\left(\frac{A}{E}\right)^n} \tag{9.17}$$

In eq. (9.17), E describes the characteristic energy of adsorption derived from the adsorption potential A at $W/W_0 = 1/e$. The exponent n was originally considered to have an integer value and served as correction factor for surface adsorption. For instance,

$n = 1$ corresponds to adsorption on the surface, $n = 2$ corresponds to adsorption in micropores and $n = 3$ to adsorption in ultramicropores.

If the Dubinin–Astakhov equation [310] is rearranged (eq. (9.18)), it is possible to determine n and E, as far as $W_{MP,0}$ is known. Therefore, the left-hand side of eq. (9.18) needs to be plotted against ln A, resulting in linear form and thus the parameter can be estimated either by graphical means or by linear regression. $W_{MP,0}$ is estimated from the maximum load capacity of the adsorbent that correspond with the micropore volume:

$$\ln\left[\ln\left(\frac{W_{MP,0}}{W_{MP}}\right)\right] = n \cdot (\ln A - \ln E) \tag{9.18}$$

Table 9.5 represents several characteristics values of $W_{MP,0}$, n and E for the adsorption on MSC-5A [291].

Tab. 9.5: $W_{MP,0}$, n and E values of different adsorbates for adsorption on carbon molecular sieve MSC-5A [291].

Adsorbate	$W_{MP,0}$ (cc g^{-1})	n	E (cal mol^{-1})
Nitrogen	0.170	2.6	2,800
Carbon dioxide	0.168	2.3	2,700
Oxygen	0.185	2.3	2,200
Methane	0.175	2.8	3,200
Ethylene	0.175	3.0	3,700
Ethane	0.175	2.9	4,000
Propylene	0.175	3.0	5,100
n-Hexane	–	2.8	7,300
Ethylacetate	–	3.1	6,600
Trichloroethene	–	3.2	7,500
Acetone	–	2.8	5,000
Methanol	–	2.7	2,600
Ethanol	–	2.7	4,100
Acetic acid	–	3.0	5,000

9.4 Transport and dispersion mechanism

Figure 9.5 depicts a section of a porous adsorbent particle to illustrate pore size distribution and the nature and location of individual transport and dispersion mechanism (=similar to catalytic reaction). The mass transport in an adsorbent is considered to be a combination of several mechanisms that can roughly be divided into intraparticle transport mechanism and extraparticle transport and dispersion mechanism. Intraparticle transport mechanism are pore diffusion, surface diffusion and reaction kinetics at phase boundaries. Film diffusion can be classified to extraparticle transport and dispersion mechanism. Which diffusion mechanism becomes dominant depends on the adsorbent structure, while sometimes two or three of them can either compete or cooperate. Adsorption condition such as temperature and concentration range and the combination of adsorbate and adsorbent also affect the transport mechanism. [291, 296]

Diffusion of adsorbed molecules occurs due to a concentration distribution in the particle. The phenomenological aspect of diffusional mass transfer is described by the diffusion coefficient defined by Fick's first law taking an appropriate concentration gradient as driving force:

$$J = -D_{\mathrm{i}}(c_{\mathrm{i}}) \cdot \frac{\mathrm{d}c_{\mathrm{i}}}{\mathrm{d}x} \tag{9.19}$$

A more correctly description of the driving force would be the description by means of the chemical potentials. However, eq. (9.19) provides a correct representation of adsorption system as long as diffusivity is allowed to be a function of the adsorbate concentration [297].

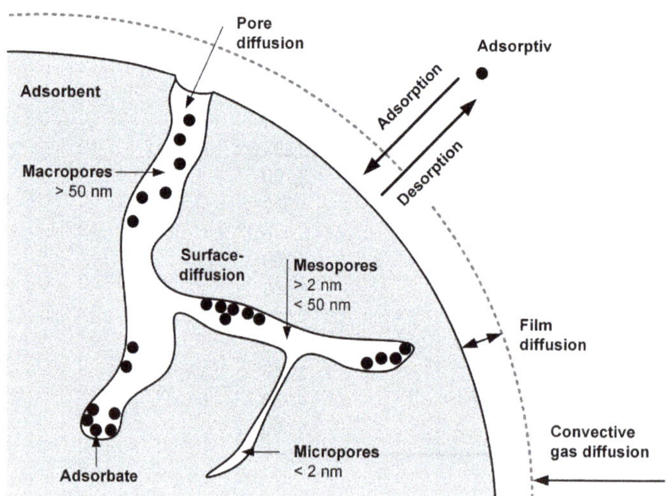

Fig. 9.5: Schematic drawing of the section of a porous adsorbent particle illustrating pore size distribution and transport and dispersion mechanism.

9.4.1 Pore diffusion

Pore diffusion occurs because of pores that are sufficiently large (= macropores) that the adsorbing molecules slip through the force field of the adsorbent surface. Macropores act like a highway for the adsorbing molecules to reach the inward of the adsorbent particle. This is typical for bidispersed pore structure, exemplarily found in activated carbon. Molecular diffusion or Knusden diffusion takes place in this case. The driving force can be approximated by the gradient in concentration of the diffusing species within the pores, if the molar concentration is constant [297].

9.4.2 Surface diffusion

Surface diffusion occurs in pores that are sufficiently small. The adsorbing molecules never escape the force field of the adsorbent surface. Strongly adsorbed molecules tend to form two or more layers and the adsorbed molecules are mobile on the adsorbent surface, for example, volatile hydrocarbons on activated carbon. Transport may occur due to an activated process involving jumps between adsorption sites while adsorbed molecules migrate quickly due to their high density relative to the fluid phase counterparts. Migration of the molecules adsorbed prevails pore diffusion to intraparticle diffusion and thus may contribute to transport of the adsorbates into the particle. When the micropore size is similar to the molecule size of the adsorbate, diffusion of the molecules is limited and the rate of adsorption in the micropore significantly affects the overall adsorption rate. The adsorbate properties are thus crucial. The concentration gradient of the species in its adsorbed state can be used to approximate the driving force for the process [288, 291, 297].

9.4.3 Film diffusion

Film diffusion occurs at the external surface of the adsorbent particle. The driving force for mass transfer is the concentration difference across the boundary layer that surrounds the particles. The boundary layer is affected by the hydrodynamic conditions of the bulk phase (= outside the particles). The diffusion of the molecules from the gas core to the solid surface is driven by a concentration gradient, equally to absorption processes [311].

9.5 Heat of adsorption

Adsorption processes are always accompanied by the release of heat. The exothermic character of adsorption processes can be explained by the lower energetical state of the adsorbate on the adsorbent surface than in the fluid carrier phase. Vapor molecules usually condense on the surface of the adsorbent if the saturation temperature is lower than the critical temperature of the transferred component, whereby it comes to an additional release of condensation heat [311]. The heat of adsorption is a measure of the energy needed for regeneration (desorption), therefore low values are preferable [288]. It also states the temperature increase that can be expected caused by adsorption under adiabatic conditions.

The heat of adsorption ΔH_{Ad} in kJ kg^{-1} for gas phase adsorption can be defined as the difference of enthalpies of the adsorptive in the fluid phase and the adsorbent fixed solute. In practical application the differential heat of adsorption Q_{diff} and the isosteric heat of adsorption Q_{st} are specified. The differential heat of adsorption Q_{diff} is the amount of heat emitted adiabatically during adsorption per mole of adsorbate. It is measured calorimetrically. The isosteric heat of adsorption Q_{st} is derived from isotherms at different temperature. The correlation between the isosteric heat of adsorption Q_{st} and the differential heat of adsorption Q_{diff} is given with the volume work equivalent to $p \cdot V$ (= $R \cdot T$), as shown in the following equation [291]:

$$Q_{st} = Q_{diff} + R \cdot T \tag{9.20}$$

Based on Van't Hoff eq. (9.21), Q_{st} can be calculated with experimentally obtained isotherms at the temperatures T_1 and T_2 as shown in the following equation:

$$Q_{st} = R \cdot T^2 \cdot \left(\frac{\mathrm{d}\ln P}{\mathrm{d}T}\right) = R \cdot \frac{\mathrm{d}\ln P}{\mathrm{d}\left(\frac{1}{T}\right)} = R \cdot T^2 \cdot \left(\frac{\mathrm{d}\ln c}{\mathrm{d}T}\right) + \frac{RT}{2} \tag{9.21}$$

$$Q_{st} = R \cdot \frac{\ln p_1 - \ln p_2}{\frac{1}{T_1} - \frac{1}{T_2}} \tag{9.22}$$

For the Henry and the Langmuir equation, Q_{st} can be described/calculated with the equilibrium constant K according to the following equations:

$$Q_{st} = R \cdot \frac{\mathrm{d}(\ln K)}{\mathrm{d}\left(\frac{1}{T}\right)} \tag{9.23}$$

or

$$K = K_0 \cdot e^{\frac{Q_{st}}{R \cdot T}} \tag{9.24}$$

In case of energetically homogenous adsorption sites without interaction between adsorbate molecules the heat of adsorption is independent of the adsorbed amount. When interaction between adsorbed molecules cannot be neglected or in case of different energy levels arising on the surface, the heat of adsorption varies with the surface coverage. The variation of the heat of adsorption can either be described with eq. (9.25), defining the spectral density of adsorption sites, or as a function of the amount adsorbed $Q(q)$:

$$\int_0^{Q_{max}} f(Q)\mathrm{d}Q = 1 \tag{9.25}$$

The relation between $f(Q)$ and $Q(q)$ is given with the following equation:

$$f(Q) = \frac{q(Q)}{q_0} \tag{9.26}$$

The description of the total isotherm equation can be simplified according to the following equation:

$$q = q_0 \cdot \int_{Q'}^{Q_{max}} f(Q) \mathrm{d}Q \tag{9.27}$$

where

$$Q' = -R \cdot T \cdot \ln(K_0 \cdot p) \tag{9.28}$$

and

$$f(Q) = -\frac{\frac{1}{R \cdot T} \cdot \mathrm{d}q}{\mathrm{d}(\ln p)_{\mathrm{T}}} \tag{9.29}$$

At higher coverage the isosteric heat of adsorption decreases with increasing amount adsorbed, as shown in the following equation:

$$Q = Q_{st} - Q_0 \cdot \ln q \tag{9.30}$$

The magnitude of the temperature rise $\Delta\theta$ can be calculated with a simple energy balance under equilibrium conditions as given with eq. (9.31):

$$\Delta\theta = \theta_{max} - \theta_0 = \frac{\frac{q \cdot \Delta H_{Ad}}{Cp_{CG}}}{\frac{q}{Y_0} - \frac{Cp_{ads}}{Cp_{CG}}} \tag{9.31}$$

with Cp_{CG} being the specific heat of the carrier gas in kJ kg^{-1} K^{-1}, Cp_{ads} the specific heat of the adsorbate in kJ kg^{-1} K^{-1} and Y_0 is the adsorbate load at the beginning of adsorption in kg kg^{-1} (comment: in the following exercises we will denote Y_0 as Y_{in}).

Cp_{ads}/Cp_{CG} can be neglected if $Cp_{ads}/Cp_{CG} \ll q/Y_0$. The energy balance simplifies as follows:

$$\Delta T = Y_0 \cdot \frac{\Delta H_{Ad}}{Cp_{CG}} \tag{9.32}$$

The magnitude of the heat of adsorption ΔH_{Ad} is about 1,000–4,000 kJ kg^{-1} and Cp_{CG} is 1 kJ kg^{-1} K^{-1} for most carrier gases. With a representative feed concentration of $Y_0 = 0.01$ kg kg^{-1} the temperature rise is about $\Delta\theta = 25$ °C, which is common in gas phase adsorption. Due to the inverse correlation of Y and the total pressure, $\Delta\theta$ will decrease with an increase in total pressure. At $P_{tot} > 5$ MPa isothermal conditions are obtained. Carrier gas with high specific heat will suppress temperature rise [292].

9.6 Multicomponent adsorption

In multicomponent adsorption, two or more adsorbable substances can occupy the same adsorption sites. Thus, isotherm relationships and their description become more complex. Extension of the Langmuir equation is the simplest approach by supposing no interaction between adsorbing molecules. For two component systems, the extended Langmuir isotherm according to Markham and Benton [312] is given as follows:

$$q_1 = \frac{q_{0,1} \cdot K_1 \cdot p_1}{1 + K_1 \cdot p_1 + K_2 \cdot p_2} \tag{9.33a}$$

$$q_2 = \frac{q_{0,2} \cdot K_2 \cdot p_2}{1 + K_1 \cdot p_1 + K_2 \cdot p_2} \tag{9.33b}$$

The extended Langmuir equation gives access to quick estimation of equilibrium relations for multicomponent adsorption since it suffices to determine the Langmuir parameters from the single-component isotherm of each component [291]. The separation factor for a mixture of two components a_{12} is given by the ratio of the single-component equilibrium constants, as shown in the following equation:

$$a_{12} = \frac{K_1}{K_2} \tag{9.34}$$

Unfortunately, the simple extended Langmuir equation has limited applicability, especially for liquid phase adsorption. This is due to the fact that even single-component isotherms in liquid phase are rarely described by the Langmuir equation. Fritz and Schlünder [313] have proposed the following equation for the description of multicomponent adsorption isotherms:

$$q_i = \frac{a_i \cdot c_i^{b_i + b_{ii}}}{c_i^{b_{ii}} + a_{ij} \cdot c_j^{b_{ij}}} \tag{9.35}$$

This equation also has its flaw as it is inconsistent with single component isotherm data and lacks thermodynamic background. Since there is a large number of empirical parameters, eq. (9.35) satisfies for a final fit with experimental data [291].

9.7 Mass balances in adsorption towers

Off-gas purification via adsorption is frequently realized in adsorption towers. The adsorbent particles are therefore packed within the tower and the off-gas, containing one or more components of adsorbate, flow through the adsorbent bed. The pressure loss is an essential technical variable if the adsorbent forms a closed layer either as a solid or moving bed. Since the adsorption rate is diffusion-controlled, small adsorbent particles and preferable short paths are advantageous but inevitably lead to higher flow resistance and pressure drop respectively. The economically optimal adsorbent particle size will therefore always be a compromise.

Three different concentration zones along a fixed adsorbent bed may be considered, schematically shown in Fig. 9.6.

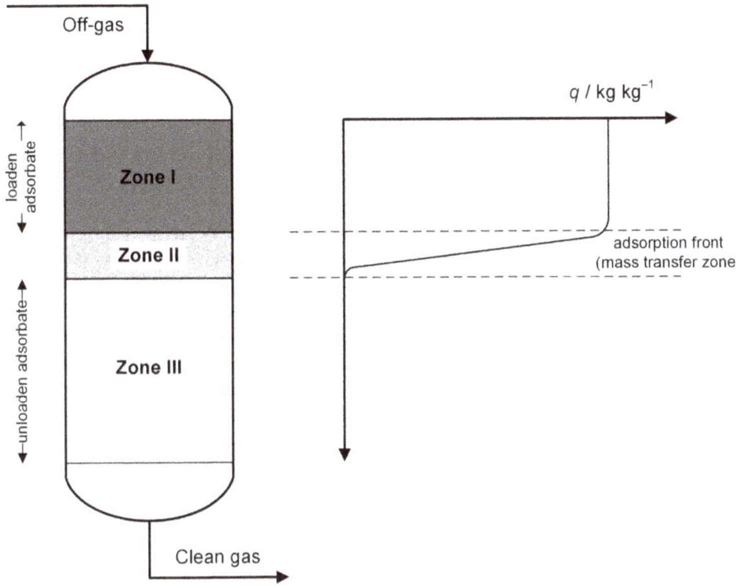

Fig. 9.6: Different zones of adsorbent load along the adsorbent bed in an adsorption column.

Zone I indicates the area in which the adsorption equilibrium between the feed concentration of the gas and the adsorbent is reached and no adsorption takes place. In this zone the adsorbent is already saturated.

Zone II is called the *mass transfer zone*, where the actual adsorption of specific species from the gas to the adsorbent occurs. Characteristically for this zone is that the equilibrium is not reached. The mass transfer zone moves slowly from the gas inlet to the outlet of the column. When it reaches the outlet of the bed, the so-called breakthrough will be obtained, and the adsorbate needs to be regenerated or exchanged. Ideally, the mass transfer zone is narrowly distributed and clearly defined. When the diffusion rate of adsorption is small compared to the gas flow rate, the mass transfer zone will be wide and blurred [290].

In Zone III of the column, the adsorbent is unloaded. It can be completely free of pollutants if it is new and has never been loaded before.

9.7.1 Type I isotherm systems

9.7.1.1 Adsorption

For type I isotherms, it can be assumed that the adsorption front penetrating the adsorber bed is a sharply defined discontinuity (= no axial dispersion) and migrate with the velocity indirectly proportional to $\Delta q/\Delta Y$ [298]. In the case of desorption, the rate of migration of the desorption front changes continuously indirectly proportional to $\mathrm{d}q/\mathrm{d}Y$. The contrast in adsorption and desorption fronts strongly affects adsorption characteristics and is the main reason why adsorption for type I isotherm systems proceeds fast, while desorption leads to tailing and proceeds slowly.

Mass balance (eq. (9.36)) is determined with the assumption of dispersion-free discontinuity by equating the cumulative amount of adsorptive fed and the mass of adsorbent required according to equilibrium conditions:

- Mass of adsorptive fed for the time t: $\Delta Y \cdot G_{CG} \cdot A \cdot t$
- Mass of adsorbent required up to position z: $\Delta q \cdot \rho_{Ad} \cdot A \cdot z$
  resulting in the following equation:

$$\Delta q \cdot \rho_{Ad} \cdot z = \Delta Y \cdot G_{CG} \cdot t \tag{9.36}$$

with t being the time in s, A is the cross sectional area of the adsorber in m^2, z is the height of the adsorbent layer in m, ρ_{Ad} is the adsorbent bulk density in kg m^{-3}, G_{CG} is the specific mass flow rate of the carrier fluid in kg m^{-2} s^{-1}, Y is the mass load of adsorptive in kg kg^{-1} and q is the mass load of adsorbate to the adsorbent mass in kg kg^{-1}.

The mass balance can be rearranged according to eq. (9.37), solving for the rate of migration of the adsorption front $v_{Ad,front} = z/t$, with the density of the carrier fluid ρ_{CG} in kg m^{-3} and the volumetric flow rate of the carrier gas $F_{V,CG}$ in m^3 s^{-1}:

$$v_{Ad,\,front} = \frac{z}{t} = \frac{G_{CG}}{\frac{\rho_{Ad} \cdot \Delta q}{\Delta Y}} = \frac{\rho_{CG} \cdot F_{V,\,CG}}{\frac{\rho_{Ad} \cdot \Delta q}{\Delta Y}} \tag{9.37}$$

The rate of migration of the desorption front $v_{De,front}$ can be derived according to eq. (9.38) with $F_{V,PG}$ being the volumetric flow rate of the purge gas (= desorbent) and ρ_{PG} its density, respectively:

$$v_{De,\,front} = \frac{z}{t} = \frac{G_{CG}}{\frac{\rho_{Ad} \cdot \mathrm{d}q}{\mathrm{d}Y}} = \frac{\rho_{PG} \cdot F_{V,\,PG}}{\frac{\rho_{Ad} \cdot \mathrm{d}q}{\mathrm{d}Y}} \tag{9.38}$$

The minimum specific amount of adsorbent is determined by the velocity of the adsorption front:

$$\frac{z}{t} = \frac{\rho_{CG} \cdot F_{V,\,CG}}{\rho_{Ad} \cdot \frac{\Delta q}{\Delta Y}} \tag{9.39}$$

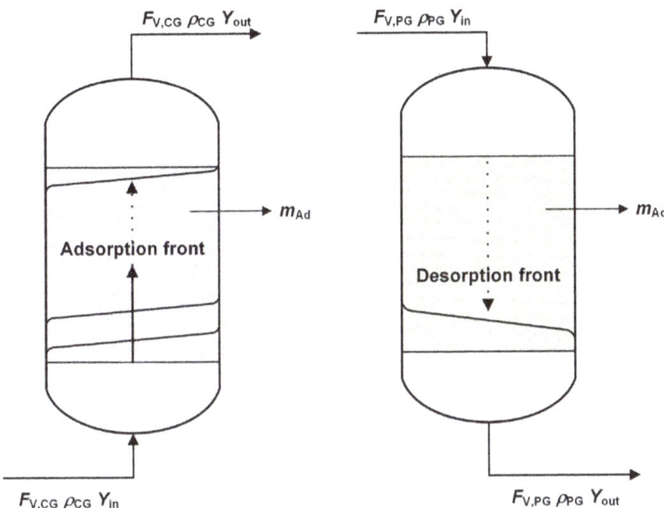

Fig. 9.7: Schematic drawing of the adsorption and desorption front along the adsorber bed.

By multiplying eq. (9.39) with the cross-sectional area, the ratio of required mass of adsorbent to the carrier fluid can be calculated with the following equation:

$$\frac{\text{Adsorbent}}{\text{Carrier gas}} = \frac{\Delta Y}{\Delta q} = \frac{Y_{\text{in}} - Y_{\text{out}}}{q_{\text{in}} - q_{\text{out}}} \tag{9.40}$$

Breakthrough of adsorption is indicated by a simultaneous breakthrough of all concentrations between feed Y_{In} and corresponding initial bed load q_1 according to $\Delta q/\Delta Y$ = constant. The discontinuous breakthrough curve can be prepared according to eq. (9.41). L is the adsorbent bed height:

$$\frac{\Delta q}{\Delta Y} = \frac{\rho_{\text{CG}} \cdot F_{\text{V,CG}} \cdot \text{t}}{\rho_{\text{Ad}} \cdot L} \tag{9.41}$$

9.7.1.2 Desorption

The purge gas consumption needed for desorption can be derived similarly to the adsorption solver [298]:

$$\frac{\text{Mass of purge gas}}{\text{Mass of adsorbent}} = K \tag{9.42}$$

Breakthrough is evaluated in the same way. Every isotherm slope $\mathrm{d}q/\mathrm{d}Y$ corresponds to a particular fluid load Y, according to the following equation:

$$\left(\frac{\mathrm{d}q}{\mathrm{d}Y}\right)_Y = \frac{\rho_{PG}\cdot F_{V,PG}\cdot t}{\rho_{Ad}\cdot L} \tag{9.43}$$

The desorption time t_{des} is then

$$t_{des} = \frac{\rho_{Ad}\cdot L\cdot K}{\rho_{PG}\cdot F_{V,PG}} \tag{9.44}$$

and the minimum amount of purge gas requirement is $(\rho_{PG}\cdot F_{V,PG})_{min} = K$ given in $\mathrm{kg_{purge}\ kg_{bed}}^{-1}$ [298].

9.7.2 Type III isotherm systems

Type III isotherms can be considered as inversion of the type I isotherms. As a consequence, adsorption leads to an expanding front and the rate of migration varies inversely with the isotherm slope:

$$\text{Rate of adsorption: } \frac{z}{t} = \frac{\rho_{CG}\cdot F_{V,CG}}{\rho_{Ad}\cdot \frac{\mathrm{d}q}{\mathrm{d}Y}} \tag{9.45}$$

$$\text{Break through time: } t_{ads} = \frac{\rho_{Ad}\cdot k\cdot z}{\rho_{CG}\cdot F_{V,CG}} \tag{9.46}$$

$$\text{Minimum bed requirement: } (A\cdot z\cdot \rho_{Ad})_{min} = \frac{1}{K} \tag{9.47}$$

$$\text{Rate of desorption: } \frac{z}{t} = \frac{\rho_{PG}\cdot F_{V,PG}}{\rho_{Ad}\cdot \frac{\Delta q}{\Delta Y}} \tag{9.48}$$

$$\text{Desorption time: } t_{des} = \frac{\rho_{Ad}\cdot z\cdot \frac{\Delta q}{\Delta Y}}{\rho_{PG}\cdot F_{V,PG}} \tag{9.49}$$

The purge requirement is given with $(\rho_{PG}\cdot F_{V,PG})_{min} = \frac{\Delta q}{\Delta Y}$ in $\mathrm{kg_{purge}\ kg_{bed}}^{-1}$ [298].

Final Exercise 9.2: Volatile organic compounds control

A roller line has to be defatted with toluene (C_7H_8). The plant is in operation for 3 months (90 days) 24 h a day. Every 3 months operation is stopped for a 2-day general revision. Because of safety reasons the whole line must be vented with air continuously. Any technical measure must guarantee a clean gas volatile organic compounds (VOC) concentration of VOC = 50 mg kg^{-1} in venting air emitted to the ambient environment.

Operation data:

- Off-gas (carrier-gas): 2,000 $\mathrm{m^3}_{\mathrm{STP,dry}}\ \mathrm{h^{-1}}$
- Temperature: 20 °C
- Pressure: 1,013 hPa
- Toluene concentration: 1 g m^{-3}
- Emission limit: 50 mg m^{-3}

Your job is to prepare off-gas treatment proposals for a feasibility study. Basically, recycling technologies would be preferred, but you shall not limit the technology study to a specific route. You organize a brainstorming session with several members of your engineering group. From the outcome you select high potential technology candidates with the help of selected physical properties of toluene, shown in Tab. 9.6.

Tab. 9.6: Selected physical properties of toluene and air.

Specification	Unit	Toluene
Antoine parameters for toluene (hPa)		
A		7.076
B		1,342.310
C		219.187
θ_{bp} (1,000 hPa)	°C	110.6
θ_{mp}	°C	−95.0
$MM_{toluene}$	g mol^{-1}	92
ΔH_v	kJ kg^{-1}	360
LCV	kJ kg^{-1}	41,000
Cp_l	kJ kg^{-1} K^{-1}	1.9
Cp_v	kJ kg^{-1} K^{-1}	1.4
ρ_l	kg m^{-3}	839
$\rho_{vap,\ STP}$	kg m^{-3}	4.11
$\rho_{vap,\ bp}$	kg m^{-3}	2.9
Carrier gas		Dry air
MM_{air}	g mol^{-1}	28.84
$\rho_{air,\ STP}$	kg m^{-3}	1.286
Cp_{air}	kJ kg^{-1} K^{-1}	1

Finally, you decide to prepare proposals for comparison of:
(a) Off-gas combustion
(b) Condensing of toluene
(c) Adsorption technology routes

Solver:

(a) Off-gas combustion

Reaction (combustion equation): $C_7H_{8(g)} + 9\ O_{2(g)} \rightleftharpoons 7\ CO_{2(g)} + 4\ H_2O_{(g)}$

In Exercise 7.14 the different solvers for shortcut estimations for the energy demand of off-gas combustion at θ = 800 °C have been discussed:

(1) We assume that no pollutant is contained in the off-gas and that the off-gas is not preheated.
(2) We assume that no pollutant is contained in the off-gas, but this time we preheat the off-gas to 400 °C.
(3) We consider combustion of the pollutant toluene. The off-gas is preheated to 400 °C.
(4) We consider combustion of the pollutant toluene. The off-gas is preheated to 400 °C. Off-gas heat is internally used.

Ad (1) Off-gas combustion:
Our off-gas contains a mass of toluene of: $m_{toluene}$ = 2,000 [m^3_{STP} h^{-1}] · 0.001 [kg m^{-3}] = 2 kg h^{-1}. We ignore the volume of toluene in air because of the low concentration and convert it into

$$Y_{Toluene} = 2\ [\text{kg h}^{-1}]/2{,}000\ [\text{m}^3_{STP}\ \text{h}^{-1}]/1.286\ [\text{kg m}^{-3}] \cdot 1{,}000 = 0.778\ \text{g kg}^{-1}.$$

The energy demand is $\dot{Q} = F_{m,carrier\text{-}gas} \cdot Cp \cdot \Delta T = 2{,}000 \cdot 1.286 \cdot 1 \cdot (800-20) = 2{,}006{,}887.2$ kJ h^{-1}, equivalent to 59.9 m^3 h^{-1} LNG or equivalent 557.5 kWh (= 278.8 kWh per $kg_{toluene}$).

Ad (2) Off-gas combustion with preheating to $\theta = 400$ °C:

The energy demand is $\dot{Q} = F_{m,carrrier\text{-}gas} \cdot Cp \cdot \Delta T = 2{,}000 \cdot 1.286 \cdot 1 \cdot (800-400) = 1{,}028{,}800$ kJ h^{-1}, equivalent to 30.7 m^3 h^{-1} LNG or equivalent 285.9 kWh (= 143 kWh per $kg_{toluene}$).

Ad (3) Off-gas combustion with preheating to $\theta = 400$ °C and consideration of the combustion energy of the pollutant: When toluene of our off-gas is incinerated it contributes an amount of energy of $\dot{Q} = 2{,}000 \cdot 1/1{,}000 = 2 \cdot 41{,}000$ kJ h^{-1}. The net energy demand is then $\dot{Q} = 1{,}028{,}800-82{,}000 = 946{,}800$ kJ h^{-1}, equivalent to 28.3 m^3 h^{-1} LNG or equivalent 263 kWh (= 131.5 kWh per $kg_{toluene}$).

Ad (4) Off-gas combustion with preheating to $\theta = 400$ °C and consideration of the combustion energy of toluene. The off-gas heat is internally used. The final off-gas temperature is $\theta = 90$ °C.

In extension to case (3) we make use of the off-gas energy between $\theta = 400$ °C and $\theta = 90$ °C, corresponding to its $\dot{Q} = F_{m,carrier\text{-}gas} \cdot Cp \cdot \Delta T = 2{,}000 \cdot 1.286 \cdot 1 \cdot (400-90) = 797{,}320$ kJ h^{-1}, leaving a final energy demand of $Q = 946{,}800-797{,}320 = 149{,}480$ kJ h^{-1}, equivalent to 4.5 m^3 h^{-1} LNG or equivalent to 41.5 kWh (= 21 kWh per $kg_{toluene}$).

Discussion: If we decide for off-gas combustion the clean gas VOC level will be far below VOC = 50 mg kg^{-1}, but operation of the process will be very expensive. Without checking for the water dew point we could easily use the off-gas heat for any purposes to minimize the auxiliary fuel consumption. We may make use of the energy demand for case (4) of 41.5 kWh. We also have to consider that toluene is lost. We may rate the energy equivalent of toluene $\dot{Q}_{equivalent} = m \cdot LCV = 2 \cdot 41{,}000 = 82{,}000$ kJ h^{-1} = 22.8 kWh. The overall energy demand is 41.5 kWh for combustion + $\lambda \cdot 22.8$ kWh (toluene loss equivalent). It is recommended to decide for the multiple $\lambda = 2$ to consider the sales value of toluene, resulting in an overall energy demand of $\dot{Q}_{tot} = 87$ kWh. This benchmark value may help for comparing technologies.

(b) Toluene separation by condensing

We have to check whether it is possible to separate toluene from the off-gas by direct condensing.

Solver:

We have rearranged the inflow concentration of $c_{toluene,\ in} = 1$ g m^{-3} into

$Y_{toluene,in} = 2$ [kg h^{-1}]/2,000 [$m^3{}_{STP}$ h^{-1}]/1.286 [kg m^{-3}] · 1,000 = 0.778 g kg^{-1}, and $Y_{toluene,out} = 4.25 \cdot 10^{-5}$ kg kg^{-1}.

The off-gas specification shows the Antoine constants of toluene, enabling us to determine the vapor pressure of toluene. We just need a link of the vapor pressure and the toluene load $Y_{toluene}$. We may make a loan from Section 2.5.2, eqs. (2.8) and (2.10), but adjust them to our needs:

$$\theta[^\circ\mathrm{C}] = \frac{B}{A - \log(p_{toluene,\theta,0})} - C \qquad (2.8)$$

$$Y = \frac{p_{toluene}}{(P_{tot} - p_{toluene})} \cdot \frac{MM_{toluene}}{MM_{CG}} \qquad (2.10)$$

We have to keep in mind that industrial-scale condensation via heat exchange shall not go below $\theta = -30$ °C. We may have in mind direct condensation with evaporating liquid nitrogen in the off-gas, a frequently applied technology. The latter technology would give us access to nearly any temperature between $\theta = -30$ °C and the boiling point temperature of nitrogen $\theta_{bp} = -195.8$ °C. However, we are recommended to stay above $\theta_{m,toluene} = -95$ °C. The enthalpy of N_2 evaporation is $\Delta h_{evap} = 199$ kJ kg^{-1} [28]. We draw a sketch for the technologies we want to discuss according to Fig. 9.8.

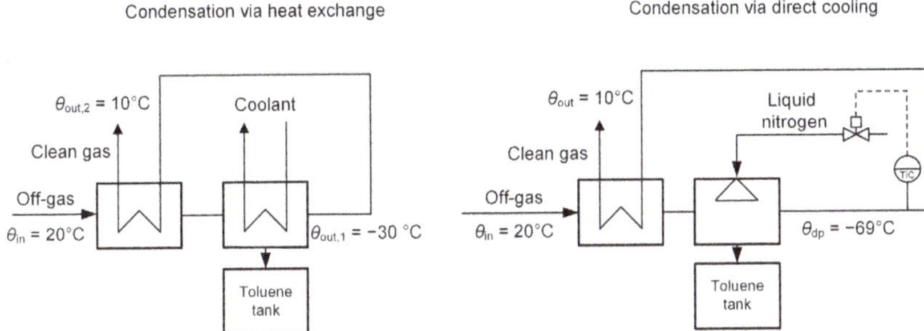

Fig. 9.8: Sketch of the considered technologies.

We have to consider that the simple technologies of Fig. 9.8 are limited to "dry off-gas" without any water vapor. For selected toluene load we calculate $p_{toluene}$ after rearranging eq. (2.10) and determine the dew point temperature θ_{bp} with eq. (2.8). We collect these data in Tab. 9.7 and Fig. 9.9.

Tab. 9.7: Dew point temperature of toluene in air for different loads $Y_{toluene}$.

$Y_{toluene}$ (g m^{-3}_{STP})	$Y_{toluene}$ (g m^{-3}_{STP})	$Y_{toluene}$ (kg kg^{-1})	$p_{toluene}$ (hPA)	θ (°C)
100,000	$1.0 \cdot 10^{-1}$	$7.77 \cdot 10^{-2}$	24.097	16.6
10,000	$1.0 \cdot 10^{-2}$	$7.77 \cdot 10^{-3}$	2.462	−18.4
1,000	$1.0 \cdot 10^{-3}$	$7.77 \cdot 10^{-4}$	0.247	−44.5
0.100	$1.0 \cdot 10^{-4}$	$7.77 \cdot 10^{-5}$	0.025	−64.6
0.010	$1.0 \cdot 10^{-5}$	$7.77 \cdot 10^{-6}$	0.002	−80.6

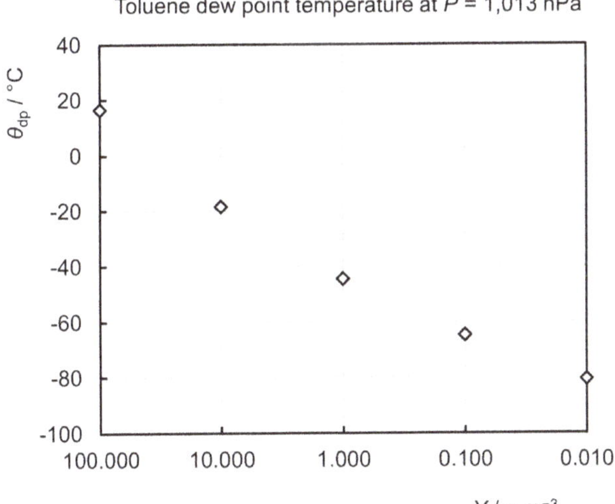

Fig. 9.9: Dew point temperature of toluene in air at P = 1,013 hPa.

Table 9.7 and Fig. 9.9 clearly show that even for the inflow toluene load of 1 g $m^{-3}{}_{STP}$ we are outside the operation range for condensing toluene within the range of standard cooling systems.

However, the engineering community always has an alternative route available, encouraging us to have a look on direct cooling with liquid nitrogen. With Fig. 9.8 in mind, we develop a solver for a first estimate of energy consumption. Who is hiding behind "a first estimate"? We assume constant physical properties, specified for $T = 293$ K, over the whole temperature span. (You may be punished for making use of that when performing detailed energy balances with these simplifications.)

The sketch of Fig. 9.8 suggests looking at the energy demand for covering the off-gas/clean gas temperature difference of $\Delta\theta = 10$ °C, with

$\Delta\dot{Q}_{cooling} = 2{,}000 \cdot 1.286 \cdot 1 \cdot (20\text{–}10) = 25{,}720$ kJ h^{-1}

Evaporation of liquid nitrogen at an estimated temperature must mainly cover the energy demand of toluene condensation. We check temperature level and energy demand for the clean gas specification of $Y_{toluene} = 4.25 \cdot 10^{-5}$ kg kg^{-1}. With eqs. (2.10) and (2.8) we identify a temperature level of $\theta_{cond} = -69.1$ °C. We have to condense 2 kg h^{-1} of toluene with $\Delta H_v = 360$ kJ kg^{-1} according to our data from Tab. 9.6:

$\dot{Q}_{cond} = m_{Toluene} \cdot \Delta H_v = 2\ [\text{kg h}^{-1}] \cdot 360\ [\text{kJ kg}^{-1}] = 720$ kJ h^{-1}

The total energy demand will then sum up to $\dot{Q}_{tot} = \dot{Q}_{cooling} + \dot{Q}_{cond} = 25{,}720\ [\text{kJ h}^{-1}] + 720\ [\text{kJ h}^{-1}] =$ 26,440 kJ h^{-1} equivalent $\dot{Q}_{tot} = 7.35$ kWh, equivalent a nitrogen demand of 133 kg h^{-1}.

In route (a) we rated the toluene loss equivalent with $\dot{Q}_{equ} = 2 \cdot 22.8$ kWh, resulting in an economic benefit of $\dot{Q}_{econ.} = 7.35\ [\text{kWh}] - 45.6\ [\text{kWh}] = -38.2$ kWh (= 19 kWh per $kg_{toluene}$).

! Attention: the negative signum indicates an economic benefit. But do not present these numbers when you have to report the outcome of your technology analysis, although the numbers of the outcome suggest a very attractive route. They will happily pick up these numbers and punish you when the outcome is less beneficial. You do have in mind that you made some significant simplifications such as zero water humidity, constant physical properties and ignorance of energy losses. Anyhow, you have to complete the job with discussing adsorption technologies.

(c) Adsorption technologies

We start with the simple part of the job. It is the adsorbent specification. We decide for activated carbon GAC with the main properties:

- GAC pellets: $d = 4$ mm
- Bulk density: $\rho_{Ad} = 400$ kg m^{-3}
- Spec. mass transfer area: 1,000 m^2 g^{-1}
- Cp_{Ad}: 0.8 kJ kg^{-1} K^{-1}

Your supplier of GAC provides you with the adsorption equilibrium data, expressed in terms of the Langmuir equation, shown in Fig. 9.10. (Just a background recommendation: Try to get into touch with your AC supplier. They are very helpful. They do have plenty of experience with AC, and they willingly share this experience with their customers.)

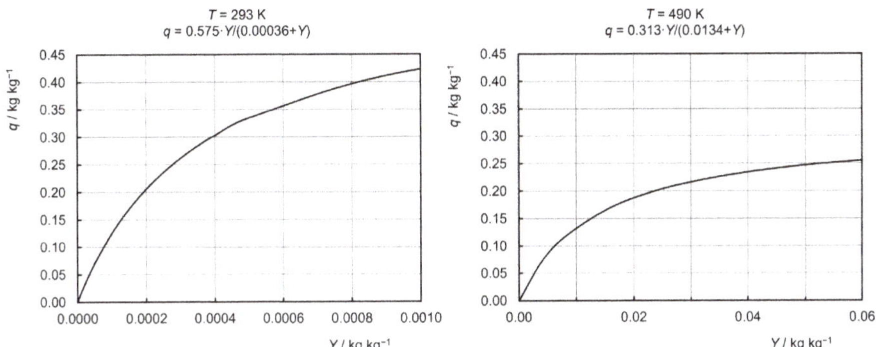

Fig. 9.10: Gas adsorbent isotherms for the system toluene/air/GAC for $T = 293$ K and $T = 490$ K.

The problem that may arise from the basic idea of applying adsorption technology is that we do not have just one technology applicable. We have to limit our analysis to a selection of technologies for not to get lost in space. We decide for similar setups as shown in Fig. 9.1.

In the basics of adsorption we have discussed the setup and the process principle of adsorption and desorption as shown in Fig. 9.7. The setup is basically a cylindric container, packed with GAC. The design of an adsorption tower may be very different from cylindric design, for example, cuboidal design. Figure 9.11 shows a cuboidal shape of the GAC packing with alternating paths for the off-gas and the clean gas. To some extent the design of the packing may be adjusted to onsite construction necessities.

We assume laminary plug flow of the adsorption front in the packed bed. We may load the packed bed with the solvent at adsorption temperature (e.g., $T_{ads} = 293$ K) as long as the concentration or load of the solute in the clean gas does not exceed the emission limit (e.g., $Y_{toluene,out} = 4.25 \cdot 10^{-5}$ kg kg^{-1}). The spent GAC then has to be regenerated in a separate process, for example, with nitrogen at elevated temperature (e.g., $T_{des} = 490$ K). We run the desorption process as long as we need to obtain a chosen minimum load at the exit (e.g., X_{out} (= q_{out}) = 0.061 kg kg^{-1}). After cooling the GAC-bed it is ready for the next cycle. Now, we have to discuss the process setup. We limit the discussion to the setup proposals shown in Figs. 9.11 and 9.12. We may also decide for external GAC regeneration. For not to get lost in design and process flexibility discussion we consider two routes:

(1) Route 1:
We install an adsorption tower of cuboidal shape as shown in Fig. 9.11. The GAC packing is designed for 3 months of operation. The spent GAC is regenerated externally.

The gas velocity is set to $v_g = 0.05$ m s^{-1}:

$Y_{toluene,\ in} = 7.8 \cdot 10^{-4}$ kg kg$^{-1}{}_{air}$

$Y_{toluene,\ out} = 4.25 \cdot 10^{-5}$ kg kg$^{-1}{}_{air}$

$q_{GAC} = 0.393$ kg kg$^{-1}{}_{GAC}$

GAC demand: 12,293 kg/service time

GAC demand: 5.7 kg h^{-1}

$m_{toluene\ adsorbed} = 4{,}083$ kg (3 months, 30 days a month, 24 h a day)

From eqs. (9.38) and (9.39) we deduce the rate of adsorption front penetration $v_{Ad.\ front} = 1.3 \cdot 10^{-4}$ m h^{-1}, respectively.

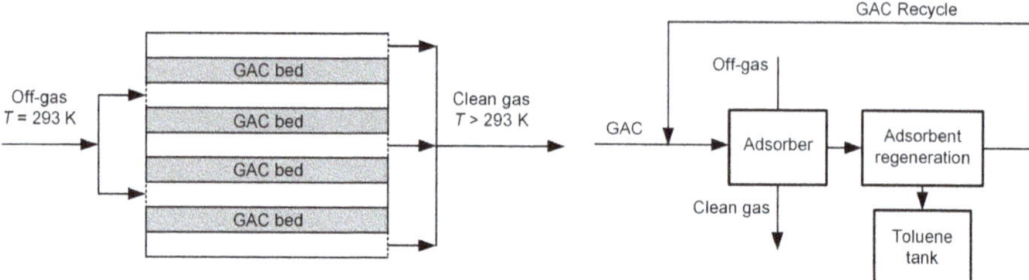

Fig. 9.11: Adsorber setup and process scheme.

(2) Route 2:

Continuous operation of the adsorption tower with internal regeneration of spent GAC by high temperature desorption with nitrogen. Figure 9.12 shows the process setup.

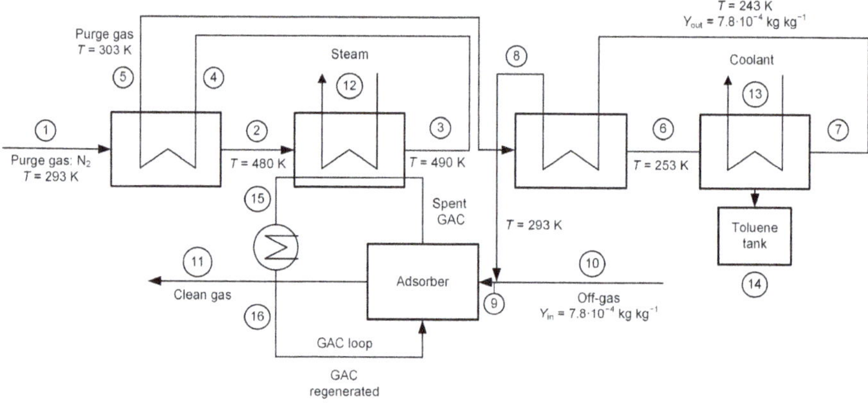

Fig. 9.12: Setup for adsorption with internal GAC regeneration.

The off-gas is fed to the adsorption tower. Adsorption is performed at $T = 293$ K. The spent GAC is continuously regenerated with nitrogen at a temperature of $T = 490$ K. To enable operation at high condensation temperature the purge gas is fed to the off-gas to minimize the loss of toluene. The desorption unit is fueled with superheated steam. The condenser is powered with a standard cooling system for operation at minimum temperature of $T = -60$ °C. Actually the condenser temperature is set to $T_{condenser} = -30$ °C. Table 9.8 shows the operation data and results of calculations. (Be careful with the numbers in line 11. We only considered the enthalpy of condensation. Actually heat of adsorption is much higher.)

Tab. 9.8: Operation data for continuous adsorption and of toluene and onsite regeneration of spent GAC according to Fig. 9.12.

Number	Specification	T (K)	Cp (kJ kg^{-1} K^{-1})	F_m (kg h^{-1})	$F_{m,toluene}$ (kg h^{-1})	ΔQ (kJ h^{-1})
1	Purge gas in	293	1.0	52.1	–	–
2	Purge gas preheat	480	1.0	52.1	–	9,737
3	Purge gas final	490	1.0	52.1	2.045	521
4	Purge gas out	490	1.0	52.1	2	521
5	Purge gas precool 1	303	1.0	52.1	2.045	9,737
6	Purge gas precool 2	253	1.0	52.1	2.045	2,604
7	Purge gas lean	243	1.0	52.1	0.167	1,241
8	Purge gas reheat	293	1.0	52.1	0.167	2,604
9	Mixed off-gas	293	1.0	2,627.1	2.167	0
10	Off-gas	293	1.0	2,575.0	2.000	0
11	Clean gas	293.3	–	2,627.1	0.112	720
12	Steam	–	–	–	–	1,687
13	Coolant	–	–	–	–	1,200
14	Toluene	–	–	2.0	1.888	179
15	Loop out	490	–	6.0	6.032	–
16	GAC cool	293	0.8	6.0	6.032	951

From the outcome we may again deduce the specific energy demand per kilogram of toluene from balance points 12 through 14 with weighing the energy equivalence of toluene again with $\lambda = 2$. From these data we deduce a specific energy demand of $\dot{Q}_{toluene} = 0.801$ kWh per $kg_{toluene}$.

Discussion: It is recommended to investigate alternative routes of off-gas treatment when possible. We decided to investigate and compare the off-gas treatment technologies combustion, condensation and adsorption. We assumed a project status of technology evaluation. For comparing technologies we limited our analysis to the comparison of specific cost of operation. The outcome shows that recycling of toluene by combination of adsorption/desorption and condensation seemingly makes sense. However, we did not make any analysis of the investment cost, which will be appreciable when just counting the number of heat exchangers in the proposed setup of Fig. 9.12. !

10 Some final remarks

Let us start with congratulations. If we did attract your attention for off-gas purification purposes, this book achieved 100% efficiency. Off-gas purification is indeed a long-term challenge, and it will need our attention for another couple of decades. We do have great technologies available, but we are still off the quality level for a balanced harmony between our economic needs and a healthy environment. At present, we work hard on developing solutions for the carbon dioxide emission problem. In this important project we are still on the level of brainstorming. However, keeping in mind that we had to go through similar phases of confusion when we developed successful strategies for getting state-of-the-art pollutants under control, we may be confident to finally get the CO_2 problem under control too. Actually there is not much of an alternative to succeeding in this challenge, except slipping from a problem into a severe crisis, but that is not what we want and expect.

This book is definitely not a comprehensive summary of pollution control solutions. Any of the topics we addressed could easily stand a separate discussion in detail of same extent and more. The intention behind this book is to support you in doing first steps in this business. We also intended to prepare strategies and solvers to offer support in the analysis of problems, the development of solvers and the design of processes. Let us sum up the strategies and recommendations at the end of this "introduction into off-gas pollution control issues":

- Whenever you are faced with an off-gas problem, try to collect all information about the problem, the background, the source and peripheral aspects of the problem.
- You will rarely find general solutions for individual problems, and you will rarely find solutions off the shelf. As a consequence, focus all your attention and experience on a specific problem.
- Do not fear to draw all advantages you can collect from literature. Have in mind that someone else might already have suffered from the same problem in the past and might have worked hard on finding solutions. It would be a waste of time to ignore this experience.
- Collect all data, physical properties of all constituents and the interaction of constituents.
- Ask thermodynamics for support, but have in mind that you must ask correct questions, and you must ask them correctly. Thermodynamics is willing to answer these questions, but you cannot blame thermodynamics for stupid answers when you ask wrong questions. Thermodynamics is the best friend of the chemical engineering community. Do not hesitate to make thermodynamics be your best friend too.
- Have in mind that the magic of thermodynamics is not of help when you ignore kinetics.

https://doi.org/10.1515/9783110763928-010

- Always try to identify all (several) solver options. Never limit your flexibility to just one route.
- Teamwork is the key to success. Never try to become the lone star. You may tremendously fail.
- Never try to hide a problem. You cannot make problems vanish.
- Do not fear problems but face them.

Yes, off-gas purification is still a huge challenge, and this challenge needs your experience, and we are very confident that you will succeed.

Bibliography

[1] I. Manisalidis, E. Stavropoulou, A. Stavropoulos and E. Bezirtzoglou, "Environmental and health impacts of air pollution: A review," *Front Public Health*, vol. 8, p. 14, 2020, doi: 10.3389/fpubh.2020.00014.

[2] *Air quality in Europe: 2019 report*. Luxembourg: Publications Office of the European Union, 2019.

[3] *Air quality in Europe: 2021 report*. Luxembourg: Publications Office of the European Union, 2021.

[4] *Health impacts of air pollution in Europe, 2021*. Luxembourg: Publications Office of the European Union, 2022.

[5] *Directive (EU) 2016/2284 of the European Parliament and of the Council of 14 December 2016 on the reduction of national emissions of certain atmospheric pollutants, amending Directive 2003/35/EC and repealing Directive 2001/81/EC: CELEX 32016L2284*, 2016.

[6] *Directive 2010/75/EU of the European Parliament and the Council of 24 November 2010 on industrial emissions (integrated pollution prevention and control): CELEX 32010L0075*, 2010.

[7] *VDI-Wärmeatlas*, 10th ed. Berlin, Heidelberg, New York: Springer.

[8] Metso Outotec, *HSC 8.0 Chemistry Software*. Finland, 2020.

[9] H. Müller, "Sulfuric Acid and Sulfur Trioxide," in *Ullmann's Encyclopedia of Industrial Chemistry*, John Wiley & Sons, Ltd, 2000.

[10] F. H. Verhoff and J. T. Banchero, "Predicting dew points of flue gases," *Chemical Engineering Progress*, no. 70, pp. 71–72, 1974.

[11] B. Xiang, B. Tang, Y. Wu, H. Yang, M. Zhang and J. Lu, "Predicting acid dew point with a semi-empirical model," *Applied Thermal Engineering*, vol. 106, pp. 992–1001, 2016, doi: 10.1016/j.applthermaleng.2016.06.040.

[12] R. R. Pierce, "Estimating acid dewpoints in stack gases," *Chemical Engineering*, 84, no. pp. 125–128, 1977.

[13] R. C. West, *Handbook of Chemistry and Physics*, 55th ed. Cleveland, Ohio: CRC Press.

[14] K. Verschueren, *Handbook of Environmental Data on Organic Chemicals*, 3rd ed., New York: Van Nostrand Reinhold, 1997.

[15] *Bundes-Immissionsschutzgesetz: BImSchG*, 2021.

[16] J. G. Speight, *Perry's Standard Tables and Formulas for Chemical Engineers*. New York: McGraw-Hill, 2003.

[17] *Directive 2008/50 of the European Parliament and of the Council of 21 May 2008 on Ambient Air Quality and Cleaner Air for Europe: CELEX 32008L0050*, 2008.

[18] Bundesministerium für Wirtschaft, Familie und Jugend, *Technische Grundlage zur Berechnung und Beurteilung von Immissionen im Nahbereich kleiner Quellen: Technische Grundlage Ausbereitungsgrechnung*. [Online]. Available: www.bmwfj.gv.at

[19] R. B. Stull, *An Introduction to Boundary Layer Meteorology*. Dordrecht: Kluwer Academic Publishers, 1988.

[20] E. W. Hewson, *Industrial Air Pollution Meteorology*. University of Michigan, Ann Abor: Meteorological Laboratories of the College of Engineering, 1963.

[21] R. B. Stull, *An Introduction to Boundary Layer Meteorolgy*. Dordrecht: Kluwer Academic Publishers, 2003.

[22] *Bundes-Immissionsschutzgesetz: BImSchG*, 2002.

[23] G. A. Briggs, *Plume Rise*. Oak Ridge, Tennessee: U.S. Atomic Energy Commission, Division of Technical Information, 1969. [Online]. Available: https://books.google.de/books?id=dKjK7Eu1fq0C

[24] H. Moses and J. E. Carson, "Stack design parameters influencing plume rise," *Journal of the Air Pollution Control Association*, vol. 18, no. 7, pp. 454–457, 1968, doi: 10.1080/00022470.1968.10469155.

[25] J. E. Carson and H. Moses, "The validity of several plume rise formulas," *Journal of the Air Pollution Control Association*, vol. 19, no. 11, pp. 862–866, 1969, doi: 10.1080/00022470.1969.10469350.

https://doi.org/10.1515/9783110763928-011

[26] J. Crank, *The mathematics of diffusion*. Oxford university press, 1979.
[27] E. L. Cussler, *Diffusion mass transfer in fluid systems*. Cambridge: Cambridge University Press, 1997.
[28] *Bundes-Immissionsschutzgesetz: BImSchG, TA-Luft 86*, 1986.
[29] G. Manier, "Vergleich zwischen Ausbreitungsklassen und Temperaturgradienten," *Meteorologische Rundschau*, no. 28, pp. 6–11, 1975.
[30] D. B. Turner, "Proposed pragmatic methods for estimating plume rise and plume penetration through atmospheric layers," *Atmospheric Environment*, vol. 19, no. 7, pp. 1215–1218, 1985, doi: 10.1016/0004-6981(85)90309-9.
[31] Á. Leelőssy, F. Molnár, F. Izsák, Á. Havasi, I. Lagzi and R. Mészáros, "Dispersion modeling of air pollutants in the atmosphere: a review," *Open Geosciences*, vol. 6, no. 3, pp. 257–278, 2014, doi: 10.2478/s13533-012-0188-6.
[32] A. Podesser, *Klimaatlas Steiermark: Stahlung 1971-2000*. Kapitel 1. [Online]. Available: http://www.umwelt.steiermark.at/
[33] K. Waltraud, "Mobile Luftgütemessungen Graz-Griesplatz," Abteilung 15-Energie, Wohnbau, Technik ABT15-Lu-09-2021, Dec. 2021. [Online]. Available: http://www.umwelt.steiermark.at/
[34] E. T. Wilkins, "Air pollution and the London fog of December, 1952," *Journal of the Royal Sanitary Institute*, vol. 74, no. 1, pp. 1–21, 1954, doi: 10.1177/146642405407400101.
[35] A. Bloder, "Klimaschutz-und Energiebericht Stadt Graz," Master thesis, Insitut für Prozess- und Partikeltechnik (Technische Universitaet Graz), Karl-Franzens-Universitaet, Graz, 2010.
[36] A. Podesser, *Klimaatlas Steiermark: Windverhältnisse 1971-2000*. Kapitel 7. [Online]. Available: http://www.umwelt.steiermark.at/
[37] W. Spangl, "Messung des Windprofils mittels Akustikradar in Graz: UBA-BE-034," Wien, 1995.
[38] H. Bauer *et al.*, "Aquella Steiermark – Bestimmung von Immissionsbeiträgen in Feinstaubproben," Lu-08/07, 2007. [Online]. Available: http://ww.umwelt.steiermark.at
[39] N. Leonida and B. Paul, "EMEP/EEA air pollutant emission inventory guidebook," 2019.
[40] R. Ranau and H. Steinhart, "Identification and evaluation of volatile odor-active pollutants from different odor emission sources in the food industry," *European Food Research and Technology*, vol. 220, no. 2, pp. 226–231, 2005.
[41] R. E. Muck, "Urease activity in bovine feces," *Journal of Dairy Science*, vol. 65, no. 11, pp. 2157–2163, 1982.
[42] M. H. Abraham, J. M. R. Gola, J. E. Cometto-Muniz and W. S. Cain, "A model for odour thresholds," *Chemical Senses*, vol. 27, no. 2, pp. 95–104, 2002, doi: 10.1093/chemse/27.2.95.
[43] Y. Nagata and N. Takeuchi, "Measurement of odor threshold by triangle odor bag method," *Bulletin of Japan Environmental Sanitation Center*, vol. 17, pp. 77–89, 1990.
[44] Baldinger S., *et al.*, "Leitfaden Gerüche in Innenräumen," Wien, Feb. 2014. [Online]. Available: www.bmwfj.gv.at
[45] Arbeitsgruppe des BMLFUW "Immissionen Geruch", "Richtlinie zur Beurteilung von Geruchsimmissionen aus der Nutztierhaltung," GeruchsRL 2017, 2017. [Online]. Available: www.bmlfuw.gv.at
[46] B. Spernbauer, "Untersuchungen zu Geruchsemissionen und -immissionen aus der Tierhaltung," Master thesis, Karl-Franzens-Universitaet, Graz, 2014.
[47] E. Zentner, Ed., *Emissionen aus der Nutztierhaltung versus Anrainnerinnen und Anrainer und Raumplanung*. 25. Österreichische Wintertagung. Irdning: Höhere Bundeslehr- und Versuchsanstalt Raumberg-Gumpenstein, 2019.
[48] D. Öttl, M. Kropsch, E. Zentner, G. Bachler and R. L. Schlacher, "Geruchsemissionen aus der Tierhaltung," Lu-01-2021, 2021. [Online]. Available: http://www.umwelt.steiermark.at/
[49] G. Eva, "Beurteilung von Geruchsimmissionen aus der Tierhaltung," Habilitation thesis, Institute of Agricultural Engineering, University of Hohenheim, Stuttgart, Germany, 2011.

[50] H.-P. Hutter, *et al.*, "Medizinische Fakten zur Beurteilung von Geruchsimmissionen: Endbericht," Leitfaden im Auftrag des Landes Steiermark, Institut für Umwelthygiene, Medizinische Universität Wien, 2007.

[51] D. Öttl, H. Moshammer, M. Mandl and L. Weitensfelder, "Richtlinie zur Beurteilung von Geruchsimmissionen," Lu-08-2018, 2018. [Online]. Available: http://www.umwelt.steiermark.at/

[52] *Feststellung und Beurteilung von Geruchsimmissionen: GIRL*, 2008.

[53] Landwirtschaftskammer Niedersachsen, "Optimierung des Stallklimas in der Mastschweinehaltung: Ein Leitfaden für die Praxis," Project number 2813MDT040, 2016.

[54] *Änderung der 1. Tierhaltungsverordnung: BGBLA_2017_II_151*, 2017. [Online]. Available: https://www.ris.bka.gv.at/eli/bgbl/II/2017/151/20170606

[55] W. Gramatte, "Die Umsetzung der aktuellen DIN 18910 in die Praxis," *Bautagung Raumberg Gumpenstein*, pp. 95–100, 2009.

[56] G. Schauberger, "Die Aufgaben der Amtssachverständigen für Veterinärwesen im landwirtschaftlichen Genehmigungsverfahren von Stallgebäuden und ihre Verantwortung bei Umweltfragen," *Veterinary Medicine Austria*, no. 103, 2016.

[57] J. Grübler, *et al.*, "Luftgütemessungen in der Steiermark – Jahresbericht 2020," ABT15-Lu-08-2021, Dec. 2021. [Online]. Available: http://www.umwelt.steiermark.at/

[58] J. M. Finch, "Dust-Collector," 325,521.

[59] O. M. Morse, "Dust-collector," 403,362.

[60] M. Bohnet, "Zyklonabscheider," in *Handbuch des Umweltschutzes und der Umweltschutztechnik*, H. Brauer, Ed., Berlin, Heidelberg: Springer Berlin Heidelberg, 1996, pp. 58–88.

[61] A. C. Hoffmann and L. E. Stein, *Gas cyclones and swirl tubes: Principles, design and operation*, 2nd ed. Berlin, Heidelberg, New York, NY: Springer, 2008. [Online]. Available: http://www.loc.gov/catdir/enhancements/fy0826/2007934931-d.html

[62] W. Barth, "Berechnung und Auslegung von Zyklonabscheidern aufgrund neuerer Untersuchungen," *Brennstoff Warme Kraft*, vol. 8, 1–9-, 1956.

[63] E. Muschelknautz, "Auslegung von Zyklonabscheidern in der technischen Praxis," *Staub, Reinhaltung der Luft*, vol. 30, no. 5, pp. 187–195, 1970.

[64] G. Wozniak, K.-P. Schade, K. Wozniak and H. H. Shalaby, "Über die Auslegung und den Entwicklungsstand von Zyklon-Abscheidern," *Forsch Ingenieurwes*, vol. 71, no. 3-4, pp. 171–180, 2007, doi: 10.1007/s10010-007-0055-7.

[65] *VDI-Wärmeatlas*. Berlin: Springer Vieweg, 2019.

[66] W. Peng, *et al.*, "Flow pattern in reverse-flow centrifugal separators," *Powder Technology*, vol. 127, no. 3, pp. 212–222, 2002, doi:10.1016/S0032-5910(02)00148-1.

[67] U. Muschelknautz, "Comparing Efficiency per Volume of Uniflow Cyclones and Standard Cyclones," *Chemie – Ingenieur – Technik*, vol. 93, no. 1–2, pp. 91–107, 2021, doi: 10.1002/cite.202000149.

[68] M. Stieß, *Mechanische Verfahrenstechnik*, 2nd ed. Berlin, Heidelberg, New York, Barcelona, Budapest, Hong Kong, London, Mailand, Paris, Tokyo: Springer, 1995.

[69] A. Gorton-Hülgerth, *Messung und Berechnung der Geschwindigkeitsfelder und Partikelbahn im Gaszyklon*. Zugl.: Graz, Techn. Univ., Diss., 1998. Düsseldorf: VDI-Verl., 1999.

[70] L. Qiang, W. Qinggong, X. Weiwei, Z. Zilin and Z. Konghao, "Experimental and computational analysis of a cyclone separator with a novel vortex finder," *Powder Technology*, vol. 360, pp. 398–410, 2020, doi: 10.1016/j.powtec.2019.10.073.

[71] V. Kumar and K. Jha, "Numerical investigations of the cone-shaped vortex finders on the performance of cyclone separators," *Journal of Mechanical Science and Technology*, vol. 32, no. 11, pp. 5293–5303, 2018, doi: 10.1007/s12206-018-1028-5.

[72] L. S. Brar, R. P. Sharma and R. Dwivedi, "Effect of vortex finder diameter on flow field and collection efficiency of cyclone separators," *Particulate Science and Technology*, vol. 33, no. 1, pp. 34–40, 2015, doi: 10.1080/02726351.2014.933144.

[73] L. S. Brar and K. Elsayed, "Analysis and optimization of cyclone separators with eccentric vortex finders using large eddy simulation and artificial neural network," *Separation and Purification Technology*, vol. 207, pp. 269–283, 2018, doi: 10.1016/j.seppur.2018.06.013.

[74] B. Zhao, H. Shen and Y. Kang, "Development of a symmetrical spiral inlet to improve cyclone separator performance," *Powder Technology*, vol. 145, no. 1, pp. 47–50, 2004, doi: 10.1016/j.powtec.2004.06.001.

[75] S. Wang, H. Li, R. Wang, X. Wang, R. Tian and Q. Sun, "Effect of the inlet angle on the performance of a cyclone separator using CFD-DEM," *Advanced Powder Technology*, vol. 30, no. 2, pp. 227–239, 2019, doi: 10.1016/j.apt.2018.10.027.

[76] M. Azadi, M. Azadi and A. Mohebbi, "A CFD study of the effect of cyclone size on its performance parameters," *Journal of Hazardous Materials*, vol. 182, no. 1–3, pp. 835–841, 2010, doi: 10.1016/j.jhazmat.2010.06.115.

[77] L. S. Brar, R. P. Sharma and K. Elsayed, "The effect of the cyclone length on the performance of Stairmand high-efficiency cyclone," *Powder Technology*, vol. 286, pp. 668–677, 2015, doi: 10.1016/j.powtec.2015.09.003.

[78] J. Chen, B. Yang, Z.-A. Jiang and Y. Wang, "Effect of external cyclone diameter on performance of a two-stage cyclone separator," *ACS Omega*, vol. 4, no. 8, pp. 13603–13616, 2019, doi: 10.1021/acsomega.9b02216.

[79] N. Ebeling, Abluft und Abgas. Reinigung und Überwachung. New York, Chichester, Brisbane, Singapore, Toronto: Wiley.VCH, 1999.

[80] A. C. Hoffmann, M. de Groot and A. Hospers, "The effect of the dust collection system on the flowpattern and separation efficiency of a gas cyclone," *The Canadian Journal of Chemical Engineering*, vol. 74, no. 4, pp. 464–470, 1996, doi: 10.1002/cjce.5450740405.

[81] W. Albring, *Angewandte Strömungslehre*, 6th ed. Berlin, Boston: De Gruyter, 1978. [Online]. Available: https://www.degruyter.com/isbn/9783112612262

[82] E. Muschelknautz and M. Trefz, "Secondary flow and short circuit flow at the dust discharge end of cyclone separators, 1," in *European Symposium: Separation of Particles from Gases*, pp. 345–407, 1989.

[83] E. Muschelknautz, *Vt-Hochschulkurs 2 [zwei], Mechanische Verfahrenstechnik*. Mainz: Krausskopf, 1972.

[84] P. Vogel, "Teil 3: Mechanisches Trennen in fluider Phase," in *Verfahrenstechnische Berechnungsmethoden*, S. Weiß, Ed., 1st ed., Weinheim: VCH Verl.-Ges, p. 257, 1985.

[85] N. Kimura, "Einführung in die Entstaubungstechnik," *The Journal of the Society of Powder Technology, Japan*, vol. 12, pp. 82–93, 1975.

[86] F. Löffler, *Staubabscheiden*. Stuttgart [Germany]: Georg Thieme Verlag, 1988.

[87] O. Lodge, *The Electrical Deposition of Dust and Smoke: With Special Reference to the Collection of Metallic Fume* and *to a Possible Purification of the Atmosphere*, 1886.

[88] K. Parker, *Electrical operation of electrostatic precipitators*, IET, 2003.

[89] H. J. White, U. Richter, H. Bober, T. Hänssgen, E. Koschany, and Deutscher Verlag für Grundstoffindustrie, *Entstaubung industrieller Gase mit Elektrofiltern*: VEB Deutscher Verlag für Grundstoffindustrie, 1969. [Online]. Available: https://books.google.de/books?id=FACLtQEACAAJ

[90] G. Skodras, S. P. Kaldis, D. Sofialidis, O. Faltsi, P. Grammelis and G. P. Sakellaropoulos, "Particulate removal via electrostatic precipitators – CFD simulation," *Fuel Process Technology*, vol. 87, no. 7, pp. 623–631, 2006, doi: 10.1016/j.fuproc.2006.01.012.

[91] K. Adamiak, "Numerical models in simulating wire-plate electrostatic precipitators: A review," *Journal of Electrostatics*, vol. 71, no. 4, pp. 673–680, 2013, doi: 10.1016/j.elstat.2013.03.001.

[92] S. M. Haque, M. G. Rasul, A. V. Deev, M. Khan and N. Subaschandar, "Flow simulation in an electrostatic precipitator of a thermal power plant," *Applied Thermal Engineering*, vol. 29, no. 10, pp. 2037–2042, 2009, doi: 10.1016/j.applthermaleng.2008.10.019.

[93] J. Anagnostopoulos and G. Bergeles, "Corona discharge simulation in wire-duct electrostatic precipitator," *Journal of Electrostatics*, vol. 54, no. 2, pp. 129–147, 2002, doi:10.1016/S0304-3886(01)00172-3.

[94] L. Zhao and K. Adamiak, "Numerical simulation of the electrohydrodynamic flow in a single wire-plate electrostatic precipitator," *IEEE Transactions on Industry Applications*, vol. 44, no. 3, pp. 683–691, 2008, doi: 10.1109/TIA.2008.921453.

[95] K. Nikas, A. A. Varonos and G. C. Bergeles, "Numerical simulation of the flow and the collection mechanisms inside a laboratory scale electrostatic precipitator," *Journal of Electrostatics*, vol. 63, no. 5, pp. 423–443, 2005, doi: 10.1016/j.elstat.2004.12.005.

[96] D. Yang, B. Guo, X. Ye, A. Yu and J. Guo, "Numerical simulation of electrostatic precipitator considering the dust particle space charge," *Powder Technology*, vol. 354, pp. 552–560, 2019, doi: 10.1016/j.powtec.2019.06.013.

[97] M. Kaul, "Einsatz elektrischer Abscheider zur Minderung von Feinstaub-Emissionen in Innenstädten," Zugl.: Wuppertal, Univ., Diss., 2015, Shaker; Shaker Verlag, Aachen, 2015.

[98] H. J. White, *Entstaubung industrieller Gase mit Elektrofiltern: Mit 43 Tab*. Leipzig: Deutscher Verlag f. Grundstoffindustrie, VEB, 1969.

[99] M. Siebenhofer, "Aerosolabscheidung durch Ionisationswäscher," *Chemie Ingenieur Technik*, vol. 63, no. 9, pp. 904–910, 1991, doi: 10.1002/cite.330630905.

[100] J. Petroll, Ed., *Verfahrenstechnische Berechnungsmethoden: Mechanisches Trennen in fluider Phase*. Teil 3, 1st ed. Weinheim: VCH Verl.-Ges, 1985.

[101] F. W. Peek, *Dielectric Phenomena in High-Voltage Engineering*, 3rd ed. New York: McGraw-Hill Book Comp, 1929.

[102] P. Cooperman, "A theory for space-charge-limited currents with application to electrical precipitation," *Transactions on AIEE, Part I: Communications Electronics*, vol. 79, no. 1, pp. 47–50, 1960, doi: 10.1109/TCE.1960.6368541.

[103] S. Kalaschnikow, "Der Einfluss der Feldstärke und der Verweildauer der Gase in Elektrofiltern auf der Reinigungsgrad," *Zeitschrift fuer Technische Physik*, vol. 14, pp. 267–270, 1933.

[104] H. J. White, "Modern electrical precipitation," *Industrial & Engineering Chemistry*, vol. 47, no. 5, pp. 932–939, 1955.

[105] W. Deutsch, "Bewegung und ladung der elektrizitätsträger im zylinderkondensator," *Annalen der Physik*, vol. 373, no. 12, pp. 335–344, 1922.

[106] M. B. Awad and G. S. P. Castle, "Ozone Generation in an Electrostatic Precipitator With a Heated Corona Wire," *Journal of the Air Pollution Control Association*, vol. 25, no. 4, pp. 369–374, 1975, doi: 10.1080/00022470.1975.10470092.

[107] A. S. Viner, P. A. Lawless, D. S. Ensor and L. E. Sparks, "Ozone generation in DC-energized electrostatic precipitators," *IEEE Transactions on Industry Applications*, vol. 28, no. 3, pp. 504–512, 1992, doi: 10.1109/28.137427.

[108] D. G. Poppendieck, D. Rim and A. K. Persily, "Ultrafine particle removal and ozone generation by in-duct electrostatic precipitators," *Environmental Science & Technology*, vol. 48, no. 3, pp. 2067–2074, 2014, doi: 10.1021/es404884p.

[109] F. Lin, *et al.*, "Flue gas treatment with ozone oxidation: An overview on NO organic pollutants and mercury," *Chemical Engineering Journal*, vol. 382, p. 123030, 2020, doi: 10.1016/j.cej.2019.123030.

[110] A. Yehia, M. Abdel-Salam and A. Mizuno, "On assessment of ozone generation in dc coronas," *Journal of Physics D: Applied Physics*, vol. 33, no. 7, pp. 831–835, 2000, doi: 10.1088/0022-3727/33/7/312.

[111] J. Katz, *The Art of electrostatic precipitation*, 2nd ed. Great Neck, N.Y.: Scholium International Inc, 1981.

[112] K. C. Schifftner and H. E. Hesketh, *Wet scrubbers*, 2nd ed. Lancaster, Basel: CRC Press, 1996.

[113] A. Bianchini, M. Pellegrini, J. Rossi and C. Saccani, "Theoretical model and preliminary design of an innovative wet scrubber for the separation of fine particulate matter produced by biomass

combustion in small size boilers," *Biomass and Bioenergy*, vol. 116, pp. 60–71, 2018, doi: 10.1016/j.biombioe.2018.05.011.

[114] C. Wang and Y. Otani, "Removal of nanoparticles from gas streams by fibrous filters: A review," *Industrial Engineering and Chemical Research*, vol. 52, no. 1, pp. 5–17, 2013, doi: 10.1021/ie300574m.

[115] D. A. Vallero, "Air Pollution Control Technologies," in *Air Pollution Calculations*, Elsevier, 2019, pp. 377–428.

[116] K. B. Schnelle, R. F. Dunn and M. E. Ternes, *Air pollution control technology handbook*. Boca Raton, Fla., London, New York: CRC Press, Taylor et Francis Group, 2016.

[117] K. T. SEMRAU, C. W. Marynowski, K. E. Lunde and C. E. Lapple, "Influence of power input on efficiency of dust scrubbers," *Industrial & Engineering Chemistry*, vol. 50, no. 11, pp. 1615–1620, 1958, doi: 10.1021/ie50587a025.

[118] K. T. Semrau, "Correlation of Dust scrubber efficiency," *Journal of the Air Pollution Control Association*, vol. 10, no. 3, pp. 200–207, 1960, doi: 10.1080/00022470.1960.10467920.

[119] K. T. Semrau, "Neuere Erkenntnisse auf dem Gebiet der Naßentstauber," *Staub*, vol. 22, pp. 184–188, 1962.

[120] K. T. Semrau, "Dust scrubber design -a critique on the state of the art," *Journal of the Air Pollution Control Association*, vol. 13, pp. 587–594, 1963, doi: 10.1080/00022470.1963.10468224.

[121] K. T. Semrau, "Practical process design of particulate scrubbers," 1977.

[122] S. Calvert, "Engineering design of fine particle scrubbers," *Journal of the Air Pollution Control Association*, vol. 24, no. 10, pp. 929–934, 1974, doi: 10.1080/00022470.1974.10469990.

[123] S. Calvert and others, "Scrubber performance for particle collection," 1974.

[124] S. Calvert and N. C. Jhaveri, "Flux Force/Condensation Scrubbing," *Journal of the Air Pollution Control Association*, vol. 24, no. 10, pp. 946–951, 1974, doi: 10.1080/00022470.1974.10469994.

[125] S. Calvert and others, "How to choose a particulate scrubber," 1977.

[126] S. Calvert and others, "Upgrading existing particulate scrubbers," 1977.

[127] S. Calvert, D. Lundgren and D. S. Mehta, "Venturi scrubber performance," *Journal of the Air Pollution Control Association*, vol. 22, no. 7, pp. 529–532, 1972, doi: 10.1080/00022470.1972.10469674.

[128] W. Barth, "Grundlegende Untersuchungen über die Reinigungsleistung von Wassertropfen," *Staub*, vol. 19, pp. 175–180, 1959.

[129] G. Schuch, *Theoretische und experimentelle Untersuchungen zur Auslegung von Nassabscheidern*, 1978.

[130] S. Calvert and others, "Wet scrubber system study. Volume I. Scrubber handbook," 1972.

[131] S. Calvert, R. Parker and D. C. Drehmel, *Particulate control highlights: fine particle scrubber research*: US Environmental Protection Agency, Office of Research and Development, 1978.

[132] G. Dau, "Zur Berechnung des Abscheidegrades von Venturi-Naßentstaubern," *Chemie Ingenieur Technik*, vol. 50, no. 9, p. 713, 1978, doi: 10.1002/cite.330500918.

[133] S. Nukiyama and Y. Tanasawa, "An Experiment on the atomization of liquid.: 4th report, the effect of the properties of liquid on the size of drops," *Transactions of the Japan Society of Mechanical Engineers. A*, vol. 5, pp. 136–143, 1938.

[134] R. H. Boll, "Particle collection and pressure drop in Venturi scrubbers," *Industrial and Engineering Chemistry Fundamentals*, vol. 12, no. 1, pp. 40–50, 1973, doi: 10.1021/i160045a008.

[135] B. J. Azzopardi and A. H. Govan, "The modelling of Venturi scrubbers," *Filtration & Separation*, vol. 21, no. 3, pp. 196–200, 1984.

[136] B. J. Azzopardi, S. Teixeira, A. H. Govan and T. R. Bott, "Improved model for pressure drop in Venturi scrubbers," *Process Safety & Environmental Protection*, vol. 69, no. 4, pp. 237–245, 1991.

[137] B. J. Azzopardi, "Liquid distribution in Venturi scrubbers: The importance of liquid films on the channel walls," *Chemical Engineering Science*, vol. 48, no. 15, pp. 2807–2813, 1993, doi:10.1016/0009-2509(93)80191-R.

[138] B. J. Azzopardi, "Gas-liquid flows in cylindrical Venturi scrubbers: Boundary layer separation in the diffuser section," *Chemical Engineering Journal*, vol. 49, no. 1, pp. 55–64, 1992, doi:10.1016/0300-9467(92)85025-5.

[139] S. Viswanathan, "Examination of liquid film characteristics in the prediction of pressure drop in a Venturi scrubber," *Chemical Engineering Science*, vol. 53, no. 17, pp. 3161–3175, 1998, doi:10.1016/S0009-2509(98)00123-7.

[140] N. V. Ananthanarayanan and S. Viswanathan, "Predicting the liquid flux distribution and collection efficiency in cylindrical Venturi scrubbers," *Industrial Engineering and Chemical Research*, vol. 38, no. 1, pp. 223–232, 1999, doi: 10.1021/ie9803321.

[141] A. Rahimi, A. Niksiar and M. Mobasheri, "Considering roles of heat and mass transfer for increasing the ability of pressure drop models in Venturi scrubbers," Chemical Engineering and Processing - Process Intensification, vol. 50, no. 1, pp. 104–112, 2011, doi: 10.1016/j.cep.2010.12.003.

[142] N. V. Ananthanarayanan and S. Viswanathan, "Effect of nozzle arrangement on Venturi scrubber performance," *Industrial Engineering and Chemical Research*, vol. 38, no. 12, pp. 4889–4900, 1999, doi: 10.1021/ie9902131.

[143] S. I. Pak and K. S. Chang, "Performance estimation of a Venturi scrubber using a computational model for capturing dust particles with liquid spray," *Journal of Hazardous Materials*, vol. 138, no. 3, pp. 560–573, 2006, doi: 10.1016/j.jhazmat.2006.05.105.

[144] F. Ahmadvand and M. R. Talaie, "CFD modeling of droplet dispersion in a Venturi scrubber," *Chemical Engineering Journal*, vol. 160, no. 2, pp. 423–431, 2010, doi: 10.1016/j.cej.2010.03.030.

[145] J. Manzano, B. M. de Azevedo, G. V. Do Bomfim, A. Royuela, C. V. Palau and V. d. A. Thales, "Design and prediction performance of Venturi injectors in drip irrigation," *Revista Brasileira de Engenharia Agricola e Ambiental*, vol. 18, no. 12, pp. 1209–1218, 2014.

[146] V. G. Guerra, R. Béttega, J. A. S. Gonçalves and J. R. Coury, "Pressure drop and liquid distribution in a Venturi scrubber: Experimental data and CFD simulation," *Industrial Engineering and Chemical Research*, vol. 51, no. 23, pp. 8049–8060, 2012, doi: 10.1021/ie202871q.

[147] M. Lehner, "Aerosol separation efficiency of a Venturi scrubber working in self-priming mode," *Aerosol Science and Technology*, vol. 28, no. 5, pp. 389–402, 1998, doi: 10.1080/02786829808965533.

[148] E. Muschelknautz, G. Hägele and U. Muschelknautz, "Naßabscheider," in *Handbuch des Umweltschutzes und der Umweltschutztechnik*, H. Brauer, Ed., Berlin, Heidelberg: Springer Berlin Heidelberg, 1996, pp. 203–229.

[149] I. B. Stechkina and N. A. Fuchs, "Studies on fibrous aerosol filters – I. Calculation of diffusional deposition of aerosols in fibrous filters," *Annals of Work Exposures and Health*, vol. 9, no. 2, pp. 59–64, 1966, doi: 10.1093/annhyg/9.2.59.

[150] K. W. Lee and B. Y. H. Liu, "Theoretical study of aerosol filtration by fibrous filters," *Aerosol Science and Technology*, vol. 1, no. 2, pp. 147–161, 1982, doi: 10.1080/02786828208958584.

[151] S. Kuwabara, "The Forces experienced by randomly distributed parallel circular cylinders or spheres in a viscous flow at small Reynolds numbers," *Journal of the Physical Society of Japan*, vol. 14, no. 4, pp. 527–532, 1959, doi: 10.1143/JPSJ.14.527.

[152] J. Happel, "Viscous flow relative to arrays of cylinders," *AIChE Journal*, vol. 5, no. 2, pp. 174–177, 1959, doi: 10.1002/aic.690050211.

[153] R. Hiller, "Der Einfluss von Partikelstoss und Partikelhaftung auf die Abscheidung in Faserfiltern," Zugl.: Karlsruhe, Univ., Fak. für Chemieingenieurwesen, Diss., 1980, VDI-Verlag; Verein Deutscher Ingenieure, Düsseldorf, 1981.

[154] M. Benarie, "Einfluss der Porenstruktur auf den Abscheidegrad in Faserfiltern," *Staub, Reinhaltung der Luft*, vol. 29, no. 2, pp. 74–78, 1969.

[155] H. Jodeit and F. Löffler, "Real influences on the collection of solid particles in technical filtermats," *Journal of Aerosol Science*, no. 15, pp. 311–317, 1984.

[156] L. Prandtl and K. Oswatitsch and Wieghardt, *Strömungslehre*. Braunschweig: Vieweg, 1965.

[157] B. P. LeClair and A. E. Hamielec, "Viscous Flow through particle assemblages at intermediate Reynolds numbers. Steady-state solutions for flow through assemblages of cylinders," *Industrial and Engineering Chemistry Fundamentals*, vol. 9, no. 4, pp. 608–613, 1970, doi: 10.1021/i160036a014.

[158] C. R. Holland and E. Rothwell, "Model studies on fabric dust filtration: 2. A study on the phenomenon of cake collapse," *Filtration Separation*, pp. 224–231, 1977.

[159] D. Leith and M. W. First, "Performance of a Pulse-Jet filter at high filtration velocity I. Particle collection," *Journal of the Air Pollution Control Association*, vol. 27, no. 6, pp. 534–539, 1977, doi: 10.1080/00022470.1977.10470452.

[160] D. Leith and M. W. First, "Performance of a Pulse-Jet Filter at high filtration velocity III. Penetration by fault processes," *Journal of the Air Pollution Control Association*, vol. 27, no. 8, pp. 754–758, 1977, doi: 10.1080/00022470.1977.10470486.

[161] F. Loeffler, H. Dietrich and W. Flatt, *Staubabscheidung mit Schlauchfiltern und Taschenfiltern*. Braunschweig: Vieweg Verlag.

[162] R. Dennis and H. A. Klemm, "Modeling concepts for pulse jet filtration," *Journal of the Air Pollution Control Association*, vol. 30, no. 1, pp. 38–43, 1980, doi: 10.1080/00022470.1980.10465912.

[163] S. Strangert, "Predicting performance of bag filters," *Filtration and Separation*, no. 15, pp. 42–55, 1970.

[164] M. Pérez-Fortes, J. A. Moya, K. Vatopoulos and E. Tzimas, "CO_2 capture and utilization in cement and iron and steel industries," *Energy Procedia*, vol. 63, pp. 6534–6543, 2014, doi: 10.1016/j.egypro.2014.11.689.

[165] K. de Ras, R. van de Vijver, V. V. Galvita, G. B. Marin and K. M. van Geem, "Carbon capture and utilization in the steel industry: Challenges and opportunities for chemical engineering," *Current Opinion in Chemical Engineering*, vol. 26, pp. 81–87, 2019, doi: 10.1016/j.coche.2019.09.001.

[166] International Energy Agency, "Global energy review: CO_2 emissions in 2021," Mar. 2022.

[167] H. Lee, S.-M. Yi, T. M. Holsen, Y.-S. Seo and E. Choi, "Estimation of CO_2 emissions from waste incinerators: Comparison of three methods," *Waste management (New York, N.Y.)*, vol. 73, pp. 247–255, 2018, doi: 10.1016/j.wasman.2017.11.055.

[168] M. Lehner, M. Ellersdorfer, R. Treimer, P. Moser, V. Theodoridou and H. Biedermann, "Carbon Capture and Utilization (CCU) – Verfahrenswege und deren Bewertung," *Berg Huettenmaenn Monatsh*, vol. 157, no. 2, pp. 63–69, 2012, doi: 10.1007/s00501-012-0056-1.

[169] M. Bailera, P. Lisbona, B. Peña and L. M. Romeo, "A review on CO_2 mitigation in the Iron and Steel industry through Power to X processes," *Journal of CO_2 Utilization*, vol. 46, p. 101456, 2021, doi: 10.1016/j.jcou.2021.101456.

[170] D. Pant, *Advances in carbon capture and utilization*. Singapore: Springer Singapore Pte. Limited, 2021.

[171] I. Ghiat and T. Al-Ansari, "A review of carbon capture and utilisation as a CO_2 abatement opportunity within the EWF nexus," *Journal of CO_2 Utilization*, vol. 45, p. 101432, 2021, doi: 10.1016/j.jcou.2020.101432.

[172] C. Bartels, *Carbon Capture and Storage: Verfahren zur Reduzierung von CO_2-Emissionen in Kraftwerken*. Hamburg: Diplomica Verlag, 2015. [Online]. Available: http://www.diplomica-verlag.de/

[173] S. Sun, H. Sun, P. T. Williams and C. Wu, "Recent advances in integrated CO_2 capture and utilization: a review," *Sustainable Energy Fuels*, vol. 5, no. 18, pp. 4546–4559, 2021, doi: 10.1039/D1SE00797A.

[174] T. Lecomte, *et al.*, "Best Available Techniques (BAT) reference document for large combustion plants," Luxembourg, Industrial Emissions Directive EUR 28836 EN, 2017.

[175] JOINT RESEARCH CENTRE, "Best Available Techniques (BAT) reference document for the ferrous metals processing industry," Industrial Emissions Directive 2010/75/EU (Integrated Pollution Prevention and Control), 2021.

[176] A. L. Kohl and R. B. Nielsen, Eds., *Gas Purification,* 5th ed., Houston: Elsevier, 2015.

[177] H. Falcke *et al.*, "Best Available Techniques (BAT) reference document for the production of large volume organic chemicals," Industrial Emissions Directive 2010/75/EU, 2017.

[178] J. J. Zhang, Y. Wei and Z. Fang, “Ozone pollution: a major health hazard worldwide,” *Show All Details*, vol. 10, p. 2518, 2019, doi: 10.3389/fimmu.2019.02518.

[179] D. R. Woods, *Rules of thumb in engineering practice*. Weinheim: WILEY-VCH, 2007.

[180] H. S. Fogler, *Elements of chemical reaction engineering*. Boston, Columbus, New York, San Francisco, Amsterdam, Cape Town, Dubai, London, Madrid, Milan, Munich, Paris, Montreal, Toronto, Delhi, Mexico City, Sa͂o Paulo, Sydney, Hong Kong, Seoul, Singapore, Taipei, Tokyo: Pearson, 2020.

[181] O. Levenspiel, *Chemical reaction engineering*, 3rd ed. Hoboken, NJ: Wiley, 1999.

[182] K. R. Westerterp, W. P. M. van Swaaij and A. A. C. M. Beenackers, *Chemical reactor design and operation*, 2nd ed. Chichester: Wiley, 2001.

[183] W. Wittenberger and W. Fritz, *Physikalisch-chemisches Rechnen mit einer Einführung in die höhere Mathematik*. Vienna, s.l.: Springer Vienna, 1980.

[184] P. W. Atkins and J. de Paula, *Atkins' physical chemistry*. Oxford: Oxford University Press, 2014.

[185] E. R. Cohen, E. R. Cohen and I. Mills, Eds., *Quantities, units and symbols in physical chemistry*, 3rd ed. Cambridge: RSC Publ, 2008.

[186] J. Rumble, Ed., *Handbook of Chemistry and Physics*, 103rd ed.: CRC Press, Taylor & Francis Group, 2022. Accessed: Nov. 11 2022.

[187] *VDI-Wärmeatlas*. Berlin: Springer Vieweg, 2019.

[188] J. Karl, *Dezentrale Energiesysteme: Neue Technologien im liberalisierten Energiemarkt*, 3rd ed. München: Oldenbourg, 2012.

[189] Metso Outotec, *HSC 8.0 Chemistry Software*. Finland, 2020.

[190] B. Roduit, A. Baiker, F. Bettoni, J. Baldyga and A. Wokaun, “3-D modeling of SCR of NO_x by NH_3 on vanadia honeycomb catalysts,” *AIChE JOURNAL*, vol. 44, no. 12, pp. 2731–2744, 1998, doi: 10.1002/aic.690441214.

[191] G. Ertl, H. Knzinger, F. Schth and J. Weitkamp, *Handbook of heterogeneous catalysis: 8 volumes*, 2nd ed. Weinheim: WILEY-VCH, 2008.

[192] T. V. Johnson, “Review of Selective Catalytic Reduction (SCR) and related technologies for mobile applications,” in *Urea-SCR Technology for deNOx After Treatment of Diesel Exhausts*, Springer, New York, NY, 2014, pp. 3–31.

[193] L. Han, *et al.*, “Selective catalytic reduction of NO_x with NH_3 by using novel catalysts: State of the art and future prospects,” *Chemical Reviews*, vol. 119, no. 19, pp. 10916–10976, 2019, doi: 10.1021/acs.chemrev.9b00202.

[194] Z. LI, L. SHEN, W. Huang and K. Xie, “Kinetics of selective catalytic reduction of NO by NH_3 on Fe-Mo/ZSM-5 catalyst,” *Journal of Environmental Sciences*, vol. 19, no. 12, pp. 1516–1519, 2007, doi:10.1016/S1001-0742(07)60247-2.

[195] R. Schmid and V. N. Sapunov, *Non-formal kinetics: In search for chemical reaction pathways*. Weinheim: Verl. Chemie, 1982.

[196] Z. Hong, Z. Wang and X. Li, “Catalytic oxidation of nitric oxide (NO) over different catalysts: an overview,” *Catalysis Science and Technology*, vol. 7, no. 16, pp. 3440–3452, 2017, doi: 10.1039/C7CY00760D.

[197] H. Tsukahara, T. Ishida and M. Mayumi, “Gas-phase oxidation of nitric oxide: Chemical kinetics and rate constant,” *Nitric Oxide*, vol. 3, no. 3, pp. 191–198, 1999, doi: 10.1006/niox.1999.0232.

[198] S. Mulla, *et al.*, “Reaction of NO and O_2 to NO_2 on Pt: Kinetics and catalyst deactivation,” *Journal of Catalysis*, vol. 241, no. 2, pp. 389–399, 2006, doi: 10.1016/j.jcat.2006.05.016.

[199] O. Carlowitz, “Grundlagen der thermischen Abgasreinigung,” *Technische Mitteilung*, no. 5, pp. 325–332, 1989.

[200] H. Wetzler, Ed., *Kennzahlen der Verfahrenstechnik*, 1970.

[201] D. W. Green and R. H. Perry, Eds., *Perry`s chemical engineers` handbook*, 8th ed., McGraw-Hill Book Comp, 2007.

[202] R. E. Treybal, *Mass-transfer operations*, 3rd ed. New York: McGraw-Hill, 2004.

[203] J. M. Thomas and W. J. Thomas, *Principles and practice of heterogeneous catalysis*. Weinheim, München: Wiley-VCH Verlag GmbH & Co. KGaA; Ciando, 2015.
[204] C. H. Bartholomew and R. J. Farrauto, *Fundamentals of industrial catalytic processes*, 2nd ed. Hoboken, NJ: Wiley-Interscience, 2010.
[205] International Union of Pure and Applied Chemistry, *IUPAC Compendium of Chemical Terminology – the Gold Book*, 2012.
[206] S. Kozuch and J. M. L. Martin, ""Turning Over" definitions in catalytic cycles," *ACS Catalysis*, vol. 2, no. 12, pp. 2787–2794, 2012, doi: 10.1021/cs3005264.
[207] M. Schultes, *Abgasreinigung – Verfahrensprinzipien, Berechnungsgrundlagen, Verfahrensvergleich.* Berlin Heidelberg: Springer-Verlag, 1996.
[208] M. Votsmeier, T. Kreuzer, J. Gieshoff, G. Lepperhoff and B. Elvers, "Automobile Exhaust Control," in *Ullmann's Encyclopedia of Industrial Chemistry*, John Wiley & Sons, Ltd, 2000, pp. 1–19.
[209] N. Salahudeen, "A review on Zeolite: Application, synthesis and effect of synthesis parameters on product properties," *Chemistry Africa*, pp. 1–18, 2022, doi: 10.1007/s42250-022-00471-9.
[210] R. W. Broach, D.-Y. Jan, D. A. Lesch, S. Kulprathipanja, E. Roland and P. Kleinschmit, "Zeolites," *Ullmann's Encyclopedia of Industrial Chemistry*, 2012, doi: 10.1002/14356007.a28_475.pub2.
[211] J. Hagen, Ed., *Industrial catalysis: A practical approach.* Weinheim, Germany: WILEY-VCH, 2015.
[212] T. Xue and L. Yang, "Zeolite-based materials for the catalytic oxidation of VOCs: A mini review," *Frontiers in Chemistry*, vol. 9, p. 751581, 2021, doi: 10.3389/fchem.2021.751581.
[213] A. W. Petrov, D. Ferri, F. Krumeich, M. Nachtegaal, J. A. van Bokhoven and O. Kröcher, "Stable complete methane oxidation over palladium based zeolite catalysts," *Nature Communications*, vol. 9, no. 1, p. 2545, 2018, doi: 10.1038/s41467-018-04748-x.
[214] Z. G. Liu, D. R. Berg and J. J. Schauer, "Effects of a zeolite-selective catalytic reduction system on comprehensive emissions from a heavy-duty diesel engine," *Journal of the Air & Waste Management Association*, vol. 58, no. 10, pp. 1258–1265, 2008, doi: 10.3155/1047-3289.58.10.1258.
[215] R. Zhang, N. Liu, Z. Lei and B. Chen, "Selective transformation of various Nitrogen-containing exhaust gases toward N_2 over zeolite catalysts," *Chemical Reviews*, vol. 116, no. 6, pp. 3658–3721, 2016, doi: 10.1021/acs.chemrev.5b00474.
[216] E. Schmidt *et al.*, "Air, 7. Waste Gases, Separation and Purification," in *Ullmann's Encyclopedia of Industrial Chemistry*, Weinheim, Germany: Wiley-VCH Verlag GmbH & Co. KGaA, 2000.
[217] R. Koppmann, Ed., *Volatile organic compounds in the atmosphere*, 1st ed. Oxford, Ames, Iowa: Blackwell Pub, 2007.
[218] *TA Luft*, 2021.
[219] *Waste gas cleaning – Methods of thermal waste gas cleaning*, VDI 2442, VDI/DIN-Kommission Reinhaltung der Luft (KRdL) – Normenausschuss, Feb. 2014.
[220] V. Veselý, "Performance of an afterburner chamber with a natural gas burner," *Fuel*, vol. 75, no. 11, pp. 1271–1273, 1996, doi:10.1016/0016-2361(96)00109-3.
[221] NIST, *NIST webbook* (accessed: Nov. 21 2022).
[222] A. M. Kanury, *Introduction to combustion phenomena: (for fire, incineration, pollution and energy applications)*, 8th ed. New York: Gordon and Breach Science Publ, 1994.
[223] M. Lackner, "Combustion," *Ullmann's Encyclopedia of Industrial Chemistry*, 2011, doi: 10.1002/14356007.b03_14.pub3.
[224] M. Lackner, Ed., *Handbook of combustion*. Weinheim: WILEY-VCH, 2010.
[225] B. Lewis, *Combustion, flames and explosions of gases*, 3rd ed. Orlando: Academic Press, 1987.
[226] C. E. Baukal, Ed., *The John Zink combustion handbook*. Boca Raton, Fla., London: CRC Press, 2001.
[227] Environmental Protection Agency, "BAT guidance note on best available techniques for oil and gas refineries," 2008.
[228] A. Bahadori, "Blow-Down and Flare Systems," in *Natural Gas Processing*, A. Bahadori, Ed., 1st ed., Boston, MA: Gulf Professional Publishing; Safari, 2014, pp. 275–312.

[229] L. Zigan, "Overview of electric field applications in energy and process engineering," *Energies*, vol. 11, no. 6, p. 1361, 2018, doi: 10.3390/en11061361.

[230] A. Starikovskiy and N. Aleksandrov, "Plasma-assisted ignition and combustion," *Progress in Energy and Combustion Science*, vol. 39, no. 1, pp. 61–110, 2013, doi: 10.1016/j.pecs.2012.05.003.

[231] W. F. Maier and J. W. Schlangen, "Efficiency of air cleaning by catalytic afterburning with electric heaters," *Catalysis Today*, vol. 17, 1–2, pp. 225–233, 1993, doi:10.1016/0920-5861(93)80027-X.

[232] J. Warnatz, *Combustion: Physical and chemical fundamentals, modelling and simulation, experiments, pollutant formation*. Berlin, Heidelberg: Springer, 1996.

[233] D. D. Reible, *Fundamentals of environmental engineering*. Boca Raton, Florida, London, [England], New York: CRC Press, 1999.

[234] J.-K. Lai and I. E. Wachs, "A perspective on the Selective Catalytic Reduction (SCR) of NO with NH_3 by supported V_2O_5–WO_3/TiO_2 catalysts," *ACS Catalysis*, vol. 8, no. 7, pp. 6537–6551, 2018, doi: 10.1021/acscatal.8b01357.

[235] D. Damma, P. Ettireddy, B. Reddy and P. Smirniotis, "A review of low temperature NH3-SCR for removal of NOx," *Catalysts*, vol. 9, no. 4, p. 349, 2019, doi: 10.3390/catal9040349.

[236] J.-H. Kim, *et al.*, "Reduction of NOx emission from the cement industry in South Korea: A review," *Atmosphere*, vol. 13, no. 1, p. 121, 2022, doi: 10.3390/atmos13010121.

[237] K. Sattler, Ed., *Thermische Trennverfahren*. Weinheim: WILEY-VCH, 2001.

[238] J. Draxler, *Verfahrenstechnik in Beispielen: Problemstellungen, Lösungsansätze, Rechenwege*. Wiesbaden: Springer Vieweg, 2014.

[239] P. Grassmann, *Einführung in die thermische Verfahrenstechnik*, 3rd ed. Berlin: De Gruyter, 2011.

[240] P. C. Wankat, *Separation process engineering: Includes mass transfer analysis*. Boston, Columbus, Indianapolis: Prentice Hall, 2017.

[241] H. J. Kandiner, *Schwefel: Gmelin Handbook of Inorganic and Organometallic Chemistry – 8th Edition Ser*, 8th ed. Berlin, Heidelberg: Springer Berlin / Heidelberg, 1974.

[242] T. G. Maple, *Gmelins Handbuch der Anorganischen Chemie*, 8th ed. Berlin, Heidelberg: Springer Berlin / Heidelberg, 1937.

[243] S. Austin and A. Glowacki, "Hydrochloric Acid," in *Ullmann's Encyclopedia of Industrial Chemistry*, Weinheim, Germany: Wiley-VCH Verlag GmbH & Co. KGaA, 2000.

[244] A. L. Kohl and R. B. Nielsen, Eds., *Gas Purification (fith Edition)*. Houston: Elsevier, 2015.

[245] S. Arrhenius, "XXXI. On the influence of carbonic acid in the air upon the temperature of the ground," *The London, Edinburgh and Dublin Philosophical Magazine and Journal of Science*, vol. 41, no. 251, pp. 237–276, 1896, doi: 10.1080/14786449608620846.

[246] P. Gabrielli, M. Gazzani and M. Mazzotti, "The role of carbon capture and utilization, carbon capture and storage and biomass to enable a Net-Zero-CO_2 emissions chemical industry," *Industrial Engineering and Chemical Research*, vol. 59, no. 15, pp. 7033–7045, 2020, doi: 10.1021/acs.iecr.9b06579.

[247] N. Koukouzas, *et al.*, "Current CO_2 capture and storage trends in Europe in a view of social knowledge and acceptance. A short review," *Energies*, vol. 15, no. 15, p. 5716, 2022, doi: 10.3390/en15155716.

[248] Institut für Energieforschung, *Weltweite Innovationen bei der Entwicklung von CCS-Technologien und Möglichkeiten der Nutzung und des Recyclings von CO_tn2: Studie; Endbericht*. Jülich: Forschungszentrum Zentralbibliothek, 2010. [Online]. Available: http://hdl.handle.net/2128/3733

[249] G. Baldauf-Sommerbauer, S. Lux and M. Siebenhofer, "Sustainable iron production from mineral iron carbonate and hydrogen," *Green Chemistry*, vol. 18, no. 23, pp. 6255–6265, 2016, doi: 10.1039/C6GC02160C.

[250] A. Loder, M. Siebenhofer and S. Lux, "The reaction kinetics of CO_2 methanation on a bifunctional Ni/MgO catalyst," *Journal of Industrial and Engineering Chemistry*, vol. 85, pp. 196–207, 2020, doi: 10.1016/j.jiec.2020.02.001.

[251] A. Brandl and A.-M. Fischer, “Method for selective desulphurisation of a synthetic raw gas,” DE 10 2012 016 643 A.

[252] W.-D. Latzin, “Linde-Verfahren zur Entschwefelung von Gasen mit Wertstoffgewinnung,” *Chemie – Ingenieur – Technik*, vol. 62, no. 4, p. 323, 1990, doi: 10.1002/cite.330620418.

[253] B. Schreiner, “Der Claus-Prozess. Reich an Jahren und bedeutender denn je,” *Chemie in Unserer Zeit*, vol. 42, no. 6, pp. 378–392, 2008, doi: 10.1002/ciuz.200800461.

[254] G. Krass and P. Grauer, “Method and device for binding gaseous CO_2 and for treating flue gases with sodium carbonate Compounds,” WO/2008/110405.

[255] E. Erich, A. Berry and S. Telge, “Alkalicarbonatwäsche für die Entfernung von Kohlendioxid aus Rauchgasen fossil befeuerter Kraftwerke als robuste Alternative zu Aminwäschen,” AIF-Vorhaben-Nr.: 15653, 2010.

[256] F. Seel, *Grundlagen der analytischen Chemie: Unter besonderer Berücksichtigung der Chemie in wäßrigen Systemen*, 7th ed. Weinheim: Verl. Chemie, 1979.

[257] P. W. Atkins, *Physikalische Chemie*, 5th ed. Weinheim: WILEY-VCH, 2013.

[258] J. G. Speight, *Perry’s standard tables and formulas for chemical engineers*. New York: McGraw-Hill, 2003.

[259] R. Sander, “Compilation of Henry’s law constants (version 4.0) for water as solvent,” *Atmospheric Chemistry and Physics*, vol. 15, no. 8, pp. 4399–4981, 2015, doi: 10.5194/acp-15-4399-2015.

[260] R. K. Freier, *Aqueous Solutions / Wässrige Lösungen*. Berlin, New York: Walter de Gruyter, 1976. [Online]. Available: https://www.degruyter.com/isbn/9783110834598

[261] R. C. West, *Handbook of Chemistry and Physics*, 55th ed. Cleveland, Ohio: CRC Press.

[262] K. Bliefert, *ph-Wert-Berechnungen: Claus Bliefert; Unter Mitarb. von*. Weinheim usw.: Verl. Chemie, 1978.

[263] Metso Outotec, *HSC 8.0 Chemistry Software*. Finland, 2020.

[264] D. W. Green and R. H. Perry, Eds., *Perry`s chemical engineers` handbook*, 8th ed.: McGraw-Hill Book Comp, 2007.

[265] NIST, *NIST webbook* (accessed: Nov. 21 2022).

[266] *VDI-Wärmeatlas*. Berlin: Springer Vieweg, 2019.

[267] J. Crank, *The mathematics of diffusion*, Oxford university press, 1979.

[268] Y. Kong and J. Vysoky, “Comparison of sodium bicarbonate and trona for multi-pollutant control at a coal-fired power plant,” Rosemont, Illinois, 2009.

[269] B. Heiting, “Grundlagen der Absorptionstechnik bei Rauchgaswäschern,” Essen.

[270] M. Luckas, *Thermodynamik der Elektrolytlösungen: Eine Einheitliche Darstellung der Berechnung Komplexer Gleichgewichte*. Berlin, Heidelberg: Springer Berlin / Heidelberg, 2001. [Online]. Available: https://ebookcentral.proquest.com/lib/kxp/detail.action?docID=6438439

[271] B. E. Poling, J. M. Prausnitz and J. P. O’Connell, *Properties of Gases and Liquids*, 5th ed., New York, N.Y.: McGraw-Hill Education; McGraw Hill, 2020. [Online]. Available: https://www.accessengineeringlibrary.com/content/book/9780070116825

[272] J. F. Zemaitis, D. M. Clark, M. Rafal and N. C. Scrivner, *Handbook of aqueous electrolyte thermodynamics: Theory & application*. New York: Wiley, 2010. [Online]. Available: http://onlinelibrary.wiley.com/book/10.1002/9780470938416

[273] B. Kolbe, J. Gmehling, M. Kleiber and J. Rarey, *Chemical thermodynamics for process simulation*. Weinheim: WILEY-VCH, 2019.

[274] L. A. Bromley, “Thermodynamic Properties of Strong Electrolytes in Aqueous Solutions,” *AIChE JOURNAL*, vol. 19, no. 2, pp. 313–320, 1973, doi: 10.1002/aic.690190216.

[275] Roth, *Sodium sulfite, safety data sheet, version 2.0*.

[276] VGB-M, *VGB Instrcution Sheet ‘Analysis of FGD gypsum’: 701e*.

[277] R. Lefet and C. Carta Sata, *Collegium Nr. 147, 311*.

[278] A. Lancia, D. Musmarra and F. Pepe, “Uncatalyzed heterogeneous oxidation of calcium bisulfite,” *Chemical Engineering Science*, vol. 51, no. 16, pp. 3889–3896, 1996, doi:10.1016/0009-2509(96)00222-9.
[279] A. Lancia, D. Musmarra, M. Prisciandaro and M. Tammaro, “Catalytic oxidation of calcium bisulfite in the wet limestone–gypsum flue gas desulfurization process,” *Chemical Engineering Science*, vol. 54, 15–16, pp. 3019–3026, 1999, doi:10.1016/S0009-2509(98)00483-7.
[280] S. Bengtsson and I. Bjerles, “Catalytic oxidation of sulphite in diluted aqueous solutions,” *Chemical Engineering Science*, vol. 30, no. 11, pp. 1429–1435, 1975, doi: 10.1016/0009-2509(75)85076-7.
[281] M. Souders and G. G. Brown, “Design of fractionating columns I. Entrainment and capacity,” *Industrial & Engineering Chemistry*, vol. 26, no. 1, pp. 98–103, 1934, doi: 10.1021/ie50289a025.
[282] K. Hoppe and H. Mittelstrass, *Grundlagen der dimensionierung von Kolonnenböden*. Dresden: Steinkopff, 1967.
[283] H. E. O’Connel, “Tray efficiency of fractionating columns and absorbers,” *Transactions of the American Institute of Chemical Engineers*, no. 42, pp. 741–775, 1946.
[284] T. H. Chilton and A. Colburn, “Distillation and absorption in packed columns a convenient design and correlation method,” *Industrial & Engineering Chemistry*, vol. 1935, 255–260.
[285] E. L. Cussler, *Diffusion Mass Transfer in Fluid Systems*. Cambridge: Cambridge University Press, 1997.
[286] R. Beck, *Ein neues Verfahren zur Berechnung von Füllkörpersäulen*, 1969.
[287] A. L. Kohl and R. B. Nielsen, Eds., *Gas Purification*, 5th ed., Houston: Elsevier, 2015.
[288] L. F. Albright, Ed., *Albright‘s Chemical Engineering Handbook*, CRC Press, Taylor & Francis Group, 2009.
[289] S. Weiss, *Verfahrenstechnische Berechnungsmethoden*. Weinheim: VCH-Verl.-Ges, 1985–.
[290] N. Ebeling, *Abluft und Abgas*: WILEY-VCH, 1999.
[291] M. Suzuki, *Adsorption engineering*. Tokyo, Amsterdam, New York: Kodansha; Elsevier, 1990.
[292] K. E. Noll, V. Gounaris and W. Hou, *Adsorption technology for air and water pollution control*. Chelsea, Mich.: Lewis Publishers, 1992.
[293] C. Tien, *Adsorption calculations and modeling*. Boston: Butterworth-Heinemann, 1994.
[294] H. Kienle and E. Bäder, *Aktivkohle und ihre industrielle Anwendung (Activated Carbon and its Industrial Application)*. Stuttgart, Weinheim: Ferdinand Enke Verlag; WILEY-VCH, 2001.
[295] K. Sattler, Ed., *Thermische Trennverfahren*. Weinheim: WILEY-VCH, 2001.
[296] International Union of Pure and Applied Chemistry, *IUPAC Compendium of Chemical Terminology – the Gold Book*, 2012.
[297] D. W. Green and R. H. Perry, Eds., *Perry‘s chemical engineers‘ handbook*, 8th ed.: McGraw-Hill Book Comp, 2007.
[298] D. Basmadjian, *The little adsorption book: A practical guide for engineers and scientists*. [Place of publication not identified]: CRC Press, 2018.
[299] S. Brunauer, P. H. Emmett and E. Teller, “Adsorption of gases in multimolecular layers,” *Journal of the American Chemical Society*, vol. 60, no. 2, pp. 309–319, 1938, doi: 10.1021/ja01269a023.
[300] A. Schönbucher, *Thermische Verfahrenstechnik: Grundlagen und Berechnungsmethoden für Ausrüstungen und Prozesse*. Berlin: Springer Berlin, 2002.
[301] S. Brunauer, L. S. Deming, W. E. Deming and E. Teller, “On a Theory of the van der Waals Adsorption of Gases,” *Journal of the American Chemical Society*, vol. 62, no. 7, pp. 1723–1732, 1940, doi: 10.1021/ja01864a025.
[302] D. G. Kinniburgh, “General purpose adsorption isotherms,” *Environmental Science & Technology*, vol. 20, no. 9, pp. 895–904, 1986, doi: 10.1021/es00151a008.
[303] I. Langmuir, “The adsorption of gases on plane surface of glass, mica and platinum,” *Journal of the American Chemical Society*, vol. 40, no. 9, pp. 1361–1403, 1918, doi: 10.1021/ja02242a004.
[304] D. O. Cooney, *Adsorption design for wastewater treatment*. Boca Raton, Fl.: Lewis Publishers, 1999.
[305] H. Freundlich, “Colloid and capillary chemistry,” *Journal of the Society of Chemical Industry*, vol. 65, no. 1672, pp. 40–41, 1926, doi: 10.1002/jctb.5000454407.

[306] C. J. Radke and J. M. Prausnitz, “Adsorption of organic solutes from dilute aqueous solution of activated carbon,” *Industrial and Engineering Chemistry Fundamentals*, vol. 11, no. 4, pp. 445–451, 1972, doi: 10.1021/i160044a003.

[307] R. H. Fowler and E. A. Guggenheim, Ed., *Statistical Thermodynamics*. New York, Cambridge: The University Press, 1939.

[308] D. P. Valenzuela and A. L. Myers, *Adsorption equilibrium data handbook*. Englewood Cliffs: Prentice Hall, op. 1989.

[309] M. M. Dubinin, “The potential theory of adsorption of gases and vapors for adsorbents with energetically nonuniform surfaces,” *Chemical Reviews*, vol. 60, no. 2, pp. 235–241, 1960, doi: 10.1021/cr60204a006.

[310] Dubinin M. M. and V. A. Astakhov, Eds., *Description of Adsorption Equilibria of Vapors on Zeolites over Wide Ranges of Temperature and Pressure*, 1971.

[311] M. Schultes, *Abgasreinigung – Verfahrensprinzipien, Berechnungsgrundlagen, Verfahrensvergleich*. Berlin Heidelberg: Springer-Verlag, 1996.

[312] E. C. Markham and A. F. Benton, “The adsorption of gas mixtures by silica,” *Journal of the American Chemical Society*, vol. 53, no. 2, pp. 497–507, 1931, doi: 10.1021/ja01353a013.

[313] W. Fritz and E. U. Schluender, “Simultaneous adsorption equilibria of organic solutes in dilute aqueous solution on activated carbon,” *Ceramic Engineering and Science Proceedings*, vol. 29, no. 5, pp. 1279–1282, 1974.

[314] *VDI-Wärmeatlas*. Berlin: Springer Vieweg, 2019.

Index

https://doi.org/10.1515/9783110763928-012

www.ingramcontent.com/pod-product-compliance
Lightning Source LLC
Chambersburg PA
CBHW080652220326
41598CB00033B/5181

* 9 7 8 3 1 1 0 7 6 3 9 0 4 *